X-ray Free Electron Lasers

Sébastien Boutet • Petra Fromme
Mark S. Hunter
Editors

X-ray Free Electron Lasers

A Revolution in Structural Biology

Springer

Editors
Sébastien Boutet
SLAC National Accelerator Laboratory
Menlo Park, CA, USA

Petra Fromme
School of Molecular Sciences
Arizona State University
Tempe, AZ, USA

Mark S. Hunter
SLAC National Accelerator Laboratory
Menlo Park, CA, USA

ISBN 978-3-030-00550-4 ISBN 978-3-030-00551-1 (eBook)
https://doi.org/10.1007/978-3-030-00551-1

Library of Congress Control Number: 2018962550

This Springer imprint is published by the registered company Springer Nature Switzerland AG
The registered company address is: Gewerbestrasse 11, 6330 Cham, Switzerland

Preface

"I was captured for life by chemistry and by crystals."
Dorothy Crowfoot Hodgkin

"The limits of the possible can only be defined by going beyond them into the impossible."
Arthur C. Clarke

"I was taught that the way of progress is neither swift nor easy."
Marie Curie

"It's kind of fun to do the impossible."
Walt Disney

X-ray free electron lasers (FELs) are revolutionary tools in the world of X-ray science. The new capabilities that they have already provided, and potentially can enable, have led to a growing revolution in structural biology. Biology using X-ray FELs is a nascent field, with the very first results of X-ray diffraction from protein crystals using an X-ray FEL published in 2011. This book represents the first collection of contributions from scientists in the field who are specifically devoted to this topic, focusing on the most emerging techniques as well as current and future challenges. It is our sincere hope that all readers enjoy reading about the revolutionary techniques developed, the obstacles that had to be overcome, and the breakthrough scientific discoveries that were enabled by X-ray FEL technology.

The reader will learn about the discoveries through the voices of innovators and pioneers: the adventurous scientists who conceived the ideas, invented the techniques, and were involved with the innovations from the very beginning. The contributing authors represent a mix of senior and young scientists who have worked together to make these experiments a reality, working tirelessly in their home laboratories and at X-ray FEL facilities, thoroughly preparing for and executing the experiments. They developed the techniques and made possible the discoveries that are described in the book. They have written their chapters to provide the reader with a sense of the adventure involved in their technology developments and discoveries, including the hurdles and difficulties that they have faced. They share not only the results of their own work and the work of others, but also give an overview of the most recent discoveries in the field and the lessons learned along the way.

This book appeals to a very broad audience of students and scientists of all levels of experience. It provides introductory background information to those new to the field and those who may have just read a newspaper article about a scientific discovery enabled by X-ray FEL technology. To the expert in the field, this book hopefully represents a rich source of teaching material about X-ray FEL technology and its application to biology.

At times, it seems like X-ray free electron lasers were made specifically to overcome the dilemma of using X-rays for biological measurements: the destruction of a biological sample by the very same X-rays needed to probe it. This fact is well known (everybody knows, e.g., that a person cannot have a chest X-ray every week since the amount of radiation would be damaging), and the problem of X-ray damage has plagued X-ray imaging techniques since their discovery by Roentgen in 1895. Ever since the first demonstration of X-ray diffraction by a physicist named William Bragg more than 100 years ago, X-ray crystallography on radiation-sensitive materials (primarily organic and biological material) has suffered from the fact that the object of desire (here the molecule of which we seek to discover the detailed structure) is damaged from the interrogation by the X-rays. With passing time and more powerful X-ray sources, the problem became even more severe, so that crystal freezing was developed to limit (but not overcome) radiation damage.

X-ray FELs are so powerful that, when focused, the generated X-ray beam destroys any solid material. It may seem at first glance to be counterintuitive to use such a destructive force to overcome radiation damage as a new tool for structural biology. The reader will learn how these powerful X-ray pulses have been generated and how they are used to outrun the traditional radiation damage issues encountered when using X-rays for structural analysis. A history of the technological advances that occurred to get us to this point with X-ray generation will set the stage for why X-ray FELs have such a potential to revolutionize the biological sciences.

A revolution means a sudden or fundamental change in a way of thinking or doing things. In science, it involves breaking new ground, exploring new territory, and conducting experiments that were deemed impossible just years prior. Indeed, most of the breakthrough experiments in the biological sciences using X-ray FELs were deemed impossible before many scientists, including those who share their experience and expertise with you in this book, developed techniques to perform the "impossible" with the new sources. A revolution also often means that many textbook paradigms that were valid before are now in question. As an example, for the last 100 years, X-ray crystallographers have worked hard to grow large, well-ordered single crystals for X-ray structure analysis. When you read this book, you may be surprised to learn that "small is beautiful" and that imperfect crystals (which were historically the bane of the traditional crystallographer's work and effort) may yet be useful and perhaps have even more to reveal than "perfect crystals."

However, the land is new and unexplored when one reaches new shores. For biology with X-ray FELs, this means that new and nearly endless opportunities exist. Scientific treasures are waiting to be unearthed, but the way is rocky with many obstacles and challenges that had, and remain, to be overcome. With X-ray FELs, one can now use nanocrystals with a few hundred unit cells for structure determination. But this raises a new question: how can one rapidly screen for, identify, and characterize these "invisible" crystals that are so small that one cannot detect them even with the best light microscope?

Each X-ray FEL shot typically destroys the sample, or at least the parts of the sample that were illuminated. Electrons are stripped from the atoms that make up the sample (e.g., biological molecules in a crystal), ultimately leading to the sample being vaporized. Understanding the radiation damage physics and its implication for solving structures or observing dynamics is critical to developing methodology to minimize the ill effects. Many experiments are designed to better understand the constraints that the powerful X-ray FELs place on data collection.

One might wonder how it can be possible to collect X-ray data under such destructive conditions. Achieving this required abandoning methods of conventional data collection where a full data set is collected on one (or a few) large crystal(s) rotated through the X-ray beam. The reader will learn about new methods that were developed to bring the crystals in their native environment to the X-ray FEL beam and replenish them between each shot, from flying crystals in a jet to rapid moving of fixed mounts that allow for X-ray data to be collected with 120 images/second at room temperature. As a result of the new techniques, data can be collected at room temperature. Reactions can be triggered "on the fly," leading toward motion pictures of biomolecules at work. X-ray FELs thereby open a new avenue in structural biology.

However, sample replenishment in a controlled state was not the only challenge to be mastered to reach the ultimate treasure of measuring the damage-free (dynamic) structure of a molecule. One of the next big obstacles in the way was the data mountain. With X-ray FELs, data are collected in a serial fashion, so multiple ultrashort diffraction snapshots are collected from thousands of crystals in random orientation. A very large number of images are coming in a stream, with some crystal hits and some crystal misses.

The first three hard X-ray FELs built in the world (the LCLS in the USA, SACLA in Japan, and PAL-XFEL in South Korea) provide between 10 and 120 X-ray shots per second, leading to a huge data mountain containing millions of images every day. Not every shot hits a crystal, and the task of finding the crystal hits in all these images, followed by finding their relative orientation and assembling accurate structure factors from the patchwork of diffraction snapshots, is monumental indeed. New data evaluation programs and algorithms were developed that are explained in this book.

One of the fundamental challenges in X-ray crystallography is that the phase of the diffracted beams is lost in the data collection process, the so-called phase problem. Data collection with X-ray FELs is no different in that regard, but X-ray FEL beams, with their short pulses, very high intensity, and high coherence, allow new avenues to be explored for determining the phases of the diffracted X-rays. Novel methods, ranging from making use of the finite size of the crystals to the idea of directly solving the phases by continuous diffraction from imperfect crystals, are actively being studied.

What scientific discoveries are now enabled by X-ray FEL technology? The reader of the book may wonder how this new technology can be applied to his/her favorite biological problem. This book features four chapters that show examples of breakthrough discoveries enabled by the X-ray FELs. The reader will be excited to learn more about the discoveries in the field of G-protein-coupled receptors, which are the targets of 50% of all current drugs, enabled by X-ray FELs. This project is especially challenging, as GPCRs are membrane proteins and are crystallized in a lipidic environment that mimics the native membrane. Unfortunately, this lipidic cubic phase (LCP) has a consistency of toothpaste, and when the idea was first proposed, experts in the field did not believe that one could "get the toothpaste to fly." Introducing LCP to the interaction region was considered a serious challenge. When you read this book, you will learn how this challenge was overcome, and you may be surprised to learn that sample delivery in highly viscous media is now forming the basis for bringing X-ray FEL technology to the conventional X-ray sources such as synchrotrons to allow for serial data collection at room temperature.

Traditionally, collecting data from a crystal and seeing the Bragg spots terminate at low resolution while diffuse scattering persists to higher resolution was deflating for the scientists working hard to produce large, well-ordered crystals. New approaches to handling the diffuse scattering have indicated that the data are useful and may allow the diffraction of the individual molecule, without augmentation from the crystal lattice, to be measured. These approaches combine methods of crystallography and coherent diffractive imaging and may allow a hybrid approach to structural analysis.

Biological processes are highly dynamic. However, classical X-ray structures provide only a static picture of a molecule. One of the most exciting developments enabled by X-ray FELs is time-resolved methods that allow scientists to capture time points of a biological reaction. The reader will learn about the first pioneering studies on time-resolved femtosecond ($1\ \text{fs} = 10^{-15}\ \text{s}$) crystallography and how these pave the way to visualizing reactions driven by light in photosensors. Subsequent (and ongoing) studies using the technique have explored light-driven biological reactions such as vision and photosynthesis, among others.

Since the majority of biological processes are not triggered by the absorption of photons, finding ways to extend the ability to follow the time course of biological reactions to all enzymes is very important. Recently, the first enzymatic reactions were studied by time-resolved crystallography at X-ray FELs, enabled by novel rapid mixing technology. This mixing-based time-resolved crystallography has already lead to active discussions in the field on how to expand the technology even more and, for example, introduce oxygen gas to study the process of oxygen transport or respiration with the X-ray FEL.

Time-resolved studies are not limited to X-ray diffraction, but also include advanced X-ray spectroscopic techniques that can probe electronic transitions and detect oxidation changes at the heart of one of the most important processes on earth, photosynthesis, which converts the light from the sun into chemical energy and produces all the oxygen that we breathe. Diffraction data, especially crystallographic data, are fairly insensitive to oxidation state or excitation level changes in molecules, and X-ray spectroscopic techniques are a powerful tool to explore the local chemical environments of metalloproteins, for instance.

So far, the "holy grail" of X-ray FEL technology for biology remains elusive, the dream to solve atomic resolution structures without the need for crystals, using single molecules in a native, noncrystalline environment. While that still a work in progress, this book discusses the advancements made on both single-particle imaging, in which diffraction data are collected from individual molecules, and solution scattering with X-ray FELs, in which diffraction data are collected from ensembles of molecules in solution. Both methods are challenging to accomplish at X-ray FELs in practice, but great strides have been made to identify necessary areas of improvement on the quest toward making the methods a reality. These results will be useful as the adventure continues to the use of superconducting accelerators in the next generation of X-ray FELs.

We cannot pause in our quest for new technique development as the mountain to climb and conquer becomes even higher with the development of new X-ray FEL technology, which is briefly summarized as an outlook in the last chapter of the book. The European XFEL just started operating in 2017 and will provide up to 27,000 pulses per second, and LCLS-II, which will start operating in 2020, will reach 1 million pulses per second. New technology must be developed rapidly to make use of these new sources, ranging from sample delivery to the collection, transfer, and storage of data with these high repetition rates. What is now already clear is that these new sources will allow for further exciting scientific developments and discoveries, with new challenges ahead of us.

Menlo Park, CA, USA Sébastien Boutet
Tempe, AZ, USA Petra Fromme
Menlo Park, CA, USA Mark S. Hunter

Contents

Contributors

Roberto Alonso-Mori Linac Coherent Light Source, SLAC National Accelerator Laboratory, Menlo Park, CA, USA

Zina Al-Sahouri Center for Applied Structural Discovery at the Biodesign Institute, School of Molecular Sciences, Arizona State University, Tempe, AZ, USA

Andrew Aquila Linac Coherent Light Source, SLAC National Accelerator Laboratory, Menlo Park, CA, USA

Kartik Ayyer Center for Free-Electron Laser Science, DESY, Hamburg, Germany

Anton Barty Centre for Free-Electron Laser Science, DESY, Hamburg, Germany

Alexander Batyuk Linac Coherent Light Source, Stanford Linear Accelerator Center (SLAC) National Accelerator Laboratory, Menlo Park, CA, USA

Sébastien Boutet Linac Coherent Light Source, SLAC National Accelerator Laboratory, Menlo Park, CA, USA

Martin Caffrey Membrane Structural and Functional Biology Group, School of Medicine and School of Biochemistry and Immunology, Trinity College Dublin, Dublin, Ireland

Carl Caleman Department of Physics and Astronomy, Uppsala University, Uppsala, Sweden

Center for Free-Electron Laser Science, Deutsches Elektronen-Synchrotron, Hamburg, Germany

Henry N. Chapman Center for Free-Electron Laser Science, DESY, Hamburg, Germany

Department of Physics, University of Hamburg, Hamburg, Germany

Centre for Ultrafast Imaging, University of Hamburg, Hamburg, Germany

Joe P. J. Chen Department of Physics, Arizona State University, Tempe, AZ, USA

Vadim Cherezov Department of Chemistry, Bridge Institute, University of Southern California, Los Angeles, CA, USA

Jesse Coe Linac Coherent Light Source, SLAC National Accelerator Laboratory, Menlo Park, CA, USA

Jacques-Philippe Colletier Institute of Structural Biology, University Grenoble, Alpes, CEA, CNRS, Grenoble, France

Sebastian Doniach Departments of Applied Physics, Physics, and Photon Science, Stanford University, Stanford, CA, USA

SLAC National Accelerator Center, Menlo Park, CA, USA

Mike Dunne Linac Coherent Light Source, SLAC National Accelerator Laboratory, Menlo Park, CA, USA

Petra Fromme School of Molecular Sciences, Arizona State University, Tempe, AZ, USA

Biodesign Center for Applied Structural Discovery, The Biodesign Institute, Arizona State University, Tempe, AZ, USA

Gye Won Han Department of Chemistry, Bridge Institute, University of Southern California, Los Angeles, CA, USA

Mark S. Hunter Linac Coherent Light Source, SLAC National Accelerator Laboratory, Menlo Park, CA, USA

Biosciences Division, SLAC National Accelerator Laboratory, Menlo Park, CA, USA

Andrii Ishchenko Department of Chemistry, Bridge Institute, University of Southern California, Los Angeles, CA, USA

So Iwata RIKEN SPring-8 Center, Sayo-gun, Hyogo, Japan

Department of Cell Biology, Graduate School of Medicine, Kyoto University, Sakyo-ku, Kyoto, Japan

Linda Johansson Department of Chemistry, Bridge Institute, University of Southern California, Los Angeles, CA, USA

Richard A. Kirian Department of Physics, Arizona State University, Tempe, AZ, USA

Christopher Kupitz Physics Department, University of Wisconsin-Milwaukee, Milwaukee, WI, USA

Ming-Yue Lee Center for Applied Structural Discovery at the Biodesign Institute, School of Molecular Sciences, Arizona State University, Tempe, AZ, USA

Dianfan Li State Key Laboratory of Molecular Biology, National Center for Protein Science Shanghai, Shanghai Science Research Center, CAS Center for Excellence in Molecular Cell Science, Shanghai Institute of Biochemistry and Cell Biology, Chinese Academy of Sciences, Shanghai, China

Wei Liu Center for Applied Structural Discovery at the Biodesign Institute, School of Molecular Sciences, Arizona State University, Tempe, AZ, USA

Andrew V. Martin School of Science, RMIT University, Melbourne, VIC, Australia

ARC Centre of Excellence for Advanced Molecular Imaging, University of Melbourne, Melbourne, VIC, Australia

Alke Meents Center for Free-Electron Laser Science, German Electron Synchrotron DESY, Hamburg, Germany

Eriko Nango RIKEN SPring-8 Center, Sayo-gun, Hyogo, Japan

Department of Cell Biology, Graduate School of Medicine, Kyoto University, Sakyo-ku, Kyoto, Japan

Dominik Oberthuer Center for Free-Electron Laser Science, German Electron Synchrotron DESY, Hamburg, Germany

Alexandra Ros School of Molecular Sciences, Arizona State University, Tempe, AZ, USA

Giorgio Schirò Institute of Structural Biology, University Grenoble, Alpes, CEA, CNRS, Grenoble, France

Marius Schmidt Physics Department, University of Wisconsin-Milwaukee, Milwaukee, WI, USA

Robert W. Schoenlein Linac Coherent Light Source, SLAC National Accelerator Laboratory, Menlo Park, CA, USA

Raymond G. Sierra Hard X-Ray Department, LCLS, SLAC National Accelerator Laboratory, Menlo Park, CA, USA

John C. H. Spence Department of Physics, Arizona State University, Tempe, AZ, USA

Benjamin Stauch Department of Chemistry, Bridge Institute, University of Southern California, Los Angeles, CA, USA

Michihiro Sugahara RIKEN SPring-8 Center, Sayo-gun, Hyogo, Japan

Department of Cell Biology, Graduate School of Medicine, Kyoto University, Sakyo-ku, Kyoto, Japan

Uwe Weierstall Department of Physics, Arizona State University, Tempe, AZ, USA

Martin Weik Institute of Structural Biology, University Grenoble, Alpes, CEA, CNRS, Grenoble, France

Thomas A. White Center for Free-Electron Laser Science, German Electron Synchrotron DESY, Hamburg, Germany

Makina Yabashi RIKEN Spring-8 Center, Sayo-gun, Japan

Junko Yano Molecular Biophysics and Integrated Bioimaging Division, Lawrence Berkeley National Laboratory, Berkeley, CA, USA

Oleksandr M. Yefanov Center for Free-Electron Laser Science, DESY, Hamburg, Germany

Chun Hong Yoon Linac Coherent Light Source, SLAC National Accelerator Laboratory, Menlo Park, CA, USA

Chapter 1
X-Ray Free Electron Lasers and Their Applications

Sébastien Boutet and Makina Yabashi

X-ray free electron lasers (FELs) represent the latest generation of X-ray sources, with unique properties and capabilities that present novel opportunities in the study of matter in unique forms as well as the study of interactions and dynamics on ultrafast timescales. For the purpose of this book focused on the use of X-ray FEL beams for the study of biological materials, the story begins with the availability of these novel sources to the scientific community as user facilities. Let us however take a quick step back and provide a brief historical background on what has led to the advent of X-ray FEL sources. This will be followed by a short description of the principles of operation of X-ray FELs and the breadth of their scientific use.

1.1 X-Rays and Their Applications

When Wilhelm Conrad Röntgen discovered X-rays in 1895, despite scientists all over the World not immediately knowing their true nature, there was a rapidly growing excitement in the scientific community and the general population regarding the potential uses of these new particles, this new form of radiation. It became quickly well known across the globe that X-rays have the power to see inside matter, inside the body of patients requiring medical care, for example. This became the first

S. Boutet (✉)
Linac Coherent Light Source, SLAC National Accelerator Laboratory, Menlo Park, CA, USA
e-mail: sboutet@slac.stanford.edu

M. Yabashi
RIKEN Spring-8 Center, Sayo-gun, Japan
e-mail: yabashi@spring8.or.jp

S. Boutet et al. (eds.), *X-ray Free Electron Lasers*,
https://doi.org/10.1007/978-3-030-00551-1_1

obvious application for these unknown rays. In this particular case, the particle-like behavior of X-rays is exploited via the absorption of energy by the material, allowing shadow images representative of how absorbing, or equivalently, how dense a material is.

Following two decades of intense scientific research, discovery and debate about the apparent dual nature of X-rays, at times seen to behave as particles and at times as waves, an increased understanding of X-rays eventually improved both the methods by which they are generated and the ways to utilize them for probing matter. In the early 1910s, the seminal works of Max von Laue, William Henry Bragg, and his son, William Lawrence Bragg led to the realization that X-rays are ideal tools to study the atomic structure of matter due to their electromagnetic wave nature, which allows them to probe features at scales similar to or longer than their wavelengths. Electromagnetic waves, including the visible light we see, can in principle exist at any wavelength. Visible light interacts with structures of size comparable to the wavelength of the light or larger, allowing us to see things. Similarly, X-rays can "see" objects as small as atoms due to their wavelength approaching atomic scales. In the continuum of wavelengths of the electromagnetic spectrum, X-rays such as those discovered by Röntgen now define a range of wavelengths spanning roughly from 0.1 to 100 Å, with the typical spacing between atoms in solids and liquids in the middle of this range, making them ideal probes of the atomic structure of matter.

Today, X-rays are recognized to possess a dual nature, as all matter does according to quantum mechanical theory. X-rays are electromagnetic waves with discrete energy units behaving as particles under some interactions with matter. Both aspects of this dual nature of X-rays are widely exploited by scientists around the World today. X-rays are used to understand the fundamental structure of matter and what happens at the atomic and electronic levels in the universe we live in, not only from a fundamental point of view but also with very practical applications. The study of biological systems represents an important aspect of the many uses of X-rays in scientific research. The reader new to the field of X-ray methods is referred to the excellent introduction by Jens Als-Nielsen [5].

1.1.1 X-Ray Methods

X-ray methods have been broadly developed over the last century and offer many diverse scientific communities a variety of tools optimized for particular samples or questions to answer. Many texts and general reviews of X-ray methods exist [5, 32, 73] and we will here only briefly discuss the evolution of the key methods now employed at X-ray FEL facilities for the study of biology.

1.1.1.1 Scattering

The natural match between the wavelength of X-rays and the typical spacing of atoms in solid and liquid matter has made the use of X-rays as a tool for research a rapidly growing field in the twentieth century. X-rays primarily interact with the electrons in matter. The simplest interaction of X-rays with matter is via elastic scattering, where X-rays essentially "bounce off" the electronic structure of matter without losing energy but their direction of travel is changed. The directional dependence of the elastic scattering from these electrons can be measured and primarily reveals the location of concentrated electronic density, typically located around the nuclei of atoms. Elastic scattering extracts information on the correlation between the location of atoms, that is the likelihood of an electron being located at a certain distance or direction from another, which can be used to deduce the detailed structure of the material or sample at hand [5].

Solid matter often arranges itself naturally in a crystalline structure where the atoms or molecules self-organize into a repeating pattern. This creates a repeating pattern of electrons where the correlation between their positions also becomes periodic. Shining an X-ray beam on a crystalline material will lead to X-rays elastically scattered from this repeating electronic density pattern to produce a corresponding repeating diffraction pattern of sharp bright spots in the measured scattered intensity. Following the first observation of X-rays diffracted by the periodic atomic arrangement in a crystal by Walther Friedrich and Paul Knipping in 1912, at the suggestion of von Laue, X-rays were soon used to map the structure of the majority of naturally occurring crystalline materials [3] as well as most man-made materials. Crystallographic methods of various forms are now regularly used to understand the structure of novel materials and some of their more unique properties such as superconductivity. The foundations of X-ray crystallographic methods are thoroughly discussed in the International Tables for Crystallography [32] for the interested reader.

Crystallographic methods, which will be featured prominently in this book, were eventually adapted to the study of biological molecules. Life is supported by a multitude of molecules with specific roles and tasks to perform for each organism. To begin to understand the details of life, and perhaps to help enhance it by, for example, curing diseases, requires an understanding of the interactions of these molecules with each other and with their environment. This is a very daunting challenge that first involves obtaining a detailed understanding of the structure of these molecules—that is the 3D arrangement of the atoms comprising them. The initial demonstration of macromolecular crystallography on the protein myoglobin took place in 1958 [35]. Since then, macromolecular structure determination using X-rays has become a very important scientific method, with wide-ranging impact in the medical and life science fields. Today, many thousands of three-dimensional structures of biological molecules are determined each year using X-ray diffraction and this trend is still increasing as shown in Fig. 1.1. X-ray FEL facilities have

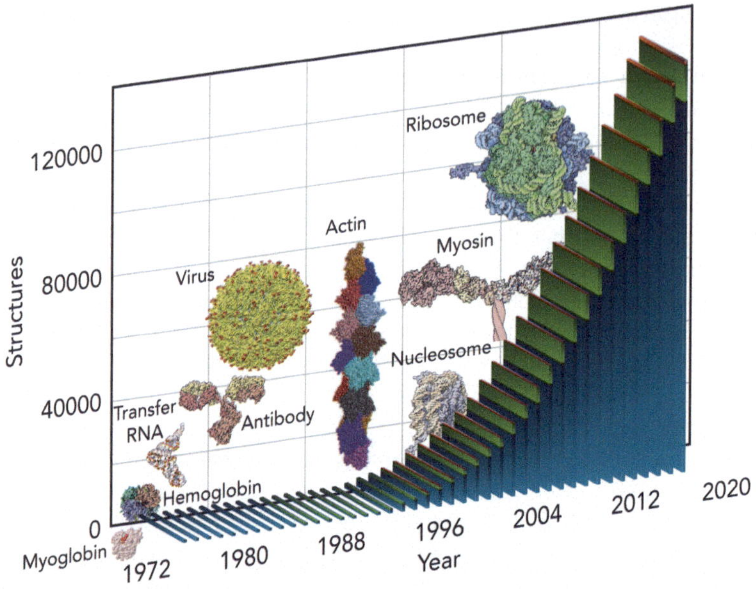

Fig. 1.1 Cumulative number of structures in the Protein Data Bank [6] used by different techniques. The Protein Data Bank is an international repository of macromolecular structures used to disseminate structural information. Figure adapted from [1]. Modified by Gregory Stewart and Terry Anderson with permission of the International Union of Crystallography

begun using their unique capabilities to contribute to the wealth of structural knowledge available. Chapter 2 of this book offers a thorough review of the current state of crystallographic methods using X-ray FELs. However, for the majority of macromolecules, a crystalline state is not a naturally forming state. The challenges of producing crystals of biomolecules of sufficient quality to allow interpreting their diffraction patterns to yield a structure exist no matter what X-ray source is used. The challenges to crystal preparation specific to X-ray FELs will be presented in Chaps. 3 and 4.

In principle, it is possible to use elastic scattering to obtain an image of a nonperiodic or noncrystalline sample to a high resolution that is only limited by the wavelength of the X-rays used. Such a possibility would obviate one of the biggest challenges in structural biology, the need to grow high-quality crystals of a molecule of interest. These relatively new imaging methods can be grouped together under the umbrella of coherent diffractive imaging (CDI) methods that were extensively reviewed recently [47].

In the case of coherent diffractive imaging, a coherent X-ray illumination, where the X-ray wave field illuminating the sample has a well-defined phase relationship across the illumination, is necessary. In simple terms, this means that the X-rays illuminating the sample look more like a long single-break wave hitting the shore rather than be composed of multiple waves colliding, leading to areas of the beach with a big surf and others more calm. Coherent diffraction methods form an image of the sample by combining the measured angle-dependent scattering amplitudes with computationally retrieved phases of the scattered waves. Mathematically, the amplitude of the elastic scattering pattern from a single object under such coherent illumination will be proportional to the amplitude of something called a Fourier transform of the illuminated electron density. The interested reader is referred to optics and X-ray physics textbooks for a detailed mathematical treatment [5, 26].

The phases of the scattered waves must be known to form an image, but these phases cannot be measured by X-ray detectors, which only sense the deposited energy in a pixel and cannot resolve the extremely rapid oscillations of the electromagnetic waves. This leads to the so-called phase problem in both imaging and crystallography, where the phases are unknown from the measurement itself. In the CDI method, iterative calculations under specific boundary conditions, by applying constraints to the image or solution based on some previous knowledge of the sample, are performed to retrieve the phase. Compared with conventional imaging methods using lenses, CDI methods can improve the resolution, because they are free from the limitations of X-ray imaging optics. Solutions to the phase problem in X-ray FEL crystallography will be discussed in Chaps. 8 and 9. The use of coherent beams recently available with brighter X-ray sources, and especially X-ray FELs, opens the door to new capabilities in the study of single particles via imaging and will be the subject of Chap. 14.

Other scattering methods are broadly used in many areas of science. For example, inelastic scattering uses the difference in the energy of incident and scattered photons to extract information on the dynamics in materials. Incident X-rays can transfer energy to the sample via inelastic interactions, where excited states with energy levels lower than the incident photon energy can be produced by absorption of X-ray photons and instantaneous reemission of photons at a lower energy. This inelastic scattering process couples incident X-rays to available excited states of the system and can therefore probe the available energy levels within the sample. Measuring inelastic scattering spectra as a function of scattering angles can reveal the dispersion relation of many dynamic processes such as phonons (lattice dynamics), magnons (spin dynamics), plasmons, and excitons (electron dynamics). Inelastic scattering methods are used more broadly in materials research than in biology and have seen limited use in biology at X-ray FELs to date. The topics covered in this book exploit elastic and coherent scattering techniques exclusively, and inelastic scattering methods will therefore not be discussed further. It is however expected that higher repetition rate X-ray FELs in the near future will make inelastic scattering methods more practical and this will benefit the biological sciences. Such potential future applications and the future of X-ray FELs will be discussed in the last chapter of this book.

1.1.1.2 Spectroscopy

Beyond the use of scattering methods to probe the primarily static structure of matter, spectroscopic methods that utilize the particle-like behavior of X-rays are commonly used. It was realized early after the discovery of X-rays that specific X-ray energies were emitted by specific materials and these differed for different materials. This is directly equivalent to the different colors emitted by neon light signs in the visible range, where different gases produce different colors. It is now known that atoms can emit X-rays when an unoccupied electronic orbital (typically an inner shell orbital for X-ray emission) gets filled by an electron from a higher orbital or by a free electron within the system, such as in metals, for example. The energy of the emitted X-ray is representative of the binding energy of the core electron orbital that got filled and its difference in energy compared to the previously filled orbital or the free electron. High atomic number elements emit harder X-rays (shorter wavelength) due to the higher energy of their core electrons, which are more tightly bound to a more highly charged atomic nucleus.

Not only does the wavelength of typical X-rays match interatomic spacings, but the energy of these X-rays also match well the energy levels of electronic orbitals of core electrons. Therefore, X-rays represent a probe capable of element sensitivity, differentiating between different atoms in the sample. Since the energy levels of electrons are perturbed by their local environment, the occupancy of valence electronic levels, as well as spin states, accurate measurements of the spectrum of emitted X-rays from a sample can be used to deduce accurate information on the electronic states of specific elements in a sample. Crystallographic and imaging techniques are not very sensitive to the finer details of the electronic structure of the samples being measured, making spectroscopy a powerful complementary tool.

X-rays of sufficiently high photon energy can be used to excite atoms by knocking out core electrons, which can be followed by X-ray emission from this core electronic level being filled shortly afterward. Methods of X-ray emission spectroscopy (XES) are powerful tools to understand, for example, oxidation states, which is particularly relevant in biocatalysts. Many important biological functions involve the binding of oxygen to one or many metal atoms in a molecule or the exchange of electrons from the molecular environment to the metal atom (oxidation). For example, the molecule Photosystem II is critical to the photosynthetic process of splitting water molecules to produce the oxygen (O_2) molecules we all breath to maintain life on Earth as we know it. X-ray emission spectroscopy can probe the oxidation state of such a metal center by measuring the spectrum of emitted X-rays and how this is modified by small changes in the electronic state of a particular atom and its local environment. Additionally, X-ray absorption spectroscopy (XAS), where electronic energy levels are probed by directly measuring the absorption (comparing an incident and transmitted spectrum), is also broadly used in X-ray science in general and beginning to find applications using X-ray FELs. Spectroscopic applications will be discussed at length in Chap. 13.

1.1.2 The Evolution of X-Ray Sources

For the first two thirds of the twentieth century, the method to generate X-rays changed little. Of course, technological improvements made sources brighter and more easily usable for experiments as X-ray generation became better understood. However, the generation of X-rays remained fundamentally the same for decades. X-ray tubes use high voltages to accelerate electrons from a cathode to an anode. As the electrons interact with the anode, a deceleration leads to emitted X-rays in a broad spectrum known as Bremsstrahlung radiation. Beyond this broad spectrum, high-energy electrons can remove core electrons from the atoms in the anode. As previously discussed, the core holes get filled rapidly and can generate emitted X-rays of a particular energy. This leads to a relatively narrow band (small range of wavelength centered around the emission line of the material) radiation emitted by the X-ray tube, however, with a high divergence and polychromaticity set by the bandwidth of the emission process. This limits the achievable brightness of the beam. Most of the energy used in generating X-rays in this fashion results in heat in the anode and cooling limitations set a limit to how much X-ray energy can be generated from an X-ray tube.

The 1970s brought the first revolution to X-ray sources. Decades of development of particle and especially electron accelerators led to more powerful machines such as cyclotrons and synchrotrons dedicated to particle physics or high-energy physics applications. It was known that such machines, with their mostly circular design, would generate emitted photon beams as the electrons are accelerated inward by the magnetic fields keeping them within a defined, mostly circular, orbit. With higher energy machines, the range of energies of the emitted photons reached the X-ray regime and a few places around the world built the capabilities to exploit, parasitically at first, this unavoidable radiation. The great initial successes of these parasitic operations, due to the very large increase in brightness compared to X-ray tubes, was rapidly sufficient to justify investment in dedicated electron storage rings built specifically for the production of X-ray beams to be used for photon science. These facilities are known as second-generation synchrotron sources, and starting in the early 1980s they provided dedicated user facility access to high-brightness X-ray beams produced from the circular trajectory of electrons through the bending magnets.

The so-called third-generation synchrotron radiation facilities made their appearance in the 1990s. These facilities are distinguished from second-generation facilities by their design specifically intended to make use of the straight sections between bending magnets to generate even brighter X-ray beams. In these straight sections, long arrays of magnets of alternating polarities called undulators are installed to send the electron beam on a rapidly oscillating sinusoidal path. The oscillating motion around a straight path leads to a much more intense, more collimated output of radiation compared to the fan of radiation from a bending magnet and this leads to much brighter sources of X-rays. These sources allow more challenging experiments and measurements, studying smaller or more dilute samples for spectroscopy, for example. Third-generation light sources are now prevalent across the globe, with hundreds of beamlines with dedicated instrumentation for a broad set of scientific fields. The contribution of these facilities to scientific knowledge and technical developments is undeniable, certainly in the life sciences but also across many fields of science. Many of the third-generation sources are now undergoing upgrades to further improve their brightness by reducing their beam emittance, a measure of the size and angular spread of the electron beam, to values approaching the theoretical minimum possible for a diffraction-limited light source. This is achieved via the use of multi-bend achromat magnets to replace double magnet benders per cell. The gentler bend in the electron beam thus afforded reduces the horizontal spread of the electron beam, increasing the beam brightness.

Following on the rough trend of a new generation of X-ray sources every decade, the 2000s saw the beginning of the X-ray FEL era, with the construction of the first FEL user facilities culminating with the start of the first ever hard X-ray FEL. The next section will describe X-ray FELs in more detail. More detailed information about the history of synchrotron sources can be found here [7].

1.2 X-Ray FELs as User Facilities

Free electron lasers have only recently become well known to the broad scientific community with the construction and start of operations of facilities open to use by scientists. Free electron lasers have, however, been around for a few decades, starting with an initial theoretical conceptualization by Madey in 1971 [44]. This was followed by a demonstration of the principles a few years later, where an optical cavity was used to generate infrared radiation [15]. The concepts and technologies eventually leading to the feasibility of FELs in the X-ray regime were developed during the 1980s and 1990s, culminating with the realization of linear accelerator-based single-pass machines dedicated to the production of short-pulsed photon beams of wavelength approaching, and eventually reaching, iteratomic spacings of a few Ångstroms or less. The first such machine built specifically to be made available to users was the Free electron LASer in Hamburg (FLASH) starting in 2005 [2, 65], followed by the SPring-8 Compact SASE Source (SCSS) [61, 74]. Both of these sources operated in the ultraviolet regime, approaching the soft X-ray range. A few years later, hard X-ray FELs became a reality with the Linac Coherent Light Source (LCLS) in the USA in 2009 [16] and the SPring-8 Angstrom Compact free electron LAser (SACLA) in Japan [31] shortly after. As of 2018, the European XFEL in Germany, SwissFEL in Switzerland, and PAL-XFEL in South Korea have joined the ranks of operating X-ray FELs. For the interested reader, the history of FELs is beautifully recounted by Pellegrini [54]. Below, the physics behind free electron lasers will be briefly described.

1.2.1 The Physics of Free Electron Lasers

Many excellent articles and reviews describe in great detail the technology and physics behind X-ray FEL radiation generation [27, 51, 59, 60]. Of particular interest should be the recent review article by Pellegrini *et al.* [55]. For the purpose of this book, we will only briefly summarize FEL physics in the hopes of stimulating the curiosity of the reader.

X-ray FELs are at times classified as fourth-generation synchrotron sources. This nomenclature is not universal due to the noncircular design of FEL facilities and other developments such as diffraction-limited storage rings and energy recovery linacs sometimes being also referred to as fourth-generation sources. Nevertheless, X-ray FELs take previous technologies from third-generation sources a large step further. Third-generation sources were based on the intended use of undulator technologies and X-ray FELs entirely rely on extending the use of undulators to devices more than one order of magnitude larger. The linear design of an X-ray FEL is not by choice since a circular design allowing multiple FEL sources would be much more desirable. It is required by the electron beam quality needed for the lasing process, which cannot be accomplished with a circular design.

An undulator causes an electron beam with energy γmc^2 to oscillate in a nearly sinusoidal fashion leading to emitted radiation of wavelength:

$$\lambda_r = \frac{\lambda_u}{2\gamma^2}\left(1 + \frac{K^2}{2}\right) = \frac{2\pi c}{\omega_r} \tag{1.1}$$

where $K \equiv eB_0\lambda_u/(2\pi mc)$ is the undulator strength parameter, B_0 is the peak magnetic field strength, λ_u is the undulator period, e is the electron charge, c is the speed of light, m is the electron mass, and ω_r is the fundamental undulator frequency. This equation holds true for the radiation emitted from any undulator system including those at conventional synchrotron sources. What distinguishes X-ray FEL radiation from synchrotron radiation is the brightness of the electron beam required, along with the length of the undulator, which allows for self-amplified spontaneous emission (SASE), an exponential growth in the radiated intensity via the interaction of the electron bunch with the previously emitted X-ray field as they co-propagate along the undulator. Achieving SASE is quite challenging and requires the electron beam to be of sufficient quality, *i.e.*, a low emittance, high peak current, and a small energy spread that are only achievable with a linear accelerator. The parameters basically control how similar all the electrons are. The lasing process is an enhancement of emitted energy by placing as many electrons as possible in the same state so that they emit in harmony (in phase).

The electron trajectory in the undulator must be sufficiently straight (on the order of 5 μm deviation over the ∼100-m-long undulator path) to maintain the spatial overlap of the electron beam and the co-propagating X-rays. If these conditions are met, it leads to a microbunching process in the electron beam in which the X-ray field slows down the faster electrons and speeds up the slower ones. The

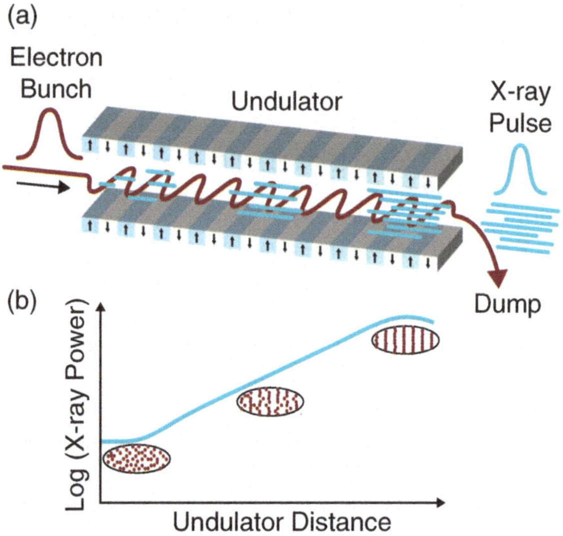

Fig. 1.2 In the self-amplified spontaneous emission (SASE) process, microbunching increases as the electron beam travels down the undulator length causing the radiation power to grow. Reproduced with permission from [10]

microbunches created behave like a single massive charge since the electrons in the bunch oscillate in phase, leading to an increased X-ray emission by a factor of N^2 along the axis of the electron beam, where N is the number of electrons in the microbunch. The process is schematically illustrated in Fig. 1.2. It eventually reaches saturation when space charge effects (electrons becoming too close and repelling each other) and the same forces that cause the microbunching in the first place eventually start to rip the microbunches apart. The SASE process also generates higher harmonics, integer multiples of the fundamental energy, at the roughly 1% intensity level of the fundamental.

The end result for an optimized accelerator and undulator system is a highly transversely coherent X-ray beam. However, the electron beam is typically comprised of many microbunches for a SASE FEL. These microbunches are uncorrelated and there is a natural spread of energy within the bunch, leading to only moderate longitudinal coherence and a typical bandwidth of the beam on the order of 0.2%. The amplification starting from random fluctuations in the initial electron beam gives rise to appreciable fluctuations in essentially all relevant parameters including pulse energy, average wavelength, and the photon energy spectrum [30, 67, 78], as well as the spatial and temporal profiles [25]. The SASE process produces short pulses, typically on the order of the few tens of femtoseconds (fs), with the potential longer pulses in the few hundred fs range or shorter pulses in the attosecond range. The X-ray pulse duration is controlled by the length of the electron bunch that possesses sufficient "quality" to produce lasing. This can be achieved by simply controlling the overall electron bunch length or possibly by intentionally producing an electron bunch where only a small part possesses the characteristics required to produce lasing. The interested reader is again referred to the review of Pellegrini for more detail [55]. The amplification process produces radiation in

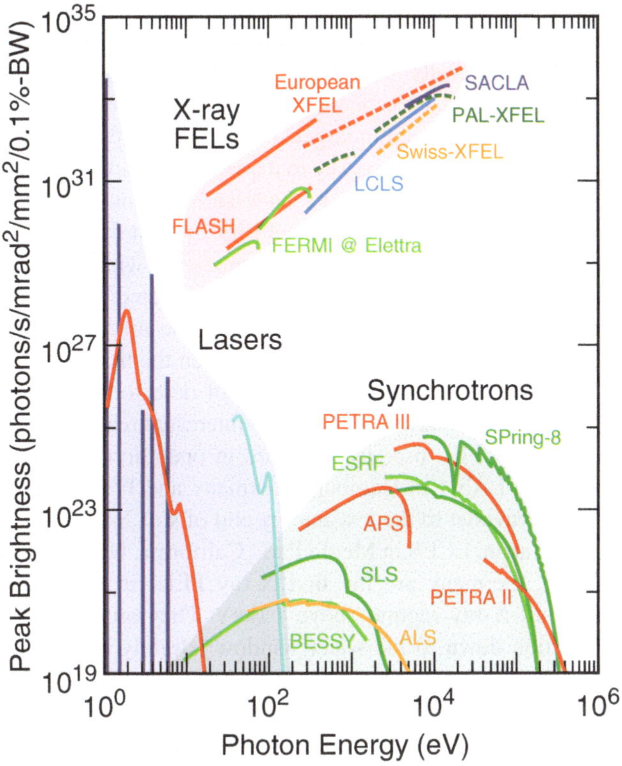

Fig. 1.3 Comparison of peak brightness as a function of photon energy between conventional lasers and higher harmonic generation sources, synchrotron sources, and X-ray free electron lasers. Modified by Gregory Stewart and Terry Anderson from [69]

a very narrow cone, with lower divergence than spontaneous radiation and with a narrower bandwidth. This ultimately results in a peak emitted brightness 9 to 10 orders of magnitude higher than the spontaneous radiation from third-generation sources. A comparison of typical performance of FELs and synchrotron sources is shown in Fig. 1.3.

The fluctuating nature of X-ray FEL beams creates a need for diagnostics to measure the pulse-to-pulse fluctuation of beam parameters that could influence the scientific measurement. For example, the technique of absorption spectroscopy, where the changes between the transmitted spectrum and the incident spectrum can reveal information on the fine details of the electronic structure of the sample, requires an accurate knowledge of the spectrum on a single pulse. Such requirement, among others, has led to the creation of single-shot spectrometer diagnostics based on bent crystal concepts where the X-rays are spatially dispersed based on their energy and measured by an area detector [78]. Other methods such as time-resolved diffraction require the measurement of very fine changes in intensity between a sample in its ground and excited states, or in the case of a sample excited via an optical laser, the ground (unpumped) and pumped state. Accurate, single-shot capable, nondestructive intensity monitors were required to achieve better than 0.1% accuracy in time-resolved measurements. These intensity measurements are based on Compton backscattering from thin targets [19, 66]. Diagnostics on the properties of any beam used to probe a sample is key to a quantitative scientific understanding. Most X-ray sources prior to X-ray FELs have stable beams that do not require constant monitoring and measurement. X-ray FELs, by their fluctuating nature, have required creativity in developing the required diagnostics, with many highlighted in the proceedings of a recent conference dedicated to photon diagnostics [14].

Beyond diagnostics, the instantaneous nature of the arrival of all X-rays at a detector and the relatively high repetition rates of even the first X-ray FELs creates a need for novel detector technology. A review of detector technology in use at LCLS is recommended as a starting point to the interested reader [9].

As of mid-2017, five FEL user facilities are in operation at or near the X-ray range. The first two, FLASH in Hamburg, Germany and FERMI in Trieste, Italy [4], operate in the ultraviolet to the low-energy end of soft X-rays. The other three, SACLA in Harima, Japan, LCLS in Menlo Park, California, USA, and the European XFEL in Hamburg, Germany are the first X-ray FELs in operation capable of operating in the hard X-ray regime above 10 keV. They are also capable of soft X-ray FEL generation down to the water window (slightly below the oxygen K-edge of 533 eV and above the carbon K-edge of 282 eV). The FEL-based results presented in this book will be entirely from four of these operating facilities, with the European XFEL being too recently operational to present results.

Other FELs have now demonstrated lasing and are undergoing early commissioning or are very near completion and will begin user programs in the late 2017 or in 2018. These include the PAL-XFEL in Pohang, South Korea and the SwissFEL in Villigen, Switzerland. Upcoming and potential new facilities as well as the novel science they will allow will be discussed more thoroughly in Chap. 16.

1.3 The Scientific Applications of X-Ray FELs

X-ray FELs produce beams of X-rays that have properties unseen before and the most successful use of these beams is via the exploitation of these unique capabilities, which can provide information unobtainable via other methods. Along with the many unique opportunities afforded by FEL beams come equivalently unique challenges. We will briefly explore the breadth of scientific exploration using X-ray FELs from the perspective of their unique capabilities and highlight how these were exploited with a few examples, ultimately leading into how X-ray FELs are valuable tools for biological studies.

1.3.1 Using the Time Resolution

The typical duration of an X-ray FEL pulse is roughly three orders of magnitude shorter than a typical shortest pulse from a synchrotron (but contains approximately the same number of photons that a synchrotron would produce in 1 s). With pulse lengths typically shorter than 30 fs, this opens a new area of ultrafast science. X-ray FELs combine these very short pulses with short wavelengths, allowing simultaneous high spatial and temporal resolution like never before. Many of the fundamental interactions in matter involving electrons and nuclei happen on the few tens of femtosecond timescales and X-ray FELs are ideal tools to study these interactions. Such fast dynamics must be triggered in a controlled and reproducible manner to extract information on these ultrafast timescales. This trigger is a brief stimulus applied to the sample, with optical pumping with an ultrafast laser being the main tool to initiate dynamics to be probed by the X-ray FEL beam. Light stimulation is currently the only reliable method to initiate dynamics with a few tens of femtosecond accuracy or better.

Pump-probe methods are ubiquitous in X-ray FEL experiments with the majority of experiments employing a laser illuminating the sample to study its dynamics. These methods are applied to all fields of science. In material science, optical pump X-ray probe methods are used to study lattice vibrations and phonon dynamics [68], material properties [33], strongly correlated systems and quantum materials [38], spin dynamics [37] as well as catalytic interfaces [53], to name just a few examples. In chemistry, ultrafast bond breaking and formation initiated by an optical laser can be observed to create a molecular movie of a reaction [36, 49]. Various scattering and spectroscopic methods can also be used to better understand charge transfer in metal complexes [12, 43, 77], eventually helping us to understand systems with potential energy-harvesting applications.

In high energy density (HED) research, it is of interest to create warm or hot states of matter that have thermal energy well beyond what is typically found on Earth but with high density typical of solid matter. These states can be created with very powerful lasers impinging on the sample and depositing a lot of energy in a short time. These states are very short lived as they will expand and cool rapidly after laser illumination. This allows, for example, conditions found in the center of planets, where high densities are sustained at high temperatures and pressures, to be reached under laboratory conditions. X-ray FEL beams are ideal to penetrate the dense material and probe the transient structure of these states of matter [18]. It is also possible to use longer pulsed lasers to initiate a shock in materials and study material failure, among other things [11, 22, 48, 63, 76].

Other variations of pump-probe techniques can involve using the X-ray beam as the pump, for example, to heat materials evenly throughout their volume (isochorically) [72]. In this case, an optical laser probe or the two-pulse capabilities of X-ray FELs [24, 46] can be used to probe the X-ray-induced dynamics in the sample. A few examples of X-ray pumped dynamics are referenced for the interested reader [17, 28, 40]. Two-pulse applications of X-ray FELs are increasing in use in many fields, including biology, to study the ultrafast processes that lead to radiation damage in the sample.

Measuring time-resolved dynamics with such high temporal resolution is a challenging endeavor due to the nature of the FEL beam. The amplification from noise that gives rise to SASE pulses causes essentially every beam parameter to fluctuate, including the arrival time of the X-rays. Diagnostics to measure this arrival time relative to the pump laser are necessary. Such diagnostics were developed with the advent of X-ray FEL beams. The interaction of the intense X-ray pulse changes the optical properties of a target, which is then probed with a small fraction of the pump laser beam intensity. The ultrafast change in the index of refraction of the target material leads to changes in the optical reflectivity and transmission through this target and can provide arrival time information for the X-rays relative to the laser [8, 25, 34]. These arrival time measurements can be used to sort data into more accurate time bins. The fluctuations in the arrival time, combined with a suitable diagnostic, can be utilized to more rapidly sample time points and sort the data in post-analysis. In this way, a challenging aspect of FEL beams can be turned into a useful advantage by just letting the beam jitter to fill the time bins and sorting that data later.

For soft X-ray and UV FEL beams, it is possible to manipulate the electron bunch with laser beams to generate shorter pulses. A device called XLEAP (X-ray Laser Enhanced Attosecond Pulse Generation) at LCLS will bring attosecond-scale dynamics into the realm of possibilities [45]. At the FERMI FEL, the very accurate synchronization between two pulses from the FEL has allowed the measurement of the beam temporal characteristics on the attosecond timescale [70] and the use of these exquisitely timed pulses to measure attosecond dynamics [56].

In biology, the high temporal-resolution capabilities of X-ray FEL is exploited for the study of light-sensitive proteins and enzymes as will be described in Chap. 11. Dynamics can also be initiated by other nonoptical methods such as rapid mixing but with lower time resolution due to the less precise initiation of dynamics. This will be the topic of Chap. 12.

1.3.2 Using the High Peak Intensity

As described above, X-ray FELs will typically produce X-ray pulses roughly three orders of magnitude shorter than conventional light sources, and these pulses will contain roughly as many X-rays as delivered in 1 s at a synchrotron. This leads to the potential for extremely high intensities when the beam is tightly focused. This can be exploited in a few ways.

1.3.2.1 Diffraction-Before-Destruction

Diffraction-before-destruction is a concept described in 2000 by Neutze *et al.* [52]. With sufficiently high intensities and sufficiently short pulses, it is possible to mitigate and possibly overcome conventional radiation damage limits that exist for longer, continuous measurements. Some of the X-rays incident on the sample probe its structure via scattering as described earlier but the majority of the incident X-rays that interact with the sample deposit their energy in the sample via X-ray absorption. This deposited energy eventually damages the sample, but this process is not instantaneous. If the pulse duration can be kept shorter than the damage dynamics, then it is in principle possible to pack more X-rays in this short time than the damage limit would allow at a synchrotron. Probing a mostly undamaged sample with a higher number of X-rays would yield a higher signal and a potentially undamaged higher-resolution structure. Diffraction-before-destruction is critical to the majority of biological applications at an X-ray FEL due to the radiation-sensitive nature of biological samples. It will be discussed further in Chap. 2 as part of the discussion on X-ray FEL crystallography, as well as Chap. 6 where radiation damage will be discussed in detail.

1.3.2.2 Using FELs to Create New States of Matter

As mentioned before, the X-ray FEL beam can be used as a pump to trigger the start of a dynamic process. Here, the very high intensities afforded by a tightly focused FEL beam provide the capability to isochorically heat a solid material to very high and uniform temperature. The penetrating power of X-rays leads to uniform illumination, even for dense samples, compared to mostly surface heating from optical lasers. This capability is of great interest to the fields of warm and hot dense matter, but also fundamental research in the effects of these extreme X-ray intensities is necessary to understand the limitations of the assumptions of the diffraction-before-destruction idea. For the smallest biological samples, a tightly focused beam is required to maximize the illumination of the sample and maximize the signal and some level of damage or structural change may occur during the pulse. Understanding the dynamics involved in short-lived hot dense states of matter during the pulse duration or shortly thereafter is of fundamental and applied interest. Radiation damage and how it affects biological samples in X-ray FEL measurements will be further discussed in Chap. 6.

1.3.2.3 Nonlinear X-Ray Physics

The high X-ray fields produced by focused X-ray FEL beams can open the door to novel methods such as nonlinear optics or nonlinear spectroscopies. A few examples are the observation of anomalous Compton scattering [20], nonsequential two-photon absorption [21, 64] as well as stimulated emission [57, 75]. The interaction of multiple X-rays with the same molecule over short times leads to novel nonlinear effects only observable using an X-ray FEL. Nonlinear spectroscopies have the potential to be very sensitive probes of the fine details of electronic states and could reveal important details on enzymatic activities, for example. Spectroscopic applications of X-ray FELs in biology will be discussed in Chap. 13.

1.3.3 Using the Coherence

Laser radiation is highly coherent, and X-ray FELs are no exception. The coherence properties of existing X-ray FELs were measured for both soft and hard X-rays at multiple facilities [23, 29, 39, 41, 50, 58, 71]. An increased level of coherence will improve the signal level and the quality of the data for scattering measurements that rely on this coherence. An example is X-ray Photon Correlation Spectroscopy (XPCS), where the intrinsic dynamics of a sample are probed by tracking the correlation of the intensities in the speckle pattern as a function of time [42, 62]. While such a method shows great promise for understanding the dynamics of untriggered processes, it has limited application in biology due to the need to measure the same sample over time, where radiation damage becomes an issue.

Coherent diffractive imaging (CDI) methods greatly benefit from the increased coherence of an FEL beam compared to a synchrotron and these have been put to use in combination with time resolution for pump-probe studies of phonons in nanoparticles, for example [13]. For biological coherent imaging, the small size of a typical reproducible sample leads to weak signals which requires a highly intense focused beam, which in turn leads to rapid damage and complete destruction of the sample on a single pulse. Therefore, biological imaging applications using an X-ray FEL beam must combine the short pulse and high intensity in a diffraction-before-destruction mode to leverage the increase in coherence of the beam. Chapter 14 will expand further on the status of imaging applications for X-ray FELs in biology.

1.4 Summary

X-ray free electron lasers are the latest X-ray sources that arise from over 120 years of X-ray science and development. They are the result of the last five decades of accelerator-based X-rays source development that have provided extremely valuable tools located at centralized facilities available to broad scientific communities. X-ray FELs push X-ray science into new directions of unprecedented peak intensities in short X-ray bursts that can be exploited for research in atomic, molecular, and optical sciences, material science, ultrafast chemistry, high-energy-density matter, and the study of soft matter and its dynamics. As is the subject of this book, they are also proving their worth in biological studies where they can overcome some of the radiation damage limitations that plague some important samples. X-ray FELs add ultrafast dynamic capabilities, in combination with high spatial resolution, to the tool set for biologists and structural biologists as they aim to understand the mechanisms of life.

X-ray FELs available today and in the near future span a spectral range from UV to soft and to hard X-rays. Hard X-rays are primarily used for seeing where the atomic nuclei are located, either in a static structure or a dynamic structure using time-resolved methods. They are also used to probe the electronic states of metal centers with absorption edges in the hard X-ray range in important metalloenzymes or proteins.

Softer X-rays are used for imaging methods where the higher scattering cross section can be beneficial for an increased signal level. They are also useful tools for spectroscopic studies of the fine electronic structure of lighter atoms with absorption edges at lower energies.

This book, via its multiple chapters on the status of multiple specific aspects of X-ray FEL experiments in biology, is hoped to provide the reader with a general introduction to the field while still providing sufficient details for those looking to contribute to the field.

References

1. Abad-Zapatero, C. (2012, May). Notes of a protein crystallographer: On the high-resolution structure of the PDB growth rate. *Acta Crystallographica Section D, 68*(5), 613–617.
2. Ackermann, W., et al. (2007). Operation of a free electron laser from the extreme ultraviolet to the water window. *Nature Photonics, 1*, 336.
3. Ahrens, T. J. (2013). *Mineral physics & crystallography: A handbook of physical constants*. Washington, DC: American Geophysical Union.
4. Allaria, E., Appio, R., Badano, L., Barletta, W. A., Bassanese, S., Biedron, S. G., et al. (2012). Highly coherent and stable pulses from the Fermi seeded free-electron laser in the extreme ultraviolet. *Nature Photonics, 6*, 699.
5. Als-Nielsen, J., & McMorrow, D. (2011). *Elements of modern X-ray physics*. New York: Wiley.
6. Berman, H. M., Westbrook, J., Feng, Z., Gilliland, G., Bhat, T. N., Weissig, H., et al. (2000). The protein data bank. *Nucleic Acids Research, 28*(1), 235–242.
7. Bilderback, D. H., Elleaume, P., & Weckert, E. (2005). Review of third and next generation synchrotron light sources. *Journal of Physics B: Atomic, Molecular and Optical Physics, 38*(9), S773.
8. Bionta, M. R., Hartmann, N., Weaver, M., French, D., Nicholson, D. J., Cryan, J. P., et al. (2014). Spectral encoding method for measuring the relative arrival time between X-ray/optical pulses. *Review of Scientific Instruments, 85*(8), 083116.
9. Blaj, G., Caragiulo, P., Carini, G., Carron, S., Dragone, A., Freytag, D., et al. (2015, May). X-ray detectors at the Linac Coherent Light Source. *Journal of Synchrotron Radiation, 22*(3), 577–583.
10. Bostedt, C., Boutet, S., Fritz, D. M., Huang, Z., Lee, H. J., Lemke, H. T., et al. (2016, March). Linac coherent light source: The first five years. *Reviews of Modern Physics, 88*, 015007.
11. Briggs, R., Gorman, M. G., Coleman, A. L., McWilliams, R. S., McBride, E. E., McGonegle, D., et al. (2017, January). Ultrafast X-ray diffraction studies of the phase transitions and equation of state of scandium shock compressed to 82 GPa. *Physical Review Letters, 118*, 025501.

12. Canton, S. E., Kjar, K. S., Vanko, G., van Driel, T. B., Adachi, S., Bordage, A., et al. (2015). Visualizing the non-equilibrium dynamics of photoinduced intramolecular electron transfer with femtosecond X-ray pulses. *Nature Communications, 6*, 6359.
13. Clark, J. N., Beitra, L., Xiong, G., Higginbotham, A., Fritz, D. M., Lemke, H. T., et al. (2013). Ultrafast three-dimensional imaging of lattice dynamics in individual gold nanocrystals. *Science, 341*(6141), 56–59.
14. Cocco, D., Moeller, S., Ploenjes, E., & Zangrando, M. (2018, January). PhotonDiag2017 workshop: Introductory overview. *Journal of Synchrotron Radiation, 25*(1), 1–2.
15. Deacon, D. A. G., Elias, L. R., Madey, J. M. J., Ramian, G. J., Schwettman, H. A., & Smith, T. I. (1977). First operation of a free-electron laser. *Physical Review Letters, 38*, 892.
16. Emma, P., Akre, R., Arthur, J., Bionta, R., Bostedt, C., Bozek, J., et al. (2010). First lasing and operation of an Ångstrom-wavelength free-electron laser. *Nature Photonics, 4*(9), 641–647.
17. Ferguson, K. R., Bucher, M., Gorkhover, T., Boutet, S., Fukuzawa, H., Koglin, J. E., et al. (2016). Transient lattice contraction in the solid-to-plasma transition. *Science Advances, 2*(1), e1500837.
18. Fletcher, L. B., Lee, H. J., Dppner, T., Galtier, E., Nagler, B., Heimann, P., et al. (2015). Ultrabright X-ray laser scattering for dynamic warm dense matter physics. *Nature Photonics, 9*, 274–279.
19. Fritz, D. M., Cammarata, M., Aymeric, R., Caronna, C., Lemke, H. T., Zhu, D., et al. (2011). A single-shot intensity-position monitor for hard X-ray FEL sources. *Proceedings of SPIE, 8140*:8140-1–8140-6.
20. Fuchs, M., Trigo, M., Chen, J., Ghimire, S., Shwartz, S., Kozina, M., et al. (2015). Anomalous nonlinear X-ray Compton scattering. *Nature Physics, 11*, 964–970.
21. Ghimire, S., Fuchs, M., Hastings, J., Herrmann, S. C., Inubushi, Y., Pines, J., et al. (2016, October). Nonsequential two-photon absorption from the *k* shell in solid zirconium. *Physical Review A, 94*, 043418.
22. Gleason, A. E., Bolme, C. A., Lee, H. J., Nagler, B., Galtier, E., Milathianaki, D., et al. (2015). Ultrafast visualization of crystallization and grain growth in shock-compressed SiO_2. *Nature Communications, 6*, 8191.
23. Gutt, C., Wochner, P., Fischer, B., Conrad, H., Castro-Colin, M., Lee, S., et al. (2012). Single shot spatial and temporal coherence properties of the SLAC linac coherent light source in the hard X-ray regime. *Physical Review Letters*, 108(2):024801.
24. Hara, T., Inubushi, Y., Katayama, T., Sato, T., Tanaka, H., Tanaka, T., et al. (2013). Two-colour hard X-ray free-electron laser with wide tunability. *Nature Communications, 4*, 2919.
25. Harmand, M., Coffee, R., Bionta, M. R., Chollet, M., French, D., Zhu, D., et al. (2013). Achieving few-femtosecond time-sorting at hard X-ray free-electron lasers. *Nature Photonics, 7*(3), 215–218.
26. Hecht, E. (2002). *Optics*. Reading, MA: Addison-Wesley.
27. Huang, Z., & Kim, K.-J. (2007). A review of X-ray free-electron laser theory. *Physical Review Special Topics – Accelerators and Beams, 10*, 034801.
28. Inoue, I., Inubushi, Y., Sato, T., Tono, K., Katayama, T., Kameshima, T., et al. (2016). Observation of femtosecond X-ray interactions with matter using an X-ray-X-ray pump-probe scheme. *Proceedings of the National Academy of Sciences of the United States of America, 113*, 1492.
29. Inoue, I., Tono, K., Joti, Y., Kameshima, T., Ogawa, K., Shinohara, Y., et al. (2015). Characterizing transverse coherence of an ultra-intense focused X-ray free-electron laser by an extended young's experiment. *IUCrJ, 2*, 620.
30. Inubushi, Y., Tono, K., Togashi, T., Sato, T., Hatsui, T., Kameshima, T., et al. (2012). Determination of the pulse duration of an X-ray free electron laser using highly resolved single-shot spectra. *Physical Review Letters, 109*, 144801.
31. Ishikawa, T., Aoyagi, H., Asaka, T., Asano, Y., Azumi, N., Bizen, T., et al. (2012). A compact X-ray free-electron laser emitting in the sub-Ångström region. *Nature Photonics, 6*, 540.
32. IUCr, & Wilson, A. J. C. (2016). *International tables for crystallography*. Dordrecht: Kluwer Academic Publishers.

33. Jiang, M. P., Trigo, M., Savić, I., Fahy, S., Murray, D., Bray, C., et al. (2016, July). The origin of incipient ferroelectricity in lead telluride. *Nature Communications, 7*, 12291.
34. Katayama, T., Owada, S., Togashi, T., Ogawa, K., Karvinen, P., Vartiainen, I., et al. (2016). A beam branching method for timing and spectral characterization of hard X-ray free-electron lasers. *Structure and Dynamics, 3*, 034301.
35. Kendrew, J. C., Bodo, G., Dintzis, H. M., Parrish, R. G., Wyckoff, H., & Phillips, D. C. (1958). A three-dimensional model of the myoglobin molecule obtained by X-ray analysis. *Nature, 181*(4610), 662–666.
36. Kim, K. H., Kim, J. G., Nozawa, S., Sato, T., Oang, K. Y., Kim, T. W., et al. (2015). Direct observation of bond formation in solution with femtosecond X-ray scattering. *Nature, 518*, 385.
37. Kubacka, T., Johnson, J. A., Hoffmann, M. C., Vicario, C., de Jong, S., Beaud, P., et al. (2014). Large-amplitude spin dynamics driven by a THz pulse in resonance with an electromagnon. *Science, 343*(6177), 1333–1336.
38. Lantz, G., Mansart, B., Grieger, D., Boschetto, D., Nilforoushan, N., Papalazarou, E., et al. (2017, January). Ultrafast evolution and transient phases of a prototype out-of-equilibrium Mott–Hubbard material. *Nature Communications, 8*, 13917.
39. Lee, S., Roseker, W., Gutt, C., Fischer, B., Conrad, H., Lehmkühler, F., et al. (2013). Single shot speckle and coherence analysis of the hard X-ray free electron laser LCLS. *Optics Express, 21*(21), 24647–24664.
40. Lehmann, C. S., Picón, A., Bostedt, C., Rudenko, A., Marinelli, A., Moonshiram, D., et al. (2016, July). Ultrafast X-ray-induced nuclear dynamics in diatomic molecules using femtosecond X-ray-pumpˇX-ray-probe spectroscopy. *Physical Review A, 94*, 013426.
41. Lehmkühler, F., Gutt, C., Fischer, B., Schroer, M. A., Sikorski, M., Song, S., et al. (2014). Single shot coherence properties of the free-electron laser SACLA in the hard X-ray regime. *Scientific Reports, 4*, 5234.
42. Lehmkühler, F., Kwaśniewski, P., Roseker, W., Fischer, B., Schroer, M. A., Tono, K., et al. (2015). Sequential single shot X-ray photon correlation spectroscopy at the SACLA free electron laser. *Scientific Reports, 5*, 17193.
43. Lemke, H. T., Kjær, K. S., Hartsock, R., van Driel, T. B., Chollet, M., Glownia, J. M., et al. (2017, May). Coherent structural trapping through wave packet dispersion during photoinduced spin state switching. *Nature Communications, 8*, 15342.
44. Madey, J. (1971). Stimulated emission of bremsstrahlung in a periodic magnetic field. *Journal of Applied Physics, 42*, 1906.
45. Marinelli, A., Coffee, R., Vetter, S., Hering, P., West, G. N., Gilevich, S., et al. (2016, June). Optical shaping of X-ray free-electron lasers. *Physical Review Letters, 116*, 254801.
46. Marinelli, A., Ratner, D., Lutman, A. A., Turner, J., Welch, J., Decker, F.-J., et al. (2015). High-intensity double-pulse X-ray free-electron laser. *Nature Communications, 6*, 6369.
47. Miao, J., Ishikawa, T., Robinson, I. K., & Murnane, M. M. (2015). Beyond crystallography: Diffractive imaging using coherent X-ray light sources. *Science, 348*(6234), 530–535.
48. Milathianaki, D., Boutet, S., Williams, G. J., Higginbotham, A., Ratner, D., Gleason, A. E., et al. (2013). Femtosecond visualization of lattice dynamics in shock-compressed matter. *Science*, 342(6155):220–223.
49. Minitti, M. P., Budarz, J. M., Kirrander, A., Robinson, J. S., Ratner, D., Lane, T. J., et al. (2015, June). Imaging molecular motion: Femtosecond X-ray scattering of an electrocyclic chemical reaction. *Physical Review Letters, 114*, 255501.
50. Mitzner, R., Siemer, B., Neeb, M., Noll, T., Siewert, F., Roling, S., et al. (2008). Spatio-temporal coherence of free electron laser pulses in the soft X-ray regime. *Optics Express, 16*(24), 19909.
51. Murphy, J. B., & Pellegrini, C. (1990). *Introduction to the physics of the free electron laser*. Amsterdam: North-Holland.
52. Neutze, R., Wouts, R., van der Spoel, D., Weckert, E., & Hajdu, J. (2000). Potential for biomolecular imaging with femtosecond X-ray pulses. *Nature, 406*(6797), 752–757.

53. Ostrom, H., Oberg, H., Xin, H., LaRue, J., Beye, M., Dell'Angela, M., et al. (2015). Probing the transition state region in catalytic CO oxidation on RU. *Science, 347*(6225), 978–982.
54. Pellegrini, C. (2012). The history of X-ray free-electron lasers. *European Physical Journal H, 37*(5), 659–708.
55. Pellegrini, C., Marinelli, A., & Reiche, S. (2016, March). The physics of X-ray free-electron lasers. *Reviews of Modern Physics, 88*, 015006.
56. Prince, K. C., Allaria, E., Callegari, C., Cucini, R., De Ninno, G., Di Mitri, S., et al. (2016). Coherent control with a short-wavelength free-electron laser. *Nature Photonics, 10*, 176–179.
57. Rohringer, N., Ryan, D., London, R. A., Purvis, M., Albert, F., Dunn, J., et al. (2012). Atomic inner-shell X-ray laser at 1.46 nanometres pumped by an X-ray free-electron laser. *Nature, 481*(7382), 488–491.
58. Ruiz-Lopez, M., Faenov, A., Pikuz, T., Ozaki, N., Mitrofanov, A., Albertazzi, B., et al. (2017, January). Coherent X-ray beam metrology using 2D high-resolution Fresnel-diffraction analysis. *Journal of Synchrotron Radiation, 24*(1), 196–204.
59. Saldin, E. L., Schneidmiller, E. A., & Yurkov, M. V. (2000). *The physics of free electron lasers*. Berlin: Springer.
60. Schmüser, P., Dohlus, M., Rossbach, J., & Behrens, C. (2014). *Free-electron lasers in the ultraviolet and X-ray regime: Physical principles, experimental results, technical realization*. Springer Tracts in Modern Physics. Cham: Springer.
61. Shintake, T., Tanaka, H., Hara, T., Tanaka, T., Togawa, K., Yabashi, M., et al. (2008). A compact free-electron laser for generating coherent radiation in the extreme ultraviolet region. *Nature Photonics, 2*(9), 555–559.
62. Shpyrko, O. G. (2014, September). X-ray photon correlation spectroscopy. *Journal of Synchrotron Radiation, 21*(5), 1057–1064.
63. Swinburne, T. D., Glavicic, M. G., Rahman, K. M., Jones, N. G., Coakley, J., Eakins, D. E., et al. (2016, April). Picosecond dynamics of a shock-driven displacive phase transformation in Zr. *Physical Review B, 93*, 144119.
64. Tamasaku, K., Shigemasa, E., Inubushi, Y., Katayama, T., Sawada, K., Yumoto, H., et al. (2014). X-ray two-photon absorption competing against single and sequential multiphoton processes. *Nature Photonics, 8*, 313.
65. Tiedtke, K., Azima, A., von Bargen, N., Bittner, L., Bonfigt, S., Düsterer, S., et al. (2009). The soft X-ray free-electron laser flash at DESY: Beamlines, diagnostics and end-stations. *New Journal of Physics, 11*(2), 023029.
66. Tono, K., Togashi, T., Inubushi, Y., Sato, T., Katayama, T., Ogawa, K., et al. (2013). Beamline, experimental stations and photon beam diagnostics for the hard X-ray free electron laser of SACLA. *New Journal of Physics, 15*(8), 083035.
67. Tono, K., Togashi, T., Inubushi, Y., Sato, T., Katayama, T., Ogawa, K., et al. (2013). Beamline for X-ray free electron laser of SACLA. *New Journal of Physics, 12*, 083035.
68. Trigo, M., Fuchs, M., Chen, J., Jiang, M. P., Cammarata, M., Fahy, S., et al. (2013). Fourier-transform inelastic X-ray scattering from time- and momentum-dependent phonon-phonon correlations. *Nature Physics, 9*(12), 790–794.
69. Ullrich, J., Rudenko, A., & Moshammer, R. (2012). Free-electron lasers: New avenues in molecular physics and photochemistry. *Annual Review of Physical Chemistry, 63*(1), 635–660. PMID: 22404584.
70. Usenko, S., Przystawik, A., Jakob, M. A., Lazzarino, L. L., Brenner, G., Toleikis, S., et al. (2017). Attosecond interferometry with self-amplified spontaneous emission of a free-electron laser. *Nature Communications, 8*, 15626.
71. Vartanyants, I. A., Singer, A., Mancuso, A. P., Yefanov, O. M., Sakdinawat, A., Liu, Y., et al. (2011). Coherence properties of individual femtosecond pulses of an X-ray free-electron laser. *Physical Review Letters, 107*(14), 144801.
72. Vinko, S. M., Ciricosta, O., Cho, B. I., Engelhorn, K., Chung, H. K., Brown, C. R., et al. (2012). Creation and diagnosis of a solid-density plasma with an X-ray free-electron laser. *Nature, 482*(7383), 59–62.

73. Warren, B. E. (1969). *X-ray diffraction*. Addison-Wesley series in metallurgy and materials engineering. Mineola, NY: Dover Publications.
74. Yabashi, M., Tanaka, H., Tanaka, T., Tomizawa, H., Togashi, T., Nagasono, M., et al. (2013). Compact XFEL and AMO sciences: SACLA and SCSS. *Journal of Physics B, 46*(16), 164001.
75. Yoneda, H., Inubushi, Y., Nagamine, K., Michine, Y., Ohashi, H., Yumoto, H., et al. (2015). Atomic inner-shell laser at 1.5-angstrom wavelength pumped by an X-ray free-electron laser. *Nature, 524*, 446–449.
76. Zastrau, U., Gamboa, E. J., Kraus, D., Benage, J. F., Drake, R. P., Efthimion, P., et al. (2016). Tracking the density evolution in counter-propagating shock waves using imaging X-ray scattering. *Applied Physics Letters, 109*(3), 031108.
77. Zhang, W., Alonso-Mori, R., Bergmann, U., Bressler, C., Chollet, M., Galler, A., et al. (2014). Tracking excited-state charge and spin dynamics in iron coordination complexes. *Nature, 509*(7500), 345–348.
78. Zhu, D., Cammarata, M., Feldkamp, J., Fritz, D. M., Hastings, J., Lee, S., et al. (2013). Design and operation of a hard X-ray transmissive single-shot spectrometer at LCLS. *Journal of Physics Conference Series, 425*(5), 052033.

Chapter 2
Serial Femtosecond Crystallography (SFX): An Overview

Mark S. Hunter and Petra Fromme

2.1 Background

Understanding the processes of life at the molecular and atomic levels is a major goal of the biological and life sciences. A challenge in this endeavor is that the length and time scales of interest for studying biological processes span many (10+) orders of magnitude. The components of life are as small as nanometers (size of atoms and molecules) to make a meter-scale living being such as a human. In the temporal domain, molecular dynamics occur on ultrafast time scales (nanoseconds or faster) leading to concerted dynamics on a much slower scale on the order of a second or more at the molecular level, to sustain a human life for up to 100 years, for example. No individual technique has proven to be the panacea to study the processes in biology at these disparate length and time scales. X-ray crystallography has traditionally provided information at high spatial resolution (atomic to molecular resolution) but information of limited temporal resolution, affecting the ability to study the dynamics of biological macromolecules. Since crystallography relies on producing crystals with repeating units of molecules that are (mostly) identical and oriented in a well-ordered manner, there is a pragmatic upper limit to the length scale that can be studied. Growing well-ordered crystals

M. S. Hunter (✉)
Linac Coherent Light Source, SLAC National Accelerator Laboratory, Menlo Park, CA, USA

Biosciences Division, SLAC National Accelerator Laboratory, Menlo Park, CA, USA
e-mail: mhunter2@slac.stanford.edu

P. Fromme
School of Molecular Sciences, Arizona State University, Tempe, AZ, USA

Biodesign Center for Applied Structural Discovery, the Biodesign Institute, Arizona State University, Tempe, AZ, USA
e-mail: PFromme@asu.edu

S. Boutet et al. (eds.), *X-ray Free Electron Lasers*,
https://doi.org/10.1007/978-3-030-00551-1_2

of large macromolecular complexes or complexes with large flexible regions is very difficult. Serial femtosecond crystallography (SFX) is utilizing the unique properties of X-ray free electron lasers (FELs) to expand the reach of crystallography to targets that are more difficult to crystallize. SFX also allows for time-resolved X-ray crystallography studies at room temperature to study the dynamics of biological molecules.

SFX is a growing field with many important experiments done to date. As such, several excellent review articles have been written to date about SFX [1–3] and the interested reader is directed to them for different perspectives on the technique.

2.2 The Need for SFX

Crystallizing macromolecules is very different from crystallization of small molecules. Crystallization of larger biomolecules involves slower transport properties in solution. Larger biomolecules are often very fragile and maintain their function and structural integrity only in a narrow window of experimental conditions such as solvent, ionic strength, pH, temperature as well as precipitants—the very parameters that can be used to lower the solubility of the molecule. Membrane protein crystallization is even more complex as they reside in the membrane and are amphiphilic and therefore must be crystallized either in form of protein-detergent micelles or in complex phases of aqueous solvent, such as lipidic cubic phases [4].

Generally, protein crystals have much higher solvent content in the crystal than small molecules, feature a small ratio of crystal contact to total surface area of the molecule and are also grown at much higher levels of supersaturation [5]. Potentially as a direct result of these differences, the growth of large single crystals of proteins can be challenging. Long range disorder effects accumulate and often lead to poor order or even termination of crystal growth. Therefore, the observations of small, nanocrystalline or microcrystalline showers during initial sparse-matrix screening is common [6] and suggests that it may in some cases be easier to generate well-ordered microcrystals than to grow large, well-ordered macrocrystals.

A technique that could make use of microcrystals would be incredibly valuable, as years can be spent optimizing the crystallization conditions for interesting macromolecules such as many of the membrane proteins and macromolecular complexes. Chapter 3 of this book is focused on the growth and characterization of nanocrystals and microcrystals for X-ray FEL studies. The challenge with using microcrystals for high-resolution structure determination is that the very act of collecting data on the samples with X-rays will damage the sample. In essence there is a competition between the X-ray dose required to accumulate high-resolution data and how rapidly the sample decays in the beam. To overcome radiation damage in order to obtain high resolution data sets typically requires spreading the dose to as many molecules as possible, often achieved with large, well-ordered crystals illuminated at a low dose per molecule or shifted during data collection to illuminate fresh spots of the sample. The damage is caused by X-ray induced photoionization

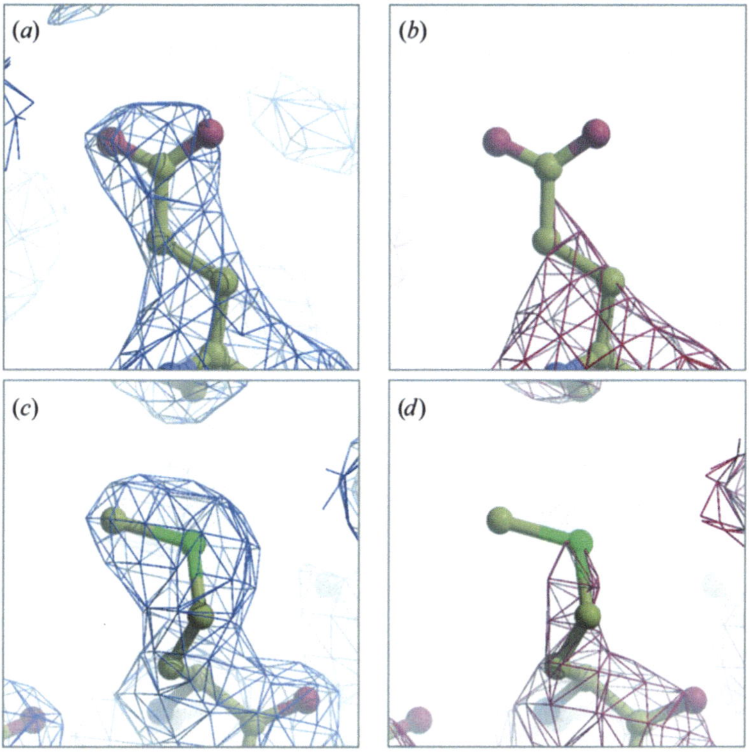

Fig. 2.1 Ill effects of radiation damage. Specific structural damage inflicted on a cryocooled crystal of apoferritin during sequential data sets collected on beamline ID14-4 at ESRF. (**a**) 2Fo—Fc map of Glu63 contoured at 0.2 e Å^{-3} after a dose of 2.5 MGy and (**b**) after 50 MGy. (**c**) 2Fo—Fc map of Met96 contoured at 0.2 e Å^{-3} after a dose of 2.5 MGy and (**d**) after 50 MGy, showing loss of electron density around the disordered atoms. Figure and caption reproduced from [7]

followed by diffusion of the radicals causing photoreduction, which leads to rapid reduction of metals, bond breaking and finally destruction of the biomolecule and the crystalline order. In the diffraction data these damage processes will manifest at both a global and local level. Global damage will appear as an overall reduction in scattering power from the sample and can manifest as large-scale destruction of a crystal, whereas local damage will lead to a reduction in interpretability of the electron density maps near certain components of the sample, as shown in Fig. 2.1 [7]. X-ray induced radiation damage will be covered in more depth in Chap. 6.

As discussed in Chap. 1, X-ray FELs have unique properties compared to synchrotron X-ray sources, namely substantially higher peak power and ultrashort pulse durations on the order of tens of femtoseconds (1 fs $= 10^{-15}$ s). Based on molecular dynamics simulations of the process of Coulomb explosion of a biomolecule irradiated by an X-ray FEL pulse in vacuum, Neutze et al. predicted that if the X-ray FEL pulse duration is shorter than the damage processes in the molecule, the diffraction data from single molecules or crystals could be collected from a damage-free structure, a concept later called diffraction-before-destruction [8]. Early experiments at FLASH (*F*ree-Electron *LAS*er in *H*amburg), the soft X-ray FEL at DESY in Hamburg, hinted at the promise of using diffraction-before-destruction for damage-free structure determination by successfully reconstructing an image of a nanofabricated artificial object from a single diffraction pattern measured by exposure to a single FLASH pulse [9]. Further tests of the diffraction-before-destruction concept were performed using FLASH. For example, a time evolution of sample damage was imaged using two illuminations of the sample with a controlled delay time between them. A novel method called time-delay holography was used to reflect the pulse back onto the exploding sample to probe the damage generated by the FLASH pulse as a function of time, confirming some of the theoretical predictions [10].

The excitement quickly built from the early FLASH results and the anticipation of the eventual impact of X-ray FELs on macromolecular crystallography (MX). However, questions remained about the viability of using submicron macromolecular crystals for structure determination, which was considered one of the most promising aspects of using X-ray FELs for MX. During the buildup to the first SFX experiments, powder diffraction measurements using liquid injection of nanocrystals of the integral membrane protein photosystem I (PSI) into the X-ray interaction region were performed at beam line 9.0.1 of the Advanced Light Source at Lawrence Berkeley National Laboratory [11]. PSI is a large membrane protein complex consisting of 36 proteins and 381 cofactors, and is a challenging test case for serial crystallography methods. Ultimately, the experiments were able to measure powder diffraction from PSI crystals that were restricted in size to <220 nm along an edge (very weak diffraction was even recorded from <100 nm PSI crystals) [11, 12]. The experiments demonstrated that submicron crystals of a membrane protein could be effectively delivered in vacuo to the X-ray interaction region and retain their integrity and diffraction power. The results suggested that these submicron crystals might be useful for structure determination, if radiation damage was not a concern, but the idea could not be further tested until MX could be done at an X-ray FEL. These experiments, along with demonstration experiments at FLASH paved the way for SFX at the Linac Coherent Light Source (LCLS), the first high energy X-ray FEL. However, adapting protein crystallography to serial data collection at X-rays FELs proved challenging.

The nature of the X-ray FEL pulses makes conventional crystallography methods challenging and therefore new methods had to be invented, which range from nanocrystal growth and characterization, sample delivery, all the way to new data evaluation algorithms. In crystallography, the reconstructed electron density is a Fourier transform of the structure factors F_{hkl}, where the amplitude of F_{hkl} is correlated to the Bragg peak intensities I_{hkl} in measured diffraction patterns. In order to determine the intensities of the individual Bragg peaks, the reciprocal lattice points associated with the structure factors must meet the Bragg conditions, that is, at a given orientation only a partial subset of the reflections can be measured (for more detailed information please see any of the excellent X-ray crystallography textbooks that are available [13–15]). A full data set is typically measured through rotating the crystal during data collection. Typically in conventional crystallography, rotation series from individual (or a small number of) macromolecular crystals are collected in order to measure full reflections from each reciprocal lattice peak [16] and data are collected at cryogenic temperatures to reduce, but not evade, X-ray damage. For large crystals, diffraction data at X-ray FELs can be collected using a highly attenuated beam and moving the crystal between shots [17–19]. Generally, the high peak intensity of an X-ray FEL pulse makes multiple measurements on any single particle or microcrystal impossible; thereby diffraction from each microcrystal can only be measured once before being destroyed. Consequently, for microcrystals, new sample introduction methods to the X-ray beam have been developed as well as novel data analysis methods that allow for accurate determination of the structure factor amplitudes from millions of random single crystal diffraction snapshots.

2.2.1 Sample Preparation and Characterization for SFX

As with traditional macromolecular crystallography, sample preparation and characterization are still key to the successful outcome of an experiment. SFX allows microcrystals to be used for high-resolution structure determination but new methods had to be developed to grow microcrystals in sufficient quality and quantity for SFX studies.

While most crystals grown for conventional crystallography are grown by vapor diffusion methods, these are not suitable for most SFX experiments due to the difficulty in scaling them up to meet the high sample consumption requirements of SFX arising from the need for a new crystal on every pulse. Crystallization methods have been adapted and developed for growth of nanocrystals and microcrystals for SFX. In addition, methods have been developed where nanocrystals and small microcrystals are grown by expression of the protein in living insect cells [20, 21] or bacterial cells. Nanocrystals require new methods for detection and characterization and have to be discriminated from non-crystalline aggregates that are not useful for SFX. Unfortunately, nanocrystals are similar in size to the wavelength of visible light and can therefore not be identified with a light microscope. Even for small microcrystals (here defined as between 1 μm and 5 μm in size) the sharp edges typically associated with a crystal are not easily discernible with a light microscope. To make matters worse, checking for birefringence, a powerful method for identifying protein crystals larger than a few micrometers, becomes very challenging, if not impossible, for nanocrystals and small microcrystals [22]. Any additional scattering from the crystallization drop, such as resulting from a PEG (polyethylene glycol) skin, will make optical identification of microcrystals very difficult, and the in meso phases can be notoriously challenging [23]. Fortunately, a lot of knowledge has been gained and new methods have been developed for growth and characterization of the nanocrystals and small microcrystals used for SFX, which are discussed in detail in Chap. 3.

Through the years of rapid development in SFX, numerous techniques have been shown to help in the identification and characterization of microcrystal samples. The crystallinity of submicron particles can be tested using electron microscopy [24, 25] but the throughput is low. Dynamic light scattering and Nanosight tracking (Malvern instruments) can be used to quickly determine size distributions for submicron to micrometer sized proteins, but do not relay information about crystallinity. DLS can identify the size of particles up to about 5 μm (depending on the buffer components), while Nanosight tracking is limited to sizes <500 nm but has the advantage to also allow for identification of the particle density.

A set of methods has been developed and is now available to identify nanocrystals and microcrystals from other protein-rich or crystalline phases. Second-order nonlinear imaging of chiral crystals (SONICC) [26, 27] is a two photon microscopy technique [28] and shows strong signal for protein crystals down to hundreds of nanometers, and can be greatly enhanced by chromophores present in the protein [29]. SONICC signal depends on the symmetry of the sample and the orientation of crystals in the field of view, and can give false positives for crystals of certain chiral inorganic and organic molecules and ligands [30]. UV-fluorescence, achieved from two-photon or standard methods, can be used to determine the presence of tryptophan in the small aggregates, allowing protein-rich phases to be discriminated from salt crystals, PEG skins, etc., but the contrast between the fluorescence signal of crystals and other aggregates may make unambiguous identification of macromolecular crystals difficult [31]. Testing for reversibility of precipitation conditions could also be used to identify potential nanocrystallization and microcrystallization conditions [6]. X-ray powder diffraction can assess the difference in diffraction quality between crystal batches, and can easily distinguish macromolecular crystals from other aggregates and salt crystals, and can be used for all space groups, but the resolution will generally be limited compared to diffraction results from an X-ray FEL [2].

2.2.2 Sample Introduction for SFX

If a microcrystal can only be measured once with an X-ray FEL beam before being destroyed, new and effective methods had to be developed to deliver the sample to the X-ray interaction region. We provide here an overview, while Chap. 5 describes these sample delivery methods in more detail.

The primary method developed for crystallography at X-ray FELs has been the serial crystallography approach, in which the crystals are delivered to the X-ray interaction region in a serial way. In the serial approach, the sample is moved through the interaction region during data collection and the data are taken from many copies of similar samples. SFX was developed for X-ray FEL sources but the techniques have recently been applied to serial crystallography data collection at room temperature with millisecond X-ray exposures of larger crystals (10–50 μm) at synchrotron sources [32–34]. While primary X-ray damage cannot be outrun at synchrotron sources, the millisecond exposure time reduces secondary damage effects.

The sample must be delivered and replenished at faster rates than the X-ray FEL frequency to avoid X-ray damage from the previous pulse. While current sample delivery systems have been optimized for the current X-ray FELs that operate at 10–120 Hz high repetition rates, the European X-ray FEL (XFEL.EU), which began user operation in the fall 2017, and LCLS-II (under construction) will approach or exceed MHz peak rates by 2020, creating a new challenge for both sample delivery and detector read out rates.

The high solvent content typical of macromolecular crystals means that the crystals need to be treated delicately in order to avoid mechanical damage. One compelling option to deliver the sample is to use a liquid stream or droplets to transport a suspension of microcrystals in their mother liquor to the X-ray interaction region, with the stream allowing replenishment of the sample while removing "used" sample. During initial development, several requirements needed to be met by liquid jets in order to be considered suitable: (1) avoid damage or alteration of the sample during deliver, (2) be compatible with in vacuo work due to soft X-ray operations and the constraints of existing X-ray FEL instruments, (3) produce thin jets in order to minimize the background from the crystallization mother liquors, and (4) use minimal amounts of sample due to the low abundance of many of the samples.

Rayleigh jets are easy to make and use, but were found to be unsuitable for sample injection because they use a large amount of sample, are prone to ice formation when running protein buffers in vacuum, and smaller, <10 μm diameter Rayleigh nozzles tend to clog quickly [35]. Gañán-Calvo showed that by replacing the glass walls of a Rayleigh nozzle with a laminar accelerating gas stream a thin liquid jet can be formed [36]. The gas dynamic virtual nozzle (GDVN) [37], shown in Fig. 2.2, was developed for SFX sample injection based upon laminar gas acceleration through an aperture and delivered samples into vacuum with significantly reduced frequency of clogging or freezing compared to the Rayleigh jet. The GDVN was a major breakthrough because (1) it produces thin jets, producing low background from the mother liquor, (2) it facilitates sample injection in vacuum because the accelerating sheath gas reduces the evaporative cooling of the injected liquid, decreasing the likelihood of ice formation on the nozzle, and (3) the large inner capillaries that deliver the sample are less likely to get clogged than the <10 μm Rayleigh nozzles needed for comparably thin jets [38]. The GDVN injection is generally driven by a high-performance liquid chromatography (HPLC) pump, allowing for a large variety of samples to be effectively injected, from soluble proteins to membrane proteins grown in certain *in meso* phases, such as lipidic sponge phase. The major limitation to the GDVN is relatively high sample consumption, so many new sample introduction methods have been developed to compliment the GDVN, such as the lipidic cubic phase (LCP) injector [39], double flow focusing injectors [40], microfluidic electrokinetic sample holder (MESH) [41,

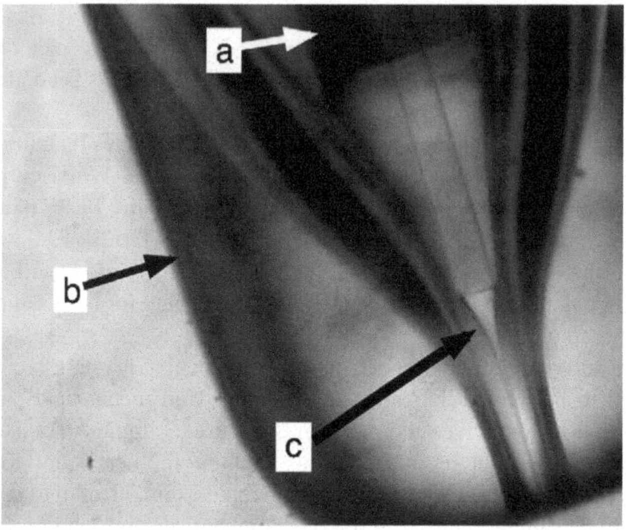

Fig. 2.2 Light Microscope image of a nozzle with liquid jet. A 360 μm OD × 40 μm ID capillary tube (**a**) with external taper injects a liquid stream into a convergent gas flow in a 1.2 mm OD capillary nozzle (**b**). This yields a rapidly tapering liquid jet (**c**) surrounded by a co-flowing gas. Figure and caption reproduced from [38]

42], and fixed target approaches [17, 19, 43, 44]. GDVN, as a fast flowing jet, remains a key tool of sample injection, especially for time-resolved SFX, and is discussed in detail, along with alternate injection methods, in Chap. 5.

2.2.3 Data Analysis for SFX

One of the major topics of discussion directly before the initial SFX experiments (and to this day!) was data processing. Unlike conventional crystallography at synchrotrons, where intensities of full reflections are recorded in a rotation series, in SFX a data set consists of tens of thousands of diffraction patterns. Each SFX diffraction pattern is essentially a "still" image (i.e., a very thin "slice" through the reciprocal lattice) and is thereby comprised of partial reflections. The integrated intensity for a reflection that is associated with the structure factors cannot be determined from a single SFX diffraction pattern [45]. To conquer the challenge, Kirian et al. [45] developed a new data evaluation method that uses Monte Carlo methods to determine accurate structure factors by averaging over the separate intensity measurements made from different crystals (in separately recorded diffraction patterns). An "average" structure factor could be calculated and used to determine the underlying structure. This "Monte Carlo" processing of the diffraction data has proven to be the workhorse for SFX data analysis, even as more sophisticated post refinement methods, such as determining the partiality of the measured Bragg peak [46, 47], are implemented.

In practice, SFX data analysis is more than simply indexing and merging of the data. After dark and gain corrections for each recorded diffraction pattern, significant effort goes into finding diffraction patterns that contain Bragg peaks (and disregard blanks), known as "hit finding." An iterative process of hit finding and indexing is usually done until a sufficiently large number of indexable diffraction patterns are identified. SFX data evaluation thereby requires assessment of data quality and quantity in real time. Furthermore intensity variations of the individual X-ray FEL pulses as well as variation of the diffraction volume of the individual crystals require data sets with high multiplicity. The amount of data required for the determination of accurate structure factors for a complete data set is a controversial point of current discussion and also depends on the resolution of the data sets, the methods used for phasing, as well as the data evaluation programs. The average SFX data set consists of more than 10,000 patterns, while large data sets over 100,000 patterns have been reported and may be required for certain time-resolved studies [48] and for experimental phasing [49]. Consequently, much work has gone into software to facilitate SFX data analysis. The major SFX analysis programs are Cheetah [50] for hitfinding, CrystFEL [51] for indexing, merging, and mtz generation, and cctbx.xfel [52] for hit finding through mtz generation. The detailed considerations for SFX data analysis are featured in Chap. 7.

2.3 Early Experiments of Serial Femtosecond Crystallography (SFX)

The first SFX experiments took place in 2009 and many advances and milestones that made SFX possible were contributed by a large and active research community. Details of the initial experiment will be discussed in the coming section but the interested reader is directed to read the original publications (cited throughout the chapter) for detailed descriptions.

2.3.1 The "Birth" of SFX

The first demonstration of SFX methods took place at LCLS simply by the nature of LCLS being the first FEL source in the X-ray regime available to users. LCLS commissioning and early user operations started in early 2009 [53]. The first protein crystallography experiments [54] were performed in December 2009 using microcrystals of Photosystem I (PSI) at the Atomic Molecular and Optical (AMO) [55] instrument in the CAMP endstation [56] using a photon energy of 1.8 keV (6.9 Å wavelength) with a 7 μm FWHM focus. The experiments also investigated

X-ray induced radiation damage by measuring diffraction data at 20, 70, and 200 fs pulse duration, with a peak power density of 10^{16} W/cm^2 and an estimated absorbed dose for the protein crystals of 700 MGy, well above the conventional X-ray dose limit of 30 MGy [57]. Two sets of detectors were used to measure the diffraction data, a "front" detector at a working distance of 68 mm that collected high spatial resolution data and a "back" detector that collected high angular resolution data.

Diffraction data from PSI nanocrystals were collected to 8.5-Å resolution at the edges of the detector, which corresponded to the resolution limit for the given experimental geometry and photon energy. In total, over 3,000,000 diffraction patterns were collected from the sample of PSI crystals between 200 and 2000 nm in size (many were blank). A total of 112,725 were identified as having over 10 Bragg peaks of which 15,445 patterns were successfully indexed and merged for the final data set and phasing was performed by molecular replacement [54, 58]. Figure 2.3 shows diffraction patterns and the electron densities derived from the SFX diffraction pattern from the PSI nanocrystals.

The 8.5-Å data and electron density maps generated from these initial SFX experiments were the first breakthrough for SFX. However, several other interesting results were obtained in these first experiments. Data collected on the back detector had high angular resolution, and interference fringes between the Bragg peaks were detected for some of the smaller crystals, as shown in Fig. 2.4. In 1952, Sayre had proposed that if one were ever able to collect X-ray diffraction data from crystals with a small number of unit cells one would see intensity between the Bragg peaks. The intensity between the peaks arises from non-delta function reciprocal lattice peaks from the finite sum of unit cells in the crystal, while interference fringes from the small crystals result from the high coherence of the beam and the faceting of the crystals, essentially representing the Fourier transform of the size and shape of the object. This was one of the first observations of intensity between the Bragg peaks in protein crystallography data. These data can be used to reconstruct the size of the object as shown in the insets of Fig. 2.4 and these particular reconstructions were accomplished using the shrinkwrap [59] algorithm. The added information could potentially be used to solve the crystallographic phase problem [60], as will be discussed in Chap. 8.

Although the electron density maps did not contain any evidence of conventional X-ray induced radiation damage, the impact of pulse duration on crystallographic data quality was explored using pulse durations of 20, 70, and 200 fs. The collected data showed a distinct reduction in (normalized) scattered intensity at larger scattering angles for the longer pulses indicating that alterations of the sample due to its interaction with the X-rays occur at longer pulse duration. The result was an important initial finding but research into radiation damage in SFX is even more complex and still ongoing. However, during the early operation of LCLS the reported X-ray pulse durations were derived from the electron bunch durations without direct measurement of the X-ray pulse duration itself. An experiment measuring photoionization of Ne atoms suggested that the X-ray pulse durations were approximately half the electron bunch duration [61]. Experiments have been

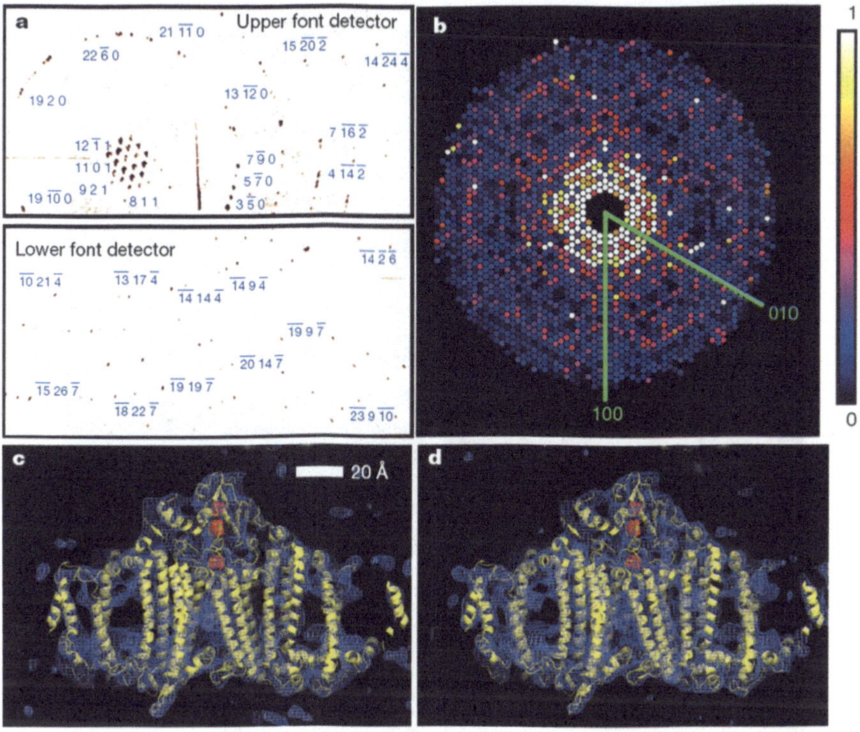

Fig. 2.3 Diffraction intensities and electron density of photosystem I. (**a**) Diffraction pattern recorded on the front pnCCDs with a single 70-fs pulse after background subtraction and correction of saturated pixels. Some peaks are labelled with their Miller indices. The resolution in the lower detector corner is 8.5 Å. (**b**) Precession-style pattern of the [001] zone for photosystem I, obtained from merging femtosecond nanocrystal data from over 15,000 nanocrystal patterns, displayed on the linear color scale shown on the right. (**c**, **d**) Region of the 2mFo—DFc electron density map at 1.0σ (purple mesh), calculated from the 70-fs data (**c**) and from conventional synchrotron data truncated at a resolution of 8.5 Å and collected at a temperature of 100 K (**d**) (Methods). The refined model is depicted in yellow. Figure and caption reproduced from [54]

dedicated to better understand this important aspect of SFX and X-ray FEL data collection in general, as discussed in later sections of this chapter and Chap. 6.

While the design and construction of the hard X-ray instruments at LCLS and the SPring-8 Angstrom Compact free electron Laser (SACLA) were still ongoing, several additional SFX experiments were done at the AMO instrument. Crystals of a photosynthetic reaction center grown in lipidic sponge phase (LSP) were introduced to the interaction point using a GDVN injector. Data from the $<10\ \mu m$ crystals were collected to the geometric limit of 7.4 Å for the experiment with the electron density reconstructed to 8.2-Å resolution [62].

Additional experiments were performed to better understand the impact of radiation damage, as the initial experiments showed a pulse duration dependence

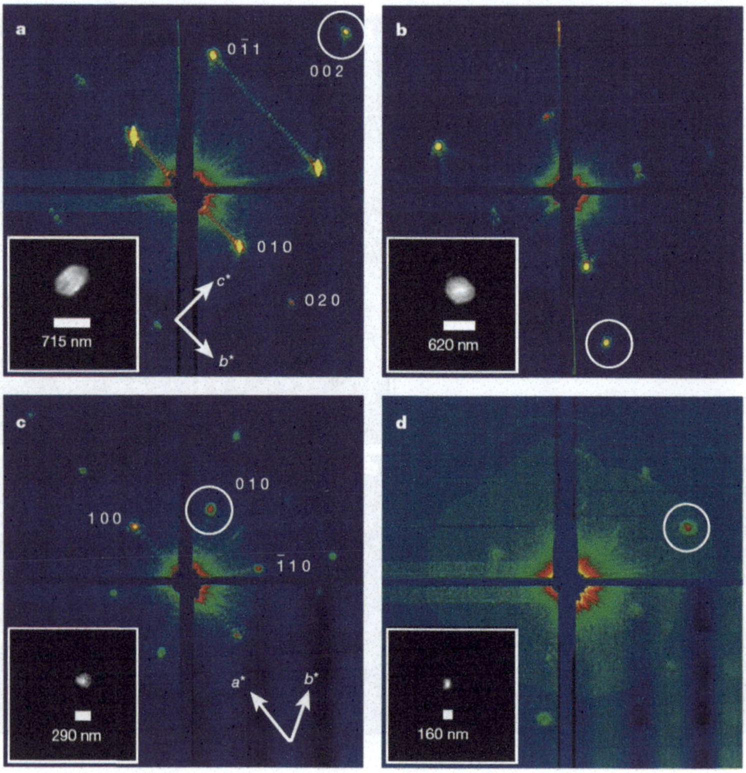

Fig. 2.4 Coherent crystal diffraction. Low-angle diffraction patterns recorded on the rear pnCCDs, revealing coherent diffraction from the structure of the photosystem I nanocrystals, shown using a logarithmic, false-color scale. The Miller indices of the peaks in (**a**) were identified from the corresponding high-angle pattern. In (**c**) we count seven fringes in the b* direction, corresponding to nine unit cells, or 250 nm. Insets, real-space images of the nanocrystal, determined by phase retrieval (using the Shrinkwrap algorithm [59]) of the circled coherent Bragg shape transform. Figure and caption reproduced from [54]

on the quality of the collected data. Detailed analysis showed that X-ray damage mechanism is different in SFX compared to synchrotron-based data collection. In synchrotron-based damage studies, the total intensity remains constant while reflections become broader and background increases until they "merge" with the background. In contrast, the first SFX damage studies show that with increased pulse duration "termination" of Bragg peaks occurs. Barty et al. [63] showed that the crystalline order of the samples would decrease during the plasma formation and onset of nuclear motion. The reduction of crystalline order will cause the Bragg diffraction to cease in a process termed "self-terminating diffraction" and shown in Fig. 2.5 [63]. Self-termination of the Bragg peaks could be interpreted as meaning that the pulse duration in SFX is inconsequential, as the Bragg diffraction terminates and the scattering of the destroyed crystals only leads to an increase in the

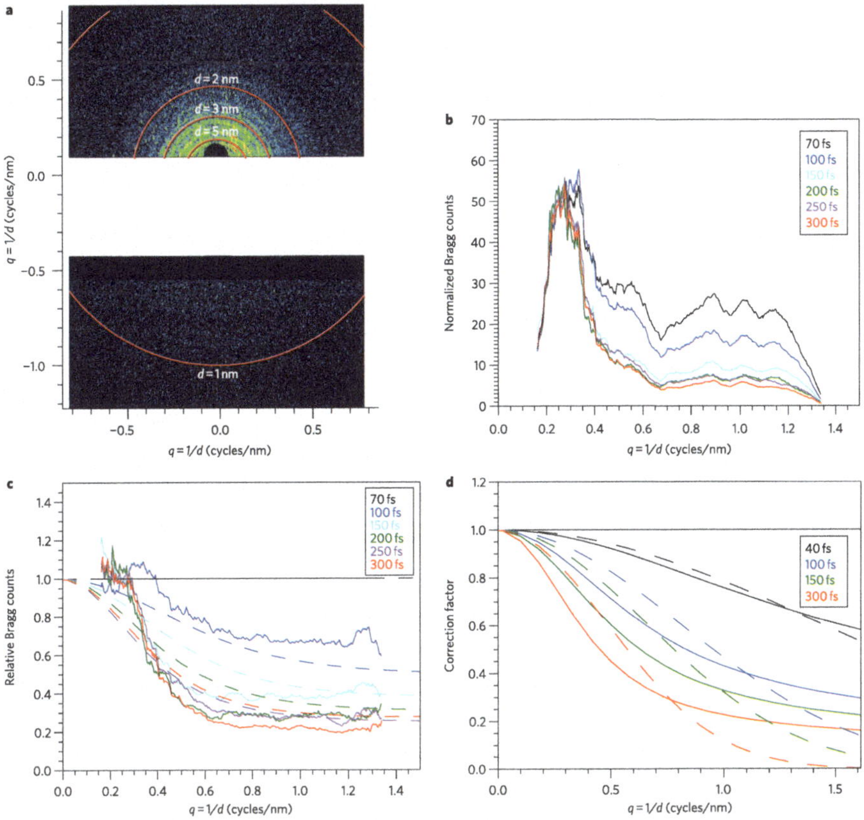

Fig. 2.5 Bragg termination observed at approximately constant X-ray pulse fluence I0T. (**a**) "Virtual powder pattern" formed by summing 3792 single-pulse patterns obtained with X-ray pulses with a duration of 300 fs. The spots in the pattern are Bragg peaks, which are visible out to the corners of the detector, corresponding to a resolution of d 1/4 0.76 nm. Because of the large unit cell size of the crystal, Debye–Scherrer rings overlap and are not resolved at $q > 0.5\ \mathrm{nm}^{-1}$. (**b**) Bragg signal I(q; T) of Photosystem I nanocrystals averaged over q shells of virtual powder patterns for nominal pulse durations T varying between 70 fs and 300 fs. (**c**) Bragg signal relative to the shortest pulses, plotted as solid lines. Dashed lines give the computed ratios of I(q; T)/I(q; T 1/4 40 fs) from the Cretin simulations. Previous experiments at LCLS indicate that the nominal "70 fs" pulses are shorter than indicated [61]. We achieve a best fit assuming these pulses have a duration of 40 fs. (**d**) Comparison of the calculated dynamic disorder factor g(q; T) (solid lines) compared to a Debye–Waller factor best-fit to the same data (dashed lines). Figure and caption reproduced from [63]

homogenous diffuse background. However, this picture is likely overly simplistic. Ionization of the atoms that results in an even distribution throughout the unit cell will lead to an overall decrease in scattered intensity in the Bragg peaks; uncorrelated ionization would lead to a background increase caused by homogenous diffuse scattering. Lomb et al. found a clear trend that the *R*-factors between data sets increase as the pulse duration increases [64]. At this point, it was clear that pulse duration had an impact on the quality of SFX data collected but the limited resolution by 2 keV X-rays was a limitation on the further exploration of the topic. These questions were further addressed as higher energy X-rays became available and are discussed in detail in Chap. 6.

2.3.2 *SFX at High Resolution*

The early SFX experiments were limited in resolution due to the photon energy usable by the available instruments at LCLS. However, in mid-to-late 2010, the X-ray pump–probe (XPP) [65] and coherent X-ray imaging (CXI) [66] instruments came online and experiments could be done using hard X-rays, with the potential to extend to atomic resolution for the first time.

In February 2011, the first SFX experiments were performed at CXI using 9.4 keV (1.33 Å wavelength) X-rays with 40 fs pulse duration using lysozyme microcrystals [67]. Data were collected to 1.9 Å resolution and the calculated difference-density maps between structures derived from SFX and low-dose synchrotron experiments showed no evident signs of radiation damage at the molecular level, as shown in Fig. 2.6. This result provided further evidence for the diffraction-before-destruction idea. The result showed the first proof-of-principle that structures can be determined by SFX at near atomic resolution without significant structural damage at room temperature from crystals that are typically too small to be used for synchrotron-based crystallography. This result has major implications for the study of membrane proteins, macromolecular complexes, and formed the basis for the first

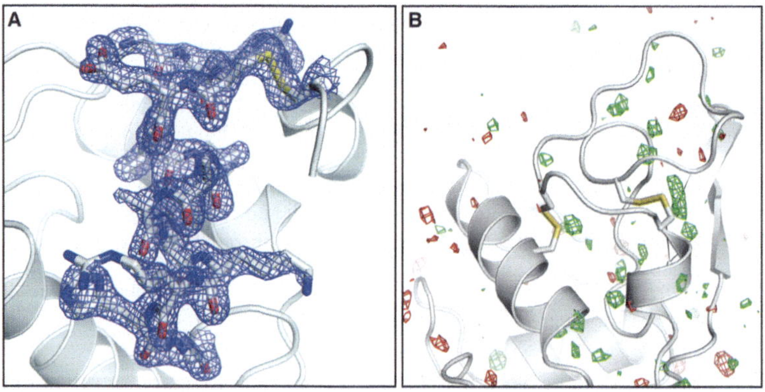

Fig. 2.6 High resolution lysozyme structure. (**a**) Final, refined 2mFobs—DFcalc (1.5σ) electron density map of lysozyme at 1.9 Å resolution calculated from 40-fs pulse data. (**b**) Fobs(40 fs)—Fobs (synchrotron) difference Fourier map, contoured at +3 σ (green) and −3 σ (red). No interpretable features are apparent. The synchrotron data set was collected with a radiation dose of 24 kGy. Figure and caption reproduced from [67]

time serial time-resolved studies at X-ray FELs. The impact of high resolution SFX was felt immediately with a major increase in demand for beamtime by the user community, still seen today with structural biology representing a large fraction of X-ray FEL use.

2.3.3 SFX of G-Protein Coupled Receptors

G-coupled protein receptors (GPCRs) are the largest class of integral membrane proteins in the human genome (>more than 600 genes code for GPRCs). They are receptors that are involved in many biochemical pathways and act as key players in signal transduction, neuronal function, hormone action, vision, sensing, mood control, pain sensing and relief and are major drug targets against diseases, with >50% of all current drugs being targeted to GPCRs [68]. However, membrane proteins are notoriously difficult to crystallize. While over 100,000 structures of proteins have been determined to date, structures for <800 unique membrane proteins have been solved to date. The intrinsic amphiphilic nature of membrane integral proteins makes them difficult to express, isolate intact from the membrane, purify and crystallize. They often form only small, weakly diffracting crystals that deteriorate rapidly when exposed to synchrotron radiation. Despite their high medicinal impact, leading to huge efforts to determine structures of GPCRs, they resisted crystallization and structure determination until a breakthrough was achieved in the field with crystallization of GPCRs in lipidic phases that mimic the native membrane [4, 69]. As discussed in Chap. 4, crystallizing GPCRs using these in meso media, such as lipidic cubic phase (LCP) or lipidic sponge phase (LSP), has been very successful to date and led to the first high-resolution GPCR structure in 2007 [70]. However, many of the crystallization trials lead to small crystals that are insufficient for high-resolution data collection at a synchrotron due to rapid deterioration in the X-ray beam even at cryogenic temperatures. Consequently, the development of new methods to enable the use of SFX for LCP-grown GPCR crystals was a very important goal. This was greatly enabled by a new sample injector capable of effectively delivering the high viscosity media (which have the consistency of tooth paste) to the X-ray FEL interaction region. These newly developed injectors are called high viscosity extruders (HVE) [32] or LCP injectors [39] and are discussed in depth in Chap. 5.

The first SFX experiments on LCP-grown GPCR crystals were done in 2012 [71]. Microcrystals of the human serotonin receptor 5-HT_{2B} bound to the agonist ergotamine were delivered using the LCP injector and a full data set was collected to 2.8-Å resolution at room temperature. The crystals used in the SFX experiments were 100× smaller in volume than the crystals required for the synchrotron experiments. This was the first X-ray structure determined from a GPCR at room temperature. The SFX-derived electron density maps were compared to synchrotron-derived electron density maps, and although the overall structures were very similar (rmsd = 0.45 Å for the backbone) significant differences were identified. Several side chains were much better resolved in the X-ray FEL structure compared to the synchrotron-based structure, and structural changes were detected in the loop regions of the proteins, side chain conformations and even part of the backbone of one of the helices was altered (it was kinked in the synchrotron structure but straight in the SFX structure). These structural differences may represent the more native conformations of the GPCR as the SFX data were collected at room temperature whereas the synchrotron-based structures are derived from data at cryogenic temperatures. The crystal size reduction is a major development for protein crystallography, as growth of large well-ordered crystals of GPCRs and other membrane protein samples can take years to finish, if ever. SFX of membrane protein targets, and especially GPCRs, represent a prominent area of current research at X-ray FELs [72, 73] and is covered in detail in Chap. 10.

The high viscosity of the LCP-grown crystal samples limits the velocity of the sample moving through the X-ray interaction region. This results in much lower sample consumption compared to GDVN (Liu et al. [71] reported 300 μg of sample used versus 10s of mg for GDVN) as the flow rate of the jet can be tightly controlled from 1 to 500 nl/min, and the jet speed is adjusted so that the sample is just replenished between the X-ray pulses. The viscous injectors have therefore been highly desired for sample delivery of crystals grown conventionally in solution by embedding crystals in LCP [74] or other viscous media. Alternate viscous media include agarose [75] or hydroxyethyl cellulose [76], grease [77], as well as high molecular weight PEG [33] matrices. The development of viscous injection methods for X-ray FELs has also recently enabled serial microsecond crystallography studies at RT at synchrotron sources [34], which highlights the strong synergistic effects of X-ray FEL developments on conventional crystallography.

2.3.4 Phasing of SFX Data

The first structures determined by SFX were all solved by use of the method of molecular replacement (in which the phases from a structurally-similar molecule are used [78]), however experimental phasing is required to solve novel structures for which no homologous structure exists. In principle the same phasing methods that are established for conventional crystallography (Single and Multiple Isomorphous Replacement (SIR and MIR), as well as Single and Multiple Anomalous Diffraction/Dispersion (SAD and MAD) can be applied to SFX. However these phasing methods are more challenging for SFX than for conventional crystallography, as the isomorphous or anomalous differences are generally small and require very accurate structure factor determination. During the Self-Amplified Spontaneous Emission (SASE) process that generates the normal lasing of an X-ray laser (see Chap. 1 for details), the X-rays are delivered with variations in pulse energy, wavelength, and pulse duration, which decrease the accuracy of individual measurements and complicate experimental phasing. Consequently, experimental phasing is easier to achieve with conventional crystallography where data can be collected on one crystal and there are smaller fluctuations of the delivered X-rays between images. These challenges can be met by a high multiplicity of the SFX data sets and several proof-of-principle studies have shown successful de novo phasing of SFX data [79, 80]. Since the initial successes using SAD phasing, SAD has been demonstrated using several the elements, such as copper [18], mercury [81], iodine [82], and. SAD phasing accounted for ~70% of all novel macromolecular structures in 2013 with the most commonly used element being selenium [83] (because selenomethionine, a derivative of methionine in which the sulfur is replaced by selenium, is easily incorporated into many proteins during expression); experiments done at LCLS [84, 85] and SACLA [86] have demonstrated Se-SAD. Additionally, native SAD using sulfur atoms has been demonstrated using lysozyme [87], thaumatin [49], and human A_{2A} adenosine receptor [80].

The X-ray FEL equivalent of MAD phasing experiments has been successfully demonstrated at SACLA [88]. In this example, two color or split beam experiments in which crystals are hit by two pulses of different color are used to record two diffraction patterns with different anomalous content for each pulse and each crystal, and can in principle improve phasing efficiency by comparing data from the same crystal. The demonstration of experimentally determining the phases of novel structures with SFX was a major milestone for the technique, both showing the accuracy of the measurements and the progress made in the analysis of the data.

It deserves mentioning that the SFX SAD/MAD phasing cases to date have been done using standard proteins, that is, proteins that would not need to have the phases determined experimentally and generally scatter very strongly (important for SAD/MAD since the relative difference signal increases with resolution). However, in the case of the mosquito larvicide protein BinAB, experimental phasing from SFX data was needed because the crystals are limited to <500 nm in size, not affording a structure from conventional crystallography. The novel structure was phased using the Multiple Isomorphous Replacement and Anomalous Scatter (MIRAS) approach, collecting data on iodine, gadolinium, and mercury derivatives of in vivo grown nanocrystals [89].

In addition to being able to continue the experimental phasing approaches developed at synchrotrons, SFX opens new avenues for phasing including phasing based on shape transforms from finite crystals [60] and direct phasing based on continuous diffraction (see below). These novel methods are discussed in Chaps. 8 and 9 in more detail.

2.3.5 Time-Resolved SFX

Time-Resolved crystallography (TR-MX) studies have been pioneered at synchrotron sources utilizing Laue diffraction methods and generally optical lasers as pumps [90–92]. However, TR-Laue-MX data collection is challenging as it requires very large crystals, which makes homogeneously pumping the sample challenging,

X-ray damage can be severe and time-resolution is limited to ~100 ps due to the pulse duration at synchrotrons [90, 93].

The ultrashort time duration of X-ray FEL pulses enables data collection at room temperature from very small nanocrystals and microcrystals, and is ideally suited to further develop the method of time-resolved crystallography. The theoretical temporal resolving power of an X-ray FEL is limited by the pulse duration (although the ultimate temporal resolution will be dependent on the sample and experimental setup) and therefore Time-Resolved SFX (TR-SFX) can explore shorter timescale phenomenon in biology than is possible using synchrotrons.

The development of the first TR-SFX experiments at X-ray FELs was driven by the motivation to study the light-driven processes in photosynthesis. The first experiments were performed using soft X-rays at the AMO beamline using a complex of Photosystem I with its electron acceptor ferredoxin [94]. With the increase in the usable X-ray FEL energy, the study of the water splitting mechanism of Photosystem II (PSII) became a very active field of X-ray FEL research at both LCLS and SACLA. The full photocycle studies can only be done at X-ray FELs as the core of the oxygen-evolving complex consists of a Mn_4O_5Ca cluster which cycles through five oxidation states during the catalytic cycles. X-ray damage of the cluster by reduction of the metals is severe, but can be outrun by X-ray FEL pulses as shown by the first undamaged dark structure of PSII determined from large frozen (1 mm) crystals, which are shifted after each X-ray FEL shot [19]. TR-SFX studies on PSII are very complex, as they require multiple laser excitations to cycle PSII through the five stages of the catalytic cycle. The resolution for TR-SFX studies has been improved from 5 Å for the first TR-SFX studies on PSII [95, 96] to 2.2–2.5 Å resolution for recent TR-SFX studies [48, 97].

Besides the ongoing PSII work, many experiments have been done establishing the ability to measure interpretable difference signals between the excited and ground states of photoactive proteins to high spatial and temporal resolution. Difference data from TR-SFX experiments using photoactive yellow protein (PYP) extended to 1.6 Å resolution and showed excellent quality and strong difference density features, as shown in Fig. 2.7 [99]. The 1-μs Difference Electron Density (DED) maps showed similar, but much stronger, features compared to 1-μs DED maps produced from synchrotron data collected using Laue diffraction.

The proof-of-principle PYP TR-SFX experiments demonstrated several substantial benefits of SFX for laser-excited time resolved experiments. For the time-resolved Laue experiments previously done using synchrotron sources, large PYP crystals were needed to achieve high resolution and several pumping cycles were needed to collect a full data set. When a ns laser is used to initiate the photocycle, the laser beam size is usually smaller than the crystal and does not penetrate fully or uniformly through the PYP crystal—the percentage of molecules photoactivated in the TR-SFX experiment was calculated at ~40%, whereas 10–15% is maximally achieved in the synchrotron experiments, leading to decreased map quality. The repeated pumping of the same crystal in the Laue experiment leads to strain in the crystal, which sets an upper limit to the laser fluence that can be used for the experiment. These issues are avoided with TR-SFX, as the crystals used were

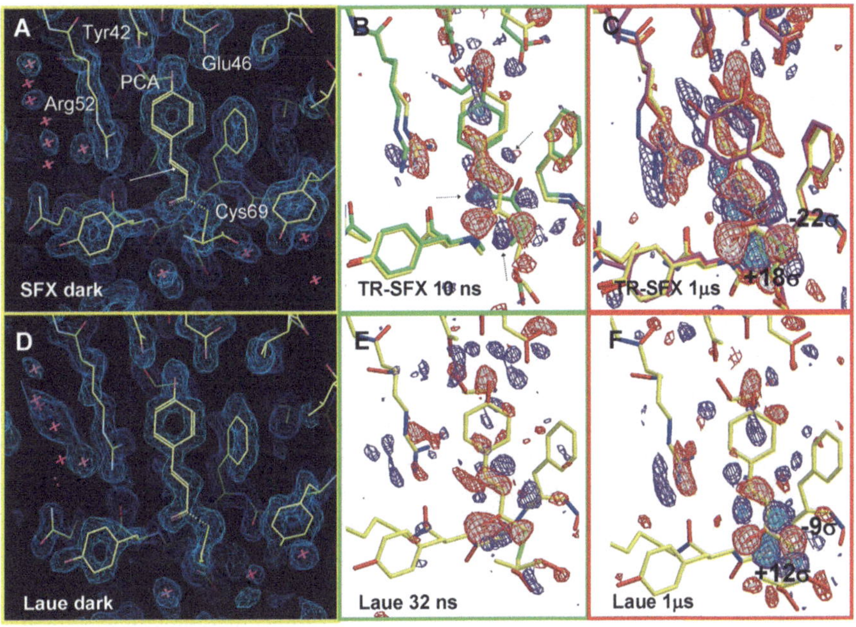

Fig. 2.7 Comparison of electron density and DED maps in the chromophore pocket obtained by TR-SFX and the Laue method. The dark state is shown in yellow in all maps. (**a**, **d**) Electron density maps for the PYP dark state obtained with TR-SFX and Laue, respectively (contour level 1.1 s, 1.6 Å resolution).The PCA chromophore and nearby residues are marked in (**a**). Arrow: Double bond in the chromophore about which isomerization occurs. (**b**) TR-SFX DED map at 10 ns. Light green structure: ICT intermediate. Features marked by dotted arrows belong to additional intermediates not shown. (**c**) TR-SFX DED map at 1 ms. Pink and red structures: structures of pR1 and pR2 intermediates, respectively. (**e**) Laue 32-ns DED map correlates best to the TR-SFX 10-ns map. (**f**) Laue 1-ms DED map. Contour levels of the DED maps: red/white −3 s/−4 s, blue/cyan +3 s/+5 s, except for (**c**) where cyan is +7 s. Figure and caption reproduced from [98]

smaller than the absorption length of the laser in the PYP crystals and each crystal was at most used for one diffraction pattern.

Several experiments have explored TR-SFX to investigate the sub-picosecond dynamics of proteins (and the chromophores typically absorbing the light energy), such as CO dissociation from myoglobin [100], cis/trans isomerization of PYP chromophore [98], and chromophore twisting of a photoswitchable fluorescent protein [101]. These experiments have shown the potential for TR-SFX to reach time resolutions below 200 fs [98], a previously inaccessible time scale. Each of these pioneering experiments required careful experimental design in order to accurately measure small differences in diffraction data. For an in-depth discussion of TR-SFX the reader is referred to Chap. 11.

The ability to use microcrystals in SFX to achieve atomic and molecular resolution also allows for reactions to be triggered by mixing, in which case a substrate/ligand is mixed with the crystals of interest [102]. The substrate diffuses into the crystal and the progression of the time-dependent changes is measured at specific time points after the mixing. This method allows for snapshots of the reaction to be recorded, which paves the way for molecular movies to be recorded in the future, when higher repetition rate X-ray FELs become operational. Reactions of biomolecules (e.g., enzymatic catalysis) could be recorded on a microsecond to second timescale. Proof-of-principle experiments on X-ray FEL-based Mix and Inject Serial Crystallography (MISC) include the study of the complex of β-lactamase and the antibiotic ceftriaxone at 2.4 Å resolution [103], and the first observation of conformational changes in riboswitches at 2.8-Å resolution [104]. An overall schematic picture of the β-lactamase experiment [103] is shown in Chap. 12. The overall design of the riboswitch MISC experiment, in which a T-junction was used for time points >10 min, is shown in Fig. 2.8a. The large differences in the riboswitch samples in the apo and ligand-bound structure are shown in Fig. 2.8b.

MISC is an exciting and novel way for X-ray FELs to impact structural biology beyond the limited field of photoactive proteins and is discussed in greater detail in Chap. 12. The high spatio-temporal resolution offered by X-ray FELs combined with the higher repetition rate machines, such as EuXFEL and (potentially) the high energy extension of LCLS-II (LCLS-II-HE, which would enable LCLS-II to produce hard X-rays) may offer a paradigm shift for structural enzymology, changing the experiments from studying a few time points that underdetermine the time course of the enzyme to a sufficiently temporally determined system. Future X-ray FEL sources and the types of experiments they will facilitate are discussed in Chap. 16.

2.3.6 Imperfect Crystals and the Push Toward Crystal-Free Imaging

Continuously modulated diffraction intensities have frequently been encountered in MX [105–110] and were generally considered a detrimental feature, although the data could be used for understanding the defects present in a crystal and potentially to infer protein dynamics within the unit cell [111]. Smaller macromolecular crystals may have fewer defects than larger crystals, and may have smaller unit cell parameter variability and mosaicity throughout [112]. In large crystals, long range order defects accumulate where disturbances during growth leads to shifts of the crystal lattice indicated by an increase in mosaicity. While these long-range growth defects can be diminished in nanocrystals and small microcrystals, other defects like dislocations caused by thermal motions in crystals with weak crystal contacts or defects caused by heterogeneities of the molecules will still be present.

SFX data that contain continuously modulated diffraction intensities from crystals of the large membrane protein complex Photosystem II (PSII) were collected at LCLS. While the Bragg peaks terminated at 4.5 Å resolution, diffuse scattering extended beyond the Bragg peaks, as shown in Fig. 2.9. While the cause of

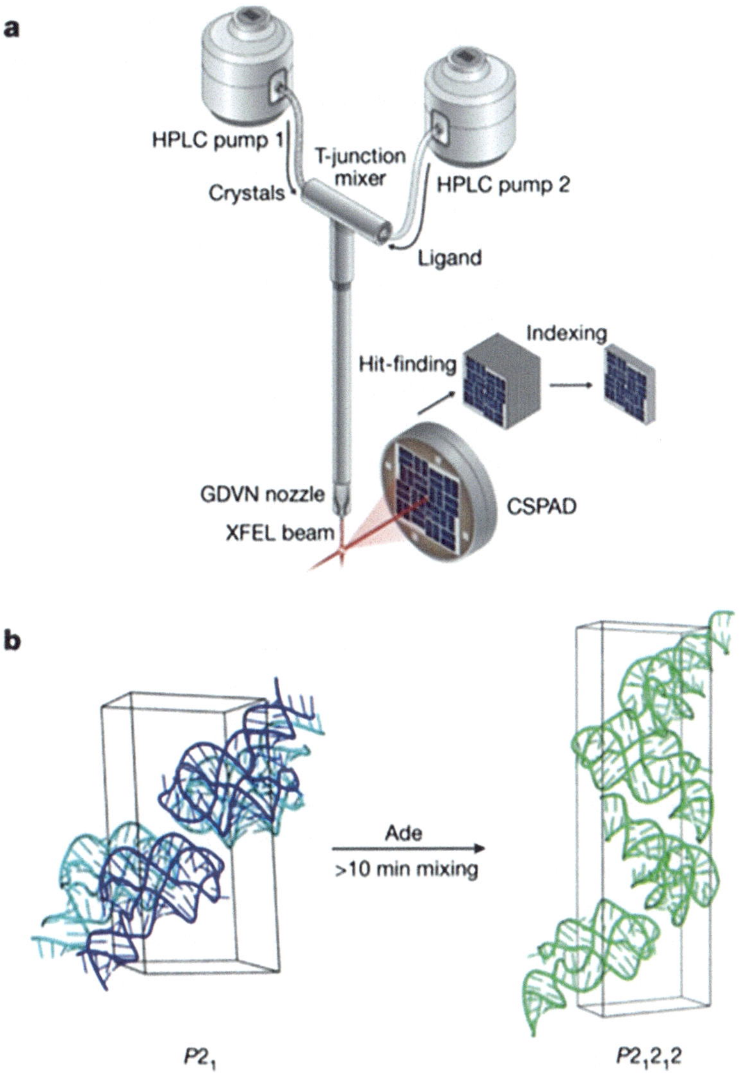

Fig. 2.8 The experimental setup of 3DX and conversion of the structure and crystal lattice after ligand mixing. (**a**) Schematic illustration of the SFX ligand-mixing experiment. For details, see the Methods. (**b**) The unit cells of the crystals of apo (P21) and ligand-bound (P21212), whose structure was converted *in crystallo* from the apo structures after >10 min mixing with adenine ligand. Figure and caption reproduced from [105]

diffuse scattering was previously discussed to be arising from flexibility within the molecule, Ayyer et al. proposed a model in which continuous diffraction is based on diffraction from rigid units of molecules within the unit cell (the PSII-dimer in this case) that are translationally displaced within the unit cell (not rotated). An

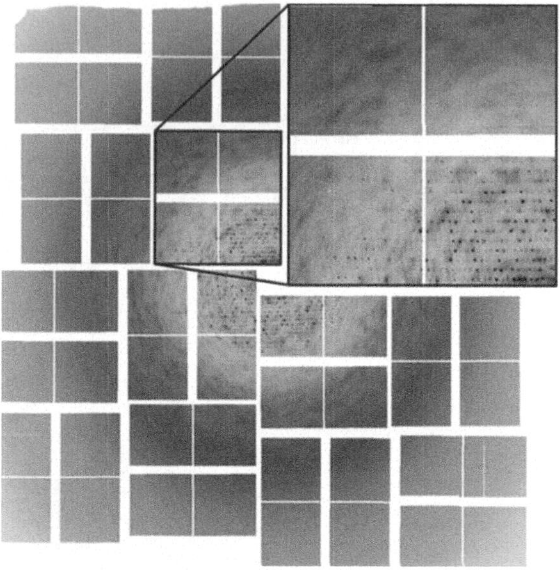

Fig. 2.9 Molecular coherent diffraction. An XFEL snapshot "still" diffraction pattern of a PSII microcrystal shows a weak speckle structure beyond the extent of Bragg peaks, which is enhanced in this figure by limiting the displayed pixel values. Figure and caption reproduced from [113]

rmsd displacement of the molecules within the unit cells by 0.8 Å could account for the Bragg peaks terminating at 5 Å resolution [114]; this displacement is smaller than the distance between a carbon–carbon bond. The diffraction from the displaced molecules would lead to a diffraction pattern that corresponds to single particle diffraction pattern of the content of the unit cell (here four PSII dimer molecules). This is the case because all unit cells are assumed in the same orientation but with no constructive interference from repeating positions, leading to an incoherent sum of the intensities from all unit cells.

A hybrid approach where Bragg and continuous diffraction data were treated as separate data sets was applied to the analysis of the diffraction patterns. The orientation of each pattern in the reciprocal space was determined by indexing the Bragg data. Then the structure was first solved based on evaluation of the Bragg data, which were phased by molecular replacement. The pixels that contain Bragg peaks were then masked to generate the continuous diffraction reciprocal space. The continuous diffraction data were then phased directly by iterative methods, where a low-resolution 9 Å molecular envelope of the complete PSII-dimer was the only input from the Bragg-diffraction-based X-ray structure. The iterative phasing of the continuous diffraction data extended the resolution range to 3.5 Å, and after convergence of the solution, the Fourier amplitudes and phases from the Bragg peaks (terminating at 4.5 Å) and the continuous diffraction data were combined to generate a 3.5 Å structure. Notably, the new electron density map showed better definition of the side chains as well as the features of the chlorophyll-cofactors when compared to the electron density maps produced solely from the Bragg data. This result has been received with much interest and controversy in the overall community. Data that were once considered reason to discard a macromolecular crystal might be useful for high-resolution structure determination, if the disorder is mainly caused by translation of the molecules in the crystal. This has sparked renewed discussion about how to interpret diffuse diffraction patterns measured in crystallography experiments as well as which metrics can be used to reflect data quality. The topic will be discussed in depth in Chap. 9.

2.3.7 *Towards Single Particle X-ray Diffraction*

The use of the continuous diffraction data along with the Bragg data is a hybrid approach combining methods from crystallography with techniques from coherent diffractive imaging (CDI). Many fields utilize CDI to study non-periodic objects, such as inorganic nanoparticles [115], frozen cells [113, 116], or even integrated circuits [117]. In fact, one of the great promises of the highly coherent, ultrashort-pulse X-ray FELs was that they would allow for Single Particle Imaging (SPI) of biological specimens, such as viruses and protein complexes, without the need to crystallize the samples to get high-resolution information.

The first SPI experiments at LCLS were performed in December 2009 collecting single particle X-ray diffraction data from mimivirus samples at the AMO [55] instrument in the CAMP endstation [56] using a photon energy of 1.8 keV (6.9 Å wavelength) with a 10 μm focus at FWHM [118]. Reconstructed single particle images from the collected data yielded 32-nm full-period resolution. Subsequent work on mimivirus data collected at LCLS allowed a 3D reconstruction to 125 nm, based upon the phase-retrieval transfer-function [119].

In early 2014, a meeting was held to discuss a roadmap for SPI to explore routes to reach atomic resolution, 3D imaging at X-ray FELs and more specifically LCLS. In response to the roadmap provided by the scientific community, LCLS started the SPI initiative [120]. The initiative, where more than 100 researchers from different groups work to explore the potential of SPI and seek solutions to the technical challenges, has proven very useful to perform highly technical proof-of-principle studies on model single particles of biological ensembles. The current status and exciting potential of SPI is discussed in greater detail in Chap. 14.

2.3.8 *Structure Determination Based on fs Small and Wide Angle Solution X-ray Scattering*

Small and Wide-Angle X-ray Scattering (SAXS/WAXS) are important tools for studying large-scale morphology and conformational changes of biomolecules in solution. The increasing use and utility of SAXS and WAXS for bioscience is indicated by dedicated beamlines being developed at most synchrotron facilities [121]. Collecting SAXS/WAXS data at an X-ray FEL allows for the method to be extended to the ultrafast time scales and, potentially for mixing experiments, analogously to MISC.

Time-Resolved SAXS/WAXS (TR-SAXS/WAXS) can be used to study important questions in light-driven biochemical processes such as how these systems cope with the energy that is absorbed by the chromophores, since the absorbed energy is similar in magnitude to the folding energy for the protein molecules themselves! In order to better understand the response of photo-activated molecules upon light excitation, the response of bacterial reaction center (RC_{vir}), a light-activated membrane protein, was studied using TR-WAXS at LCLS and showed that RC_{vir} relieved the strain induced via the photon absorption by structural deformation that propagates throughout the protein structure [122]. The propagating structural changes are called a protein quake and the changes were observed to continue for several picoseconds. The analysis indicated an ultrafast protein conformational mode that decayed with a half-life of ~44 ps and reached the half-maximum value in 1.4 ps after photoexcitation with the 800-nm pump laser. Although the study was done with higher than physiologically relevant pump energies, the data and analysis show that strain induced by the absorbed light at the RC_{vir}, is relaxed via protein structural changes that propagate more rapidly than heat.

Including SAXS measurements in the analysis of structural changes to molecules after light excitation allows structural parameters such as the radius of gyration and molecular volume to be followed in the time course, giving additional information to TR-WAXS. The TR-SAXS/WAXS method has been used to measure ultrafast dynamics in a complex of myoglobin and carbon monoxide (MbCO) by photoexciting the solution with a 538-nm pump laser and probing the sample at several time delays up to 100 ps, with a time resolution of ~500 fs [123]. The results demonstrated that Mb undergoes significant global structure rearrangement after the photolysis of the Mb-CO bond. Similar to what was shown for RC_{vir}, the light-induced structural changes occurring at the Mb-CO bond propagate through the protein as a quake-like intramolecular rearrangement, and the structural parameters of the molecules (radius of gyration and protein volume) oscillate about their equilibrium values for several picoseconds after the initial excitation. The maximum of the ultrafast change in the radius of gyration occurred in 1 ps, and with a radius of gyration of ~17 Å this would put the speed of the strain relief at approximately 20 Å/ps, consistent with the expected speed of sound in the protein.

The MbCO complex is a model system for studying ultrafast dynamics in proteins. Using TR-SAXS/WAXS at X-ray FELs is an active area of research on photo-activated proteins and appears likely to contribute novel information about protein dynamics in the future.

Another method for structure determination of non-periodic objects is through Fluctuation X-ray Scattering (FXS). The basic idea behind FXS is that if X-ray scattering data are collected from randomly oriented molecules in solution in a time frame shorter than the rotational diffusion times of the molecules, more detailed information can be extracted from the pattern than from a traditional SAXS/WAXS pattern [124]. X-ray FELs have pulse durations shorter than the rotational diffusion times of biomolecules and could thereby open a new era for FXS structure determination of biomolecules in physiological conditions. The theory and practical considerations of FXS are discussed in Chap. 15.

2.4 Pushing SFX Forward

2.4.1 Serial Operations at Beamlines for SFX at Current X-ray FELs

Access to beamtime at X-ray FELs has been a hindrance to the SFX community and increased access to the X-rays would be very beneficial. In addition to multiplexing instruments using upstream monochromators to remove small portions of the SASE spectrum [125], another method to increase access is through refocusing the unused beam from SFX experiments to a downstream, serial experiment. The CXI instrument of LCLS is already conducting serial SFX operations [84], and the SPB/SFX instrument of XFEL.EU has designs for serial operations starting in the early 2020s.

2.4.2 New Experimental Endstations for X-ray FEL Experiments

All initial SFX experiments were performed in vacuum, which set strict limitations to the way samples could be introduced into the X-ray FEL. These sample delivery methods are the workhorse for SFX, however the throughput achieved strongly depends on the crystal density, which is restricted by the clogging challenge. The DAPHNIS (Diverse Application Platform for Hard X-ray diffraction in SACLA) [126] endstation at SACLA and the MFX beamline at LCLS, which operate in air or in a He atmosphere (which can be humidified) have paved the way for the development of new sample delivery methods for SFX based on the fixed target approach.

Two very promising new systems have been developed for measurements at atmospheric pressure: one novel system is based in acoustic droplet generation onto a conveyer belt that delivers the crystals within the droplet one-by one to the X-ray beam [127]. As clogging is not an issue with the tape drive, high hit rates can be achieved and sample consumption is minimized. However, the system is currently only used at 10 Hz but in principle the system could also be adjusted to 120 Hz data collection rates in the future. This tape drive system also allows for combination of X-ray emission spectroscopy (XES) with X-ray diffraction and has been successfully used for combined TR-SFX and TR-XES experiments at LCLS [97]. Chapter 13 will discuss XES spectroscopy using this setup.

The second system, known as the roadrunner system, is a goniometer-based system for fixed target measurements [44]. This system features a silicon chip with 120,000 windows that allows for data collection with 120 Hz at up to 50% hit rate, where >50% of the windows are covered by crystals. In this setup the windows of the chips are aligned to the X-ray interaction region using an inline microscope. This system has recently been used for the high-resolution structure determination of a virus [44]. Further developments are on the way to enable TR-SFX experiment as well as combination with spectroscopy in parallel to X-ray diffraction on the chip using the roadrunner. These represent just two examples of sample delivery methods that were developed specifically for X-ray FELs and other sample delivery methods in use at SACLA and LCLS will also be discussed in Chap. 5.

2.4.3 Higher Repetition Rate X-ray FELs

One very promising way to increase the number of experiments and enable new types of experiments that can be conducted using biological samples at X-ray FELs is the increase of the repetition rate of the X-ray FELs. Two new X-ray FELs, the European XFEL and LCLS-II will deliver X-ray pulses at much higher repetition rates than the current maximum of 120 Hz at LCLS. The European XFEL differs from current X-ray FELs as it delivers pulses as ten bunch trains per second. For the first operation cycle, the bunch train length will be 60 μs with 1–60 equally spaced pulses per bunch. In the future, the train length can be increased to 600 μs and it is planned that eventually up to 27,000 pulses can be delivered in 1 s.

While this increase in repetition rate can in principle dramatically increase the amount of data collected per time, it poses new challenges for the application of the SFX method, mainly with respect to sample delivery and data collection and evaluation. Sample replenishment and destruction of the flow in the jet will be an increasing challenge as pulses in the pulse train will come with a very rapid rate with spacing of only 1 μs between the pulses (for comparison the LCLS pulses are spaced 8.3 ms between pulses). Imaging of jet explosions have shown the impact of the focused beam on the flowing jet and raise a key concern about the ultimate limitation from sample delivery at high repetition rates [128]. Fixed target approaches as well as viscous jets will not be able to achieve MHz scale sample replenishment and likely only very rapid jets such as the GDVN can currently meet these goals. New experimental approaches using droplet or jet on demand techniques may prove useful for the higher repetition rate machines, although the continuous waveform machines above ~100 kHz, such as LCLS-II, will favor jets. However, the pulse train structure also has advantages, especially for time resolved studies as reactions can be induced in the time between pulses and then up to 60 (and later 2700) images of the reaction can be recorded in the pulse train, thereby getting close to the goal of molecular movies of biomolecules at work.

A further step in this direction will be LCLS-II, which is currently under construction and will start operating in 2020. The unique feature of LCLS-II is that it will reach extremely high repetition rates of close to one million pulses/s. While LCLS-II will operate in the soft to tender X-ray energy regime (up to 5 keV for the first harmonic), which limits the achievable resolution, a proposal for a high energy upgrade (to >12 keV) has already been developed, which will allow for high resolution data to be collected at ultrahigh repetition rates. With the maximal rate of one million pulses/s, the pulses will come with a frequency of $\sim$1 pulse per microsecond and therefore at a similar rate to the $\sim$1 pulse per 230 ns pulse train at XFEL.EU. Therefore, sample delivery systems that are being developed to meet the speed requirement at the XFEL.EU can directly be used for the data collection at LCLS-II. As LCLS-II will deliver the pulses equally spaced the sample would not run to waste. Huge multiplicities of data could be achieved, which would allow for extremely accurate structure factor determination for SFX. In addition, the high repetition rate will allow a high number of images for single particle diffraction, which may significantly increase the resolution limit.

With high repetition rate machines X-ray spectroscopy and solution scattering experiments could take only few seconds instead of hours, which would bring a huge increase in data quality and quantity and allow for more time points to be measured for time resolved experiments for on single crystals and by solution scattering.

However, the very high repetition rates of the XFEL.EU and LCLS-II also lead to formidable novel challenges concerning detectors, data acquisition, data evaluation and data storage. An overview of the ongoing developments of upcoming X-ray FEL facilities will be given in Chap. 16.

2.5 Outlook

Since the start of user operations at X-ray FELs, new methods have been developed to study structure and dynamics in biological samples and systems, with SFX being a major contributor. SFX results to date have shown that the hope of structures being determined while avoiding the ill effects of conventional radiation damage were not overly optimistic. The unique capabilities of X-ray FELs have allowed new frontiers of structural biology to be pursued, pushing further into the ultrafast processes that drive photobiology and studying biological reactions of enzymes with an unprecedented combination of spatial and temporal resolutions. Although the results to date have been exciting, the high-repetition rate X-ray FELs will further enable bioscience at high spatial and temporal resolutions. The new instruments and techniques may continue a revolution in structural biology, transforming high-resolution information from (mostly) static structures into information about the structural dynamics and, ultimately, to a combined view containing structural information as it evolves in time.

Acknowledgements This chapter describes the work of a very large number of people at SLAC National Accelerator Laboratory, the users of LCLS, and the wider community. Use of the LCLS is supported by the US Department of Energy, Office of Science, Office of Basic Energy Sciences under Contract no. DE-AC02-76SF00515.

References

1. Aquila, A., Barty, A., Bostedt, C., Boutet, S., Carini, G., dePonte, D., et al. (2015). The linac coherent light source single particle imaging road map. *Structural Dynamics, 2*(4), 041701. https://doi.org/10.1063/1.4918726.
2. Aquila, A., Hunter, M. S., Doak, R. B., Kirian, R. A., Fromme, P., White, T. A., et al. (2012). Time-resolved protein nanocrystallography using an X-ray free-electron laser. *Optics Express, 20*(3), 2706–2716. https://doi.org/10.1364/OE.20.002706.
3. Arnlund, D., Johansson, L. C., Wickstrand, C., Barty, A., Williams, G. J., Malmerberg, E., & Neutze, R. (2014). Visualizing a protein quake with time-resolved X-ray scattering at a free-electron laser. *Nature Methods, 11*(9), 923–926. https://doi.org/10.1038/nmeth.3067 https://www.nature.com/articles/nmeth.3067#supplementary-information.
4. Ayyer, K., Yefanov, O. M., Oberthür, D., Roy-Chowdhury, S., Galli, L., Mariani, V., et al. (2016). Macromolecular diffractive imaging using imperfect crystals. *Nature, 530*(7589), 202–206. https://doi.org/10.1038/nature16949.
5. Barends, T. R., Foucar, L., Botha, S., Doak, R. B., Shoeman, R. L., Nass, K., et al. (2014). De novo protein crystal structure determination from X-ray free-electron laser data. *Nature, 505*(7482), 244–247. https://doi.org/10.1038/nature12773.
6. Barends, T. R. M., Foucar, L., Ardevol, A., Nass, K., Aquila, A., Botha, S., et al. (2015). Direct observation of ultrafast collective motions in CO myoglobin upon ligand dissociation. *Science*. https://doi.org/10.1126/science.aac5492.
7. Barty, A., Caleman, C., Aquila, A., Timneanu, N., Lomb, L., White, T. A., et al. (2012). Self-terminating diffraction gates femtosecond X-ray nanocrystallography measurements. *Nature Photonics, 6*(1), 35–40. http://www.nature.com/nphoton/journal/v6/n1/abs/nphoton.2011.297.html#supplementary-information.
8. Barty, A., Kirian, R. A., Maia, F. R., Hantke, M., Yoon, C. H., White, T. A., & Chapman, H. (2014). *Cheetah* : Software for high-throughput reduction and analysis of serial femtosecond X-ray diffraction data. *Journal of Applied Crystallography, 47*(Pt 3), 1118–1131. https://doi.org/10.1107/S1600576714007626.
9. Batyuk, A., Galli, L., Ishchenko, A., Han, G. W., Gati, C., Popov, P. A., et al. (2016). Native phasing of x-ray free-electron laser data for a G protein–coupled receptor. *Science Advances, 2*(9). https://doi.org/10.1126/sciadv.1600292.
10. Botha, S., Nass, K., Barends, T. R. M., Kabsch, W., Latz, B., Dworkowski, F., et al. (2015). Room-temperature serial crystallography at synchrotron X-ray sources using slowly flowing free-standing high-viscosity microstreams. *Acta Crystallographica Section D, 71*(2), 387–397. https://doi.org/10.1107/S1399004714026327.
11. Boutet, S., Lomb, L., Williams, G. J., Barends, T. R., Aquila, A., Doak, R. B., et al. (2012). High-resolution protein structure determination by serial femtosecond crystallography. *Science, 337*(6092), 362–364. https://doi.org/10.1126/science.1217737.
12. Caffrey, M. (2003). Membrane protein crystallization. *Journal of Structural Biology, 142*(1), 108–132. https://doi.org/10.1016/S1047-8477(03)00043-1.
13. Caspar, D. L. D., Clarage, J., Salunke, D. M., & Clarage, M. (1988). Liquid-like movements in crystalline insulin. *Nature, 332*(6165), 659–662.
14. Chapman, H. N., Barty, A., Bogan, M. J., Boutet, S., Frank, M., Hau-Riege, S. P., et al. (2006). Femtosecond diffractive imaging with a soft-X-ray free-electron laser. *Nature Physics, 2*(12), 839–843.
15. Chapman, H. N., Fromme, P., Barty, A., White, T. A., Kirian, R. A., Aquila, A., et al. (2011). Femtosecond X-ray protein nanocrystallography. *Nature, 470*(7332), 73–77. https://doi.org/10.1038/nature09750.
16. Chapman, H. N., Hau-Riege, S. P., Bogan, M. J., Bajt, S., Barty, A., Boutet, S., et al. (2007). Femtosecond time-delay X-ray holography. *Nature, 448*(7154), 676–679.

17. Cho, H. S., Dashdorj, N., Schotte, F., Graber, T., Henning, R., & Anfinrud, P. (2010). Protein structural dynamics in solution unveiled via 100-ps time-resolved x-ray scattering. *Proceedings of the National Academy of Sciences, 107*(16), 7281–7286. https://doi.org/10.1073/pnas.1002951107.
18. Chollet, M., Alonso-Mori, R., Cammarata, M., Damiani, D., Defever, J., Delor, J. T., et al. (2015). The X-ray pump–probe instrument at the Linac coherent light source. *Journal of Synchrotron Radiation, 22*(3), 503–507. https://doi.org/10.1107/S1600577515005135.
19. Cohen, A. E., Soltis, S. M., González, A., Aguila, L., Alonso-Mori, R., Barnes, C. O., et al. (2014). Goniometer-based femtosecond crystallography with X-ray free electron lasers. *Proceedings of the National Academy of Sciences, 111*(48), 17122–17127. https://doi.org/10.1073/pnas.1418733111.
20. Colletier, J.-P., Sawaya, M. R., Gingery, M., Rodriguez, J. A., Cascio, D., Brewster, A. S., & Eisenberg, D. S. (2016). De novo phasing with X-ray laser reveals mosquito larvicide BinAB structure. *Nature, 539*, 43. https://doi.org/10.1038/nature19825 https://www.nature.com/articles/nature19825#supplementary-information.
21. Colletier, J. P. (n.d.). Chromophore twisting in the excited state of a photoswitchable fluorescent protein captured by time-resolved serial femtosecond crystallography. *Nature Chemistry*.
22. Conrad, C. E., Basu, S., James, D., Wang, D., Schaffer, A., Roy-Chowdhury, S., et al. (2015). A novel inert crystal delivery medium for serial femtosecond crystallography. *IUCrJ, 2*(Pt 4), 421–430. https://doi.org/10.1107/S2052252515009811.
23. DePonte, D. P., Doak, R. B., Hunter, M., Liu, Z., Weierstall, U., & Spence, J. C. H. (2009). SEM imaging of liquid jets. *Micron, 40*(4), 507–509. https://doi.org/10.1016/j.micron.2008.12.009.
24. DePonte, D. P., Weierstall, U., Schmidt, K., Warner, J., Starodub, D., Spence, J. C. H., & Doak, R. B. (2008). Gas dynamic virtual nozzle for generation of microscopic droplet streams. *Journal of Physics D: Applied Physics, 41*(19), 195505.
25. Dörner, K., Martin-Garcia, J. M., Kupitz, C., Gong, Z., Mallet, T. C., Chen, L., et al. (2016). Characterization of protein Nanocrystals based on the reversibility of crystallization. *Crystal Growth & Design, 16*(7), 3838–3845. https://doi.org/10.1021/acs.cgd.6b00384.
26. Doucet, J., & Benoit, J. P. (1987). Molecular dynamics studied by analysis of the X-ray diffuse scattering from lysozyme crystals. *Nature, 325*(6105), 643–646.
27. Drenth, J. (2007). *Principles of protein X-ray crystallography* (3rd ed.). New York: Springer-Verlag.
28. Echalier, A., Glazer, R. L., Fulop, V., & Geday, M. A. (2004). Assessing crystallization droplets using birefringence. *Acta Crystallographica Section D, 60*(4), 696–702. https://doi.org/10.1107/S0907444904003154.
29. Ekeberg, T., Svenda, M., Abergel, C., Maia, F. R. N. C., Seltzer, V., Claverie, J.-M., et al. (2015). Three-dimensional reconstruction of the Giant Mimivirus particle with an X-ray free-Electron laser. *Physical Review Letters, 114*(9), 098102.
30. Emma, P., Akre, R., Arthur, J., Bionta, R., Bostedt, C., Bozek, J., et al. (2010). First lasing and operation of an ångstrom-wavelength free-electron laser. *Nature Photonics, 4*(9), 641–647. https://doi.org/10.1038/nphoton.2010.176.
31. Faure, P., Micu, A., Perahia, D., Doucet, J., Smith, J. C., & Benoit, J. P. (1994). Correlated intramolecular motions and diffuse x-ray scattering in lysozyme. *Nature Structural & Molecular Biology, 1*(2), 124–128.
32. Ferguson, K. R., Bucher, M., Bozek, J. D., Carron, S., Castagna, J.-C., Coffee, R., et al. (2015). The atomic, molecular and optical science instrument at the Linac coherent light source. *Journal of Synchrotron Radiation, 22*(3), 492–497. https://doi.org/10.1107/S1600577515004646.
33. Foadi, J., Aller, P., Alguel, Y., Cameron, A., Axford, D., Owen, R. L., et al. (2013). Clustering procedures for the optimal selection of data sets from multiple crystals in macromolecular crystallography. *Acta Crystallographica Section D: Biological Crystallography, 69*(Pt 8), 1617–1632. https://doi.org/10.1107/S0907444913012274.

34. Frank, M., Carlson, D. B., Hunter, M. S., Williams, G. J., Messerschmidt, M., Zatsepin, N. A., et al. (2014). Femtosecond X-ray diffraction from two-dimensional protein crystals. *IUCrJ, 1*(2), 95–100. https://doi.org/10.1107/S2052252514001444.
35. Fromme, R., Ishchenko, A., Metz, M., Chowdhury, S. R., Basu, S., Boutet, S., et al. (2015). Serial femtosecond crystallography of soluble proteins in lipidic cubic phase. *IUCrJ, 2*(5), 545–551. https://doi.org/10.1107/S2052252515013160.
36. Fukuda, Y., Tse, K. M., Nakane, T., Nakatsu, T., Suzuki, M., Sugahara, M., et al. (2016). Redox-coupled proton transfer mechanism in nitrite reductase revealed by femtosecond crystallography. *Proceedings of the National Academy of Sciences, 113*(11), 2928–2933. https://doi.org/10.1073/pnas.1517770113.
37. Fuller, F. D., Gul, S., Chatterjee, R., Burgie, E. S., Young, I. D., Lebrette, H., & Yano, J. (2017). Drop-on-demand sample delivery for studying biocatalysts in action at X-ray free-electron lasers. *Nature Methods, 14*(4), 443–449. https://doi.org/10.1038/nmeth.4195 http://www.nature.com/nmeth/journal/v14/n4/abs/nmeth.4195.html#supplementary-information.
38. Gañán-Calvo, A. M. (1998). Generation of steady liquid microthreads and Micron-sized Monodisperse sprays in gas streams. *Physical Review Letters, 80*(2), 285–288.
39. Garman, E. (2010). Radiation damage in macromolecular crystallography: What is it and why should we care? *Acta Crystallographica Section D, 66*(4), 339–351. https://doi.org/10.1107/S0907444910008656.
40. Ginn, H. M., Brewster, A. S., Hattne, J., Evans, G., Wagner, A., Grimes, J. M., et al. (2015). A revised partiality model and post-refinement algorithm for X-ray free-electron laser data. *Acta Crystallographica Section D, 71*(6), 1400–1410. https://doi.org/10.1107/S1399004715006902.
41. Gorel, A., Motomura, K., Fukuzawa, H., Doak, R. B., Grünbein, M. L., Hilpert, M., et al. (2017). Multi-wavelength anomalous diffraction de novo phasing using a two-colour X-ray free-electron laser with wide tunability. *Nature Communications, 8*(1), 1170. https://doi.org/10.1038/s41467-017-00754-7.
42. Graewert, M. A., & Svergun, D. I. (2013). Impact and progress in small and wide angle X-ray scattering (SAXS and WAXS). *Current Opinion in Structural Biology, 23*(5), 748–754. https://doi.org/10.1016/j.sbi.2013.06.007.
43. Gualtieri, E. J., Guo, F., Kissick, D. J., Jose, J., Kuhn, R. J., Jiang, W., & Simpson, G. J. (2011). Detection of membrane protein two-dimensional crystals in living cells. *Biophysical Journal, 100*(1), 207–214. https://doi.org/10.1016/j.bpj.2010.10.051.
44. Guizar-Sicairos, M., Johnson, I., Diaz, A., Holler, M., Karvinen, P., Stadler, H.-C., et al. (2014). High-throughput ptychography using Eiger: Scanning X-ray nano-imaging of extended regions. *Optics Express, 22*(12), 14859–14870. https://doi.org/10.1364/OE.22.014859.
45. Hendrickson, W. A. (2014). Anomalous diffraction in crystallographic phase evaluation. *Quarterly Reviews of Biophysics, 47*(1), 49–93. https://doi.org/10.1017/S0033583514000018.
46. Hunter, M. S., DePonte, D. P., Shapiro, D. A., Kirian, R. A., Wang, X., Starodub, D., et al. (2011). X-ray diffraction from membrane protein Nanocrystals. *Biophysical Journal, 100*(1), 198–206. https://doi.org/10.1016/j.bpj.2010.10.049.
47. Hunter, M. S., Yoon, C. H., DeMirci, H., Sierra, R. G., Dao, E. H., Ahmadi, R., & Boutet, S. (2016). Selenium single-wavelength anomalous diffraction de novo phasing using an X-ray-free electron laser. *Nature Communications, 7*, 13388. https://doi.org/10.1038/ncomms13388 http://www.nature.com/articles/ncomms13388#supplementary-information.
48. Johansson, L. C., Arnlund, D., White, T. A., Katona, G., DePonte, D. P., Weierstall, U., & Neutze, R. (2012). Lipidic phase membrane protein serial femtosecond crystallography. *Nature Methods, 9*(3), 263–265. http://www.nature.com/nmeth/journal/v9/n3/abs/nmeth.1867.html#supplementary-information.

49. Jung, Y. O., Lee, J. H., Kim, J., Schmidt, M., Moffat, K., Šrajer, V., & Ihee, H. (2013). Volume-conserving trans–cis isomerization pathways in photoactive yellow protein visualized by picosecond X-ray crystallography. *Nature Chemistry, 5*(3), 212–220. http://www.nature.com/nchem/journal/v5/n3/abs/nchem.1565.html#supplementary-information.
50. Kam, Z., Koch, M. H. J., & Bordas, J. (1981). Fluctuation X-ray scattering from biological particles in frozen solution by using synchrotron radiation. *Proceedings of the National Academy of Sciences of the United States of America, 78*(6), 3559–3562.
51. Kern, J., Tran, R., Alonso-Mori, R., Koroidov, S., Echols, N., Hattne, J., & Yachandra, V. K. (2014). Taking snapshots of photosynthetic water oxidation using femtosecond X-ray diffraction and spectroscopy. *Nature Communications, 5*, 4371. https://doi.org/10.1038/ncomms5371 https://www.nature.com/articles/ncomms5371#supplementary-information.
52. Kirian, R. A., Wang, X., Weierstall, U., Schmidt, K. E., Spence, J. C. H., Hunter, M., et al. (2010). Femtosecond protein nanocrystallography—Data analysis methods. *Optics Express, 18*(6), 5713–5723. https://doi.org/10.1364/OE.18.005713.
53. Kirian, R. A., White, T. A., Holton, J. M., Chapman, H. N., Fromme, P., Barty, A., et al. (2011). Structure-factor analysis of femtosecond microdiffraction patterns from protein nanocrystals. *Acta Crystallographica Section A, 67*(2), 131–140. https://doi.org/10.1107/S0108767310050981.
54. Kissick, D. J., Gualtieri, E. J., Simpson, G. J., & Cherezov, V. (2010). Nonlinear optical imaging of integral membrane protein crystals in Lipidic Mesophases. *Analytical Chemistry, 82*(2), 491–497. https://doi.org/10.1021/ac902139w.
55. Kissick, D. J., Wanapun, D., & Simpson, G. J. (2011). Second-order nonlinear optical imaging of Chiral crystals. *Annual Review of Analytical Chemistry (Palo Alto, Calif.), 4*, 419–437. https://doi.org/10.1146/annurev.anchem.111808.073722.
56. Koopmann, R., Cupelli, K., Redecke, L., Nass, K., DePonte, D. P., White, T. A., & Duszenko, M. (2012). In vivo protein crystallization opens new routes in structural biology. *Natural Methods, 9*(3), 259–262. http://www.nature.com/nmeth/journal/v9/n3/abs/nmeth.1859.html#supplementary-information.
57. Kupitz, C., Basu, S., Grotjohann, I., Fromme, R., Zatsepin, N. A., Rendek, K. N., & Fromme, P. (2014). Serial time-resolved crystallography of photosystem II using a femtosecond X-ray laser. *Nature, 513*(7517), 261–265. https://doi.org/10.1038/nature13453 http://www.nature.com/nature/journal/v513/n7517/abs/nature13453.html#supplementary-information.
58. Kupitz, C., Olmos, J. L., Holl, M., Tremblay, L., Pande, K., Pandey, S., & Schmidt, M. (2017). Structural enzymology using X-ray free electron lasers. *Structural Dynamics, 4*(4), 044003. https://doi.org/10.1063/1.4972069.
59. Landau, E. M., & Rosenbusch, J. P. (1996). Lipidic cubic phases: A novel concept for the crystallization of membrane proteins. *Proceedings of the National Academy of Sciences, 93*(25), 14532–14535.
60. Levantino, M., Schirò, G., Lemke, H. T., Cottone, G., Glownia, J. M., Zhu, D., & Cammarata, M. (2015). Ultrafast myoglobin structural dynamics observed with an X-ray free-electron laser. *Nature Communications, 6*, 6772. https://doi.org/10.1038/ncomms7772 https://www.nature.com/articles/ncomms7772#supplementary-information.
61. Liang, M., Williams, G. J., Messerschmidt, M., Seibert, M. M., Montanez, P. A., Hayes, M., et al. (2015). The coherent X-ray imaging instrument at the Linac coherent light source. *Journal of Synchrotron Radiation, 22*(Pt 3), 514–519. https://doi.org/10.1107/S160057751500449X.
62. Liu, W., Wacker, D., Gati, C., Han, G. W., James, D., Wang, D., et al. (2013). Serial femtosecond crystallography of G protein-coupled receptors. *Science, 342*(6165), 1521–1524. https://doi.org/10.1126/science.1244142.
63. Lomb, L., Barends, T. R. M., Kassemeyer, S., Aquila, A., Epp, S. W., Erk, B., et al. (2011). Radiation damage in protein serial femtosecond crystallography using an x-ray free-electron laser. *Physical Review B, 84*(21), 214111.

64. Madden, J. T., DeWalt, E. L., & Simpson, G. J. (2011). Two-photon excited UV fluorescence for protein crystal detection. *Acta Crystallographica Section D, 67*(10), 839–846. https://doi.org/10.1107/S0907444911028253.
65. Malkin, A. J., Kuznetsov, Y. G., Land, T. A., DeYoreo, J. J., & McPherson, A. (1995). Mechanisms of growth for protein and virus crystals. *Nature Structural & Molecular Biology, 2*(11), 956–959.
66. Marchesini, S., He, H., Chapman, H. N., Hau-Riege, S. P., Noy, A., Howells, M. R., et al. (2003). X-ray image reconstruction from a diffraction pattern alone. *Physical Review B, 68*(14), 140101.
67. Martin-Garcia, J. M., Conrad, C. E., Nelson, G., Stander, N., Zatsepin, N. A., Zook, J., et al. (2017). Serial millisecond crystallography of membrane and soluble protein microcrystals using synchrotron radiation. *IUCrJ, 4*(4), 439–454. https://doi.org/10.1107/S205225251700570X.
68. Moffat, K. (2001). Time-resolved biochemical crystallography: A mechanistic perspective. *Chemical Reviews, 101*(6), 1569–1582. https://doi.org/10.1021/cr990039q.
69. Nakane, T., Hanashima, S., Suzuki, M., Saiki, H., Hayashi, T., Kakinouchi, K., et al. (2016). Membrane protein structure determination by SAD, SIR, or SIRAS phasing in serial femtosecond crystallography using an iododetergent. *Proceedings of the National Academy of Sciences of the United States of America, 113*(46), 13039–13044. https://doi.org/10.1073/pnas.1602531113.
70. Nakane, T., Song, C., Suzuki, M., Nango, E., Kobayashi, J., Masuda, T., et al. (2015). Native sulfur/chlorine SAD phasing for serial femtosecond crystallography. *Acta Crystallographica Section D, 71*(12), 2519–2525. https://doi.org/10.1107/S139900471501857X.
71. Nass, K., Meinhart, A., Barends, T., Foucar, L., Gorel, A., Aquila, A., et al. (2016a). Protein structure determination by single-wavelength anomalous diffraction phasing of X-ray free-electron laser data. *IUCrJ, 3*, 1–12.
72. Nass, K., Meinhart, A., Barends, T. R. M., Foucar, L., Gorel, A., Aquila, A., et al. (2016b). Protein structure determination by single-wavelength anomalous diffraction phasing of X-ray free-electron laser data. *IUCrJ, 3*(3), 180–191. https://doi.org/10.1107/S2052252516002980.
73. Nave, C. (1998). A description of imperfections in protein crystals. *Acta Crystallographica Section D, 54*(5), 848–853. https://doi.org/10.1107/S0907444998001875.
74. Neutze, R., Wouts, R., van der Spoel, D., Weckert, E., & Hajdu, J. (2000). Potential for biomolecular imaging with femtosecond X-ray pulses. *Nature, 406*(6797), 752–757. https://doi.org/10.1038/35021099.
75. Nogly, P., James, D., Wang, D., White, T. A., Zatsepin, N., Shilova, A., et al. (2015). Lipidic cubic phase serial millisecond crystallography using synchrotron radiation. *IUCrJ, 2*(2), 168–176. https://doi.org/10.1107/S2052252514026487.
76. Oberthuer, D., Knoška, J., Wiedorn, M. O., Beyerlein, K. R., Bushnell, D. A., Kovaleva, E. G., et al. (2017). Double-flow focused liquid injector for efficient serial femtosecond crystallography. *7*, 44628. https://doi.org/10.1038/srep44628 http://dharmasastra.live.cf.private.springer.com/articles/srep44628#supplementary-information
77. Owen, R. L., Rudiño-Piñera, E., & Garman, E. F. (2006). Experimental determination of the radiation dose limit for cryocooled protein crystals. *Proceedings of the National Academy of Sciences of the United States of America, 103*(13), 4912–4917. https://doi.org/10.1073/pnas.0600973103.
78. Padayatti, P., Palczewska, G., Sun, W., Palczewski, K., & Salom, D. (2012). Imaging of protein crystals with two-photon microscopy. *Biochemistry, 51*(8), 1625–1637. https://doi.org/10.1021/bi201682q.
79. Pande, K., Hutchison, C. D. M., Groenhof, G., Aquila, A., Robinson, J. S., Tenboer, J., et al. (2016). Femtosecond structural dynamics drives the trans/cis isomerization in photoactive yellow protein. *Science, 352*(6286), 725–729. https://doi.org/10.1126/science.aad5081.
80. Perez, J., Faure, P., & Benoit, J.-P. (1996). Molecular rigid-body displacements in a tetragonal Lysozyme crystal confirmed by X-ray diffuse scattering. *Acta Crystallographica Section D, 52*(4), 722–729. https://doi.org/10.1107/S0907444996002594.

81. Rasmussen, S. G. F., Choi, H.-J., Rosenbaum, D. M., Kobilka, T. S., Thian, F. S., Edwards, P. C., & Kobilka, B. K. (2007). Crystal structure of the human [bgr]2 adrenergic G-protein-coupled receptor. *Nature, 450*(7168), 383–387. http://www.nature.com/nature/journal/v450/n7168/suppinfo/nature06325_S1.html.
82. Redecke, L., Nass, K., DePonte, D. P., White, T. A., Rehders, D., Barty, A., et al. (2013). Natively inhibited *Trypanosoma brucei* Cathepsin B structure determined by using an X-ray laser. *Science, 339*(6116), 227–230. https://doi.org/10.1126/science.1229663.
83. Roedig, P., Ginn, H. M., Pakendorf, T., Sutton, G., Harlos, K., Walter, T. S., et al. (2017). High-speed fixed-target serial virus crystallography. *Natural Methods.*, advance online publication, *14*, 805–810. https://doi.org/10.1038/nmeth.4335 http://www.nature.com/nmeth/journal/vaop/ncurrent/abs/nmeth.4335.html#supplementary-information.
84. Rossmann, M. G. (1990). The molecular replacement method. *Acta Crystallographica. Section A, 46*(Pt 2), 73–82.
85. Rupp, B. (2013). Macromolecular crystallography: Overview. In G. C. K. Roberts (Ed.), *Encyclopedia of biophysics* (pp. 1346–1353). Berlin, Heidelberg: Springer Berlin Heidelberg.
86. Schlichting, I. (2015). Serial femtosecond crystallography: The first five years. *IUCrJ, 2*(Pt 2), 246–255. https://doi.org/10.1107/S205225251402702X.
87. Schlichting, I., & Miao, J. (2012). Emerging opportunities in structural biology with X-ray free-electron lasers. *Current Opinion in Structural Biology, 22*(5), 613–626. https://doi.org/10.1016/j.sbi.2012.07.015.
88. Schmidt, M. (2013). Mix and inject: Reaction initiation by diffusion for time-resolved macromolecular crystallography. *Advances in Condensed Matter Physics, 2013*, 10. https://doi.org/10.1155/2013/167276.
89. Schotte, F., Soman, J., Olson, J. S., Wulff, M., & Anfinrud, P. A. (2004). Picosecond time-resolved X-ray crystallography: Probing protein function in real time. *Journal of Structural Biology, 147*(3), 235–246. https://doi.org/10.1016/j.jsb.2004.06.009.
90. Seibert, M. M., Ekeberg, T., Maia, F. R. N. C., Svenda, M., Andreasson, J., Jonsson, O., et al. (2011). Single mimivirus particles intercepted and imaged with an X-ray laser. *Nature, 470*(7332), 78–81.
91. Shapiro, D., Thibault, P., Beetz, T., Elser, V., Howells, M., Jacobsen, C., et al. (2005). Biological imaging by soft x-ray diffraction microscopy. *Proceedings of the National Academy of Sciences, 102*(43), 15343–15346. https://doi.org/10.1073/pnas.0503305102.
92. Shapiro, D. A., Chapman, H. N., DePonte, D., Doak, R. B., Fromme, P., Hembree, G., et al. (2008). Powder diffraction from a continuous microjet of submicrometer protein crystals. *Journal of Synchrotron Radiation, 15*(6), 593–599. https://doi.org/10.1107/S0909049508024151.
93. Sierra, R. G., Gati, C., Laksmono, H., Dao, E. H., Gul, S., Fuller, F., et al. (2016). Concentric-flow electrokinetic injector enables serial crystallography of ribosome and photosystem II. *Nature Methods, 13*(1), 59–62. https://doi.org/10.1038/nmeth.3667.
94. Sierra, R. G., Laksmono, H., Kern, J., Tran, R., Hattne, J., Alonso-Mori, R., et al. (2012). Nanoflow electrospinning serial femtosecond crystallography. *Acta Crystallographica Section D, 68*(11), 1584–1587. https://doi.org/10.1107/S0907444912038152.
95. Spence, J. C. H., & Chapman, H. N. (2014). The birth of a new field. *Philosophical Transactions of the Royal Society B: Biological Sciences, 369*(1647). https://doi.org/10.1098/rstb.2013.0309.
96. Spence, J. C. H., Kirian, R. A., Wang, X., Weierstall, U., Schmidt, K. E., White, T., et al. (2011). Phasing of coherent femtosecond X-ray diffraction from size-varying nanocrystals. *Optics Express, 19*(4), 2866–2873. https://doi.org/10.1364/OE.19.002866.
97. Stagno, J. R., Liu, Y., Bhandari, Y. R., Conrad, C. E., Panja, S., Swain, M., & Wang, Y. X. (2016). Structures of riboswitch RNA reaction states by mix-and-inject XFEL serial crystallography. *Nature.*, advance online publication, *541*(7636), 242–246. https://doi.org/10.1038/nature20599 http://www.nature.com/nature/journal/vaap/ncurrent/abs/nature20599.html#supplementary-information.

98. Stan, C. A., Milathianaki, D., Laksmono, H., Sierra, R. G., McQueen, T. A., Messerschmidt, M., & Boutet, S. (2016). Liquid explosions induced by X-ray laser pulses. *Nature Physics, 12*(10), 966–971. https://doi.org/10.1038/nphys3779http://www.nature.com/nphys/journal/v12/n10/abs/nphys3779.html#supplementary-information.
99. Stevens, R. C., Cherezov, V., Katritch, V., Abagyan, R., Kuhn, P., Rosen, H., & Wuthrich, K. (2013). The GPCR network: A large-scale collaboration to determine human GPCR structure and function. *Nature Reviews. Drug Discovery, 12*(1), 25–34.
100. Stevenson, H. P., DePonte, D. P., Makhov, A. M., Conway, J. F., Zeldin, O. B., Boutet, S., & Cohen, A. E. (2014a). Transmission electron microscopy as a tool for nanocrystal characterization pre- and post-injector. *Philosophical Transactions of the Royal Society B: Biological Sciences, 369*(1647). https://doi.org/10.1098/rstb.2013.0322.
101. Stevenson, H. P., Makhov, A. M., Calero, M., Edwards, A. L., Zeldin, O. B., Mathews, I. I., et al. (2014b). Use of transmission electron microscopy to identify nanocrystals of challenging protein targets. *Proceedings of the National Academy of Sciences, 111*(23), 8470–8475. https://doi.org/10.1073/pnas.1400240111.
102. Strüder, L., Epp, S., Rolles, D., Hartmann, R., Holl, P., Lutz, G., et al. (2010). Large-format, high-speed, X-ray pnCCDs combined with electron and ion imaging spectrometers in a multipurpose chamber for experiments at 4th generation light sources. *Nuclear Instruments and Methods in Physics Research Section A: Accelerators, Spectrometers, Detectors and Associated Equipment, 614*(3), 483–496. https://doi.org/10.1016/j.nima.2009.12.053.
103. Suga, M., Akita, F., Hirata, K., Ueno, G., Murakami, H., Nakajima, Y., et al. (2015). Native structure of photosystem II at 1.95 a resolution viewed by femtosecond X-ray pulses. *Nature, 517*(7532), 99–103. https://doi.org/10.1038/nature13991.
104. Suga, M., Akita, F., Sugahara, M., Kubo, M., Nakajima, Y., Nakane, T., et al. (2017). Light-induced structural changes and the site of O=O bond formation in PSII caught by XFEL. *Nature, 543*(7643), 131–135. https://doi.org/10.1038/nature21400.
105. Sugahara, M., Mizohata, E., Nango, E., Suzuki, M., Tanaka, T., Masuda, T., & Iwata, S. (2015). Grease matrix as a versatile carrier of proteins for serial crystallography. *Natural Methods, 12*(1), 61–63. https://doi.org/10.1038/nmeth.3172 http://www.nature.com/nmeth/journal/v12/n1/abs/nmeth.3172.html#supplementary-information.
106. Sugahara, M., Nakane, T., Masuda, T., Suzuki, M., Inoue, S., Song, C., et al. (2017). Hydroxyethyl cellulose matrix applied to serial crystallography. *Scientific Reports, 7*, 703. https://doi.org/10.1038/s41598-017-00761-0.
107. Tenboer, J., Basu, S., Zatsepin, N., Pande, K., Milathianaki, D., Frank, M., et al. (2014). Time-resolved serial crystallography captures high-resolution intermediates of photoactive yellow protein. *Science, 346*(6214), 1242–1246. https://doi.org/10.1126/science.1259357.
108. Thibault, P., Elser, V., Jacobsen, C., Shapiro, D., & Sayre, D. (2006). Reconstruction of a yeast cell from X-ray diffraction data. *Acta Crystallogr. Sect. A, 62*(4), 248–261. https://doi.org/10.1107/S0108767306016515.
109. Tono, K., Nango, E., Sugahara, M., Song, C., Park, J., Tanaka, T., et al. (2015). Diverse application platform for hard X-ray diffraction in SACLA (DAPHNIS): Application to serial protein crystallography using an X-ray free-electron laser. *Journal of Synchrotron Radiation, 22*(3), 532–537. https://doi.org/10.1107/S1600577515004464.
110. Van Benschoten, A. H., Liu, L., Gonzalez, A., Brewster, A. S., Sauter, N. K., Fraser, J. S., & Wall, M. E. (2016). Measuring and modeling diffuse scattering in protein X-ray crystallography. *Proceedings of the National Academy of Sciences, 113*(15), 4069–4074. https://doi.org/10.1073/pnas.1524048113.
111. Wall, M. E., Ealick, S. E., & Gruner, S. M. (1997). Three-dimensional diffuse x-ray scattering from crystals of staphylococcal nuclease. *Proceedings of the National Academy of Sciences, 94*(12), 6180–6184.

112. Wampler, R. D., Kissick, D. J., Dehen, C. J., Gualtieri, E. J., Grey, J. L., Wang, H.-F., et al. (2008). Selective detection of protein crystals by second harmonic microscopy. *Journal of the American Chemical Society, 130*(43), 14076–14077. https://doi.org/10.1021/ja805983b.
113. Wanapun, D., Kestur, U. S., Kissick, D. J., Simpson, G. J., & Taylor, L. S. (2010). Selective detection and Quantitation of organic molecule crystallization by second harmonic generation microscopy. *Analytical Chemistry, 82*(13), 5425–5432. https://doi.org/10.1021/ac100564f.
114. Weierstall, U., Doak, R. B., Spence, J. C. H., Starodub, D., Shapiro, D., Kennedy, P., et al. (2008). Droplet streams for serial crystallography of proteins. *Experiments in Fluids, 44*(5), 675–689. https://doi.org/10.1007/s00348-007-0426-8.
115. Weierstall, U., James, D., Wang, C., White, T. A., Wang, D., Liu, W., et al. (2014). Lipidic cubic phase injector facilitates membrane protein serial femtosecond crystallography. *Nature Communications, 5*, 3309–3309. https://doi.org/10.1038/ncomms4309.
116. Welberry, T. R., Heerdegen, A. P., Goldstone, D. C., & Taylor, I. A. (2011). Diffuse scattering resulting from macromolecular frustration. *Acta Crystallographica Section B, 67*(6), 516–524. https://doi.org/10.1107/S0108768111037542.
117. White, T. A. (2014). Post-refinement method for snapshot serial crystallography. *Philosophical Transactions of the Royal Society B: Biological Sciences, 369*(1647). https://doi.org/10.1098/rstb.2013.0330.
118. White, T. A., Kirian, R. A., Martin, A. V., Aquila, A., Nass, K., Barty, A., & Chapman, H. N. (2012). CrystFEL: A software suite for snapshot serial crystallography. *Journal of Applied Crystallography, 45*(2), 335–341. https://doi.org/10.1107/s0021889812002312.
119. Woolfson, M. M. (1997). *An introduction to X-ray crystallography* (2nd ed.). Cambridge: Cambridge University Press.
120. Xiong, G., Moutanabbir, O., Reiche, M., Harder, R., & Robinson, I. (2014). Coherent X-ray diffraction imaging and characterization of strain in silicon-on-insulator nanostructures. *Advanced Materials, 26*(46), 7747–7763. https://doi.org/10.1002/adma.201304511.
121. Yamashita, K., Kuwabara, N., Nakane, T., Murai, T., Mizohata, E., Sugahara, M., et al. (2017). Experimental phase determination with selenomethionine or mercury-derivatization in serial femtosecond crystallography. *IUCrJ, 4*(5), 639–647. https://doi.org/10.1107/S2052252517008557.
122. Yamashita, K., Pan, D., Okuda, T., Sugahara, M., Kodan, A., Yamaguchi, T., et al. (2015). An isomorphous replacement method for efficient de novo phasing for serial femtosecond crystallography. *Scientific Reports, 5*, 14017. https://doi.org/10.1038/srep14017.
123. Yoon, C. H., DeMirci, H., Sierra, R. G., Dao, E. H., Ahmadi, R., Aksit, F., et al. (2017). Se-SAD serial femtosecond crystallography datasets from selenobiotinyl-streptavidin. *Scientific Data, 4*, 170055. https://doi.org/10.1038/sdata.2017.55.
124. Young, I. D., Ibrahim, M., Chatterjee, R., Gul, S., Fuller, F. D., Koroidov, S., & Yano, J. (2016). Structure of photosystem II and substrate binding at room temperature. *Nature, 540*(7633), 453–457. https://doi.org/10.1038/nature20161 http://www.nature.com/nature/journal/v540/n7633/abs/nature20161.html#supplementary-information.
125. Young, L., Kanter, E. P., Krässig, B., Li, Y., March, A. M., Pratt, S. T., et al. (2010). Femtosecond electronic response of atoms to ultra-intense X-rays. *Nature, 466*, 56. https://doi.org/10.1038/nature09177.
126. Zeldin, O. B., Brewster, A. S., Hattne, J., Uervirojnangkoorn, M., Lyubimov, A. Y., Zhou, Q., & Brunger, A. T. (2015). Data exploration toolkit for serial diffraction experiments. *Acta Crystallographica Section D, 71*(2), 352–356. https://doi.org/10.1107/S1399004714025875.
127. Zhang, H., Han, G. W., Batyuk, A., Ishchenko, A., White, K. L., Patel, N., & Cherezov, V. (2017a). Structural basis for selectivity and diversity in angiotensin II receptors. *Nature.*, advance online publication, *544*(7650), 327–332. https://doi.org/10.1038/nature22035 http://www.nature.com/nature/journal/vaop/ncurrent/abs/nature22035.html#supplementary-information.

128. Zhang, X., Zhao, F., Wu, Y., Yang, J., Han, G. W., Zhao, S., & Xu, F. (2017b). Crystal structure of a multi-domain human smoothened receptor in complex with a super stabilizing ligand. *Nature Communications, 8*, 15383. https://doi.org/10.1038/ncomms15383 https://www.nature.com/articles/ncomms15383#supplementary-information.
129. Zhu, D., Feng, Y., Stoupin, S., Terentyev, S. A., Lemke, H. T., Fritz, D. M., et al. (2014). Performance of a beam-multiplexing diamond crystal monochromator at the Linac coherent light source. *Review of Scientific Instruments, 85*(6), 063106. https://doi.org/10.1063/1.4880724.

Chapter 3
Small Is Beautiful: Growth and Detection of Nanocrystals

Jesse Coe and Alexandra Ros

3.1 Introduction

With the advent of X-Ray free electron lasers (FELs), the field of serial femtosecond crystallography (SFX) was borne, allowing a stream of nanocrystals to be measured individually and diffraction data to be collected and merged to form a complete crystallographic data set. This allows submicron to micron crystals to be utilized in an experiment when they were once, at best, only an intermediate result towards larger, usable crystals. SFX and its variants have opened new possibilities in structural biology, including studies with increased temporal resolution, extending to systems with irreversible reactions, and minimizing artifacts related to local radiation damage. Perhaps the most profound aspect of this newly established field is that "molecular movies," in which the dynamics and kinetics of biomolecules are studied as a function of time, are now an attainable commodity for a broad variety of systems, as discussed in Chaps. 11 and 12. However, one of the historic challenges in crystallography has always been crystallogenesis and this is no exception when preparing samples for serial crystallography methods. In the following chapter, we focus on some of the specific characteristics and considerations inherent in preparing a suitable sample for successful serial crystallographic approaches.

While this chapter's title directly refers to "nano-"crystals, the following is also applicable to small crystals that may not be strictly submicron. Synchrotron serial

J. Coe (✉)
Linac Coherent Light Source, SLAC National Accelerator Laboratory, Menlo Park, CA, USA
e-mail: jdcoe@SLAC.Stanford.EDU

A. Ros
School of Molecular Sciences, Arizona State University, Tempe, AZ, USA
e-mail: Alexandra.Ros@asu.edu

S. Boutet et al. (eds.), *X-ray Free Electron Lasers*,
https://doi.org/10.1007/978-3-030-00551-1_3

crystallography at a micro-focused beamline is also a highly effective technique that is very similar to nano-crystallography at X-Ray FELs but requires larger crystals than the minimum needed for SFX. Furthermore, depending on specifics of a given experiment, larger crystals may also be preferable at an X-Ray FEL, sometimes up to a few tens of microns in the largest dimension, which is common for membrane proteins as discussed in Chap. 4. Other times, growth of crystals larger than a couple microns tends to be elusive at best, a challenge common with G-protein coupled receptors (covered in depth in Chap. 10). In any case, it is common within the field for the terms "nanocrystal" and "microcrystal" to be somewhat synonymous, with "nanocrystal" appearing for a broad size regime in the literature. For the sake of brevity, we will refer to all small crystals suitable for serial crystallography as "nanocrystals" hereto forth.

3.2 Nanocrystallogenesis

When approaching nanocrystallogenesis, the parameters governing growth remain largely the same from macrocrystallography, namely thermodynamics, kinetics, and solubility. The notable difference is that the objective occupies a different region of the phase space. To additionally optimize diffraction quality, one must also consider and control crystal size and size homogeneity. These parameters have an impact upon data collection in serial crystallography and failure to optimize can cause malfunction of sample introduction and/or an increase in time and sample needed to complete a data set. It is also very important to consider and prioritize characteristics for a given experiment, especially in the case of a serial experiment aimed at something more complex than a single static structure.

In general, crystallogenesis can be thought about as a multidimensional phase space consisting of any and all conditions experienced by the protein in solution. A simplified depiction of this can be seen in Fig. 3.1 where a 2D slice is shown between protein concentration and a generic precipitant. However, in practice a comprehensive phase space is highly complicated and can be sensitive to multiple additives, temperature, protein homogeneity, time, pH, etc., essentially anything that comprises part of the crystallization environment. While each protein will have its own unique phase space, there are generalities that can help guide optimization and avoid a brute force approach past initial screening (even this is not strictly brute force as most commercial screens rely on empirical successes). It should also be noted that all the traditional pre-screening optimizations (purification, configurational and oligomeric homogeneity, detergent screening, mutagenic engineering, etc.) will still play an immense role in the ability to obtain quality samples. We will not go into the details about these here but would refer you to texts by Scopes [1] or Doublie [2] for additional reading on the topic.

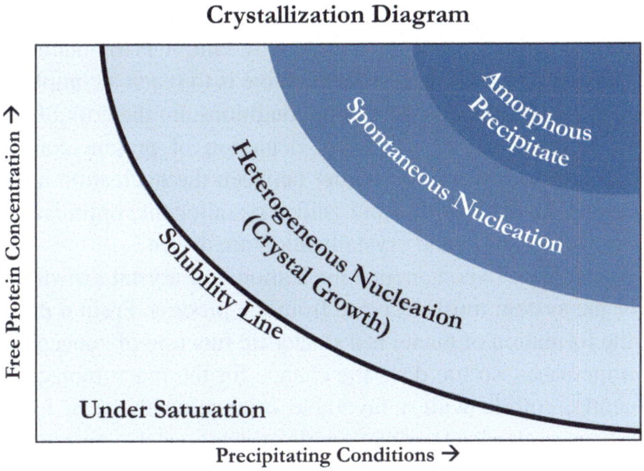

Fig. 3.1 Generalized solubility phase diagram for crystallization. Illustrates the typical qualitative relationship between solubility and concentrations of protein and precipitating conditions (e.g., salt concentration, pH, temperature). As one or both of these concentrations are increased, the solubility tends to decrease until a supersaturated state is achieved where nucleation is able to occur. The segmented regions in the supersaturated phase qualitatively increase in their propensity for nuclei to form as they move away from the solubility curve. Once supersaturation becomes very large, disordered (non-crystalline) nuclei tend to be favored, illustrated by the amorphous precipitate region in the top right

3.2.1 Thermodynamics of Solubility and Nucleation

In its most fundamental sense, the formation of crystals is nothing more than a controlled precipitation of a molecule or molecular system such that translational symmetry is achieved through free energy minima. The process is driven by thermodynamics through solubility and intermolecular forces. This pertains to any size of crystal and just like macrocrystallogenesis of macromolecules, the first step is to find a foothold within the broad parameter phase space governing a particular sample of interest. This typically involves trying many conditions in a broad screen to look for trends and patterns in solubility by sparsely sampling phase space [3]. It is important to remember that in order to find precipitating conditions, one must consider a plethora of variables including not only the concentration and composition of protein and mother liquor but also the pH, temperature, time and method among others. While Fig. 3.1 is an example of a generic crystallization diagram, a useful tool in visualizing phase space, it should be stressed that this is simplified and, as mentioned above, many variables govern the actual phase space, making it n-dimensional. The "precipitating condition" represented on the x-axis can either be thought of as representing the entire suite of precipitating variables other than the concentration of the protein itself, or as a single variable, equating to simply a "slice" of the n-dimensional phase space.

In contrast to macrocrystallogenesis, where a single or few large crystals are desired, the main goal for a serial crystallography sample is to obtain *many* crystals that are *small* in size. That is to say, the objective is to reach a completely different area in phase space. Highly oversaturating conditions are the crux of this, allowing the formation of many nuclei and rapid depletion of protein concentration. In Fig. 3.1 this equates to nearing the border between the nucleation and amorphous precipitate regions. To achieve this and guide crystallogenic optimization, we must first consider how nucleation and crystallization are driven.

For the favorability of spontaneous nucleation and crystal growth to occur, the free energy of the system must decrease from the process. From a physiochemical perspective, the formation of nuclei is a stochastic function of concentration. As the concentration increases, so too does the chance for the macromolecules to collide and, subsequently, collide with a favorable orientation that can lower the local energy (but collisions also lead to unfavorable orientations that eventually dissipate). This can be written in terms of the free energy of crystallization as

$$\Delta G_{\text{crystal}} = \Delta H_{\text{crystal}} - T\left(\Delta S_{\text{crystal}} + \Delta S_{\text{solvent}}\right) \tag{3.1}$$

Since a crystal is inherently ordered, the entropy from the macromolecular crystal will always be negative (equating to a positive contribution to the free energy). Thereby, the loss of degrees of freedom for the protein needs to be compensated with an entropic gain from disrupted solvent shells (there may also be an enthalpic contribution but it has generally been shown to be minimal in comparison [4]). Even upon such an event, the solution is still dynamic and subsequent collisions can cause the nuclei to increase in size or disperse. General phase transitions (e.g., nanocrystallization) intrinsically are a competition between the destabilizing interfacial (surface area) free energy components and the stabilizing bulk (volume) free energy components. For simplicity, consider an ideal case where the surface energy does not vary with orientation (i.e., a spherical nucleus). This leads to a free energy for the nucleus being the Gibbs-Thomson equation for a condensed droplet as a function of radius, namely

$$\Delta G_{\text{nucleation}} = \frac{4}{3}\pi r^3 \nu k_B T \ \ln(S) + 4\pi r^2 \gamma \tag{3.2}$$

where k_B, T, and S are the Boltzmann constant, temperature, and entropy, respectively, the ν term represents a molar volume element for an additional molecule with respect to the packing, r is the radius of the droplet, and the γ term representing the specific energy of the surface [5]. Certainly, in the real nucleation case, the γ term will have an anisotropic dependence upon facet composition but the fundamental form of the equation remains a competition between a positive surface component (acting similar to an activation barrier) and a negative (stabilizing) volumetric component that dominates as r becomes larger. This means that a critical radius occurs when we set the derivative of Eq. (3.2) to zero:

$$\frac{d\Delta G_{\text{nucleation}}}{dr} = 4\pi r^2 \nu k_B T \ln(S) + 8\pi r\gamma = 0 \tag{3.3}$$

$$\therefore r_{\text{critical}} = -\frac{2\gamma}{\nu k_B T \ \ln(S)} \tag{3.4}$$

In essence, a critical sized nucleus is formed when the contribution to the overall energy from the nucleus volume overcomes the surface energy increase as an additional molecule is added [5]. However, in order to get to this point, growth of smaller, quasi-stable nuclei must continue to grow against an uphill energy barrier until a critical size is reached.

For any amount of supersaturation, critical nuclei are possible given enough time but at low levels of supersaturation, deterioration of the sub-critical nuclei dominates. This region is typically represented as the heterogeneous nucleation or growth zone (see Fig. 3.1). As can be qualitatively understood, higher degrees of supersaturation lead to more collision events and shorter, more feasible time scales for critical nuclei to occur. This is represented as the spontaneous nucleation zone. In extremely elevated levels of supersaturation through increased protein concentration, collisions will be so frequent as to overcome the preference for energetically favorable orientations necessary for quasi-stable nuclei, as the unstable nuclei lifetimes are outcompeted by collisions. Alternatively, increasingly precipitating conditions also lead to higher levels of supersaturation that change the potential energy surface of collisions, allowing slower relaxation times. Both of these can lead to critical radii being achieved without the molecules exhibiting translational symmetry, resulting in an amorphous precipitate. It should be noted that other than the solubility line, the zone separations represented in crystallization phase diagrams are not well defined and are, without a conventional metric, somewhat arbitrary. This does not exclude their usefulness, especially in conveying and differentiating phase space trajectories experienced in different methods. For any set of sample characteristics, fine screening solution content and physical parameters is crucial. One of the most impactful factors in attaining your goal lies in the chosen method, for which the following section is dedicated.

3.2.2 Methods

It has been almost 60 years since the first protein crystal structure was published [6] and even longer since the first protein crystals were observed in 1840 [7]. In the time since, many ways to produce macromolecular crystals have been devised. Vapor diffusion, batch, free interface diffusion, and dialysis methods are among the most historically popular ways to make macrocrystals, each having their own benefits and challenges (these and other methods have been covered extensively elsewhere [8, 9], the reader is referred to these publications for detailed review of these and other methods). Again, when aiming for nanocrystals, we are simply trying to access a different area of phase space and so, in most cases, we can just adapt these techniques to suit our purpose. Another pervading theme in sample production for serial crystallography is the sheer mass needed for a successful data set, sometimes requiring hundreds of milligrams of protein! This is certainly something to consider and, when possible, increasing crystallization setup volumes should be considered as it can lead to consistency throughout data collection on top of reducing tediousness of sample preparation.

It is important to remember that crystals are not grown statically and the method and implementation will largely affect experimental crystallogenesis results. Once a foothold condition is found, fine screening around it should be done with multiple methods. This is especially important for nanocrystals due to the multitude of parameters that need to be simultaneously optimized. While one method may give the best diffraction, this must sometimes be weighed against size or yield or even growth time. It is important to keep in mind the specifics of the experiment at hand. If it is the case that the goal is simply to obtain a novel structure from a protein that resists the formation of large, well-ordered crystals then diffraction quality will of course take precedence. However, this is not always the case. For example, in a kinetic study with substrate mixing there are multiple variables to simultaneously optimize. In this type of experiment, as reaction time regimes become short, size and size homogeneity become increasingly important to maintain temporally homogeneous data sets along a reaction timeline. Of course, enhancing resolution is still very important but so long as the resolution is sufficient to see conformational changes in a particular system, the other parameters may become more beneficial to focus ones efforts.

3.2.2.1 Adaptation of Existing Conditions

While obtaining structures on samples that only seem to form small crystals is certainly a benefit of SFX, this represents only a fraction of targets. The ability to "outrun" radiation damage, the high temporal resolving power and the extreme brilliance found at an X-Ray FEL lend it to completely new areas of study and the ability to overcome some of the shortcomings of other light sources. For example, the structures of metalloproteins can be determined without site specific radiation damage [10, 11], irreversible reactions and conformational homogeneity in transient states can be probed [12, 13], and ultrafast time regimes (sub-ps) are now accessible to study dynamics [14, 15]. This opens many avenues for progression on proteins that already have a static crystal structure and, subsequently, already have well-established crystallization conditions. With this in mind, a primary goal of this chapter is to guide the translation of existing macrocrystallogenic conditions into those suitable for SFX and initial conditions are assumed.

By far, the most common and effective way to get a foothold on possible crystallization conditions is mass sparse-matrix screening, and this remains true for nanocrystallization. Interpreting the results differs as one searches for conditions giving nanocrystals but most times the large crystal "hit" in a screen can be optimized into nanocrystal conditions in the same way that a shower of crystals can lead to macrocrystal conditions. It should also be noted that most large-scale screening is done with vapor diffusion and while this can still lead to nanocrystals, it is a method that moves through phase space because of evaporative concentration and can be a generally "slow" method with respect to inducing nucleation. In its traditional form, it is also an extremely tedious method to obtain the milligrams of sample generally necessary for a serial crystallography experiment. That is not to say that it could not be used and one can certainly imagine a volumetrically upscaled sitting drop setup, but it is rarely the most convenient or optimal technique. It is important to keep in mind that, in general, increased concentration of precipitating condition leads to smaller crystals. While this is certainly true for chemical precipitants (e.g., salt, polyethylene glycol (PEG)), it is worthwhile to think about the kinetics or time as well. For example, in a vapor diffusion experiment, the volume and concentration of the well solution controls how fast the sample cocktail concentrates via evaporation. Parameters such as temperature or viscosity can also have a large effect on the thermodynamic rate of nucleus formation. For example, a highly viscous precipitant (e.g., PEG) can slow down the process by which the protein molecules collide in solution versus a lower viscosity precipitant, which will favor a much faster diffusion rate. Or in the case of temperature, crystallogenesis at a higher temperature will increase the available energy in solution, giving faster diffusion and thus tending to favor more nuclei and smaller crystals (of course caution must be taken when varying the temperature too much to avoid possibly unwanted effects such as denaturation or expansive freezing of water). The speed of a method will depend on its crystallization diagram; the time spent in each region of phase space will dictate the overall results of a given setup.

3.2.2.2 Free Interface Diffusion

One method that has been adapted from a traditional method is that of nanocrystalline free interface diffusion (FID), originally described by Kupitz et al. 2014 [16] and shown in Fig. 3.2. In an FID setup, the protein and precipitant solutions are combined to create a layered mixture similar to the microcapillary approaches [17] used traditionally. The difference lies in the need for larger volumes and high nucleation rates. Typically, the less dense of the two solutions—which is typically the protein solution—is first aliquoted into a vessel, often a microcentrifuge tube because the sloped sides have the potential to act as a parameter by influencing the mixing region volume and concentration profile. A quick centrifugation can be helpful to remove any bubbles and create a flat interface before addition of the second solution. Once a flat interface is obtained, the denser solution is added through the center of the surface dropwise. It will form a bottom layer and some perturbation at the interface, giving a small volume mixing zone. This allows for very high concentrations of each solution at the interface, higher than can be achieved with thorough mixing (most other methods involve the need for mixing

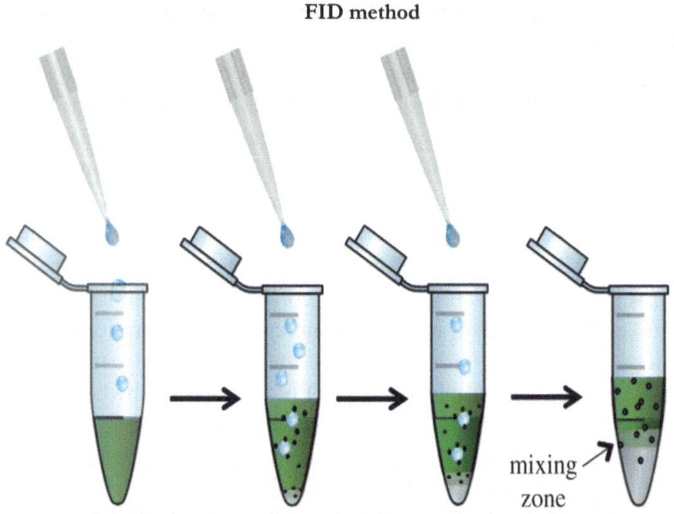

Fig. 3.2 Schematic of nanocrystal free interface diffusion (FID) setup. As the denser solution (either protein or precipitant) is dropwise added to the center of the other and a layered setup with a mixing region between occurs. The mixing region acts as an interface with high local concentrations of both protein and precipitating conditions, allowing diffusive mixing. The concentrations and volume of this will depend upon the container geometry, droplet size and viscosities of the two solutions. The sloped sides of a microcentrifuge tube (pictured here) can allow for easily varying the surface area-to-volume ratio, though this should be taken into account when attempting to upscale a setup (adapted from Kupitz et al. [16])

of a protein containing solution and a precipitant containing solution, leading to a necessary dilution of both upon mixing). The perturbation from dropping one solution through the other tends to speed up the process, favoring nucleation by inducing a minimal but necessary mixing region.

The access to high nucleation regions of phase space experienced at this interface can cause nanocrystals to form and, as they grow larger, tend to settle towards the bottom layer. In the case that the precipitant is denser, as is common with many precipitants containing high salt or PEG, this serves as a sort of auto-quenching effect as the crystals settle away from the free protein layer and into the precipitant rich layer. Different volumetric ratios, total volume and even droplet radius can have a profound effect on results from this method and volumetrically conserved upscaling tends to eventually break down reproducibility and quality. The limit of this is highly specific to a given protein/mother liquor and is likely a function of the mixing region profile which depends upon surface area and volume ratios, viscosities and perturbation. Gentle centrifugation is a way to enhance gravitational settling and can also expedite crystal formation and uniform growth. As can be expected, over extended periods full mixing of the two layers can occur and to prevent any loss of quality or even dissolution, crystals should be harvested and/or quenched prior to complete diffusion (the time sensitivity of this will be a function

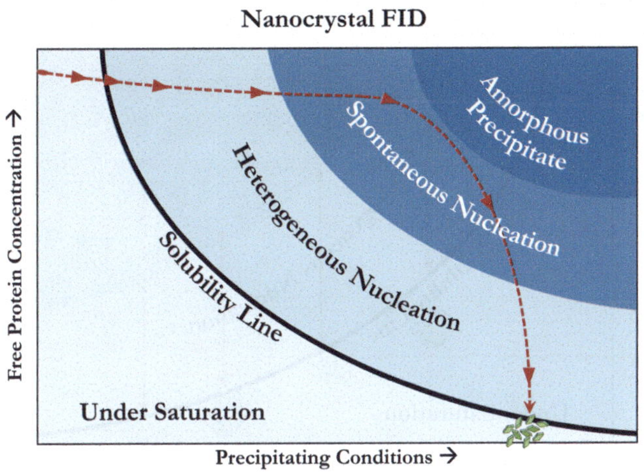

Fig. 3.3 Nanocrystal FID phase diagram. The red line shows a conceptual path experienced by proteins moving through the interface into the mixing region and finally depositing as part of a crystal. The spacing of the arrows denote a likely temporal trajectory as diffusion slowly allows it to enter the mixing region and then more rapidly adsorb onto a nucleus before settling as crystal growth is achieved

of miscibility between the layers as well as volume–interfacial surface area ratio). This method can be particularly useful when either or both precipitant and protein concentrations are constrained by solubility in other methods and a smaller crystal size is desired. This is due to the ability to have saturated concentrations at the interface. It should be noted, however, that reproducibility is sensitive to even minor differences in the setup of this method due to the many affecting variables involved such as volume, ratio, drop size, position of the perturbation in the interface, and container (Fig. 3.3).

3.2.2.3 Batch Nanocrystallization

Batch crystallization is a "cocktail" approach where the protein and precipitating conditions are homogenized. This is particularly useful in making the large volumes often necessary for SFX experiments due to its ease and simplicity. This has the added benefits of decreasing sensitivity to user technique and volumetric scaling, though caution should be taken to ensure scaling does not interfere with homogeneity. It can be seen in Fig. 3.4 that there is no movement through phase space prior to nucleation (or lack thereof). Protein stock and precipitating mother liquor are typically added to a vial or microcentrifuge tube and either homogenized by pipette mixing or using a magnetic stir rod. The stir rod allows homogeneity to be retained as protein begins to precipitate out and can allow either the protein or precipitant to be added slowly, avoiding high concentration interfaces with significant contact time. Analogously, when pipette mixing, multiple aliquots can

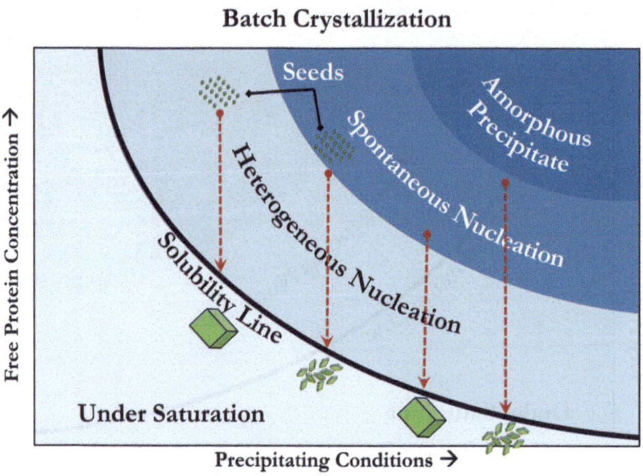

Fig. 3.4 Batch crystallization phase diagram. Illustration of the "cocktail" method where a protein and precipitant are homogenized, essentially picking a supersaturated point in phase space and letting it evolve with equilibration. As can be seen, the crystal size (among other parameters) is highly sensitive to these conditions and will tend to smaller crystals further away from the solubility curve. The use of seeding is also highlighted here as it often allows milder conditions to result in a similar result. Seeding can also encourage homogeneity and accelerate crystallogenesis

be serially pipette mixed to allow a more gradual introduction of precipitating conditions. This is perhaps the most convenient method as it is a short, simple setup that can usually be upscaled to complete experimental volumes suitable for consistent data collection. While this method can theoretically be done in any size vessel, experimental demand usually requires a few milliliters of sample for a full data set. Fine screening with a dilution gradient using a crystallization robot can help explore specific batch conditions prior to upscaling. It is also worth noting that recently, microfluidic approaches have also been successful at fine screening batch conditions with minimal volume constraints [18].

3.2.2.4 Other Methods and Sample Delivery Considerations

There are many other methods that have been successfully used to grow macromolecular nanocrystals that are thus far a little less general. Almost any method that has been used for microcrystal growth can be adapted to nanocrystal generation, the key is using parameters to enter a different region of phase space. In fact, even methods to create nanocrystals such as mechanically crushing macrocrystals have had success [19], although caution should be taken with this as homogeneity and quality can be severely impacted, the degree of which is very dependent on the crystal contacts within the crystal.

A particularly interesting method of crystallization that should be mentioned is that of in vivo over-expression. Using baculovirus-Sf9 cells, multiple proteins have now been shown to crystallize in vivo [20] with successful injecting of the un-lysed cells in one case [21]. While the mechanism behind in vivo crystallization and its general applicability is currently unclear, it is certainly worth monitoring the progress behind this phenomenon.

One method of crystallization that has been extremely successful with membrane proteins, particularly G-coupled protein receptors, is the use of the lipidic cubic phase (LCP) as both a crystallization and delivery media. LCP is a bi-continuous mesophase that can allow type I crystal packing (i.e., stacked membrane embedded planar 2D crystals lacking detergent micelles [22]) due to its membrane mimetic properties, typically leading to tighter packing and often higher resolution than obtainable with crystals grown in detergents or other surfactants (in surfo). Chapter 4 is dedicated to this method as it has been so successful when approaching some of the most challenging proteins.

One of the clear advantages of an LCP capable system for serial crystallography is the ability to create a much slower moving viscous jet, which reduces sample consumption by orders of magnitude. The reduction of sample requirements has driven the development of alternative viscous carriers in which pre-grown crystals can be embedded into a viscous media, allowing 1–2 orders of magnitude less material [23–25].

A useful tool that can be implemented with any method is seeding, the use of pre-grown crystals to "seed" nucleation. These seeds are of course obtained by some initial crystallization method but the addition to a subsequent crystallization attempt can have a profound effect. Thinking back to the crystallization phase diagrams, different areas of phase space can now be approached. Using seeds also tends to encourage homogeneity among crystal size [26] and across different batches. The most convenient side effect is that while pursuing nanocrystallogenic optimization, any sample made along the way can be used for subsequent trials and can be doped into any method easily.

3.2.3 Stability and Storage

Since serial crystallography avoids radiation damage via constantly replenishing sample during data collection, freezing techniques and optimization therein are, in general, unnecessary. Instead, one must consider how the sample is stored and handled prior to sample introduction to best preserve diffraction quality and physical characteristics that are optimized during crystallogenesis.

Crystal size is one of the most sensitive characteristics to the storage method and technique. Ideally, crystals are kept in a solution of their mother liquor or a variant thereof. This has the benefit of needing only to resuspend the sample prior to introduction and data collection (once an optimal crystal density has been determined and achieved). It is not uncommon, however, that the precipitating conditions used in initial crystallogenesis allow further growth over time, since many crystallization experiments have a slow growth phase due to supersaturated conditions in the mother liquor. This can occur from any free protein continuing to adsorb to crystal surfaces. It is therefore highly advisable to either remove uncrystallized protein as soon as the desired crystals are obtained or to further decrease solubility, though the impact is highly dependent upon the amount of time the sample must be stored and specific conditions. While in some cases an optimized method will have little remaining protein in solution, this can still be necessary. Fortunately, this can usually be achieved by allowing crystals to settle, carefully removing the supernatant and replacing with fresh mother liquor, effectively "washing" the crystals. It should be noted that this does have the effect of shifting the solution out of equilibrium and will cause the surface molecules to fall back into solution until equilibrium is obtained. For a larger size crystal this can be negligible, but for a smaller crystal it may be significant and may be advisable to try to find a minimal concentration that keeps stability or to shift to higher precipitating conditions, effectively decreasing equilibrium concentration of free protein.

Ostwald ripening, the process by which larger crystals tend to grow while smaller ones tend to dissolve due to thermodynamic favorability, may also disturb the size, homogeneity and number of individual crystals. This occurs because even though crystals have been formed, the solution is effectively in a dynamic equilibrium with solubility being low but not strictly absent. Because of the lower number of interactions with the bulk crystal for the molecules on the surface of a crystal, they tend to detach and go into solution periodically before stochastically recombining with a crystal. Due to the difference in surface area, this process favors large crystals in the long run. Certainly, the more size homogenous the sample, the slower this process will occur but absolute size homogeneity is impossible to achieve practically. One way to avoid Ostwald ripening is to further decrease the solubility of the protein, effectively quenching the exchange. Simply by "washing" the sample as described above with a higher precipitating condition can achieve this, although a dramatic change in solubility can sometimes damage the crystals and sometimes a stepwise approach is preferable. Temperature can also be used to decrease solubility, though again care must be taken to monitor unwanted effects on crystal quality.

As data are usually collected at or near biological temperatures, stability during transportation can also be a concern. Fortunately, due to the high nucleating conditions needed for a plethora of small crystals, nanocrystallogenesis is typically a fast process and oftentimes samples can be grown on site within a matter of hours or days prior to an experiment. However, if this is not the case, it is imperative to ensure the integrity of the sample is not compromised during transport by anticipating environmental perturbations such as handling or temperature variance. As automation and remote data collection in serial experiments is almost certainly inevitable, this will likely become more and more general of a consideration. Temperature secure containers, eliminating gas from sample head space, shock absorption or even embedding crystals into a viscous media (if viable) are among the approaches that can ameliorate shipping concerns. As always, testing and characterizing to optimize results and protect precious sample is vital.

3.3 Considerations and Characterization for SFX Optimization

3.3.1 Characteristics: What Is Optimal?

There are some unique characteristics that apply in serial crystallography that must be controlled and optimized for a given sample. Size homogeneity and sample density are always general concerns, affecting data collection efficiency and quality. Of course, size itself plays a vital role and sometimes bigger is better. But there are many circumstances where it is not and thoughtful selection is crucial to a successful experiment (e.g., for time-resolved studies). In fact, for time resolved studies, it can be important for both size and size distribution to be minimized. Size governs reaction homogeneity upon probing since the activation trigger (e.g., light, substrate) will have a different distribution to the different molecules in the crystal dependent upon volume. In the case of a chemical trigger, molecules towards the center would on average experience a delayed reaction initiation due to reactant diffusion within the crystal. For optically triggered reactions, molecules downstream in the direction of the pump laser propagation would experience increasing attenuation in pump power, leading to lower yields of reaction initiation. There will also be a temporal range of reaction initiation similar to the chemical trigger case but the range would be in the 10s–100s of femtoseconds for micrometer sized crystals. This still must be taken into account when exploring dynamics on these timescales and, like the diffusive case, smaller crystals are preferable with respect to homogeneous initiation. Reaction timeline homogeneity itself is of course important as the serial snapshots are merged into a data set and large distributions will broaden conformational heterogeneity. This will be covered in more detail in the later Chaps. 11 and 12.

Sample density, that is, the concentration of crystals in a suspension, is another key factor. Depending on X-Ray source size, crystal size and jet size, the optimal sample density can be calculated by approximating a Poisson distribution for the hit rate versus concentration (usually falling in the 10^9–10^{11} crystals/mL range), though a Poisson approximation will become less accurate as the crystal size becomes much larger than the beam focal spot. It is important to verify that the protein-rich phase is in fact crystalline and, especially in the submicron range, this is not always straightforward. In fact, many times small crystals may be mistaken for amorphous precipitate to even the trained eye. Techniques in microscopy and diffraction that can be used to verify crystallinity are covered in Sect. 3.3.2.

3.3.1.1 Data Analysis Considerations

In serial crystallography, data are collected as a series of diffraction "snapshots" from different crystals that are merged together to form a complete data set. As opposed to a rotation series on a goniometer, only partial reflections are measured and thus structure factors must be elucidated using Monte Carlo methods. The high multiplicities of measured reflections needed necessitate a large amount of sample even in an ideal case, where each crystal is the exact same size, morphology, quality, etc. Changes in sample homogeneity signify that even more measurements, that is, single snapshot diffraction patterns, are needed for the data to converge to reliable and comparable intensities.

In early SFX studies, $>10^5$ diffraction patterns were thought to be necessary to determine structures [27]. This has considerably decreased in the past few years and full data sets have now been obtained with under 10,000 (in some cases, under 1000! [28]) images needed [23, 29, 30]. A narrower size distribution of protein nanocrystals, however, might greatly reduce the number of diffraction patterns required for successful merging and integration. In addition, the peak intensity of Bragg reflections in the individual diffraction patterns scales with the crystal size quadratically, which may lead to significant variation in peak intensities. This means that a high size inhomogeneity not only introduces another parameter that needs to be addressed using Monte Carlo methods, practical considerations may necessitate attenuation of the beam to avoid detector damage. This can lead to a decreased signal-to-noise ratio (SNR) for patterns representing the smaller sized crystals in a sample since the beam intensity used is often determined from the strongest scattering crystals.

Another consideration for data analysis related to crystal size is experimental solutions to the phase problem, that is, phasing. There is a huge interest to improve SFX data analysis for de novo structure determination. Crystal size homogeneity may play a crucial role in this approach. For example, Spence et al. have proposed that coherent diffraction intensities between Bragg reflections of sufficiently small crystals may be used for novel phasing approaches [31]. These coherent shape-transformed Bragg reflections allow for two-dimensional projection images of the entire nanocrystal, which could be used to solve the crystallographic phase problem in SFX without prior information, crystal modifications, or resolution restrictions. It is expected that this novel approach works best for a specific crystal size range. When crystals become too small (containing too few unit cells), the inter-Bragg diffraction intensities reduce, thus an optimal intermediate size is desired [32]. The details of methods involving these properties will be discussed in Chap. 8. Whether exploiting nanocrystals discreteness for novel phasing approaches or minimizing the time and amount of sample needed for a complete data set, there is a clear motivation to obtain control over crystallogenic parameters in order to improve data quality.

3.3.1.2 Practicality: Sample and Hardware

There are some practical concerns that arise from the hardware used for SFX experiments, which must also be considered. From the sample delivery point of view, most experiments have been performed using some type of gas focused jet (sample delivery is covered in detail in Chap. 4) and to minimize wasted sample, constrictions in the hardware are often very small (nozzles usually contain a capillary with 30–100 μm inner diameter). Minimum flow rates for a stable jet (equating to minimal sample consumption) can vary depending on buffer composition, especially viscosity. With polyalcohols, such as polyethylene glycol, being a common precipitating agent, this can be a frequent concern and it is worth experimenting to try minimizing viscosity during final sample preparation. Setting up small aliquots of different concentrations and monitoring crystal integrity over time can save later frustration over hindrance of data collection due to a clogged nozzle or inconsistent jet.

Another issue that arises due to the small nature of sample delivery hardware is clogging due to the crystals themselves or other particles. Certainly, one must select a nozzle size appropriate for the employed crystals (a good rule of thumb is *at least twice the size* of the largest crystals in the batch) but many times even a sample with relatively good size homogeneity will have a few large outliers. It should be kept in mind that optimal sample densities are on the order of 10^9–10^{11} crystals/mL and it only takes one crystal to clog the nozzle. It is therefore in a user's best interest to have filtering systems in place prior to the injection hardware to avoid time loss for fixing/replacing the hardware. In-line filtering is almost always a necessity but it is oftentimes advisable to "pre-filter" the sample prior to containment in a sample delivery reservoir. Many commercially available plumbing and filter setups can be adapted to this purpose, for example standard liquid chromatography hardware. It should be noted that not all filters are created equal and different types and quoted porosities can have detrimental effects on a crystalline sample such as shearing crystals apart. It is advisable to test the effect of filtration on the sample prior to a beamtime in order to know which filter will work best for a given crystal suspension. The density at which a crystal suspension is filtered can also have an effect and, in general, it is prudent to filter at low concentrations and allow a sample to settle before removing supernatant for concentration. Filtering can also serve to improve size homogeneity during sample preparation should you experience a bimodal or multimodal distribution.

Even when a particular sample is not strongly constrained by quantity or another method of sample introduction is used that can sidestep the above-mentioned concerns, size heterogeneity can cause other concerns. In addition to data quality and efficiency concerns addressed in the previous section, saturation or even destruction of the detector must be considered. Especially with the extremely brilliant X-Ray FELs, a well diffracting crystal can easily exceed the intensity at which a detector can be damaged. Even when not damaged, saturation can occur, obscuring values for structure factors (i.e., intensities above the saturation threshold will be measured incorrectly as the threshold value). To avoid this, the beam is usually attenuated to a level that certainly avoids damage and minimizes saturation. Remembering that peak scattered intensity scales quadratically with respect to the number of unit cells illuminated by the beam, it becomes clear how attenuating to the largest crystals that are introduced to the beam can quickly limit the lower size limit for useful data collection.

3.3.1.3 Control of Homogeneity Through Post Growth Methods

Once crystals are obtained in suitable concentrations, it is important to characterize crystal size homogeneity. This can be accomplished with the methods described in Sect. 3.3.2. However, we emphasize that dynamic processes may play a significant role, requiring stringent analysis of crystals prior to crystallographic measurements. Crystals may grow, aggregate or dissolve after production and these processes need to carefully be accounted for. To reduce the amount of unwanted larger crystals, a straight forward approach of filtering may be employed. While this approach is suitable for sufficiently large amounts of crystals available, it might not be applicable when the concentration of smaller crystals is low or when the crystals are prone to decomposition through mechanical filtering approaches. In addition, as discussed above, novel phasing approaches may require specific size ranges and narrow size distributions, which require more sophisticated approaches for post growth crystal sizing.

One approach to fractionate crystals by size has recently proposed by Abdallah et al. [33] In this novel microfluidic method, crystals are deviated in their migration path while flowing through a micrometer-sized constriction and collected in various outlet reservoirs of a microfluidic device. The method relies on dielectrophoresis, where the applied dielectrophoretic force scales with the radius of the crystals to the power of three [33]. In the application demonstrated by Abdallah et al., crystals experienced negative dielectrophoresis, repulsing large objects into a center stream, while smaller particles deviate to side channels. Smaller crystals can thus be recovered in the side channels. The first realization of this dielectrophoretic crystal sorting was demonstrated with photosystem I (PSI crystals, as demonstrated in Fig. 3.5). The sorted crystals were characterized with various techniques, demonstrating that they were not decomposed during the sorting process and that they retained excellent diffraction quality even after sorting [34]. This continuous crystal sorting method was further developed for higher throughput to account for the mL volumes of crystal suspension needed in liquid jet based injection methods for SFX with X-Ray FELs [34].

3.3.2 Characterization

During the crystal screening and growth process, one needs to continually characterize the crystallogenic results to guide optimization. Compared to macrocrystals, this can be a significantly more complex task due to the necessity for high density, homogeneous crystal suspensions typical in serial crystallography, which are also generally harder or sometimes impossible to visualize with routine microscopy. This is not to say that the task is necessarily always difficult, just that it requires a thoughtful approach with a larger arsenal of tools than is traditionally necessary.

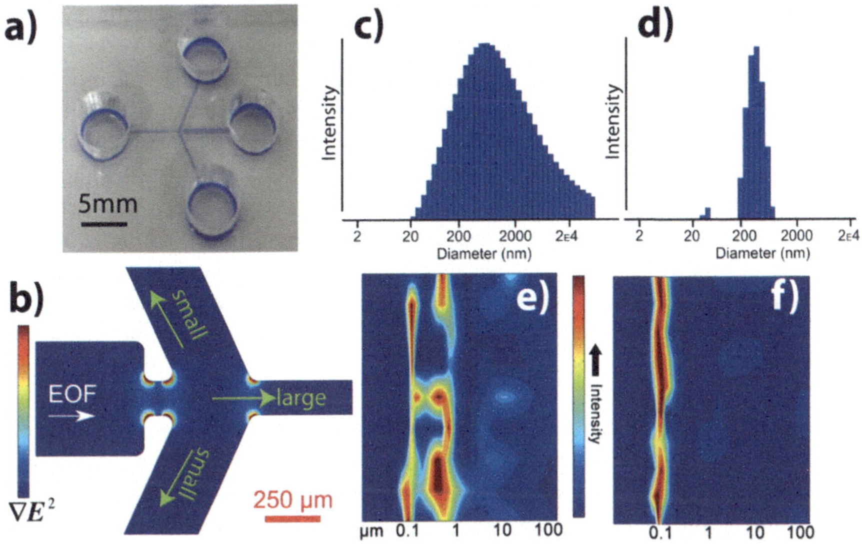

Fig. 3.5 Microfluidic Sorting: (**a**) Image of the microfluidic sorter used for high throughput crystal sorting. (**b**) Mechanism of crystal sorting in the microfluidic device near the constriction region. Electroosmotic flow (EOF) is used to deliver sample to the constriction region. The green arrows indicate the flow of large and small particles after the constriction regions, where sorting occurs. (**c**) Dynamic light scattering (DLS) analysis of PSI crystal suspension prior to sorting. (**d**) DLS analysis of PSI crystal suspension after sorting with the dielectrophoretic sorter indicating crystals <600 nm). (**e**) PSI crystal suspension prior to dielectrophoretic sorting and (**f**) after sorting in the low throughput version (adapted from: Abdallah et al. [33]) A homogeneous size fraction of about 100 nm was achieved (**f**)

3.3.2.1 Optical Detection: Visualizing Your Crystals

While it is still a good idea to first look at potential crystals under an optical microscope, it can be difficult to score results, especially as crystals approach one micron or smaller. Using a polarized filter to look for birefringence can be particularly useful to look for crystals whose sizes are near this threshold. In practice, the intensity from birefringence scales with the size of the crystals and so its usefulness does have a lower limit. Nonetheless, it can certainly enhance the ability to differentiate crystals from amorphous precipitate and can often provide enough contrast to indicate crystals below a micron. It should be noted that birefringence requires optical anisotropy and will therefore be absent or diminished in space groups with high symmetry (e.g., cubic). However, this does allow most salt crystals to be ruled out, which can be a concern with the high concentrations that can arise in nanocrystal creation.

One of the most useful ways to identify crystallinity is that of second harmonic generation (SHG) microscopy. There has been considerable effort developing this technique, specifically for identification of nanocrystals, and commercial instruments, such as Formulatrix's SONICC (second order nonlinear imaging of chiral crystals), have enjoyed success within the community. SHG works on the principle that under very intense illumination with light, two photon processes become significant and subsequent frequency doubling occurs over the fundamental wavelength. In our case, this is dependent on the polarizability of both the molecule and the crystal as a whole, constructively interfering with additional unit cells. In practice, this means that it is only observable from crystals (or other periodic objects) and will differentiate between them and amorphous material, often down to ~100 nm or smaller. It is also very sensitive to anisotropy and complete destructive interference will occur in any centrosymmetric space group. Like birefringence, this also rules out cubic space groups but prevents false positives from most salts. In addition, other space groups of high symmetry may suffer from diminished signal, albeit not strictly zero [35]. The signal also has an inherent dependence on the polarizability of the protein itself and this can lead to varying size limits or even feasibility. Commercial dyes have been developed that can intercalate through the crystal and enhance signal, a noteworthy aid for one aiming for the smaller end of the size regime. A good cross validation for potential nanocrystals observed in either SHG or polarized microscopy is UV-microscopy, which can easily highlight protein from salt by way of tryptophan fluorescence. In fact, the SONICC instrument couples an ultraviolet two-photon excitation fluorescence (UV-TPEF) method with SHG imaging, allowing overlapping images of a drop for secondary characterization. This is a great cross validation for proteins with any aromatic residues that either have a SONICC prohibitive space group (false negatives) or a precipitant that can form chiral (non-protein) crystals (false positives). The use of a multiphoton process is inherently confocal in nature and allows a decreased background and narrower depth of field, particularly important with the smallest size regime or unoptimized conditions. Excellent reviews on the principles of SHG and UV-TPEF imaging can be found in Kissick et al. [36] and Madden et al. [37], respectively.

3.3.2.2 Light Scattering Techniques for Size Determination

Light scattering methods are traditionally used to characterize particle sizes in suspension. Among those, dynamic light scattering (DLS) and nanoparticle tracking analysis (NTA) have been primarily applied for the characterization of crystal suspensions. NTA has only become recently available commercially but has found immediate application in crystallography. DLS is a well-established method used routinely in nanoparticle and microparticle analysis.

DLS takes advantage of the scattering characteristics of suspended particles in solution. It is thus not surprising that DLS has also been employed for the characterization of suspensions of small crystals required for crystallography and in particular the smaller crystals needed for SFX. In DLS, a laser is directed into the crystal suspension and the scattered light is measured at some fixed angle. Since the particles in the solution exhibit Brownian motion, the measured scattered light intensity will also vary randomly over time, typically in the microsecond time regime. If the particles are large, the time variations at the detector are slow, whereas for small particles the time variations are faster. The time fluctuations of this scattering are recorded in DLS and related to the particle size distribution in solution using suitable correlation analysis.

We may write the autocorrelation function of the scattered light intensity as [38]:

$$C(t) = \frac{\langle I(t)I(0)\rangle}{\langle I(0)I(0)\rangle} = 1 + g(t)^2 \tag{3.5}$$

where t is the time and $g(t)$ is the normalized first order time autocorrelation function. For monodiperse particles, $g(t)$ constitutes an exponential function with a time decay governed by the particle diffusion coefficient, D, and the scattering vector, q:

$$g(t) = \exp\left(-q^2 Dt\right) \tag{3.6}$$

The scattering vector is given by:

$$q = \frac{4\pi}{\lambda} \sin\frac{\theta}{2} \tag{3.7}$$

with λ the wavelength and θ the scattering angle. From Eq. (3.6) we notice the direct relation of $g(t)$ to the diffusion coefficient and thus the size of a particle. The latter is obtained through the Stokes–Einstein relation:

$$D = \frac{k_B T}{6\pi\eta r} \tag{3.8}$$

Here, k_B is the Boltzmann constant, T is the temperature, η is the viscosity, and r is the radius of a sphere. In a DLS measurement, we thus determine the diffusion coefficient of a particle corresponding to a sphere. Therefore r is replaced with the hydrodynamic diameter ($d_h = 2r$) for non-spherical particles. Suitable mathematical corrections need to be applied to obtain the size of non-spherical particles in DLS measurements.

Most particle and specifically crystal suspensions are not monodisperse. To account for polydispersity in DLS measurements, one introduces the methods of "cumulants." The decay function $g(t)$ is then assumed to consist of the sum of decay functions, where each summand accounts for a specific subset of particles. Analyzing polydispersity in DLS measurements requires suitable software, which is included in most commercial DLS instrumentation. It is also interesting to note that while DLS typically cannot distinguish between crystal and amorphous particles, a technique in which depolarized light from the scattered light is ascribed to birefringence has been developed, indicating crystallinity [39]. For a more detailed description of the DLS theoretical framework we refer the interested reader to the literature, for example a review by Pecora [38] or a book by Schmitz [40].

The size range suitable for a DLS analysis spans from as low as 1 nm well into the micron regime where optical characterization becomes available. However, a sample needs to be carefully characterized in order to avoid gravitational settling, which may compete with Brownian motion in μm-sized crystals. Moreover, the scattered intensity scales with the particle diameter to the sixth power, which signifies that a ten times larger particle scatters a million times more intensely. This relation needs to be critically viewed in DLS as the scattering of larger particles can easily overtake the much weaker scatting of smaller particles and bias the data analysis. Small crystals can thus be easily overlooked in suspension containing larger particles, such that the potentially more useful particle sizes for SFX may not be recognized.

However, DLS has become a routine tool in size characterization and it is powerful when (1) different crystallization batches are compared, (2) rapid analysis is necessary—such as at a beam time prior to injection, (3) the amount of crystal suspension is limited and not compatible with sample cell size for NTA (see below) or dilution of the sample cannot be performed for NTA, and (4) if a broad size range from several tens of nanometers up to micrometers is to be characterized.

DLS instruments offer a variety of cuvettes and thus variation in the sample volume to be analyzed. Standard measurements can be routinely carried out in volumes from 1 mL down to 50 μL. For crystallization trials, performed in small volumes, such as the hanging droplet method or even miniaturized on microfluidic platforms [38], it becomes important to perform DLS analysis in volumes below 50 μL. This can be accomplished with instruments exhibiting specialized optics, such that the DLS laser can be directed into a hanging droplet or microfluidic channel.

Alternative to DLS, particle tracking has been applied for the characterization of crystal size distributions. In particle tracking methods, the scattering or fluorescence of small objects below 1 μm is recorded by video microscopy; hence, the method is often referred to as nanoparticle tracking analysis, or NTA. A laser of certain wavelength is directed in a sample chamber and the displacement of individual particles from frame to frame is recorded. The mean square displacement of a particle, $\overline{x}$, is related to the diffusion coefficient, D, in the two-dimensional case via:

$$\overline{x} = \sqrt{4Dt} \tag{3.9}$$

where t is the time. Once D is determined it can be related to the particle size, or more precisely its hydrodynamic radius, r_H, via the Stokes–Einstein relation shown in Eq. (3.8). Be aware that D is a function of both temperature and viscosity in addition to the particle size so a new calibration must be performed whenever one or both of these variables are modified.

Suitable algorithms can track the particle motions and displacement and determine their size. In NTA, since it is a direct visualization process, particles can be counted and thus particle concentrations can be determined. This is an important additional data point for SFX experiments and the particle concentration can be adjusted to optimize hit rates. Since NTA tracks single particles, it also allows a more detailed analysis of multimodal distributions compared to DLS, an ensemble process, and is less prone to masking of smaller particles due to the augmented scattering properties of larger particles as apparent in DLS [40].

Particle tracking analysis per se is not a novel method and can be easily implemented via suitable imaging instrumentation and free software packages [18], such as available for ImageJ [41]. NTA has recently become available commercially and thus facilitated greatly for crystallography applications. The NanoSight instrument from Malvern (UK) has suitable measurement cells that allow for size distribution analysis for proteins and protein crystals in the range from ~30 nm to 2 μm. Clearly, this size range shows that crystal suspensions with larger expected particle size distributions should be analyzed by DLS. Another consideration in particle tracking analysis is the size of the measurement chamber, requiring several hundred microliters to fill the entire chamber. If crystal suspensions are limited in amount, recovery needs to be attempted or in the worst case, this analysis cannot be performed.

3.3.2.3 Transmission Electron Microscopy

Transmission electron microscopy (TEM) is a useful tool in crystallography, and has provided valuable information on the quality of micro- and nanocrystals prior to serial crystallography. TEM represents a vacuum technique, where an electron beam is directed through a thin specimen. Microcrystals and nanocrystals are typically mounted on a thin grid for TEM imaging with use of a negative stain for increased contrast. The electron beam interacts with both the electron cloud and nuclei of the atoms in the crystals leading to electron scattering. Scattered electrons pass through an objective lens, which upon focusing creates the primary image. Additional optical components are used to form a highly magnified final image of the primary image.

Obviously, as an imaging technique for nm-sized particles, a strength of TEM is to provide unequivocal information of the size of nanocrystals and microcrystals. It is the most direct visualization method of assessing the size of crystals, but is not suited for fast analysis. Analyzing the size-distribution of protein crystals with TEM is a time-consuming process, and requires the sophisticated and expensive TEM instrument as well as specialized training for the experimenter. However, provided enough crystalline material is at hand, this analysis could be automated to provide size and heterogeneity information of a particular crystallization trial. But this analysis is typically not performed in favor of faster and less cumbersome techniques such as NTA and DLS related to size-based analysis.

The major strength of TEM relates to revealing information about crystallinity. TEM can provide information about the existence of nanocrystals including variations in the crystal forms or the evaluation of diffraction quality. TEM allows for direct visualization of the crystal lattices and the Fourier transform of TEM images from protein crystals reveals their electron diffraction patterns or “Bragg” spots. The higher the order of these spot, the better—in general—the diffraction quality of protein crystals.

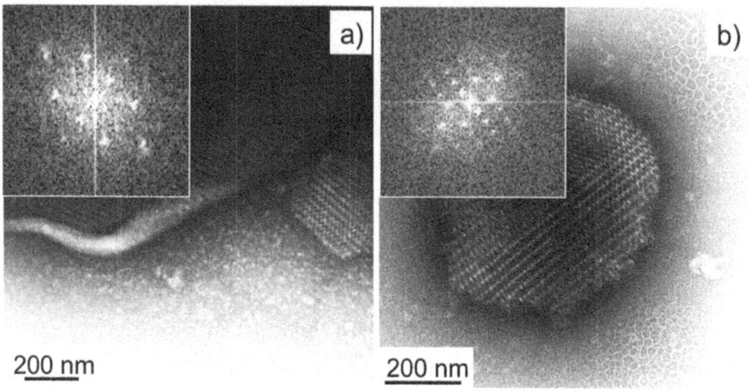

Fig. 3.6 TEM imaging of nanocrystals (**a**) TEM image of PSI crystals prior to microfluidic size fractionation and corresponding Bragg spots after Fourier transformation (inset); (**b**) TEM image obtained using negative uranyl acetate staining image of PSI crystals after microfluidic size fractionation and corresponding Bragg spots after Fourier transformation (inset). TEM analysis confirms excellent crystallinity of PSI crystals after fractionation (Reproduced with permission of the International Union of Crystallography and adapted from Stevenson et al. [44])

Stevenson et al. demonstrated that Bragg spots obtained from TEM analysis of several analyzed protein crystals are indicative of crystal quality [42]. The information gained with TEM analysis of crystals includes information related to crystal lattice variations between different crystallization methods as well as crystal pathology including lattice defects, anisotropic diffraction, and nanocrystal nuclei contamination by heavy protein aggregates.

TEM thus constitutes a useful tool to identify nanocrystals for challenging protein targets [43] and for the evaluation and optimization of crystal growth [44]. TEM has also been used to characterize the crystal quality after the fractionation of protein crystals by size using microfluidic tools. Abdallah et al. used a negative stain TEM (following a procedure previously published by Stevenson et al. [45]) to show that PSI crystals subjected to size-based sorting showed excellent lattice resolution and highly ordered Bragg spots, as demonstrated in Fig. 3.6. Indeed, SFX at an X-Ray FEL was successful with these sorted crystals [45].

3.3.2.4 Powder Diffraction

As anyone who has grown aesthetically beautiful crystals only to find that they have less than stellar diffraction can attest, you never know until you shoot them! While the size of the samples generally prohibits any single crystal diffraction at home X-Ray sources, nanocrystals are perfectly suited for powder diffraction at more accessible X-Ray sources. If possible, this should be attempted before a serial crystallography experiment. While the low comparative flux can prevent knowledge of the actual diffraction limit that might be attainable at a more powerful source, powder diffraction is the best way to test for a diffracting sample and can differentiate between batches or conditions by comparative resolution limits. It is also very easy to prepare for since typically nanocrystals solutions are already a "powder." However, at higher resolutions the powder rings become more frequent and faint, eventually causing them to be indiscernible from background, providing a limit on resolution. The powder diffraction rings can be evaluated to get a rough estimate of the crystal size contributing to the powder rings, though resolution of a reliable size estimate is likely limited to a very small size regime (less than a few hundred nanometers). One can also obtain other information such as using the spacing of the rings to obtain information about the lattice spacing in the crystal.

The easiest way to do this is to harvest an aliquot of the sample (about 10 μL of pelleted nanocrystals usually works well) and put them into an X-Ray transparent mountable capillary (e.g., MiTeGen MicroRT). This should be centrifuged to create a dense powder pellet from which the liquid should be removed to ensure density. In practice, some of the mother liquor should be implemented elsewhere in the tube to avoid drying of the pellet, which can affect resolution. Then the sealed capillary can be mounted and data collected on the dense nanocrystalline pellet. Be sure to mimic the conditions as close as possible to your actual experiment (e.g., lighting for light-sensitive proteins, temperature, mother liquor composition).

3.4 Recap

Experimental beamtime at X-Ray FELs are currently much more limited than at other light sources. The careful development and characterization of crystal samples for SFX experiments with X-Ray FELs is thus extremely important. There are unique characteristics that must be considered for each specific experiment, having a profound effect upon the quality of data that can be obtained. Generally, as many characterization methods as possible should be performed and this chapter summarized the most important and currently applied methods. The primary factor to consider for initial nanocrystallization trials are the higher concentrations generally key to converting macrocrystal conditions into nanocrystal conditions. Obtained nanocrystals need to be carefully characterized by size, size homogeneity, density, quantity, and quality, which will all play a critical role in a successful experiment. It is always helpful to carry out cross validation whenever possible while characterizing results with nanocrystals. With the further development and improvement of facilities at the existing X-Ray FEL instruments and the next generation of X-Ray FELs either coming online or in the planning stages, the available characterization facilities are expected to be considerably more abundant and state of the art. This will certainly facilitate crystal characterization prior to SFX with X-Ray FELs for any future experiment. In addition, as Chap. 8 will explore, the movement towards smaller size regimes accessible in crystallography has led to the ability to access interesting characteristics inherent to truly discrete crystals, resulting in new modes of data collection and analysis.

References

1. Scopes, R. K. (2013). *Protein purification: Principles and practice*. Berlin, Germany: Springer.
2. Doublié, S. (2007). *Macromolecular crystallography protocols* (Vol. 1). New York: Springer.
3. Jancarik, J., & Kim, S.-H. (1991). Sparse matrix sampling: A screening method for crystallization of proteins. *Journal of Applied Crystallography, 24*(4), 409–411.
4. Vekilov, P. G., Feeling-Taylor, A., Yau, S.-T., & Petsev, D. (2002). Solvent entropy contribution to the free energy of protein crystallization. *Acta Crystallographica Section D: Biological Crystallography, 58*(10), 1611–1616.
5. Garcıa-Ruiz, J. M. (2003). Nucleation of protein crystals. *Journal of Structural Biology, 142*(1), 22–31.
6. Perutz, M. F., Rossmann, M. G., Cullis, A. F., Muirhead, H., Will, G., & North, A. (1960). Structure of hæmoglobin: A three-dimensional Fourier synthesis at 5.5-Å. Resolution, obtained by X-ray analysis. *Nature, 185*(4711), 416–422.
7. Giegé, R. (2013). A historical perspective on protein crystallization from 1840 to the present day. *The FEBS Journal, 280*(24), 6456–6497.
8. McPherson, A. (2017). Protein crystallization. In *Protein Crystallography: Methods and Protocols* (pp. 17–50). New York: Springer.
9. Rupp, B. (2009). *Biomolecular crystallography: Principles, practice, and application to structural biology*. Abingdon, UK: Garland Science.
10. Cohen, A. E., Soltis, S. M., Gonzalez, A., Aguila, L., Alonso-Mori, R., Barnes, C. O., Baxter, E. L., Brehmer, W., Brewster, A. S., Brunger, A. T., Calero, G., Chang, J. F., Chollet, M., Ehrensberger, P., Eriksson, T. L., Feng, Y., Hattne, J., Hedman, B., Hollenbeck, M., Holton, J. M., Keable, S., Kobilka, B. K., Kovaleva, E. G., Kruse, A. C., Lemke, H. T., Lin, G., Lyubimov, A. Y., Manglik, A., Mathews, I. I., McPhillips, S. E., Nelson, S., Peters, J. W., Sauter, N. K., Smith, C. A., Song, J., Stevenson, H. P., Tsai, Y., Uervirojnangkoorn, M., Vinetsky, V., Wakatsuki, S., Weis, W. I., Zadvornyy, O. A., Zeldin, O. B., Zhu, D., & Hodgson, K. O. (2014). Goniometer-based femtosecond crystallography with X-ray free electron lasers. *Proceedings of the National Academy of Sciences of the United States of America, 111*(48), 17122–17127.
11. Hirata, K., Shinzawa-Itoh, K., Yano, N., Takemura, S., Kato, K., Hatanaka, M., Muramoto, K., Kawahara, T., Tsukihara, T., & Yamashita, E. (2014). Determination of damage-free crystal structure of an X-ray-sensitive protein using an XFEL. *Nature Methods, 11*(7), 734–736.
12. Aquila, A., Hunter, M. S., Doak, R. B., Kirian, R. A., Fromme, P., White, T. A., Andreasson, J., Arnlund, D., Bajt, S., Barends, T. R., Barthelmess, M., Bogan, M. J., Bostedt, C., Bottin, H., Bozek, J. D., Caleman, C., Coppola, N., Davidsson, J., DePonte, D. P., Elser, V., Epp, S. W., Erk, B., Fleckenstein, H., Foucar, L., Frank, M., Fromme, R., Graafsma, H., Grotjohann, I., Gumprecht, L., Hajdu, J., Hampton, C. Y., Hartmann, A., Hartmann, R., Hau-Riege, S., Hauser, G., Huaser, H., Hirsemann, P., Holl, J., Holton, M., Hömke, A., Johansson, L., Kimmel, N., Kassemeyer, S., Krasniqi, F., Kühnel, K.-U., Liang, M., Lomb, L., Malmerberg, E., Marchesini, S., Martin, A. V., Maia, F. R., Messerschmidt, M., Nass, K., Schlichting, I., Schmidt, C., Schmidt, K. E., Schulz, J., Seibert, M. M., Shoeman, R. L., Sierra, R., Soltau, H., Starodub, D., Stellato, F., Stern, S., Strüder, L., Timneanu, N., Ullrich, J., Wang, X., Williams, G. J., Weidenspointner, G., Weierstall, U., Wunderer, C., Barty, A., Spence, J. C. H., & Chapman, H. N. (2012). Time-resolved protein nanocrystallography using an X-ray free-electron laser. *Optics Express, 20*(3), 2706–2716.
13. Kupitz, C., Basu, S., Grotjohann, I., Fromme, R., Zatsepin, N. A., Rendek, K. N., Hunter, M. S., Shoeman, R. L., White, T. A., Wang, D., James, D., Yang, J.-H., Cobb, D. E., Reeder, B., Sierra, R. G., Liu, H., Barty, A., Aquila, A. L., Deponte, D., Kirian, R. A., Bari, S., Bergkamp, J. J., Beyerlein, K. R., Bogan, M. J., Caleman, C., Chao, T.-C., Conrad, C. E., Davis, K. M., fleckenstein, H., Galli, L., Hau-Riege, S. P., Kassemeyer, S., Laksmono, H., Liang, M., Lomb, L., Marchesini, S., Martin, A. V., Messerschmidt, M., Milathianaki, D., Nass, K., Ros, A., Roy-Chowdhury, S., Schmidt, K., Seibert, M., Steinbrener, J., Stellato, F., Yan, L., Yoon, C., Moore, T. A., Moore, A. L., Pushkar, Y., Williams, G. J., Boutet, S., Doak, R. B., Weierstall, U.,

Frank, M., Chapman, H. N., Spence, J. C. H., & Fromme, P. (2014). Serial time-resolved crystallography of photosystem II using a femtosecond X-ray laser. *Nature, 513*(7517), 261–265.
14. Pande, K., Hutchinson, C. D. M., Groenhof, G., Aquila, A., Robinson, J. S., Tenboer, J., Basu, S., Boutet, S., Deponte, D., Liang, M., White, T., Zatsepin, N., Yefanov, O., Morozov, D., Oberthuer, D., Gati, C., Subramanian, G., James, D., Zhao, Y., Koralek, J., Brayshaw, J., Kupitz, C., Conrad, C., Roy-Chowdhury, S., Coe, J., Metz, M., Paulraj Lourdu, X., Grant, T., Koglin, J., Ketawala, G., Fromme, R., Srajer, V., Henning, R., Spence, J., Ourmazd, A., Schwander, P., Weierstall, U., Frank, M., Fromme, P., Barty, A., Chapman, H., Moffat, K., Van Thor, J. J., & Schmidt, M. (2016). Femtosecond structural dynamics drives the trans/cis isomerization in photoactive yellow protein. *Science, 352*(6286), 725–729.
15. Tenboer, J., Basu, S., Zatsepin, N., Pande, K., Milathianaki, D., Frank, M., Hunter, M., Boutet, S., Williams, G. J., Koglin, J. E., Oberthuer, D., Heymann, M., Kupitz, C., Conrad, C., Coe, J., Roy-Chowdhury, S., Weierstall, U., James, D., Wang, D., Grant, T., Barty, A., Yefanov, O., Scales, J., Gati, C., Seuring, C., Srajer, V., Henning, R., Schwander, P., Fromme, R., Ourmazd, A., Moffat, K., Van Thor, J. J., Spence, J. C. H., Fromme, P., Chapman, H. N., & Schmidt, M. (2014). Time-resolved serial crystallography captures high-resolution intermediates of photoactive yellow protein. *Science, 346*(6214), 1242–1246.
16. Kupitz, C., Grotjohann, I., Conrad, C. E., Roy-Chowdhury, S., Fromme, R., & Fromme, P. (2014). Microcrystallization techniques for serial femtosecond crystallography using photosystem II from thermosynechococcus elongatus as a model system. *Philosophical Transactions of the Royal Society B: Biological Sciences, 369*(1647), 20130316.
17. Salemme, F. (1972). A free interface diffusion technique for the crystallization of proteins for X-ray crystallography. *Archives of Biochemistry and Biophysics, 151*(2), 533–539.
18. Abdallah, B. G., Roy-Chowdhury, S., Fromme, R., Fromme, P., & Ros, A. (2016). Protein crystallization in an actuated microfluidic nanowell device. *Crystal Growth & Design, 16*, 2074–2082.
19. Kupitz, C., Olmos, J. L., Jr., Holl, M., Tremblay, L., Pande, K., Pandey, S., Oberthür, D., Hunter, M., Liang, M., Aquila, A., Tenboer, J., Calvey, G., Katz, A., Chen, Y., Wiedorn, M. O., Knoska, J., Meents, A., Mariani, V., Norwood, T., Poudyal, I., Grant, T., Miller, M. D., Xu, W., Tolstikova, A., Morgan, A., Metz, M., Martin-Garcia, J. M., Zook, J. D., Roy-Chowdhury, S., Coe, J., Nagaratnam, N., Meza, D., Fromme, R., Basu, S., Frank, M., White, T., Barty, A., Bajt, S., Yefanov, O., Chapman, H. N., Zatsepin, N., Nelson, G., Weierstall, U., Spence, J., Schwander, P., Pollack, L., Fromme, P., Ourmazd, A., Phillips, G. N., & Schmidt, M. (2017). Structural enzymology using X-ray free electron lasers. *Structural Dynamics, 4*(4), 044003.
20. Redecke, L., Nass, K., DePonte, D. P., White, T. A., Rehders, D., Barty, A., Stellato, F., Liang, M., Barends, T. R., Boutet, S., Williams, G. J., Messerschmidt, M., Seibert, M. M., Aquila, A., Arnlund, D., Bajt, S., Barth, T., Bogan, M. J., Caleman, C., Chao, T. C., Doak, R. B., Fleckenstein, H., Frank, M., Fromme, R., Galli, L., Grotjohann, I., Hunter, M. S., Johansson, L. C., Kassemeyer, S., Katona, G., Kirian, R. A., Koopmann, R., Kupitz, C., Lomb, L., Martin, A. V., Mogk, S., Neutze, R., Shoeman, R. L., Steinbrener, J., Timneanu, N., Wang, D., Weierstall, U., Zatsepin, N. A., Spence, J. C., Fromme, P., Schlichting, I., Duszenko, M., Betzel, C., & Chapman, H. N. (2013). Natively inhibited *Trypanosoma brucei* cathepsin B structure determined by using an X-ray laser. *Science, 339*(6116), 227–230.
21. Sawaya, M. R., Cascio, D., Gingery, M., Rodriguez, J., Goldschmidt, L., Colletier, J. P., Messerschmidt, M. M., Boutet, S., Koglin, J. E., Williams, G. J., Brewster, A. S., Nass, K., Hattne, J., Botha, S., Doak, R. B., Shoeman, R. L., DePonte, D. P., Park, H. W., Federici, B. A., Sauter, N. K., Schlichting, I., & Eisenberg, D. S. (2014). Protein crystal structure obtained at 2.9 a resolution from injecting bacterial cells into an X-ray free-electron laser beam. *Proceedings of the National Academy of Sciences of the United States of America, 111*(35), 12769–12774.
22. Ostermeier, C., & Michel, H. (1997). Crystallization of membrane proteins. *Current Opinion in Structural Biology, 7*(5), 697–701.

23. Fromme, R., Ishchenko, A., Metz, M., Chowdhury, S. R., Basu, S., Boutet, S., Fromme, P., White, T. A., Barty, A., Spence, J. C., Weierstall, U., Liu, W., & Cherezov, V. (2015). Serial femtosecond crystallography of soluble proteins in lipidic cubic phase. *IUCrJ, 2*(5), 545–551.
24. Martin-Garcia, J. M., Conrad, C. E., Nelson, G., Stander, N., Zatsepin, N. A., Zook, J., Zhu, L., Geiger, J., Chun, E., Kissick, D., Hilgart, M. C., Ogata, C., Ishchenko, A., Nagaratnam, N., Roy-Chowdhury, S., Coe, J., Subramanian, G., Schaffer, A., James, D., Ketawala, G., Venugopalan, N., Xu, S., Corcoran, S., Ferguson, D., Weierstall, U., Spence, J. C. H., Cherezov, V., Fromme, P., Fischetti, R. F., & Liu, W. (2017). Serial millisecond crystallography of membrane and soluble protein microcrystals using synchrotron radiation. *IUCrJ, 4*(4), 439–454.
25. Weierstall, U., James, D., Wang, C., White, T. A., Wang, D., Liu, W., Spence, J. C., Doak, R. B., Nelson, G., & Fromme, P. (2014). Lipidic cubic phase injector facilitates membrane protein serial femtosecond crystallography. *Nature Communications, 5*, 3309.
26. Ibrahim, M., Chatterjee, R., Hellmich, J., Tran, R., Bommer, M., Yachandra, V. K., Yano, J., Kern, J., & Zouni, A. (2015). Improvements in serial femtosecond crystallography of photosystem II by optimizing crystal uniformity using microseeding procedures. *Structural Dynamics, 2*(4), 041705.
27. Kirian, R. A., Wang, X., Weierstall, U., Schmidt, K. E., Spence, J. C., Hunter, M., Fromme, P., White, T., Chapman, H. N., & Holton, J. (2010). Femtosecond protein nanocrystallography—Data analysis methods. *Optics Express, 18*(6), 5713–5723.
28. Uervirojnangkoorn, M., Zeldin, O. B., Lyubimov, A. Y., Hattne, J., Brewster, A. S., Sauter, N. K., Brunger, A. T., & Weis, W. I. (2015). Enabling X-ray free electron laser crystallography for challenging biological systems from a limited number of crystals. *eLife, 4*, e05421.
29. Ginn, H. M., Brewster, A. S., Hattne, J., Evans, G., Wagner, A., Grimes, J. M., Sauter, N. K., Sutton, G., & Stuart, D. (2015). A revised partiality model and post-refinement algorithm for X-ray free-electron laser data. *Acta Crystallographica Section D: Biological Crystallography, 71*(6), 1400–1410.
30. Ginn, H. M., Messerschmidt, M., Ji, X., Zhang, H., Axford, D., Gildea, R. J., Winter, G., Brewster, A. S., Hattne, J., Wagner, A., Grimes, J. M., Evans, G., Sauter, N. K., Sutton, G., & Stuart, D. I. (2015). Structure of CPV17 polyhedrin determined by the improved analysis of serial femtosecond crystallographic data. *Nature Communications, 6*.
31. Spence, J. C. H., Weierstall, U., & Chapman, H. N. (2012). X-ray lasers for structural and dynamic biology. *Reports on Progress in Physics, 75*(10), 102601.
32. Brehms, W., & Diederichs, K. (2014). Breaking the indexing ambiguity in serial crystallography. *Acta Crystallographica Section D, D70*, 101–109.
33. Abdallah, B. G., Zatsepin, N. A., Roy-Chowdhury, S., Coe, J., Conrad, C. E., Dörner, K., Sierra, R. G., Stevenson, H. P., Grant, T. D., Nelson, G., James, D. R., Calero, G., Wachter, R. M., Spence, J. C. H., Weierstall, U., Fromme, P., & Ros, A. (2015). Microfluidic sorting of protein nanocrystals by size for XFEL diffraction. *Structural Dynamics, 2*, 041719.
34. Abdallah, B. G., Chao, T. C., Kupitz, C., Fromme, P., & Ros, A. (2013). Dielectrophoretic sorting of membrane protein nanocrystals. *American Chemical Society Nano, 7*(10), 9129–9137.
35. Haupert, L. M., & Simpson, G. J. (2011). Screening of protein crystallization trials by second order nonlinear optical imaging of chiral crystals (SONICC). *Methods, 55*(4), 379–386.
36. Kissick, D. J., Wanapun, D., & Simpson, G. J. (2011). Second-order nonlinear optical imaging of chiral crystals. *Annual Review of Analytical Chemistry, 4*, 419.
37. Madden, J. T., DeWalt, E. L., & Simpson, G. J. (2011). Two-photon excited UV fluorescence for protein crystal detection. *Acta Crystallographica Section D: Biological Crystallography, 67*(10), 839–846.
38. Pecora, R. (2000). Dynamic light scattering measurement of nanometer particles in liquids. *Journal of Nanoparticle Research, 2*(2), 123–131.

39. Schubert, R., Meyer, A., Dierks, K., Kapis, S., Reimer, R., Einspahr, H., Perbandt, M., & Betzel, C. (2015). Reliably distinguishing protein nanocrystals from amorphous precipitate by means of depolarized dynamic light scattering. *Journal of Applied Crystallography, 48*(5), 1476–1484.
40. Schmitz, K. S. (1990). *An introduction to dynamic light scattering of macromolecules*. Kansas City, MO: University of Missouri.
41. Rasband, W. (1997). *ImageJ*. Bethesda, MD: US National Institutes of Health.
42. Bai, K., Barnett, G. V., Kar, S. R., & Das, T. K. (2017). Interference from proteins and surfactants on particle size distributions measured by nanoparticle tracking analysis (NTA). *Pharmaceutical Research, 34*(4), 800–808.
43. Regtmeier, J., Eichhorn, R., Duong, T. T., Anselmetti, D., & Ros, A. (2007). Dielectrophoretic manipulation of DNA: Separation and polarizability. *Analytical Chemistry, 79*, 3925–3932.
44. Stevenson, H. P., Lin, G. W., Barnes, C. O., Sutkeviciute, I., Krzysiak, T., Weiss, S. C., Reynolds, S., Wu, Y., Nagarajan, V., Makhov, A. M., Lawrence, R., Lamm, E., Clark, L., Gardella, T. J., Hogue, B. G., Ogata, C. M., Ahn, J., Gronenborn, A. M., Conway, J. F., Vilardaga, J. P., Cohen, A. E., & Calero, G. (2016). Transmission electron microscopy for the evaluation and optimization of crystal growth. *Acta Crystallographica Section D-Structural Biology, 72*, 603–615.
45. Stevenson, H. P., Makhov, A. M., Calero, M., Edwards, A. L., Zeldin, O. B., Mathews, I. I., Lin, G., Barnes, C. O., Santamaria, H., Ross, T. M., Soltis, S. M., Khosla, C., Nagarajan, V., Conway, J. F., Cohen, A. E., & Calero, G. (2014). Use of transmission electron microscopy to identify nanocrystals of challenging protein targets. *Proceedings of the National Academy of Sciences of the United States of America, 111*(23), 8470–8475.

Chapter 4
The Lipid Cubic Phase as a Medium for the Growth of Membrane Protein Microcrystals

Zina Al-Sahouri, Ming-Yue Lee, Dianfan Li, Wei Liu, and Martin Caffrey

4.1 Introduction

SFX[1] Serial femtosecond X-ray crystallography (SFX) is a relatively new method for collecting crystallographic information on small crystals fed continuously across a free-electron laser (FEL) beam composed of high-fluence X-ray bunches femtoseconds long [2, 3]. Each encounter between an X-ray bunch and a microcrystal (*hit*) ideally gives rise to a single, still diffraction pattern with greater than 15 measurable reflections. Since the crystals are randomly oriented, collecting patterns on enough crystals (thousands, typically) produces a complete data set of high redundancy for structure determination by molecular replacement (MR) and de novo phasing [4–8]. Data are collected in a sample chamber at atmospheric pressure or in vacuo at 20 °C. Despite the intensity of the X-ray bunch (typically 10^{12}

[1] Parts of this chapter are reproduced directly from Ref. [1]

Z. Al-Sahouri · M.-Y. Lee · W. Liu (✉)
Center for Applied Structural Discovery at the Biodesign Institute, School of Molecular Sciences, Arizona State University, Tempe, AZ, USA
e-mail: Zina.Al-Sahouri@asu.edu; minglee@asu.edu; w.liu@asu.edu

D. Li
State Key Laboratory of Molecular Biology, National Center for Protein Science Shanghai, Shanghai Science Research Center, CAS Center for Excellence in Molecular Cell Science, Shanghai Institute of Biochemistry and Cell Biology, Chinese Academy of Sciences, Shanghai, China
e-mail: dianfan.li@sibcb.ac.cn

M. Caffrey (✉)
Membrane Structural and Functional Biology Group, School of Medicine and School of Biochemistry and Immunology, Trinity College Dublin, Dublin, Ireland
e-mail: martin.caffrey@tcd.ie

S. Boutet et al. (eds.), *X-ray Free Electron Lasers*,
https://doi.org/10.1007/978-3-030-00551-1_4

photons/bunch), each pulse is of such short duration that the changes associated with radiation damage do not progress sufficiently before the diffracted X-rays have departed (*run*) and their structural manifest recorded. We refer to this as *"hit and run"* SFX. It is also known as *"diffraction-before-destruction"* SFX.

The SFX:LCP Marriage A fluid medium is used to ferry crystals of membrane proteins through the pulsed XFEL beam for SFX [4, 8]. Because productive interactions between X-rays and crystals in the flowing stream were so infrequent, typically, only 1 in 25,000 crystals produce a useful diffraction pattern. Thus, vast amounts of valuable membrane protein were required for data collection and most of the protein went to waste. For example, when photosystem I (PSI) crystals dispersed in a continuous liquid jet, data collection required 10 mg of protein [4]. In contrast, when photosynthetic reaction center crystals were dispersed in the more viscous but still quite fluid lipid sponge phase that they had been grown in by the in meso method, only 3 mg of protein were needed [8]. Due to the viscous nature of the mesophases, flow rate would be reduced dramatically. And if suitably high crystal densities in the LCP could be achieved, the rate of delivery of single crystals and X-rays to the interaction region could be matched for a most efficacious use of both. The realization of this idea led to the development of the LCP-SFX method.

Advantages Afforded by LCP-SFX LCP-SFX is appealing as a method because it offers the prospect of obviating some of the issues that arise with in meso-based structure determination using synchrotron X-ray radiation. With the in meso method, crystals are typically grown in a sealed glass sandwich plate. By contrast with crystals grown by the more traditional so-called in surfo methods, their in meso counterparts tend to be considerably smaller. Harvesting crystals from the viscous mesophase is a somewhat cumbersome process that can lead to substantial loss of crystals and to degradation in crystal quality, which may affect diffraction quality. Data collection at a synchrotron source is typically done at 100 K. Such a frigid temperature can stabilize conformational sub-states, particularly in the protein's side chains, that may not be physiologically relevant and may lead to potential structure–function misinterpretation [9]. Radiation damage is also a major concern when using small crystals with synchrotron radiation sources where residues, such as aspartate and glutamate, are particularly prone to undergo decarboxylation [10]. Damage can be mitigated to a degree with larger crystals, beam attenuation, and data collection at cryo-temperatures, but a complete data set often requires many tens of crystals. In this context then, LCP-SFX is attractive in that it offers an approach to in situ data collection with micrometer or nanometer-sized crystals at or close to the more physiologically relevant 20 °C and the prospect of outrunning the structural consequences of radiation damage.

The LCP as a Liquid Crystal or Mesophase The lipid cubic phase takes center stage in this work. It serves as the medium in which crystallization occurs and is, in turn, used to port those same crystals into the XFEL beam for SFX. As a lyotropic liquid crystal, it is formed most simply by mixing together the monoacylglycerol lipid, monoolein, and water, in approximately equal parts at 20 °C

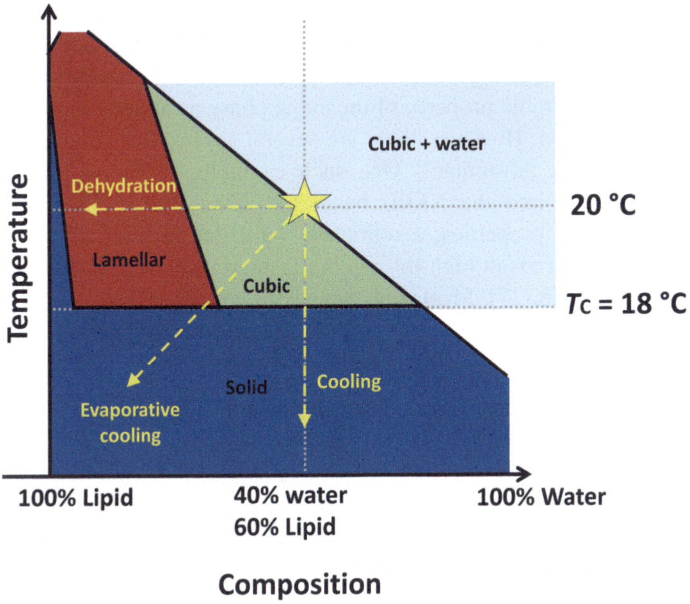

Fig. 4.1 Schematic of the equilibrium temperature–composition phase diagram for the monoolein (9.9 MAG)-water system near 20 °C. The different phases are shown as colored zones and labelled accordingly. The cubic mesophase is extruded into the evacuated sample chamber for SFX under conditions indicated by the yellow star at 20 °C and ~40% aqueous medium. Possible trajectories through the phase diagram taken upon dehydration, cooling and evaporative cooling are indicated by dashed arrows. The 20 °C isotherm is identified by a horizontal dashed line. The liquid crystal-to-solid (Lc) transition is identified by the horizontal dashed line at 18 °C. This schematic is adapted from Ref. [1]

(Fig. 4.1). The hydrophobic effect primarily drives the spontaneous self-assembly of the mesophase. As with any state of matter, mesophase behavior is dictated by Gibbs' phase rule, and is conveniently and concisely summarized in the form of a temperature–composition phase diagram (Fig. 4.1). The equilibrium phase diagram for the monoolein-water system has been mapped out thoroughly based on small- and wide-angle X-ray scattering measurements [11, 12]. Below about 18 °C, cubic phase gives way to a solid, the lamellar crystalline or Lc phase. The cubic phase consists of a single, continuous highly curved and multiply branched lipid bilayer on either side of which is a bathing aqueous channel. These two continuous channels interpenetrate but never contact one another directly because a lipid bilayer separates them. For use in in meso crystallogenesis, the mesophase is prepared typically by combining the host lipid with an aqueous solution of a pure membrane protein solubilized in detergent [13]. The most commonly used host lipids are *cis*-monoenoic monoacylglycerols (MAGs) with acyl chains 14–18 carbon atoms long [14, 15]. Originally, a lipid synthesis program in the Caffrey lab provided these MAGs in support of the in meso method of crystallization [16]. Given their success, many are now commercially available.

Viscosity A noted feature of the cubic phase is its viscosity. The older literature referred to it as the viscous isotropic or VI phase, reflecting its "challenging" rheological and non-birefringent optical properties. Viscosity has been highlighted by some as an undesirable property of the cubic phase as a medium with which to perform crystallization. However, there are several applications where the viscous nature offers distinct advantages. One such example is the use as a medium to transport crystals into the XFEL beam for SFX. Despite its viscosity and challenging handling properties, a robot was built that enables the setting of in meso crystallization trials in high-throughput fashion using miniscule quantities of mesophase and protein [17]. Most trials are set up now using anywhere from 20 to 50 nL of mesophase corresponding to ~5–20 ng protein/well. Given the success of this robot, variations on the original design are available commercially [18].

Rheology Rheology and flow as a jet is dictated to a significant degree by the make-up of the mesophase and in particular the identity of the host MAG component. Different MAGs produce mesophases, each with its own microstructure and physicochemical properties, which will impact uniquely on jet flow and behaviour as a tubular conveyor of crystals. A mesophase with undesirable characteristics might require exorbitantly high and possibly damaging pressures to induce flow. As well, it may produce an unstable jet that breaks up prematurely before reaching the interaction region or curve back out of the beam and onto the nozzle to cause nozzle blockage, interference with flow, and subsequent downtime to correct such issues. A more fluid mesophase will require higher flow rates to produce a stable jet with the added cost of wasted protein crystals as noted above for the sponge phase.

Crystallization Mechanism It is the bicontinuous nature of the LCP that is at the heart of in meso crystallogenesis. Our working hypothesis for how crystallization takes place begins with the target protein reconstituted into and uniformly distributed throughout the continuous, bilayer membrane that permeates the mesophase. Components of the precipitant stabilize a transition locally to the lamellar phase into which proteins diffuse to preferentially partition, concentrate, and subsequently nucleate giving rise to macroscopic crystals [19]. These crystals tend to be generally small, but of high diffraction quality, and considerable effort is usually required to optimize conditions that produce crystals large enough for synchrotron radiation-based data collection.

In Meso Successes to Date Despite the challenges of the method, it has been used to generate crystal structures of a number of different membrane proteins and complexes [13]. The most notable, of late, include the β_2 adrenergic receptor-Gs protein complex that was the subject of the 2012 Nobel Prize in Chemistry [20] and the rhodopsin–arrestin complex [21] (please refer to Chap. 10 for more details). To date, over 400 recorded entries in the Protein Data Bank (PDB, www.pdb.org) are attributed to the in meso method (Fig. 4.2). This corresponds to about 11% of all membrane protein structures deposited in the PDB. Attesting to the growing interest in the method, almost half the in meso PDB records has been added in the past 3 years.

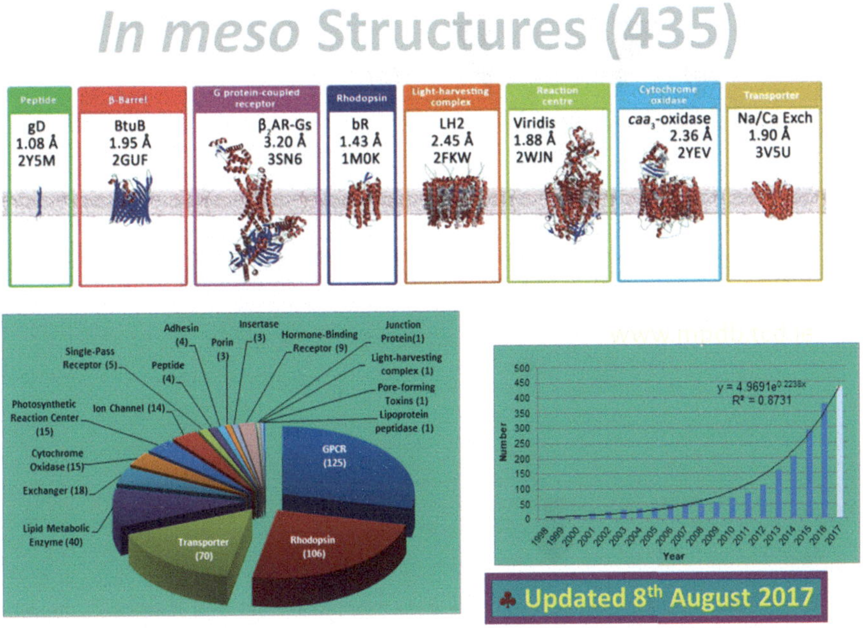

Fig. 4.2 Latest in meso stats

LCP-SFX Technical Challenges Three major technical challenges were identified in implementing LCP-SFX. These included: (1) vacuum incompatibility of the monoolein-based LCP, (2) the need to scale-up from nanoliter to microliter volumes of crystal-laden mesophase, and (3) the provision of an injector that could extrude the highly viscous mesophase in the form of a micrometer-diameter, continuous bolus into the XFEL beam.

Host Lipid, Vacuum Compatibility The first technical challenge relates to the phase behavior of the medium in which crystals are grown and then ported into the XFEL beam for SFX. As noted, data are frequently collected in an evacuated sample chamber at 20 °C. The mesophase containing well-dispersed microcrystals is extruded from the injector as a fully hydrated tubular bolus. Immediately upon entering the chamber, volatiles (water in particular) will evaporate from the surface of the bolus and the sample temperature will drop due to evaporative cooling (Fig. 4.3). Evaporation also leads to the concentration of all non-volatiles in the bolus to increase. These include lipid, detergent, protein, protein crystals, and buffer and precipitant components. Therefore, concentration gradients develop along the length and across the diameter of the cylindrically shaped mesophase bolus. The magnitude of the gradients depends on flow rate and distance along the bolus from the tip of the injector nozzle. Depending on the final concentrations reached, these assorted components can crystallize directly and/or destabilize the dispersing mesophase.

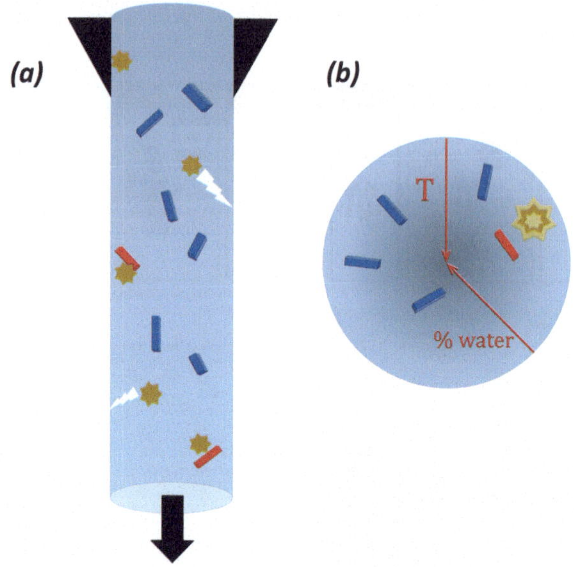

Fig. 4.3 Cartoon representation of the crystal-laden mesophase bolus as it is extruded through the nozzle (black triangles) of the LCP-injector into the evacuated sample chamber at 20 °C for serial femtosecond crystallographic measurements with an X-ray free electron laser. (**a**) Side view of the bolus where the gradient in color from Top to Bottom corresponds to the gradient in temperature and composition along the length of the bolus induced by evaporative cooling. (**b**) End on view of the bolus where the gradient in color corresponds to the gradient in temperature and composition (arrows) along the radius of the cylindrical bolus induced by evaporative cooling. Pristine, undamaged membrane protein crystals are colored blue and are shown dispersed in a light blue cubic mesophase. Stars correspond to sites where the mesophase has transformed from the cubic to the solid Lc phase that may damage the crystals (red) and introduce defects (lightning bolt) in the bolus thereby affecting flow. The enlarged star in (**b**) is drawn to suggest local heating due to the heat of fusion associated with the solidification reaction that may damage dispersed crystals nearby. This schematic is adapted from Ref. [1]

As noted, mesophase behavior is dictated by temperature and composition [11, 12]. Evaporative cooling, brought about by loss of water, will induce sample cooling as well as an increase in lipid concentration. From the equilibrium temperature–composition phase diagram for the monoolein/water system (Fig. 4.1), a reduction in water content together with a drop in temperature will result in a transition from the cubic mesophase to the solid Lc phase [11, 12]. This change in phase can have detrimental consequences such as a change in the rheological and flow characteristics of the extruded jet, creating problems with sample positioning in the XFEL beam. Because the transition has a large associated heat of fusion, wherever crystallization occurs a local "hot" spot will develop that, in turn, may impact negatively on jet flow characteristics and on membrane protein crystal quality. Further, the Lc phase itself, as a solid, may damage the delicate membrane protein crystals dispersed in the bolus. Lastly, Lc phase bolus medium contributes strong and sharp background powder diffraction at low and wide-angles. Background scatter from the Lc phase creates problems for the recovery of crystal diffraction data from recorded composite images. More importantly, it can damage some detectors, and it is partly for this reason that the incident beam is routinely attenuated some 20-fold during data collection. This is in contrast to the cubic phase, which gives rise to relatively benign diffuse scattering at wide-angles, although diffraction in the low-angle region can be strong and sharp.

It was therefore important to avoid the undesirable cubic-to-Lc transition due to evaporative cooling. An obvious way around this was to reduce the cubic-to-Lc transition temperature, Tc, which could be achieved by using an alternative host MAG to monoolein (9.9 MAG) whose Tc is 18 °C (Fig. 4.1). Separately, we had designed 7.9 MAG (nomenclature described in ref. [22]) for in meso crystallization at low temperatures [23]. The Tc of 7.9 MAG under conditions of full hydration is about 6 °C. This was deemed low enough and 7.9 MAG was chosen as the host lipid for use in a feasibility study with the diacylglycerol kinase, DgkA. It was subsequently shown to behave as expected and to prevent the formation of Lc phase under conditions of SFX data collection.

The 2.05 Å structure of DgkA was obtained using synchrotron X-rays after extensive crystallization screening and optimization [14, 24]. Final crystals were generated in 7.8 MAG at 4 °C. It was necessary therefore to rescreen and optimize in 7.9 MAG, and ideally this should be done at 20 °C, the temperature at which SFX data were to be collected. However, the crystal requirements for synchrotron radiation and for SFX data collection are entirely different. For the former, a few large single crystals suffice. For SFX, tens of microliters of mesophase containing a high density of micrometer-sized crystals are needed.

Adjusting the concentration of 2-methyl-2,4-pentanediol (MPD) in the precipitant solution provided crystals in 7.9 MAG at 20 °C that ranged from showers of microcrystals required for SFX, to isolated, relatively large single crystals suitable for synchrotron radiation data collection. The former were used successfully for SFX and provided a structure of DgkA to 2.18 Å [25]. The latter, however, diffracted at the synchrotron to no better than 6 Å. Additional rounds of optimization, performed at 4 °C, resulted in large, single crystals that provided a synchrotron radiation structure at the same resolution of 2.18 Å [25].

Scaling Up The next technical challenge required increasing the scale of crystallogenesis. For SFX, it was anticipated that tens of microliters of crystal-laden mesophase would be required to collect enough data for a structure solution. As noted, in meso crystallization screening is highly efficient and is performed typically on a 50 nL mesophase per well basis [13]. The challenge then was to scale up crystallogenesis by about a thousand-fold. Due to a scarcity of membrane proteins generally, extensive screening for optimal crystallization conditions is practically impossible on such a large scale. It is necessary therefore to identify conditions that produce a high density of microcrystals, first under standard conditions in glass sandwich plates at the 50 nL level, and then scale up by a factor of a thousand and hope that the same condition translates directly. However, for the conditions to translate it is necessary to perform the large volume crystallogenesis while maintaining, as much as possible, the same geometrical relationships between mesophase bolus and precipitant solution that prevails during nanoliter volume crystal growth. The geometry in question relates to the shape and size, especially the diameter, of the bolus in contact with the surrounding precipitant solution. Since in meso crystallization has been reasoned to depend on such factors [19], every effort should be made to replicate those conditions for large scale microcrystal production. This can be realized by carrying out crystal growth in a bolus of protein-laden mesophase approximately 15 cm long and 0.4 mm in diameter located toward the center of the barrel of a 100 μL Hamilton syringe containing precipitant solution. The composition of the precipitant solution would have been identified previously to generate the desired, high density of microcrystals in standard nanoliter-scale glass sandwich plate screening. Upon incubation for a period at 20 °C, microcrystals ideally up to 30 μm long are obtained at a suitably high density. Most of the excess precipitant solution is removed mechanically with the aid of an empty syringe and a narrow bore syringe coupler [26]. The last vestiges of residual excess precipitant are incorporated lyotropically by combining the opaque dispersion with a small volume of host MAG. This procedure generates the required bulk volume of optically clear cubic mesophase in which microcrystals (Fig. 4.4) are uniformly dispersed and ready for SFX measurements in vacuo at 20 °C.

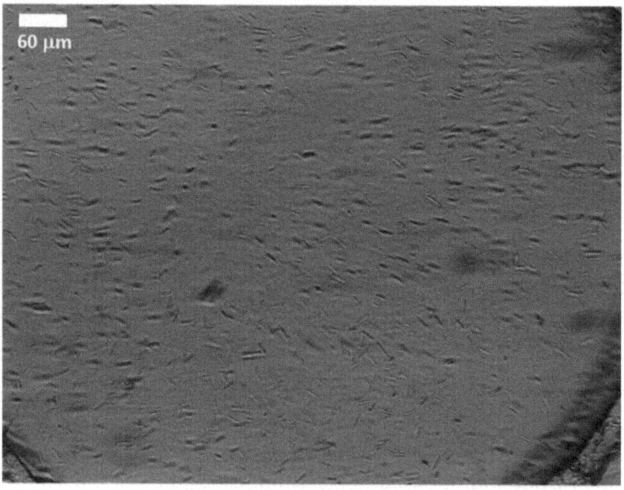

Fig. 4.4 Microcrystals of DgkA grown in the cubic mesophase with 7.9 MAG as host lipid at 20 °C in a 0.5 mL syringe. Details of sample preparation are described in Ref. [31]

LCP Injector The third challenge associated with realizing LCP-SFX was the development, building and implementation of an injector capable of delivering the highly viscous, crystal-laden mesophase at a fixed rate as a uniform, continuous, micrometer-diameter, cylindrically shaped bolus to the interaction region in an evacuated sample chamber at 20 °C. Inspired by the simple coupled syringe mixing device used for mesophase preparation and delivery in manual and robotic application of the in meso method, this was realized in the form of the LCP injector [27]. In operation, it involves the extrusion of mesophase from a 20 to 50 μL reservoir through a 6 cm-long glass capillary with an internal diameter of 20–50 μm. The tapered end of the capillary extends beyond the tip of a specially designed gas virtual nozzle which provides a co-flowing stream of gas for reliable, co-axial mesophase extrusion. Pressure, generated by a HPLC pump, is transmitted through water to the mesophase with a pair of Teflon beads separating and providing a water-tight seal between the two media. The analogy between the injector and the syringes used in the coupled syringe mixer is striking.

Growing Small Crystals for LCP-SFX The LCP-SFX method introduced in 2013 is robust and well proven. The protocol in place for generating microliter volumes of mesophase with a suitably high density of microcrystals has been described in detail in several publications. An overview of the protocol is outlined below (Fig. 4.5).

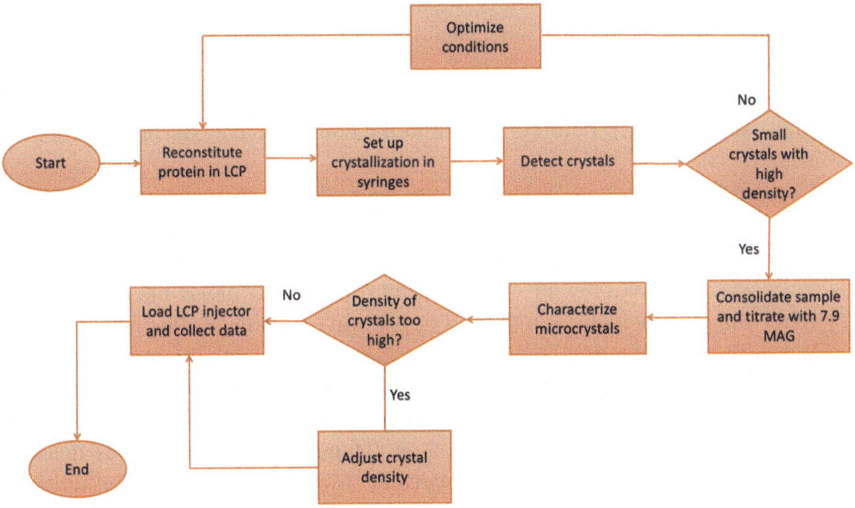

Fig. 4.5 Flowchart summarizing the process of protein reconstitution in lipid cubic phase (LCP) in order to obtain crystals via serial femtosecond crystallography

4.2 Procedure for the Preparation and Characterization of Microcrystals for LCP-SFX

4.2.1 Membrane Protein Reconstitution in LCP (Modified from Ref. [28])

1. Transfer 15 μL of molten 9.9 MAG into syringe no. 1, and 10 μL of protein solution into syringe no. 2.
2. Connect the syringes together using a syringe coupler and homogenize the sample by pushing it through the coupler back-and-forth between syringes, until a transparent LCP forms.
3. To set up crystallization in syringes, move the entire LCP sample into syringe no. 2. Disconnect the empty syringe no. 1, while keeping the coupler connected to syringe no. 2.
4. Attach a removable needle to a 100-μL syringe (no. 3) and aspirate ~60 μL of the precipitant solution.
5. Disconnect the needle from syringe no. 3, keeping the Teflon ferrule inside the syringe.
6. Connect syringe no. 3 to the coupler attached to syringe no. 2. Carefully screw and tighten the coupler.
7. Inject ~5 μL of protein-laden LCP sample from syringe no. 2 into syringe no. 3 (extruded as a continuous extended string, fully immersed in the precipitant solution).
8. Disconnect syringe no. 3 from the coupler and attach a needle stopper to it. Make sure that LCP does not adhere to the coupler needle during the coupler withdrawal.
9. Use Parafilm strips to seal the needle stopper and the plunger syringe interface to prevent dehydration.
10. Repeat steps 3–8 to set up crystallization in four additional syringes (nos. 4–7).
11. Place syringe nos. 4–7 in a Ziploc bag, and add a moist fiber-free tissue (Kimwipes) to maintain a high level of humidity. Seal the Ziploc bag and store it in a 20 °C incubator.
12. Inspect the samples directly inside syringes every 12 h, using a stereo-zoom microscope equipped with cross-polarizers. Microcrystals typically appear within 1–3 day and can be detected as a faint uniform glow or as densely packed bright dots under cross-polarizers. Microcrystals grown in syringes can be stored for several days at 20 °C. Avoid large temperature fluctuations (over 2 °C) during sample storage and inspections. Samples in syringes can be transported at this stage to the XFEL source using a Greenbox thermal management system pre-equilibrated at 20 °C.

4.2.2 Sample Consolidation and Titration With 7.9 MAG

13. Take out all the samples from the 20 °C incubator ~1 h before the expected start of LCP-SFX data collection.
14. Carefully remove Parafilm seals from syringe no. 3.
15. Replace the needle stopper with a removable needle.
16. Slowly push the plunger of syringe no. 3, squeezing out the precipitant through the needle. Push the plunger of syringe no. 3 slowly and carefully. Abrupt movement can accidentally eject some LCP along with the precipitant solution, resulting in sample loss.
17. Stop pushing the plunger when most of the precipitant solution has been removed.
18. Replace the removable needle with a needle stopper.
19. Repeat steps 13–17 with syringe nos. 4–7.
20. Remove needle stoppers from syringes no. 3 and no. 4 and connect them together using a syringe coupler.
21. Transfer all of the sample from syringe no. 4 into syringe no. 3.
22. Repeat steps 19–20 with syringe nos. 5–7 to consolidate the entire sample in syringe no. 3. Squeeze out as much precipitant as possible.
23. Transfer ~5 μL of 7.9 MAG into a clean 100-μL syringe (no. 2), and connect syringes no. 2 and no. 3 by way of a coupler. The use of 7.9 MAG is only required if LCP is extruded in vacuum for LCP-SFX data collection. When performing LCP-SFX experiments at ambient pressure, 9.9 MAG can be used in this step.
24. Homogenize the sample by moving it through the coupler back-and-forth between syringes.
25. Repeat steps 22–23 until the sample becomes fully homogeneous and transparent. As the exact amount of residual precipitant solution is unknown, 7.9 MAG is titrated in 5-μL increments.
26. Move the entire optically clear sample into syringe no. 2 and disconnect syringe no. 3.

4.2.3 Microcrystal Characterization

27. Attach an LCP injector loading needle (1 in. long, gauge 22, point style 3) to syringe no. 2.
28. Extrude ~1 μL of the sample onto a glass slide, cover it with a glass coverslip and gently press on the coverslip to sandwich the sample.
29. Take images through a high-magnification microscope in bright-field illumination mode and under cross-polarizers. If possible, taking UV fluorescence and SONICC images can help better characterize the sample.

30. Estimate crystal size and density. The minimum crystal size is ~1 μm. The minimum crystal density that will work depends on the crystal size, the size of the beam, the diameter of the injector's nozzle and the diffraction strength of the crystal. If the crystal density is too high, perform steps 30–34. Otherwise, proceed to step 35.

4.2.4 *Adjusting Crystal Density*

31. Prepare the volume of LCP as needed for dilution using 50% (vol/vol) 7.9 MAG and 50% (vol/vol) precipitant solution with two clean 100-μL syringes (washed and dried syringes no. 3 and no. 4) and a coupler, as described in step 1.
32. Move the entire 7.9 MAG LCP sample into syringe no. 3. Disconnect syringe no. 4 keeping the coupler connected to syringe no. 3.
33. Connect syringe no. 2, containing the microcrystal-laden LCP, to the coupler attached to syringe no. 3.
34. Homogenize the contents of the two syringes by moving the sample back-and-forth between syringes ~100 times.
35. Repeat steps 26–29 to re-evaluate microcrystal size and density.

4.2.5 *Loading the Sample in an LCP Injector for LCP-SFX Data Collection*

36. Attach an LCP injector loading needle (1 in. long, gauge 22, point style 3) to syringe no. 2 that contains the final LCP sample homogeneously filled with microcrystals.
37. Transfer the sample into an LCP injector.
38. Insert the LCP injector loaded with sample into the sample chamber, start the injector, adjust the LCP flow rate based on the XFEL repetition pulse rate and the detector readout rate, and collect LCP-SFX data.

Issues Issues arise with occasional large crystals present in the mesophase sample. It will/can block/clog the delivery nozzle and can halt the experiment at great cost in failure to use valuable beam time and sample. Also, strong diffraction from big crystals can damage the detector leading to permanently dead pixels. It is partly for this reason that the beam is attenuated to 5–10% of full intensity, which represents underutilization/inefficient use of a very valuable resource, XFEL photons.

Dust and lint will also clog the nozzle and require special sample treatment to avoid. This includes filtering lipid, buffer and precipitant solutions, use of lint-free paper, and working in a relatively dust-free environment.

Not all proteins produce small crystals and optimization may be needed to generate small crystals for use in LCP-SFX. There is also the need to generate a suitably high density of crystals to increase the single crystal hit rate. Unfortunately, the higher density brings with it a rise in wasted protein between shots.

Crystal Detection and Mensuration It is one thing to grow small crystals. It is another altogether to detect and visualize them. Detection is integral to the success of the downstream SFX measurement and it must be done ideally in a quantitative manner. Thus, not only are crystal dimensions needed with some accuracy, it is also desirable to know the distribution of sizes and crystal density. The methods that have been used to date for these purposes include bright field and cross-polarized light microscopy, UV-fluorescence microscopy, and second order nonlinear imaging of chiral crystals (SONICC). Each comes with its pros and cons. Bright field microscopy is simple and accurate but the size limit extends only to that of light microscopy, which is in the vicinity of 1 μm. Polarizing light microscopy can help with visualization provided the crystal is birefringent. The stronger the birefringence, the smaller the crystal that can be seen. Even still, the method is limited to a resolution of about 1 μm. Unfortunately, optically isotropic crystals with cubic space group symmetry are not birefringent. UV-fluorescence is a very powerful method with a size limit possibly extending to a little below 1 μm. However, it requires that the protein contains amino acids that fluoresce upon excitation with 280 nm light. Tryptophan is the most fluorescent of the amino acids and the more tryptophans in the protein the higher the fluorescence yield and the more sensitive the measurement. For proteins that lack tryptophan, a high tyrosine content may compensate, to a limited degree. An alternative is to trace fluorescently label the protein, which involves separate chemical and purification steps and the introduction of heterogeneity into the sample. SONICC has been used successfully to visualize small membrane protein crystals growing in meso for use in SFX. An advantage of the method is that it has a spatial resolution that extends to <1 μm. Further, it has been adapted for high-throughput measurements with standard in meso crystallization plates. However, the equipment needed to make a SONICC measurement is expensive, several space groups do not give a SONICC signal (providing false negatives) and the method suffers from false positives in that non-proteinaceous crystalline materials such as detergents and lipids can give rise to a SONICC signal. Despite the many disadvantages, the information forthcoming from SONICC is considered sufficiently valuable that an instrument has been installed at the European XFEL. Other methods, such as light scattering and electron microscopy that are used extensively in characterizing crystals of soluble proteins for SFX measurements, thus far have not found application with mesophase grown crystals.

4.3 Results

A feasibility study of the LCP-SFX method was conducted using the coherent X-ray imaging (CXI) instrument at the Linac Coherent Light Source (LCLS) over the course of seven 12-h data collection shifts in March 2013. The CXI operating conditions included: photon energy, 9.5 keV; wavelength, 1.3 Å; fluence, 10^{12} photons/pulse; average pulse energy at the sample, 0.05 mJ; bunch delivery rate, 120 Hz; pulse length, 50 fs; X-ray focus, 1.5 μm; attenuation, 3–6% of full beam. The injector operated at an effective pressure up to 10,000 psi and a constant volumetric flow rate of 170 nL/min, corresponding to a linear flow rate of 1.4 mm/min. The extruded bolus diameter was ~50 μm as defined by the 50 μm internal diameter of the nozzle capillary. The X-ray beam intersected the mesophase bolus ~100 μm from the tip of the capillary extending from the injector nozzle. The evacuated sample chamber operated at 10^{-4} Torr and 20 °C. Diffraction data were collected on Cornell-SLAC Pixel Array Detector (CSPAD) detectors at sample-to-detector distance of 122 cm. Images and diffraction data were analyzed and processed following published procedures [29–31].

The data required to solve a structure of the test protein DgkA by molecular replacement were collected using approximately 4 h of beam time, 42 μL mesophase, and 200 μg protein. Data collection was greatly facilitated by the high hit rate provided by the LCP jet. The SFX structure, at ~2.18 Å resolution, is very similar to the corresponding structures determined using synchrotron radiation at 100 K. In addition, during the beam time for this feasibility study, structures were obtained for two liganded G protein-coupled receptors (GPCRs) [14].

The use of the LCP-SFX method with GPCRs is described in detail in Chap. 10. One example, 5-HT2B, is a member of the class A GPCR superfamily. The purified 5-HT2B/ergotamine complex was prepared following the protocol described above. Conditions with magnesium sulfate were optimized further to obtain relatively large crystals for traditional microcrystallography at a synchrotron source [32], whereas conditions with magnesium chloride, which reproducibly yielded high-density microcrystals, were used to prepare samples for LCP-SFX [21, 33].

Crystals optimized for traditional crystallography were collected directly from LCP using MiTeGen micromounts and flash-frozen in liquid nitrogen. Crystallographic data were collected at the 23ID-D beamline of the Advanced Photon Source (APS; Argonne, Illinois, USA) using a 10-μm minibeam at a wavelength of 1.0330 Å and a MarMosaic 300 charge-coupled device (CCD) detector. Data from the 17 best crystals collected under cryo-conditions were merged and used to solve the structure by molecular replacement at 2.7 Å resolution [32]. LCP-SFX data were collected at the CXI end station at the LCLS, using 50-fs X-ray pulses (3×10^{10} photons/pulse) at a repetition rate of 120 Hz and a wavelength of 1.3 Å, focused to a 1.5-μm spot size by Kirkpatrick–Baez mirrors. LCP with randomly dispersed 5-HT2B/ergotamine microcrystals was extruded through a 20–50 μm-diameter nozzle into a vacuum chamber at room temperature and at a constant flow rate of 50–200 nL/min, to intersect with the XFEL beam. Single-shot diffraction images were collected using a CSPAD located at a distance of 100 mm from the sample. The structure was determined by molecular replacement to 2.8 Å resolution [32].

Prospects Clearly, the LCP-SFX method for membrane proteins works and will continue to be used with a host of important membrane proteins and complexes, particularly where sufficiently large crystals cannot be grown for synchrotron radiation diffraction experiments. While the accessibility of LCP-SFX to the scientific community is currently limited by the small number of XFEL sources worldwide, SFX and XFEL have many desirable features, such as beam intensity, reduced radiation damage, and the ability to collect diffraction data from many crystals in a high-throughput manner without the need for mounting single crystals individually. It is expected that these advantages will make use of XFELs more attractive—for soluble as well as membrane proteins. Some thoughts along the lines of how future SFX studies might be improved are presented below.

A detector with wider dynamic range would certainly be of great benefit given that the current CSPAD requires the beam to be attenuated by a factor of ~20 to prevent damage from strong and sharp reflections. The latter derive, in part, from the occasional larger membrane protein crystals. Of equal import is detector damage from the solid Lc phase induced to form by evaporative cooling of the host LCP. If an evacuated sample chamber will be used for future LCP-SFX studies then a better understanding is needed of, with a view to controlling, the conditions that prevail in the extruded bolus under data collection conditions. These include knowing the temperature and composition along the length and across the diameter of the bolus and how this impacts phase behavior of the various components therein. The focus must be on mesophase behavior and how this is affected by changes in lipid hydration and temperature, with reference to the relevant temperature–composition phase diagram. The problem of converting to the solid Lc phase can be averted by using a lipid, 7.9 MAG for example, with a lower cubic-to-solid phase transition temperature (Tc). There are other MAGs with low Tc values that can be used for this purpose. An alternative approach, implemented successfully with GPCRs, involves doping mesophase prepared with a different host lipid or lipid mixture in which microcrystals had already grown and were dispersed with the low-Tc 7.9 MAG. Monoolein containing 10 mol% cholesterol was the lipid mixture used in the GPCR application, and 7.9 MAG was added to the extent of 30 mol% post-crystal growth. The preexisting crystals apparently do not suffer any deterioration in diffraction quality as a result of the doping and mixing exercise. However, for this to be a generally applicable procedure checks on crystal quality must be performed and, as needed, doping and mixing protocols where damage is avoided or minimized must be developed. Should the post-crystal growth doping approach work with other host lipids and lipid mixtures, it will expand the space available for crystallization screening enormously in that screening will not be tied to a specific host lipid.

As noted, a synthesis program in the Membrane Structural and Functional Biology Group provides rationally designed lipids for in meso crystallization and other applications in the membrane structural and functional biology field [16, 22, 34]. Separately, MAGs similar to those employed in the field to date and other lipid types have been designed and synthesized for use in low temperature crystallogenesis that may find application in future LCP-SFX studies. To do so, they must be shown to be effective hosting lipids for in meso crystallization and to form a crystal transport medium that is stable to evaporative cooling that takes place during SFX data collection in vacuo. Alternatively, they might be used to dope microcrystal-mesophase dispersions thereby lowering the effective Tc and preventing evaporative cooling-driven solidification, as already demonstrated with GPCRs [14, 32, 35].

The cubic mesophase, with its rheological hallmark of viscosity, is integral to LCP-SFX. However, not all in meso screening efforts generate structure-quality crystals in the cubic phase. As often as not, the much more fluid, yet bicontinuous sponge phase is the medium from which final crystals emerge [13, 36]. The sponge phase evolves from the cubic phase in the presence of certain precipitant components such as PEG 400, MPD, and butanediol, and appears more prone to form with the shorter chain host MAGs [14]. It is characterized by significantly enlarged aqueous channels, long-range disorder, optical clarity, non-birefringence, and fluidity. It is the latter property that makes it inefficient as an SFX medium [8]. However, the process can be reversed to induce a sponge-to-cubic phase transition. These methods are used, for example, to facilitate crystal harvesting which is generally easier in the viscous cubic phase. The conversion is relatively easy to do when the "spongifying agent" is an additive such as MPD where, typically, reducing the spongifier concentration in the sponge phase by dilution is sufficient to recover the cubic phase. Such an approach might be taken for LCP-SFX when final microcrystals only form in the sponge phase. Presumably, the conversion would be implemented immediately prior to running the SFX measurement and only in situations where it was shown that the process did not compromise crystal quality.

It is important to recognize that the limit to the size of crystals with which quality data can be collected using synchrotron radiation continues to drop as brighter and more intense X-ray beams are produced enabling the creation of smaller beams and as better detectors become available. It is now routine to collect synchrotron data on crystals just 10–50 μm in maximum dimension. Indeed, with new and improved synchrotron beams in the offing, crystals with single digit micrometer dimensions may well provide useful samples for routine data collection.

The in meso method is used primarily for crystallizing membrane proteins. However, it works also with soluble proteins. Lysozyme, thaumatin, and insulin are cases in point [37, 38]. It makes sense therefore to explore the utility of the LCP as a viscous, slow "flowing" medium in which to port microcrystals of soluble proteins and complexes into the XFEL for efficient, high hit-rate SFX. Crystals can be grown in situ and used, essentially, as with membrane proteins. The alternative is to combine extant crystals with preformed mesophase to create a dispersion that can be loaded directly into the reservoir of the LCP injector for SFX measurement. In this latter case, the mesophase would best be prepared with the mother liquor in which the soluble protein crystals grew. As with membrane proteins, MAGs having different acyl chain characteristics and correspondingly different mesophase microstructures and rheologies should prove useful for generating and porting crystals of the widest possible range of soluble protein targets.

Mixing crystals grown by the more traditional in surfo method may be the only possibility for certain membrane proteins. If these only yield microcrystals unsuitable for use with synchrotron radiation and are in short supply where the more wasteful liquid delivery system is not practical, then using LCP-SFX may be the option of last resort. In this case the extant microcrystals, dispersed in a liquid mother liquor, can be mixed with mesophase, ideally equilibrated with an appropriate mother liquor, and the crystal-laden mesophase used for SFX. The crystals would need to be stable to such a treatment, of course, and conditions may need to be adjusted to find those suited to producing a useful crystal dispersion.

Some proteins are difficult, if not impossible, to optimize to larger size and these would naturally lend themselves to LCP-SFX work. Rhodopsin–arrestin is a particularly compelling example of this. It may also be that with certain proteins or complexes growth is prohibitively slow. Using smaller crystals that can be grown in a reasonable time period would make sense especially if structures were needed for high-throughput screening purposes such as drug design, discovery and development where too slow a growing process could render the project impractical.

Note too that any protein that crystallizes progresses from an initial nucleus followed by a small and then a bigger crystal. It is possible to stop the growth process at a size optimally suited for SFX measurements. However, it may be difficult to stop growth without damaging extant crystals. A simpler solution might be to halt the growth process by using them directly at an intermediate stage in the growth process when crystals are suitably sized for SFX. However, if SFX is to be performed at RT the option of cyropreservation at an intermediate stage of growth is not available and crystals would need to be processed by SFX immediately as they achieve the right size. This would likely mean having to grow crystals on-site and using them for measurement as they matured—a luxury available to a select few.

In the interest of completeness it is appropriate to consider the alternatives to LCP-SFX, the most obvious of which is LCP synchrotron-based serial crystallography (LCP-SSX). LCP-SSX can be done in both fixed and moving target mode. The latter involves using the LCP injector to flow crystal-laden mesophase across a synchrotron X-ray beam. The feasibility of this approach has been demonstrated at several synchrotron facilities and has been shown to work with the model protein bacteriorhodopsin [39]. As with any flowing jet methodology, the process is wasteful requiring large amounts of protein, lipid and ligand if present. The fixed target approach has been implemented in several ways. The in meso in situ serial X-ray crystallography (IMISX) method where data are collected at cryogenic temperatures (IMISXcryo) is particularly effective and routinely used in the Caffrey lab. The method dispenses with the need for the technically demanding, inefficient, and potentially damaging crystal (loop) harvesting step that is an integral part of the traditional in meso method. For the IMISX method, crystals are grown in a bolus of mesophase sandwiched between thin plastic windows. The bolus contains tens to hundreds of crystals, visible with an in-line microscope at the MX synchrotron beamline and suitably disposed for conventional or serial crystallographic data collection. Wells containing the crystal-laden boluses are removed individually from hermetically sealed glass plates in which crystallization occurs, affixed to pins on goniometer bases and excess precipitant removed from around the mesophase. The wells are snap cooled in liquid nitrogen, stored and shipped in Dewars, and manually or robotically mounted on a goniometer in a cryo-stream for diffraction data collection at 100 K as is done routinely with standard, loop-harvested crystals. The IMISX-cryo approach has been used to generate high-resolution crystal structures of a G protein-coupled receptor, α-helical and β-barrel transporters, and an enzyme as model integral membrane proteins. Insulin and lysozyme were used as test soluble proteins. The quality of data that can be generated by this method has been attested to by performing sulfur and bromine SAD phasing with two of the test proteins. By comparison with the LCP-SFX, the IMISX method can be used with microcrystals, it uses nanogram-to-single digit microgram quantities of protein, there is little if any wasted protein, there are no limitations regarding the type of hosting lipid that can be used for crystallization, it can be used at most synchrotron MX mini-beamlines, it uses materials and instrumentation available for traditional in meso work including robots for setting plates and for mounting samples in the beam, it can be used at both ambient and room temperatures, and in the latter format represents true in situ crystallography in that measurements are made where and as crystals grow.

Crystal structures with ligands bound come at a premium and are much sought after for investigations of mechanism of action and structure-based drug design. Ligands are typically added to the protein before making the mesophase and setting up crystallization trials. They can also be doped into the mesophase to provide an environment for growth that is saturated with ligand. Or crystals of the apo-form can be soaked with ligand directly in meso. In the latter case, smaller crystals may provide an advantage where the surface area to volume (SA/V) ratio is higher for more effective, rapid and complete uptake. Thus, the LCP-SFX approach may prove particularly fruitful for ligand screening in a drug discovery campaign where high-throughput and parallelism is required.

Smaller crystals lend themselves to time-resolved SFX (TRSFX) where a sudden change in condition, for example, substrate or ligand concentration, can be used to trigger a reaction or a signalling event. The smaller the crystal the larger is the SA/V ratio and the faster is complete saturation of the crystal with ligand, enabling better time resolution.

Smaller crystals also provide advantages with light-activated processes where attenuation in the crystal is less of an issue. This enables a more rapid and uniform initiation of the process throughout the crystal for a more revealing view of the process under investigation. Where this is not possible, the interrogation of multiple states by the beam and their contribution to the measured diffraction signal complicate and may render impossible signal deconvolution and subsequent kinetic analysis.

MR is the phasing method most commonly used in the area of LCP-SFX. However, experimental or de novo phasing that includes S-SAD, is possible with some examples in hand. The requirement in terms of numbers of crystals and amount of protein/ligand for effective de novo phasing is however prohibitive. Only cases where an abundance of protein and ligand is available are likely to benefit from this in the short-term.

Smaller crystals call for smaller jets to reduce background scatter and absorption from the surrounding mesophase thereby maximizing the signal-to-noise (S/N) ratio for better quality data and statistics. However, a jet size that matches that of the crystal and that maximizes S/N may not be ideal for several reasons. Firstly, the mesophase may be more prone to clogging due to the occasional rogue large crystals and contaminating lint and dust which create costly plumbing problems and instrument downtime. Secondly, because of the larger SA/V ratio, evaporation and evaporative cooling and the attendant problems, as noted above, become more extreme. Thirdly, crystals may suffer from an increased level of mechanical damage as a result of direct contact with the wall of the nozzle capillary and the surface of the jet during delivery. For these reasons using a mesophase jet diameter somewhat larger than the maximum dimension of the dispersed crystals may prove optimal.

4.4 Conclusion

The development of the mesophase and its application as a growing medium for SFX-suitable microcrystals has many advantages. Harvesting crystals is avoided, radiation damage is not a major effect on data quality, and data collection is performed at a biologically relevant temperature. Many advances have been made to use LCP as a medium for SFX, including host lipid selection, development of successful scaling up procedures, and the design of an injector ideal for LCP-SFX, making this a versatile methodology with potential because it fulfills limitations previously involved in crystallizing membrane and soluble proteins.

Acknowledgements This review was supported in part by Science Foundation Ireland (12/IA/1255, 16/IA/4435; M.C.), the National Institutes of Health grants R21 DA042298 (W.L.), R01 GM124152 (W.L.), the National Science Foundation (STC award 1231306) (M.C., W.L.), and the Flinn Foundation Seed Grant (W.L.).

References

1. Caffrey, M., Li, D., Howe, N., & Shah, S. T. A. (2014). 'Hit and run' serial femtosecond crystallography of a membrane kinase in the lipid cubic phase. *Philosophical Transactions of the Royal Society B: Biological Sciences, 369*, 20130621. https://doi.org/10.1098/rstb.2013.0621.
2. Spence, J. C. H., Weierstall, U., & Chapman, H. N. (2012). X-ray lasers for structural and dynamic biology. *Reports on Progress in Physics, 75*, 102601. https://doi.org/10.1088/0034-4885/75/10/102601.
3. Barty, A., Kupper, J., & Chapman, H. N. (2013). Molecular imaging using X-ray free-electron lasers. *Annual Review of Physical Chemistry, 64*, 415–435. https://doi.org/10.1146/annurev-physchem-032511-143708.
4. Chapman, H. N., Fromme, P., Barty, A., White, T. A., Kirian, R. A., Aquila, A., et al.\ (2011). Femtosecond X-ray protein nanocrystallography. *Nature, 470*, 73–77.
5. Boutet, S., Lomb, L., Williams, G. J., Barends, T. R., Aquila, A., Doak, R. B., et al. (2012). High-resolution protein structure determination by serial femtosecond crystallography. *Science, 337*, 362–364.

6. Kern, J., Alonso-Mori, R., Hellmich, J., Tran, R., Hattne, J., Laksmono, H., et al. (2012). Room temperature femtosecond X-ray diffraction of photosystem II microcrystals. *Proceedings of the National Academy of Sciences of the United States of America, 109*, 9721–9726. https://doi.org/10.1073/pnas.1204598109.
7. Redecke, L., Nass, K., DePonte, D. P., White, T. A., Rehders, D., Barty, A., et al. (2013). Natively inhibited Trypanosoma brucei cathepsin B structure determined by using an X-ray laser. *Science, 339*, 227–230.
8. Johansson, L. C., Arnlund, D., White, T. A., Katona, G., Deponte, D. P., Weierstall, U., et al. (2012). Lipidic phase membrane protein serial femtosecond crystallography. *Nature Methods, 9*, 263–265. https://doi.org/10.1038/nmeth.1867.
9. Fraser, J. S., van den Bedem, H., Samelson, A. J., Lang, P. T., Holton, J. M., Echols, N., et al. (2011). Accessing protein conformational ensembles using room-temperature X-ray crystallography. *Proceedings of the National Academy of Sciences of the United States of America, 108*, 16247–16252. https://doi.org/10.1073/pnas.1111325108.
10. Burmeister, W. P. (2000). Structural changes in a cryo-cooled protein crystal owing to radiation damage. *Acta Crystallographica. Section D, Biological Crystallography, 56*, 328–341. https://doi.org/10.1107/s0907444999016261.
11. Briggs, J., Chung, H., & Caffrey, M. (1996). The temperature-composition phase diagram and mesophase structure characterization of the monoolein/water system. *Journal de Physique II, EDP Sciences, 6*, 723–751.
12. Qiu, H., & Caffrey, M. (2000). The phase diagram of the monoolein/water system: Metastability and equilibrium aspects. *Biomaterials, 21*, 223–234. https://doi.org/10.1016/s0142-9612(99)00126-x.
13. Caffrey, M., Li, D., & Dukkipati, A. (2012). Membrane protein structure determination using crystallography and lipidic mesophases–recent advances and successes. *Biochemistry, 51*, 6266–6288. https://doi.org/10.1021/bi300010w.
14. Liu, W., Wacker, D., Gati, C., Han, G. W., James, D., Wang, D., et al. (2013). Serial femtosecond crystallography of G protein-coupled receptors. *Science, 342*, 1521–1524. https://doi.org/10.1126/science.1244142.
15. Xu, F., Liu, W., Hanson, M. A., Stevens, R. C., & Cherezov, V. (2011). Development of an automated high throughput LCP-FRAP assay to guide membrane protein crystallization in lipid mesophases. *Crystal Growth & Design, 11*, 1193–1201.
16. Caffrey, M., Lyons, J. A., Smyth, T., & Hart, D. J. (2009). Monoacylglycerols: The workhorse lipids for crystallizing membrane proteins in mesophases. In L. DeLucas (Ed.), *Current topic in membranes* (pp. 83–108). Burlington, NJ: Academic Press.
17. Cherezov, V., Peddi, A., Muthusubramaniam, L., Zheng, Y. F., & Caffrey, M. (2004). A robotic system for crystallizing membrane and soluble proteins in lipidic mesophases. *Acta Crystallographica. Section D, Biological Crystallography, 60*, 1795–1807. https://doi.org/10.1107/S0907444904019109.
18. Li, D., Boland, C., Walsh, K., & Caffrey, M. (2012). Use of a robot for high-throughput crystallization of membrane proteins in lipidic mesophases. *Journal of Visualized Experiments, 67*, e4000. https://doi.org/10.3791/4000.
19. Caffrey, M. (2008). On the mechanism of membrane protein crystallization in lipidic mesophases. *Crystal Growth & Design, 8*, 4244–4254.
20. Rasmussen, S. G., DeVree, B. T., Zou, Y., Kruse, A. C., Chung, K. Y., Kobilka, T. S., et al. (2011). Crystal structure of the beta2 adrenergic receptor-Gs protein complex. *Nature, 477*, 549–555.
21. Kang, Y., Zhou, X. E., Gao, X., He, Y., Liu, W., Ishchenko, A., et al. (2015). Crystal structure of rhodopsin bound to arrestin by femtosecond X-ray laser. *Nature, 523*, 561–567. https://doi.org/10.1038/nature14656.
22. Misquitta, L. V., Misquitta, Y., Cherezov, V., Slattery, O., Mohan, J. M., Hart, D., et al. (2004). Membrane protein crystallization in lipidic mesophases with tailored bilayers. *Structure, 12*, 2113–2124. https://doi.org/10.1016/j.str.2004.09.020.

23. Misquitta, Y., Cherezov, V., Havas, F., Patterson, S., Mohan, J. M., Wells, A. J., et al. (2004). Rational design of lipid for membrane protein crystallization. *Journal of Structural Biology, 148*, 169–175. https://doi.org/10.1016/j.jsb.2004.06.008.
24. Wang, C., Jiang, Y., Ma, J., Wu, H., Wacker, D., Katritch, V., et al. (2013). Structural basis for molecular recognition at serotonin receptors. *Science, 340*, 610–614.
25. Li, D., Stansfeld, P. J., Sansom, M. S. P., Keogh, A., Vogeley, L., Howe, N., et al. (2015). Ternary structure reveals mechanism of a membrane diacylglycerol kinase. *Nature Communications, 6*, 10140 https://www.nature.com/articles/ncomms10140#supplementary-information.
26. Cheng, A., Hummel, B., Qiu, H., & Caffrey, M. (1998). A simple mechanical mixer for small viscous lipid-containing samples. *Chemistry and Physics of Lipids, 95*, 11–21. https://doi.org/10.1016/S0009-3084(98)00060-7.
27. Weierstall, U., James, D., Wang, C., White, T. A., Wang, D., Liu, W., et al. (2014). Lipidic cubic phase injector facilitates membrane protein serial femtosecond crystallography. *Nature Communications, 5*, 3309. https://doi.org/10.1038/ncomms4309.
28. Liu, W., Ishchenko, A., & Cherezov, V. (2014). Preparation of microcrystals in lipidic cubic phase for serial femtosecond crystallography. *Nature Protocols, 9*, 2123–2134. https://doi.org/10.1038/nprot.2014.141.
29. White, T. A., Kirian, R. A., Martin, A. V., Aquila, A., Nass, K., Barty, A., et al. (2012). CrystFEL: A software suite for snapshot serial crystallography. *Journal of Applied Crystallography, 45*, 335–341.
30. Kirian, R. A., White, T. A., Holton, J. M., Chapman, H. N., Fromme, P., Barty, A., et al. (2011). Structure-factor analysis of femtosecond microdiffraction patterns from protein nanocrystals. *Acta Crystallographica. Section A, Foundations of Crystallography, 67*, 131–140. https://doi.org/10.1107/S0108767310050981.
31. Kirian, R. A., Wang, X., Weierstall, U., Schmidt, K. E., Spence, J. C., Hunter, M., et al. (2010). Femtosecond protein nanocrystallography-data analysis methods. *Optics Express, 18*, 5713–5723. https://doi.org/10.1364/OE.18.005713.
32. Wacker, D., Wang, C., Katritch, V., Han, G. W., Huang, X.-P., Vardy, E., et al. (2013). Structural features for functional selectivity at serotonin receptors. *Science, 340*, 615–619. https://doi.org/10.1126/science.1232808.
33. Liu, W., Wacker, D., Gati, C., Han, G., James, D., Wang, D., et al. (2013). Serial femtosecond crystallography of G protein-coupled receptors in lipidic cubic phase. *Science, 342*, 1521–1524. https://doi.org/10.1126/science.1244142.
34. Li, D., & Caffrey, M. (2011). Lipid cubic phase as a membrane mimetic for integral membrane protein enzymes. *Proceedings of the National Academy of Sciences of the United States of America, 108*, 8639–8644. https://doi.org/10.1073/pnas.1101815108.
35. Xiang, J., Chun, E., Liu, C., Jing, L., Al-Sahouri, Z., Zhu, L., et al. (2016). Successful strategies to determine high-resolution structures of GPCRs. *Trends in Pharmacological Sciences, 37*(12), 1055–1069. https://doi.org/10.1016/j.tips.2016.09.009.
36. Cherezov, V., Clogston, J., Papiz, M. Z., & Caffrey, M. (2006). Room to move: Crystallizing membrane proteins in swollen lipidic mesophases. *Journal of Molecular Biology, 357*, 1605–1618. https://doi.org/10.1016/j.jmb.2006.01.049.
37. Aherne, M., Lyons, J. A., & Caffrey, M. (2012). A fast, simple and robust protocol for growing crystals in the lipidic cubic phase. *Journal of Applied Crystallography, 45*, 1330–1333. https://doi.org/10.1107/S0021889812037880.
38. Caffrey, M. (2000). A lipid's eye view of membrane protein crystallization in mesophases. *Current Opinion in Structural Biology, 10*, 486–497. https://doi.org/10.1016/S0959-440X(00)00119-6.
39. Nogly, P., James, D., Wang, D., White, T. A., Zatsepin, N., Shilova, A., et al. (2015). Lipidic cubic phase serial millisecond crystallography using synchrotron radiation. *IUCrJ, 2*, 168–176.

Chapter 5
Sample Delivery Techniques for Serial Crystallography

Raymond G. Sierra, Uwe Weierstall, Dominik Oberthuer, Michihiro Sugahara, Eriko Nango, So Iwata, and Alke Meents

5.1 Overview

In serial femtosecond crystallography (SFX), protein microcrystals and nanocrystals are introduced into the focus of an X-ray free electron laser (FEL) beam, ideally, one-by-one in a serial fashion. The high photon density in each pulse is the double-edged sword that necessitates the serial nature of the experiments. The high photon count focused spatially and temporally leads to a *diffraction-before-destruction* snapshot, but this single snapshot is not enough for a high-resolution three-dimensional structural reconstruction. To recover the structure, more snapshots are required to sample all of reciprocal space from randomly oriented crystal diffraction, and in practice, some redundancy is necessary in these measurements. Please see Chap. 8 for more details.

R. G. Sierra (✉)
Hard X-Ray Department, LCLS, SLAC National Accelerator Laboratory, Menlo Park, CA, USA
e-mail: rsierra@slac.stanford.edu

U. Weierstall (✉)
Department of Physics, Arizona State University, Tempe, AZ, USA
e-mail: weier@asu.edu

D. Oberthuer · A. Meents
Center for Free-Electron Laser Science, German Electron Synchrotron DESY, Hamburg, Germany
e-mail: dominik.oberthuer@desy.de; Alke.Meents@desy.de

M. Sugahara · E. Nango · S. Iwata
RIKEN SPring-8 Center, Sayo-gun, Hyogo, Japan

Department of Cell Biology, Graduate School of Medicine, Kyoto University, Sakyo-ku, Kyoto, Japan
e-mail: msuga@spring8.or.jp; nango@spring8.or.jp; s.iwata@spring8.or.jp

S. Boutet et al. (eds.), *X-ray Free Electron Lasers*,
https://doi.org/10.1007/978-3-030-00551-1_5

Fixed targets were initially used to verify the *diffract-before-destruction* idea central to biological imaging with X-ray FELs and specifically serial femtosecond crystallography (SFX) experiments [1]. In most traditional optical imaging experiments, the sample is mounted onto a glass slide, goniometer pin, electron microscopy grid, or some other substrate; therefore, fixed targets seem a logical first step for X-ray FEL sample delivery. However, the repetition rate of the FEL pulses combined with their destructive power made the efficient use of fixed targets difficult. At the time of the first X-ray FEL experiments, fixed targets could neither be replenished at the X-ray repetition rate nor could they be properly protected from the adverse effects of the vacuum environment without drastically increasing background scattering. This drove the desire and need to use flowing liquid sample delivery methods. Later in the chapter, the idea of fixed targets for in-vacuum and ambient experiments is revisited.

Biological samples, which are by nature sensitive to damage by X-rays, should be replenished, at the minimum, with the X-ray FEL repetition rate, which has, until recently, ranged up to 120 Hz at X-ray facilities that have operated for a few years. If synchronized with a 120 Hz source, every sample, most commonly a crystal of a particular biological molecule, that is hit by a focused X-ray pulse is destroyed after the pulse has passed through; within 8.3 ms the debris must be cleared out and a new crystal is ideally supplied for the next pulse. With the advent of superconducting accelerators, an average repetition rate of 27 kHz (4.5 MHz peak) and up to 1 MHz will be available with the opening of the European XFEL and the LCLS-II upgrade, respectively. To make use of the peak repetition rates at these facilities, samples must be replaced every 220 ns or 1 μs, respectively. Figures 5.1 and 5.2 show the effects of the intense incident X-ray pulses on a solid substrate and a liquid water jet in vacuum.

The first structural biology experiments at an X-ray FEL were performed at soft X-ray energies [3, 4]. At these lower energies (<2 keV), X-rays interact more strongly with electrons than at the higher energies typical for X-ray crystallography (>6 keV). Due to the low photon energy, the first experiments were performed

Fig. 5.1 SEM images of a pattern etched with a focused ion beam into a 20 nm silicon nitride membrane, before (*left*) and after (*right*) of an incident FEL pulse. The damage is evident, yet the authors were able to reconstruct the original structure, thus verifying the *diffraction-before-destruction* concept. The right image demonstrates the destructive power of the incident X-rays which affect the interaction region (central 20 μm) as well as the adjacent area [1]. Reproduced with permission from Chapman et al. [1]

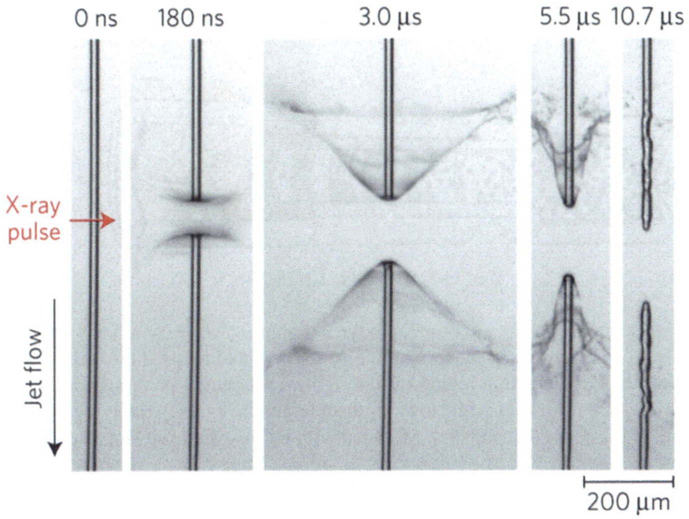

Fig. 5.2 A Rayleigh jet of water (20-µm-diameter) injected into vacuum, imaged stroboscopically, to view the effects of the incident X-ray FEL pulse on the stream of liquid. Varying time delays after the incident X-ray pulses (0.75 ± 0.08 mJ, 8.2 keV, 120 Hz) are shown. A gap forms after the X-ray pulse vaporizes the liquid explosively. Liquid from the jet forms thin conical films of water, which later collapse onto the jet, while the liquid continues to flow (downwards in the image). Reproduced with permission from Stan et al. [2]

in vacuum to prevent strong interactions between ambient gas molecules and soft X-rays. Additionally, the samples had to consist of nanometer sized crystals or single particles to prevent excessive X-rays absorption by the sample. The hydration layer around the sample had to be small, to minimize absorption from surrounding water molecules, and to minimize background scatter in the case of single particles.

Flowing liquid delivery systems were thought to be able to both sufficiently replenish the samples and keep them protected in vacuum. For single particles (e.g., viruses, whole cells), Seibert et al. [4] used an aerodynamic lens stack, similar to that shown in Fig. 5.3, in which an aerosolized sample passes through a series of chambers at decreasing pressure, separated by apertures. The pressure gradient from atmosphere to vacuum through each section acts as a focusing lens. As the aerosolized particles accelerate with the gas passing through the focusing apertures, the particles' inertia keeps them closer to the centerline while the carrier gas quickly expands, creating a focused particle beam, while simultaneously evaporating a volatile solvent. Ultimately, the particles arrive at the X-ray interaction region with minimized hydration shell, ensuring low background scattering critical to imaging single particles while still protecting the particle from the detrimental effects of vacuum. Transmission inefficiencies and low particle densities at the interaction region make the aerodynamic lens less than ideal [for SFX experiments], and aerosol injection for single particle imaging experiments is still an area of ongoing research [6–8].

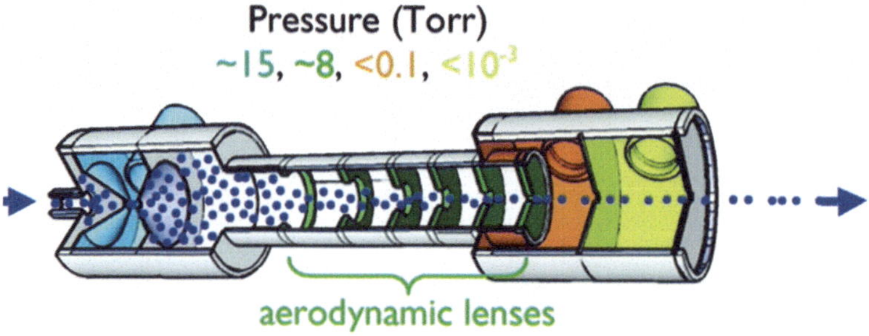

Fig. 5.3 Cross-sectional schematic of an aerodynamic lens stack. Aerosolized particles are generated (*left*) from an electrospray, for example (*not shown*), and focused through an aerodynamic lens stack towards the interaction region (to the right). Differential pumping regions are serially decreasing in pressure from left to right towards the interaction region, [with typical pressure values corresponding to the colored chambers]. Modified with permission from Bogan et al. [5]

For crystallographic applications, an increased water layer around the sample is more tolerable due to the increased scattering intensity at the Bragg peak locations. Therefore, liquid sample delivery directly into the beam is more commonly used. The most familiar liquid jet is a Rayleigh jet. Seen in faucets, garden hoses and fountains, a Rayleigh jet is formed when there is sufficient flow rate to overcome the resistive properties intrinsic in a fluid, such as its viscosity and/or surface tension. Rayleigh was able to describe the eventual breakup of the jet into droplets from small surface perturbations [9, 10]. Faubel et al. [11] and later, for the purpose of SFX, Weierstall et al. [12] demonstrated that liquid streams could be created and delivered into vacuum. Rayleigh jets use large quantities of sample, on the order of 100–1000s of μL/min. Small jet diameters with reduced flow rate and improved stability compared to Rayleigh jets were made possible by the gas dynamic virtual nozzle (GDVN) [13]. In the next section, the fluid mechanics that governs liquid injection devices is introduced and discussed. These concepts set up the boundary conditions of sample delivery for SFX and allow the reader to appreciate the nuances of the subsequent, seemingly similar, sample delivery methods.

SFX takes advantage of some of the unique capabilities of X-ray FELs by keeping protein crystals in a solvated state and close to room temperature. The crystal concentration should ideally be adjusted so that, on average, only one crystal is in the interaction region during a given X-ray pulse. For a 1–10 μm X-ray focus, typical concentrations are 10^7–10^{10} particles per milliliter and are optimized during the experiment.

The requirements for SFX sample delivery are therefore:

- Replenish the protein crystals at the X-ray interaction region as efficiently as possible
- The delivery method should be compatible with the experimental environment (e.g., vacuum, helium, air)

- No sample damage due to the injection process (compatible carrier media/support and sample, no undue shear forces, charging, etc.)
- Minimal background scattering from the carrier media or support
- Reliable operation for hours

Sample delivery for SFX experiments can be nuanced and complicated for crystallographers looking to perform an experiment for the first—or *n*-th time. The following section will serve as a primer of the underlying fluid mechanics that govern sample delivery. The problem of delivering a slurry of crystals to the incident X-ray beam in the most efficient way is not trivial and is at the crux of a successful experiment. It can be the difference between hours of frustrating, fruitless data collection and a successful experiment. By the end of this chapter, the reader should have an appreciation for the status quo of the numerous sample delivery techniques available to interface with the myriad crystallization conditions possible, while being able to identify the pros and cons of each technique and how a specific sample delivery approach might mate with their particular crystal system.

In this chapter, several different methods are presented which have been used to deliver microcrystals to an X-ray FEL beam. In Sect. 5.2 the gas dynamic virtual nozzle (GDVN) is introduced, which has been the workhorse for many SFX experiments since the start of user operations at the first hard X-ray FEL. These are, historically, handmade nozzles which help minimize clogging by the crystal solution or foreign objects, since they produce a micron sized jet (<10 μm) from a larger aperture; passing the same crystal slurry through a micron sized orifice is quite impractical due to clogging issues. To make them more reproducibly, high-resolution 3D printing is currently being employed and first results are shown. GDVNs have a relatively high flow rate and jet speed, which leads to considerable sample waste when used at low repetition rate X-ray FELs such as the Linac Coherent Light Source (LCLS) and the SPring8 Angstrom Compact free electron LAser (SACLA). High viscosity injectors, such as the lipidic cubic phase (LCP) injector, which were initially developed due to their compatibility with membrane protein crystal growth, and later were adapted for multiple sample types to reduced sample waste, are introduced in Sect. 5.3. These allow a reduction of the sample flow rate by a factor of 100 as long as the media has a high viscosity similar to LCP, a high viscosity growth medium for many membrane proteins. In Sect. 5.4 the microfluidic electrokinetic sample holder (MESH) and its variant, the concentric MESH (coMESH), is introduced, which borrows methods from electrospray and electrospinning to move the sample in a charged stream through the X-ray focus. Sample flow rates with this technique are lower than with the GDVN and can be as low as with the high viscosity injectors, therefore filling the gap between these two techniques in terms of sample consumption.

Section 5.4 introduces two variants of the GDVN, double flow focusing nozzles and mixing nozzles. Mixing nozzles are currently of great interest when studying enzyme reactions in a time-resolved fashion. Double flow focusing nozzles are more reliable than regular GDVNs since clogging events are largely avoided due to the use of an outer focusing liquid and they also allow a reduction of the sample consumption, since the sample flow can be pinched off by the outer liquid flow. Section 5.5 introduces other viscous carrier media that can be used for SFX with the high viscosity injector. In addition, it presents a drop-on-demand system synchronized with the X-ray FEL pulses for SFX, which has been developed at SACLA. The in-helium atmosphere at SACLA has different considerations for sample delivery as compared to the in-vacuum techniques initially developed to interface with LCLS. An example of a fixed target sample holder is presented in Sect. 5.7. With fixed target sample holders, thousands of microcrystals are mounted on a solid support and scanned through the X-ray beam. One microcrystal is exposed to an X-ray pulse, then the support is moved to the next crystal. The sample support has to be moved from crystal to crystal at the repetition rate of the X-ray laser. Fixed target sample delivery results in very high hit fractions since ideally every X-ray pulse hits a crystal. The chapter closes with an outlook on what is next in the field of sample delivery as mixing experiments and high repetition rate sources start to become the new norm for SFX.

5.1.1 Fluid Mechanics for Crystallographers

Any crystallographer knows the delicate balance needed between the protein and its surrounding *mother liquor*—the fluid composition that coaxes the protein out of solution and into a crystalline lattice. Finding the right conditions for crystals to grow can take years, and even optimizing them for cryo-conditions or improved resolution can take a whole career. By the end of this chapter you will appreciate that it is ultimately the protein crystals and their mother liquor that govern the performance, and thus success, of an SFX experiment using liquid sample delivery. The fluid properties of the mother liquor alone can vary from low surface tension detergents to high viscosity polymers and lipids or simple salt and water brines, and these differences can be felt as early as during pipetting in the lab. The addition of suspended protein crystals increases the apparent viscosity and introduces complexities to the experiment, such as potentially disrupting any meniscus exposed from a capillary opening, causing clogs at junctions and throughout liquid lines when transporting the sample through liquid lines, and sedimentation in reservoirs. If performing the experiment in vacuum, the evaporation of the exposed meniscus modifies the local chemistry and can cause precipitation of salts and dehydration of the sample and severely impede the experiment; the chemicals that cause your proteins to precipitate into crystals can also cause salts to precipitate once exposed to vacuum, for example. The following section will give a brief overview of some fundamental fluid mechanics that govern the different sample delivery methods and setup the physical constraints which might answer the crystallographers first question when [planning for sample delivery] for an SFX experiment: "Why do we have to do that?" A successful SFX crystallographer will be familiar with these concepts and work closely with the sample injection team to get the crystals through one of the major SFX bottlenecks.

To understand the constraints of sample injection, and thus the need for different sample delivery techniques, pressure-driven fluid flow in a pipe should be understood; more simply the Hagen–Poiseuille [14] flow equation:

$$Q = \frac{\pi \Delta \mathrm{Pr}^4}{8\mu L} = \frac{\Delta P}{R} \tag{5.1}$$

where the flow rate Q, of a fluid with dynamic viscosity μ, is driven by a pressure gradient ΔP, over a length of tubing L, with radius r. The dynamic viscosity (or simply viscosity) here, in SI units of Pa·s, is indicative of the fluid's resistance to shear stresses. This is the parameter we intuit when we say that honey is *thicker* than water; that is, honey has a higher dynamic viscosity than water. The dynamic viscosity is different from the kinematic viscosity, $\nu = \mu/\rho$, with SI units of m^2/s. The kinematic viscosity represents the ratio of the fluid's viscous force to the inertial force and indicates how fast momentum is diffused throughout the fluid. For example, the viscosity of air and water at standard atmospheric conditions are approximately 0.02 and 1 mPa·s respectively, which agrees with our expectation that water is *thicker* than air. The kinematic viscosities, however, are 1×10^{-5} m^2/s and 1×10^{-6} m^2/s, respectively, implying that momentum diffuses through air faster than in water. The kinematic viscosity is important when discussing velocity profiles within the fluid, as well as mass diffusion. An ion of hydrogen has a diffusivity of 10^{-10}–10^{-9} m^2/s in water, meaning that momentum information of the bulk water diffuses 3–4 orders of magnitude faster than the mass diffusion of hydrogen ions; for example, the parabolic velocity profile of flowing water in a pipe is established sooner than a uniform pH.

A fluid deforms continuously as a shear stress is applied to it, whereas a solid object resists this applied force. Both liquids and gases are fluids and can exhibit similar behaviors. Most of the basic fluid mechanics concepts discussed here will assume Newtonian fluids. A Newtonian fluid, such as water and oil, is one whose viscosity is a property of the fluid's state and is not affected by the applied shear rate. In contrast, toothpaste, ketchup, blood, polymer solutions, or colloidal suspensions like mayonnaise and paint, are non-Newtonian fluids. Here an increased shear rate can cause the fluid to move easier or more difficultly depending on the specific flow properties of the fluid. It is likely that certain protein crystallization conditions might lead to fluids which exhibit these complex behaviors, but the basic principles discussed here will focus on a Newtonian fluid assumption. In later sections some injection methods will be discussed where non-Newtonian viscous media, such as lipidic cubic phase of monoolein (LCP), agarose, or high molecular polymer solutions, are used as carrier media for sample delivery. See White [14] or other fluid mechanics texts for more details on shear thinning, shear thickening, Bingham plastic, or thixotropic fluids.

The fluid properties of a crystal's mother liquor might dictate the viscosity and thus cannot be readily changed. Under many circumstances, the size of the protein crystals itself begins to limit the geometries suitable for the SFX experiment. In most fluid mechanics calculations applicable to SFX, the fluid is treated as a continuum and species inside of the fluid are treated as dilute, solvated species, not causing changes in the bulk properties of the fluid. This assumption cannot be maintained when dealing with a suspension of solid protein crystals on the order of 100s of nanometers to dozens of micrometers in dimension. The crystals themselves have solvent channels that can interact with the fluid but this is beyond the scope of this discussion. The idea should be clear, though, that a crystal slurry and the same fluid without crystals will invariably behave differently, whether it be different effective viscosities, or non-Newtonian behaviors, much like the way particles suspended in water prevent ketchup from moving until a sufficiently high shear rate is applied.

Although not always physically accurate, a useful heuristic in understanding basic fluid flows is to use an electric circuitry analogy. The terms in the pipe flow equation above can be grouped to define a hydrodynamic resistance R. Much like an element in an electronic circuit has some resistance to the flow of current from an applied voltage (Ohm's law), a capillary of fluid will restrict volumetric flow with the application of a volumetric potential (i.e., the pressure (ΔP)). Ignoring the effects of evaporative cooling possible at a vacuum orifice, if water is placed inside a 50 μm inner diameter capillary tube, 1 m in length, with atmospheric pressure applied to one end and vacuum applied to the other, the water will flow at approximately 1 μL/min. This fluid flow is so low that the ensuing fluid meniscus on the vacuum side will freeze due to evaporative cooling in vacuum. Changing the fluid's properties or driving the flow much faster (creating a Rayleigh jet), would be the only way to prevent that meniscus from freezing. However, the high flow rate of a Rayleigh jet (>100 μL/min) can be prohibitive for most protein crystal slurries. The Hagen–Poiseuille equation indicates that the only ways to reduce the sample consumption through the capillary tube are to decrease the pressure difference, increase the length of the tube, decrease the tube diameter, or to increase the viscosity. We will briefly discuss the implications of a change in these parameters.

Pressure Gradient Although serial crystallography experiments do not necessarily require in-vacuum injection, many have been and continue to be done in vacuum. In the case of vacuum injection, a way to reduce the pressure gradient below one atmosphere is to apply vacuum upstream of the sample reservoir. This can lead to outgassing of the solution in the capillary, which can lead to cavitation and bubble formation and can severely limit or interrupt the jetting ability, and thus the data collection. This might also affect the stability of the suspended protein crystals as the solution chemistry is potentially affected through dehydration. At the vacuum end, differential pumping schemes can be employed to keep the exposed meniscus nearer to atmospheric pressures, thus diminishing the pressure gradient. In non-vacuum experiments, the pressure gradient is arbitrary and bounded by the injection method rather than the vacuum conditions.

Tubing Inner Diameter One of the most powerful relationships for fluid flow in a pipe is the fourth-power dependence on the inner diameter of the tube. Going from a 50-μm diameter tube to a 100-μm tube can give a 16-fold increase in flow rate or a 16-fold decrease in pressure to run the same flow rate. In practice, to maintain the same liquid jet, the pressure will stay and the flow rate will have to compensate. A gross simplification of the process of ejecting liquid streams from a liquid's surface would be to imagine that sufficient kinetic energy must be imparted onto the fluid by some sort of potential energy well, be it pneumatic, hydraulic, electrostatic, or other. Once the applied potential energy is enough to fight resistive forces in the bulk fluid and the exposed meniscus, mainly the viscosity and surface tension, then the remaining energy is used to flow the jet. In order to further focus the meniscus of the fluid into a small stream, one must add some excess kinetic energy to the system [to accelerate the flow], and herein lies one of the fundamental challenges to sample delivery. The atmosphere pressure difference, in our hypothetical system above, dictates a minimum flow rate of 1 μL/min, as most SFX experiments have historically occurred within a vacuum. Adding energy might increase this flow rate, as seen with either the GDVN, where flow rates on the order of tens of microliters per minute are necessary to make a thin stable jet, or the thicker Rayleigh jet with hundreds of microliters per minute. See Eggers and Villermaux's review on the physics of liquid jets for more details [15].

As appealing as it seems to then reduce the capillary diameter, modifications to the tubing diameter quickly come at odds with the necessary driving pressure. For example, to run at 10 μL/min (similar to flows of a nominal GDVN) through a 1 m-long capillary with an inner diameter of 25 μm, a driving pressure of over 2500 psi (~170 bar) is needed, as opposed to 160 psi (~11 bar) needed for the same flow rate through a 50 μm inner diameter capillary with the same length. These values assume the capillary meniscus is in air and does not take into account whether the flow creates a liquid jet or not. The pressure values would likely change if modified to create a jet, however, the relative pressure differential between the two capillary diameters will remain. Such a large operating pressure approaches safety limits of common fluid handling and microfluidic equipment and might cause failures, and although not prohibitive, this quickly becomes non-ideal.

The smaller the tubing diameter gets, the easier it can clog with the protein slurries as well as any foreign particulates. A general heuristic is that the largest crystal dimension should not exceed 2/5 of the inner diameter of the tubing. The more concentrated the crystal slurry gets, the higher the likelihood that even small nanocrystals can clog a 50 μm opening. A macroscopic analogy would be a crowd of people exiting a doorway. Although one person fits easily through the doorway, moving many people through quickly can cause "clogging." Unfortunately, capillary diameters are not a readily tunable parameter for sample delivery, even though a narrow range of sizes are typically employed: 30, 40, 50, 75, and 100 μm inner diameters have been used with varying success depending on the size of the suspended crystals. Increasing the diameter to accommodate for larger crystals or to prevent clogging will quickly lead to exponentially higher sample consumption rates.

Tube Length The experimental geometry can quickly become complicated with vacuum chambers as large as 1 m wide to support the necessary equipment to execute the experiment. This results in capillary lengths on the order of 1 m in order to have the sample injector close to the interaction region while reaching relevant sample injection equipment outside of the vacuum chamber. Making the capillaries longer is not as problematic; however, increases in length cause increases in the necessary driving pressure. Keep in mind that minor modifications in tubing diameters can give similar results to increases in tubing length; for example, going from 100 to 50 μm inner diameter has the same effect on the necessary pressure as increasing the larger tube's length from 1 to 16 m.

Viscosity The mother liquor's viscosity can be one of the most important differentiators between applicable SFX jet techniques. Of the four parameters outlined to diminish the flow rate, it is the only one intrinsic to the system, while the others are externally applied parameters. The viscosity of a protein crystal slurry is not readily modified due to the effects that a change in electrochemistry might have on the quality of the crystals. Modifications of the viscosity—if possible—can lead to reductions in sample consumption. The viscosity of a solution can be quickly increased by simply adding long polymer chains of large molecular weight and glycerol [16, 17].

As previously outlined, the viscosity of the fluid is one of the key parameters that capture how a fluid will resist the transfer of momentum required to make a liquid jet. As the viscosity of a fluid is increased by either dissolving additives or suspending high-density particulates, there is a need to supply more potential energy (e.g., pressure) to overcome the added resistance and drive the flow. Einstein theorized [18] and Hiemenz [14] and references therein later verified experimentally, that the viscosity of a *dilute* solution will increase by $1 + 2.5\phi$, where φ is the volume fraction of the suspended particles (no greater than 0.02). These are empirical approximations and have since been further modeled. Probstein [19] notes that volume fractions as high as 0.1 have been used in the past. For reference, Fig. 5.4b) shows typical volume fractions of protein crystals, which are generally around 0.2 but can be higher [20]. Assuming the largest volume fraction in the dilute assumption ($\varphi = 0.1$) the viscosity increases 25% thus affecting the flow conditions accordingly.

As an example, running a typical 50 μm inner diameter GDVN with water will give typically a minimum flow rate of 3–4 μL/min with a driving pressure of approximately 400 psi. Running a suspension of lysozyme crystals (similar to those in Fig. 5.4 with >10% volume fraction), with mother liquor of water, salt, and acetic acid through the same nozzle, causes the minimum flow rates to increase by dozens of microliters per minute and causes the driving pressure to increase as well. This increase is much more than the 25% increase we previously approximated from the dilute suspension assumption (<10% volume fraction). The slurry's properties create different conditions at the meniscus and in the bulk fluid, resulting in higher driving pressures to overcome the resistance from the small solid crystal chunks, which periodically pass through the capillary and meniscus. The resulting liquid jet length is now significantly shorter before breaking up. The driving pressure can increase by no more than fivefold, as many of the fittings and tubings might fail past 2000 psi. This implies that a standard 50-μm inner diameter GDVN is well suited for injection of liquids with viscosity similar to water, but might struggle as soon as viscous additives and high concentrations of crystals are added.

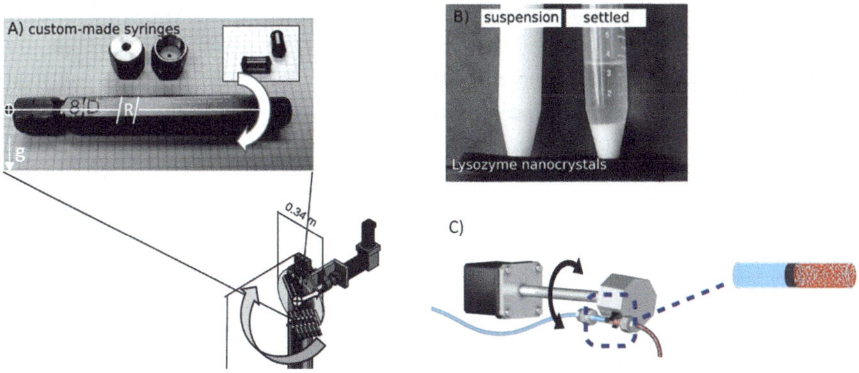

Fig. 5.4 Custom-made syringes (**a**) mounted on an anti-settling device near the sample chamber. The gravity vector, g, points down in all images; the syringe oscillates about a fixed radius of rotation, R, about the central black and white cross 0.17 m away (not to scale). The centerline of the syringe is parallel to the radial vector (perpendicular to the axis of rotation). Reproduced with permission of the International Union of Crystallography [20]. Lysozyme crystals in high concentration (**b**). The crystals on the left were resuspended just before the picture was taken, while those on the right were allowed to settle (minutes to hours); Reproduced with permission from Lomb et al. [20]. An alternative anti-settling device (**c**), where the centerline of the syringe is perpendicular to the radial vector (parallel to the axis of rotation). From the anti-settling device, the sample is brought into the interaction region, in vacuum, by means of a liquid jet; Reproduced with permission from Sierra et al. [21]

High pressure steel fittings (withstanding up to 10,000 psi) are common commercially; however, throughout many of the injection schemes we will discuss, fused silica capillaries are common, which interface with polymer fittings and polymer tubing sleeves, which might not withstand the high pressures. Also, common pressurized gas cylinders only go up to ~2000 psi meaning that HPLC pumps running at constant flow are the only option to drive the flow. The issue of high pressure is not insurmountable; it just requires additional design restrictions that might not be tolerable in some cases. The section on high viscosity extrusions shows how a pressure amplifier can achieve these high pressures to extrude viscous media through small fused silica capillaries. Although Johansson et al. [22] showed that more viscous media, such as lipidic sponge phase, can pass through a GDVN, this has proven quite difficult to reproduce. Therefore the GDVN can be used with lower water-like viscosities, and the High Viscosity Extrusion (HVE) method can handle higher viscosities, leaving a wide range of viscosities in between. (The MESH and double flow focusing nozzle (DFFN) sections will address this.)

An added complication with gas focused jets and viscosity is that the boundary condition necessary to focus the liquid meniscus with the gas sheath must have matching shear rates, a condition dependent on each fluid's viscosity. The more viscous the liquid meniscus, the higher shear rate the gas must apply, again running into an upper limit of applicable gas pressure values. A normal GDVN gas sheath mass flow rate is 20 mg/min which requires a gas pressure of about 400 psi when using a 2 m long gas tubing with 100 μm inner diameter. A 5–6 times increase is tolerable (assuming a full gas cylinder of ~2500 psig), with a higher probability of some connection failure happening before that point. The increased gas load on the vacuum chamber from the helium sheath gas might present issues for the turbo pumps or sensitive electronic equipment. This helps explain why more viscous substances will have difficulty injecting through a GDVN nozzle and why the helium sheath gas in a HVE does not focus the media and simply helps straighten the extruded media and prevent the sample from curling upon itself and the nozzle. Note that recently, a concentric flow GDVN, the dual flow focusing nozzle (DFFN) injector, was developed as a way to handle different media through a gas focused injector by focusing an intermediate fluid, such as ethanol, which in turn focuses the central fluid containing the sample [23] (as well as Sect. 5.5). A useful metric is to calculate the hydrodynamic resistance of the system, *R*, as defined in Eq. (5.1) in order to determine the liquid injection method best suited for a given sample. Estimate your system's resistance and compare it to a published injection methods estimated resistance and see if your system might be compatible.

One important restriction made on the geometry by the crystals is a minimum inner diameter to accommodate the protein crystal dimensions. The 2/5 approximation seems to hold for crystals with morphologies of low aspect ratios, such as cubes, blocks, diamond, etc. as evidenced by a 20 μm filter being used on 50 μm inner diameter GDVN geometries. High aspect ratio crystals might not obey this approximation, since the long dimension might flow align and not cause the same concern. However, this should be tested prior to the experiment to ensure there are no issues of clogging. Of course, smaller, uniform crystal sizes are better for sample injection in all sample delivery methods and lead to injection more similar to flow conditions without suspended crystals. Unfortunately, crystal size and uniformity are not always controllable parameters.

5.1.1.1 Suspended Solids

Much of the work used to describe fluid flows typically considers the bulk flow with the properties of the fluid being a continuum. Typically dissolved salts and other soluble objects are assumed dilute enough to not disturb the continuum characteristics of the properties [19]. In the case of sample injection for SFX, it is inherently a two-phase flow, with solid crystals subject to the continuous fluid they are suspended in. The best crystalline suspension is one that is uniform in size and monodisperse, making the fluid behavior less erratic due to variations in size, as well as aiding in the X-ray data collection.

The Reynolds number (Re) is an important dimensionless parameter in fluid mechanics. It represents the ratio of inertial forces to viscous forces in the media. It is calculated as $\mathrm{Re} = \rho UD/\mu$ where the mass density ρ, the characteristic bulk velocity U, the characteristic length D, and dynamic viscosity μ, are properties of the fluid and the relevant physical parameters. To describe flow through a tube, the characteristic length that dictates the flow phenomena is the wetted diameter;

flow over a flat plate would have the length of the plate as the characteristic length; flow around a small object suspended in a semi-infinite fluid would have the largest particle dimension as its characteristic length. Fluid flows in tubes with Reynolds numbers less than 2300 empirically move in laminar sheets and lack any turbulent effects [19] (and references therein). In the case of small particles suspended in solutions, they are typically at Re < 1 and are in a regime called *creep* (or Stoke's) flow, where inertial forces are completely negligible and viscous or other forces might dominate. In this flow regime, particles can readily sediment, and they will sediment at different rates, based on their morphology, size, and their alignment with the gravitational field [14]. A particle in a parabolic flow field, typical in fully developed pipe flow, will stay on a streamline unless it diffuses out due to Brownian motion (Taylor dispersion) or some external force, such as gravity [14, 19, 24]. For reference, water flowing at 10 mm/s through a 50 μm orifice has $Re = 0.5$ (Q, the volumetric flow rate ~2 μL/min); slower flows and more viscous flows will have even lower Re values. Unlike inertially dominated flows, two bodies suspended in creep flow can affect each other's flow fields significantly; for example, two identical particles 5 diameter-lengths apart can still feel the others' presence, whereas inertially dominated flows around a particle would return to the bulk velocity at such distances. Creep flow fields never have an increase in velocity as opposed to inertial flow fields, meaning nearby objects slow each other down; for example, objects sedimenting in creep flow near a wall settle more slowly than those away from the wall.

Unions and Filters High viscosity and MESH injectors typically have no more than one union connecting the capillary tip at the interaction region to the sample reservoir. GDVN and DFFN injections more commonly use tubing unions. Unions rarely connect two tubings of the same inner diameter; rather it is most common to go from larger inner diameters and interface down to smaller inner diameters, such as the final size in the liquid injector. Even when connecting two of the same capillaries, the union itself has an inner diameter which might not match (e.g., common tubing unions have a 250 or 500 μm inner diameter, connecting two capillaries of 50 μm inner diameters will go from the original capillary dimension to one 5–10 times larger, and back down to the original dimension). The crystals flowing near the periphery of the larger diameter tube cannot easily cross streamlines to get into the next smaller inner diameter line, thus causing a buildup at the union interface, which quickly culminates into clogging. Taneda showed how streamlines around a corner in low Reynolds number flow can create a vortex in the corner, where particles remain trapped [25]. The interface between capillaries and unions of differing inner diameters effectively create corners for crystals to become trapped and eventually interfere with the flow of incoming crystals, and should be minimized where possible.

In order to ensure the crystal slurry does not clog the capillary it is sometimes necessary to filter the larger crystals from the slurry. Off-the-shelf stainless steel and PEEK filtration frits are common in chromatography applications to filter out particulates of different sizes, such as 2 or 20 μm cutoffs. These filters are effective for biological fluids with objects typically much smaller and more flexible able to pass through unharmed while filtering foreign particulates. However, crystals are more rigid and might not pass through these filters. Testing should be done to understand how the crystals interact with the filtration method or devise new gentler filtration methods compatible with the crystals.

Sedimentation Except for the high viscosity injection methods, settling is one of the most prevalent problems throughout all injection techniques—given sufficient time crystals can even sediment in high viscosity media. For μm-sized crystals, it is more a matter of when, not if, the crystals will sediment. Placing a cylindrical reservoir vertically with the sample exit at the bottom seems ideal, but it will lead to an eventual clog. Most crystal slurries are in the creep flow regime, where viscous forces are left to balance the gravitational forces. Particles the size of the protein crystals reach their terminal velocity quite readily. Balancing the Stoke's drag of a particle moving at its terminal velocity with its buoyant forces results in an equation for the particle velocity

$$\vec{U} = \frac{2\rho_0 a^2}{9\mu}\left(\frac{\rho}{\rho_0} - 1\right)\vec{g}$$

of a particle with radius a, and the particle's mass density ρ, suspended in a fluid with viscosity and mass density μ and ρ_0, respectively. The velocity of sedimentation helps determine when the particles have sufficiently sedimented and will likely no longer flow through appropriate flow paths. As the particle concentration at the bottom of the reservoir continues to increase, the connected capillary will ultimately clog, despite the crystal sizes being smaller than the capillary diameter. Placing the reservoir horizontally with the exit to the side avoids the issue of clogging from sedimentation, but now the sample sediments past the streamlines that will guide the crystals into the capillary entrance towards the interaction region. A standard SFX reservoir inner diameter is approximately 5 mm and it would take approximately 90 min for 1 μm sized objects to settle past the capillary entrance (at the reservoir's midline) if it were suspended in water; 20 min for 2 μm objects (assuming an object with density similar to that of a protein $\sim$1.2 g/cm^3).

The most commonly used method to prevent sedimentation in the sample reservoirs is to rotate the reservoirs in order to modify the direction of the gravity vector. Lomb et al. [20] has shown an approach where the sample is loaded into syringe-like reservoirs and rotated. There have been some modifications made to the technique, but the basic idea of rotating the sample slowly back and forth has been used frequently in SFX experiments. It is not an infallible approach and should be tested prior to the experiment to ensure the crystal morphologies and the properties of the fluids are properly matched to the rotation speeds and frequencies of the sample shaker. Recall, it is a device to slow the sedimentation and not resuspend it. Once the particles are settled, the properties of the creep flow regime make it difficult to resuspend them by shaking or moving the plunger back and forth. Properties of laminar flow mean that the particles will flow back and forth to the same position along the same streamlines and require turbulence or mixing or some other external force to homogeneously redistribute them. This is the same physical phenomena that explains why microscopic spermatazoa swim with cork screw tail motions instead of the back and forth motion typical in humans and fish swimming at the macroscale [26].

Key Takeaways

- The crystal slurry dominates the flow physics
  - Particle size sets minimum capillary diameters (<2/5 capillary diameter)
  - Fourth order dependence on transport tube inner diameter can increase flow rates too much, setting upper and lower size limits
  - The viscosity of the mother liquor combined with the increased viscosity caused by the suspended particles means sample fluids in capillaries are more viscous than expected
- Flows are typically laminar and dominated by viscous forces rather than inertial; diffusion and viscosity limited, no turbulence.
- Settling affects numerous sample delivery methods and can be a detriment to experiments.
- In vacuum experiments have an intrinsic minimum pressure gradient not present at ambient pressure.
- Although not discussed in the chapter, proper microfluidic hygiene is important
  - Smooth perpendicular cuts to tubes and capillaries
  - Minimize flow impedances
    - Prime bubbles out of the lines, as trapped gasses can be large flow capacitances
    - Resistances such as too many unions, or small flow restrictions should be minimized
    - Minimize swept, or dead, volumes to conserve sample.
- There is no substitute for testing and characterizing the experiment as faithfully as possible, that is, injection tests with the actual crystalline samples, if possible, in the actual vacuum chamber in front of an X-ray source will minimize surprises during the beam time. Successfully delivering just the mother liquor in air or rough vacuum is a good start but can be misleading compared to [the actual] sample slurry.

5.2 Standard Liquid Injectors

In Serial Femtosecond Crystallography, every nanocrystal that is hit by an X-ray pulse is destroyed by the interaction that deposits a lot of energy in a small volume. Most of the requirements on SFX sample delivery listed in the overview can be fulfilled with a liquid jet injector. A microscopic liquid jet can be obtained in the

Fig. 5.5 Schematic view of GDVN. The end of the nozzle is beveled to allow diffracted X-rays (the red cone) to reach the detector. Green lines are gas streamlines, which accelerate and compress the liquid to produce a small jet from a larger orifice. Blue is the liquid containing protein microcrystals. The X-ray FEL pulses hit the liquid jet before it breaks up into droplets

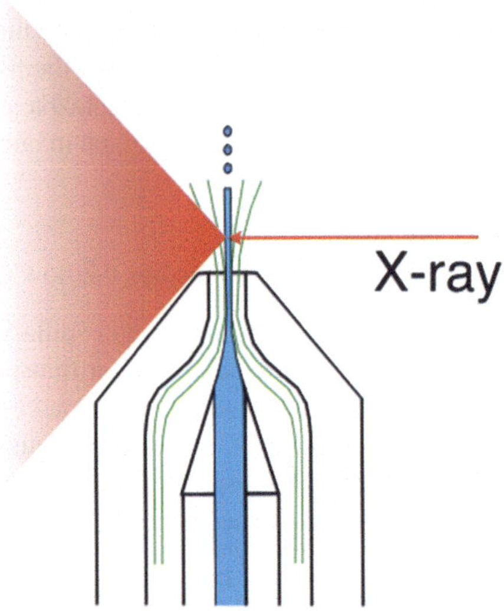

simplest way by discharging a liquid through an orifice at high liquid pressure. A continuous liquid jet emerges, with a diameter identical to the orifice diameter (Rayleigh jet) [27], which breaks up into droplets due to capillary forces. The droplets have a diameter of about twice the jet diameter and an average spacing of 4.5 times the orifice diameter [28]. In X-ray FEL experiments with liquid jets, the X-ray pulses hit the liquid stream ideally before it breaks up into droplets to assure a high efficiency (every X-ray pulse hits the liquid). However, Rayleigh jets have a few disadvantages: high flow rate (0.4–7 mL/min) and high jet speed (100 m/s), large jet diameter (10–100 μm), susceptibility to clogging for smaller orifice diameter (especially when the liquid contains protein crystals) and susceptibility to freezing in vacuum when the flow is interrupted. The liquid jet diameter for serial femtosecond crystallography should ideally be similar to the X-ray beam diameter, which is between 10 μm and 0.1 μm at LCLS for example. Liquid jets of such diameter cannot be obtained with a Rayleigh jet nozzle due to clogging issues.

These issues have been solved with the Gas Dynamic Virtual Nozzle (GDVN). This nozzle was developed based on the observation that shear and pressure forces of a sheath gas can reduce the diameter of a liquid jet to a value smaller than the orifice diameter [13, 29]. GDVN nozzles generate a liquid stream of micron to submicron diameter with flow speeds of 10–100 m/s at flow rates of 1–20 μL/min. They consist of an inner sample capillary with typically 40–75 μm inner diameter, and an outer gas capillary that provides a co-flowing gas stream that accelerates the liquid and thereby reduces the diameter of the stream (see Fig. 5.5).

The acceleration and reduction in jet diameter occurs over a very short distance at the exit aperture of the nozzle. The liquid microjet breaks eventually up into droplets due to the Rayleigh–Plateau instability [27]. The diameter of the liquid jet is reduced by a factor of 10–50 compared to the inner diameter of the sample capillary. Any particle which is smaller than the sample capillary inner diameter but larger than the jet diameter (e.g., a large crystal) will simply pass through the system and only momentarily disrupt the jet, that is, the nozzle does not clog as easily as a physical nozzle of a size similar to the GDVN jet size. The inner sample capillary is centered in the outer gas tube by laser-cut Kapton spacers. The mass flow rate of the co-flowing gas stream is 20–30 mg/min and the sheath gas used is typically Helium or Nitrogen, where the use of Helium usually results in a more reliable jet. Due to the low viscosity of the mother liquor used to grow most protein crystals, microcrystals settle in the liquid due to gravity, and therefore an anti-settling device [30], as discussed in the previous section, is required for SFX to slowly rotate the sample reservoir and prevent settling.

The X-ray beam is focused onto the contiguous part of the jet, before the droplet breakup, usually 100–200 μm away from the nozzle exit. The background scattering from the contiguous jet is lower than from the droplets, since the droplets have about twice the diameter of the jet. Scattering from the cylindrical shape of the jet leads to a streak in the diffraction pattern at low spatial frequencies if the X-ray beam is not significantly smaller than the jet.

The outside of the GDVN is beveled at the end to avoid shadowing of the diffracted X-rays. The gas focusing aperture is formed by flame polishing and variations in shape and size between different nozzles are almost unavoidable. For the liquid jet to emerge aligned with the axis of the GDVN, the sample capillary must be accurately centered on the axis of the gas-focusing aperture. To make centering easier, a square outer glass tube may be used [31] for automatic centering of the round sample capillary. But even in this case, the flame polishing procedure may not result in a gas aperture hole which is exactly on the center axis of the square glass tube. These manufacturing difficulties led to the introduction of 3D printed GDVN nozzles [32]. A high resolution two-photon 3D printer is used to print the critical GDVN endpiece with high precision at a resolution of 500 nm, and a liquid and gas capillary are glued into the 3D printed part (see Fig. 5.6 and Fig. 5.7). A keyadvantage of 3D printing technology is that more complicated shapes can be printed (e.g., mixing nozzles [33] or double flow focusing nozzles [23]) as will be discussed further in this chapter.

GDVN jets can be used at atmospheric pressure and in a vacuum environment. Using the GDVN in a vacuum environment can create challenges: High vacuum can be easily maintained in the presence of a microscopic liquid Rayleigh jet [11], but the focusing gas introduced into the vacuum raises the base pressure of the experimental chamber. Therefore, a differential pumping system is used, for example, at the Coherent X-ray Imaging (CXI) endstation at LCLS to protect the high vacuum system and the beamline optics and X-ray detector. The GDVN is mounted on a nozzle rod in a differential pumping shroud, which contains several openings for cameras and pump lasers for pump probe experiments. The nozzle rod can be retracted behind a gate valve and removed from the vacuum system for nozzle changes. Several motorized stages allow alignment of the nozzle relative to the X-ray beam [34]. An initially installed in-vacuum microscope for nozzle observation [34] has since been replaced by an optical system (microscope

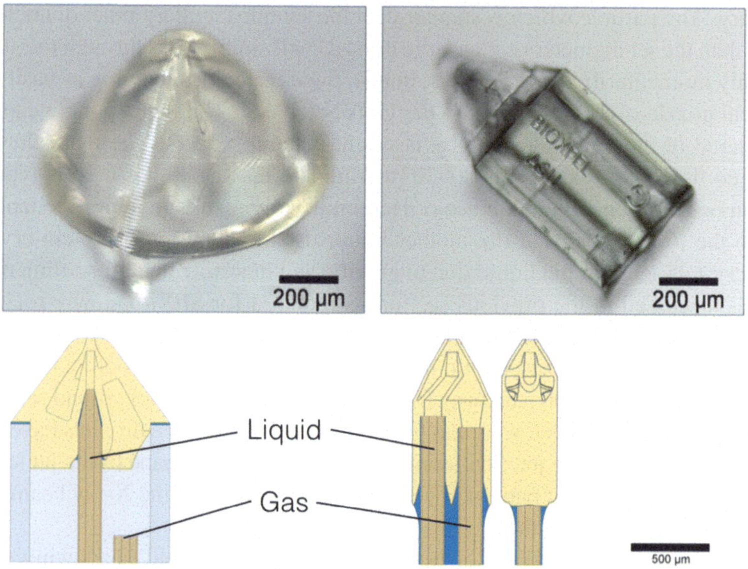

Fig. 5.6 3D printed nozzles: left: this design requires a micromanipulator and mounting stage to glue the capillary into the printed nozzle tip. Right: the new design is easier to attach to the gas and liquid capillaries. Grey: Steel or Glass, light brown: printed material, dark brown: polyimide-coated glass capillary, blue: epoxy

Fig. 5.7 3D printed nozzle in operation. Liquid line on bottom, gas line on top

objective and transfer lens in vacuum), which allows the use of more sophisticated cameras outside the vacuum chamber. This allowed the use of high-speed cameras or stroboscopic recording modes to visualize the damage inflicted onto the jet during X-ray exposures [2].

GDVNs have been successfully used in many SFX and Wide Angle X-ray Scattering (WAXS) experiments at LCLS and SACLA, as well as in electroflow-focusing mode to aerosolize single particle samples in an aerodynamic lens system (as shown in Fig. 5.3) [35]. The flow rate and jet speed of the GDVN is not optimized for the X-ray pulse repetition rates of SACLA or LCLS: at 120 Hz repetition rate and 10 m/s jet speed, the distance that the jet travels between two pulses is about 8 cm. This means most of the sample goes to waste in between pulses. This is unacceptable for many membrane proteins which can only be expressed in small quantities. Therefore, other injection methods have been explored which consume less sample, as will be described in the following chapters. For future high repetition rate X-ray FELs, sample waste with the GDVN is not expected to be a problem. At the European XFEL with bunch trains arriving at 10 Hz, where the pulse repetition rate within each bunch is 4.5 MHz, it remains to be seen if the GDVN jet speed is even fast enough to keep up with the high repetition rate pulses. This is especially a problem if shock waves from the sample explosion travel upstream in the jet and cause additional damage [2]. At LCLS-II, with a sustained repetition rate of 1 MHz, the jet speed seems to be perfectly matched to make use of the entire sample with minimal waste. This however presents a new challenge with sample debris. Currently, for a GDVN, most of the sample collects in a catcher away from the interaction region. With a high intensity, high repetition rate X-ray FEL beam, most of the sample will be vaporized by the beam, no longer collecting in a catcher away from the interaction region but rather rapidly coating every surface close to the interaction region, including the nozzle tip. The extent to which debris will be a problem with high repetition rate machines remains to be seen. Sample consumption with the GDVN can be reduced by a factor of about eight by using a double flow focusing nozzle [23], where the sample flow can be reduced by a second focusing liquid, before gas focusing occurs (see Sect. 5.5).

5.3 High Viscosity Injectors

In recent years, protein crystallization in lipidic cubic phase (LCP) has led to structures of many important human membrane proteins, which are highly important and the subject of Chaps. 4 and 11 of this book. The desire to stream protein crystals in LCP into the X-ray FEL beam for SFX led to the development of the high viscosity injector. LCP has a viscosity comparable to car grease or tooth paste (dynamic viscosity of ~48.3 Pa·s [36] compared to water at 8.9×10^{-4} Pa·s and Ethanol at 1.1×10^{-3} Pa·s) and thus cannot be used with a GDVN, since the long sample capillary with an inner diameter of tens of microns would require impractically high pressures to make the liquid flow. Therefore, a new injector had to be designed that uses only a very short sample capillary and a hydraulic stage to achieve the high pressure needed to drive the sample through this capillary. Figure 5.8 shows a schematic of one version of an LCP or high viscosity injector.

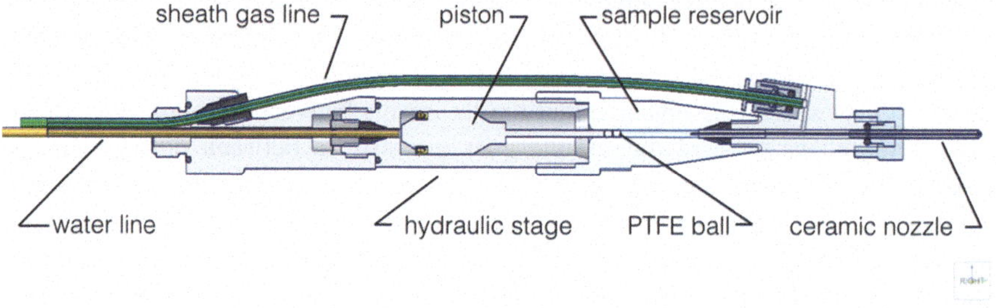

Fig. 5.8 High viscosity injector. It consists of a hydraulic stage, which is pressurized by an HPLC pump. The piston pressurizes the sample in the reservoir and drives it through the capillary. The ceramic nozzle ensures that sheath gas surrounds the extruded high viscosity material and keeps the flow on axis

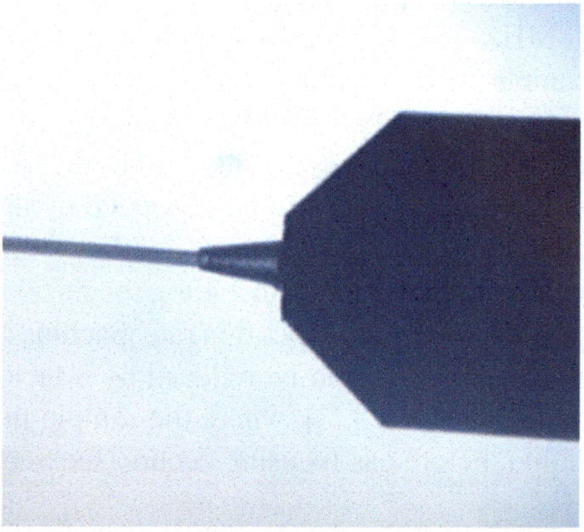

Fig. 5.9 LCP stream extruded through a 50 μm inner diameter capillary; the stream has the same diameter as the capillary inner diameter. The large shadow on the right is the ceramic gas aperture

An HPLC pump supplies water at constant flow rate and moves a plunger in a syringe body. The plunger has a large diameter on the water side and a smaller diameter on the sample side, where it applies pressure on a Teflon ball. The ratio of the two plunger diameters equals the pressure amplification of the hydraulic stage. The Teflon ball acts as a secondary plunger, it is sitting in the reservoir bore, which has a slightly smaller inner diameter then the ball diameter, thus forming a tight seal against pressures up to 15,000 psi. The LCP sample is pushed by the Teflon ball from the reservoir into a capillary with a diameter of 20 to 50 μm, selected depending on crystal size. An LCP stream emerges at the distal end of the capillary (Fig. 5.9). Different reservoir sizes can be made, for example: 25, 40 and 120 μL.

The diameter of the extruded stream is identical to the capillary inner diameter. To keep the stream straight and on axis, a co-flowing gas is surrounding the stream. The distal end of the capillary is ground in a conical shape and protrudes out of the gas aperture, which has a square inner diameter and opening to allow a path for the sheath gas (see Fig. 5.10). A ceramic injection molded gas aperture is used,

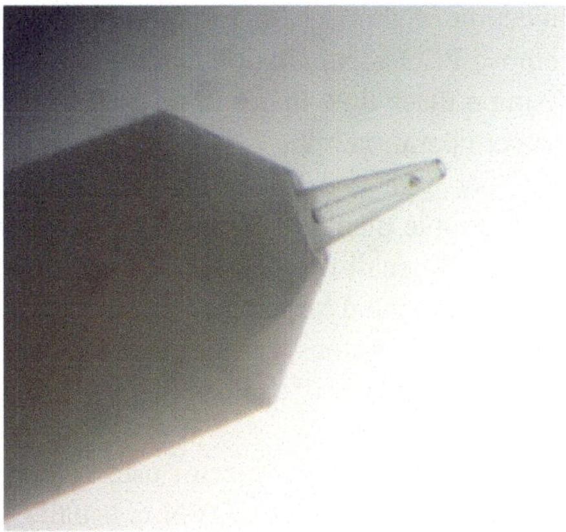

Fig. 5.10 Injection molded ceramic gas aperture with protruding coned sample capillary (50 μm inner diameter). The gas aperture hole in the ceramic part has a square shape to allow sheath gas to exit and surround the sample stream

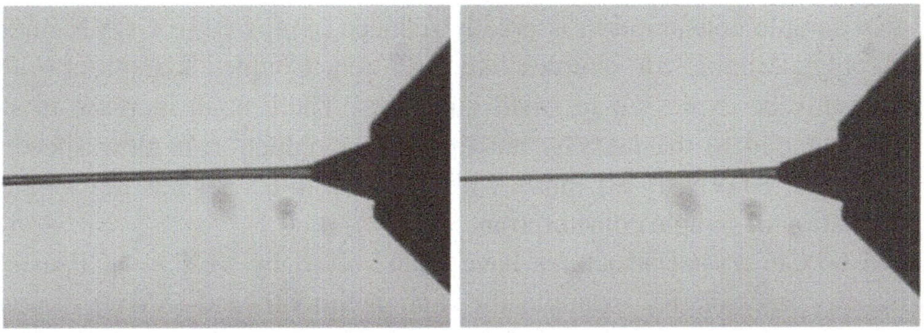

Fig. 5.11 PEO extrusion with high viscosity injector into ambient pressure. Nitrogen sheath gas exits the nozzle surrounding the capillary cone, which is protruding out of the ceramic gas aperture. Left: mass flow rate of sheath gas (Nitrogen): 5 mg/min. Right: mass flow rate of sheath gas (Nitrogen): 15 mg/min. Same liquid flow rate in both images: 90 nL/min

which has a more reproducible shape and longer life than the previously used hand melted glass tubes [37]. The high viscosity of LCP prevents crystal settling due to gravity on the time scales of a measurement. The sample flow rate can be adjusted and optimized for the X-ray FEL pulse repetition rate, so that the stream advances only the distance needed to replenish the damaged material from the last pulse.

Unlike in the GDVN, the LCP stream cannot be focused down to a smaller diameter by the sheath gas. However, another high viscosity medium does allow a reduction of the stream diameter by shear forces from the sheath gas, as seen in Fig. 5.11 for PEO, a gel polymer of high-molecular-weight poly(ethylene oxide). The PEO stream, which shows already reduced background scattering compared to LCP [38] (by a factor of about 1.5 in the diffuse ring region), can be reduced in diameter from initially 50 μm to about 20 μm, which reduces scattering background further. PEO extrusion works best at ambient pressure, for example at the MFX beamline at LCLS or the DAPHNIS endstation at SACLA [39] and at Synchrotrons. It is particularly hard to start a stream of PEO with the high viscosity injector, since PEO is very sticky and tends to ball up at the end of the nozzle. Once started, though, the PEO stream runs very reliably if the sample does not contain any gas bubbles.

Injection of short bursts of isopropanol into the gas line with a switching valve to remove PEO accumulation at the nozzle tip makes starting the stream much easier. Using a mass flow meter in the gas line also helps to find the correct gas flow for proper operation.

Other high viscosity media for crystal embedding have been developed (e.g., hydrophobic media like grease [40] and Vaseline [41]) which form emulsions with crystals in small droplets of mother liquor. These carrier media however produce high background scattering and are often incompatible with membrane proteins. Hydrophilic viscous carrier media do not have these disadvantages, and several have been described: agarose [42], hyaluronic acid [43], and two hydrophilic gelling polymers [44].

The high viscosity injector uses sample much more economically: the flow rate of the microjet from a GDVN is usually 10–30 μL/min, whereas the sample flow rate with the high viscosity injector can be adjusted from 0.01 to 2 μL/min. This means that sample consumption is greatly reduced compared to GDVN injection. This is highly desirable for samples like G-Protein Coupled Receptors (GPCRs) which can only be expressed in small quantities. There is an increase in stream diameter compared to the GDVN, but the drastic reduction in flow speed from 10 m/s with the GDVN to ~2 mm/s with the LCP injector leads nevertheless to a large reduction in sample consumption.

Several GPCR crystal structures have been solved by SFX in LCP using the high viscosity injector, for example the human Smoothened (SMO) receptor in complex with cyclopamine [45], the angiotensin receptor [46], the human δ-opioid receptor [47] and rhodopsin bound to arrestin [48] and the human glucagon receptor. Pump probe experiments with bacteriorhodpsin crystals in an LCP stream have been performed [49] as well as native phasing measurements on human A2A adenosine receptor, which after phase extension lead to a 1.9 Å structure [50]. Microcrystals of soluble proteins have been mixed into LCP and used with the high viscosity injector for structure solutions [51]. For LCP SFX experiments in vacuum, as at the LCLS CXI instrument, extra measures must be taken to prevent a phase change of monolein LCP into a lamellar phase due to water loss by desiccation and temperature reduction due to evaporative cooling. This would lead to strong diffraction rings which mask weak Bragg reflections. To that end, a shorter chain length lipid can be added to the LCP post crystallization [45] which lowers the phase transition temperature.

Serial crystallography experiments with high viscosity micro streams can also be done at synchrotron sources, albeit without the dramatic reduction in radiation damage possible at XFELs. Fast detectors allow shutterless data collection from a stream of microcrystals in its LCP growth medium with 20–100 ms exposure times. The extrusion speed of the sample and exposure time can be adjusted depending on the flux of the beamline to allow for full exposure of each crystal to the damage limit at room temperature. Serial Millisecond Crystallography has been demonstrated successfully at the Advanced Photon Source [38], the Swiss Light Source [41] and the European Synchrotron Radiation Facility [52].

5.4 The Microfluidic Electrokinetic Sample Holder

Mating the protein crystal suspension with a type of sample delivery method is one of the most critical aspects to successfully execute an experiment at an X-ray FEL. A major concern in SFX is sample consumption, since many proteins cannot be readily crystallized in arbitrary quantities. In 2009, Chapman and Seibert et al.'s [3, 4] first biological experiments at an X-ray FEL were limited to soft X-rays (<2 keV) and thus required a vacuum sample environment and the low solvent background and sample replenishment of a thin liquid jet or a solvent-less aerosol injector. To address the issue of higher sample consumptions, the microfluidic electrokinetic sample holding (MESH) technique was developed in 2011, where instead of accelerating the meniscus with aerodynamic or hydrodynamic forces, electrokinetic forces are employed. A room temperature structure of thermolysin was originally solved to better than 4 Å [53], demonstrating the MESH technique, but as better data analysis algorithms were developed, the same data could be used to extend the resolution to 2.1 Å [54]. The MESH was successfully used to solve room temperature structures of photosystem II [55–57], demonstrate minimal electronic structure changes caused by high intensity X-ray FEL pulses by collecting simultaneous X-ray emission spectra and diffraction [58, 59], collect X-ray emission spectra in the soft X-ray regime [60–62], and collect novel structures on nanocrystals [63]. In all of these experiments, the mother liquor was doped in order for the crystals to survive the vacuum injection. The capillary geometries varied from 50 to 100 μm, with the flow rate not exceeding 3 μL/min, and in good running conditions, having flow rates in the sub-microliter per minute regime.

The following sections will explain how to use the MESH technique to deliver sample for in-vacuum SFX experiments. After discussing the MESH, a new method involving a concentric flow that keeps the low flow rate and allows the MESH technique to be applied to a broader range of fluid conditions will be discussed. The chapter concludes with a discussion on how a technique like the MESH and its variants fit into the grander scheme of a suite of sample delivery techniques available for SFX.

5.4.1 The Microfluidic Electrokinetic Sample Holder

The main difference between the MESH injection and other SFX injection methods is the way it supplies energy to the microfluidic system: an applied voltage rather than applied pressures. The microfluidic electrokinetic sample holder (MESH) technique is physically identical to electrospraying or electrospinning techniques [64]; however, the MESH technique is interested in the thin liquid jet created after the Taylor cone but before the eponymous spray or spun filaments typical of the other techniques. The applied voltage causes any free charge in the solution, typically solvated ions, to migrate towards or away from the voltage source, subject to the polarities. A net excess charge builds at the exposed meniscus, which continues to repel from the like-charged voltage source. Once the applied voltage potential is high enough to overcome the opposing surface tension and viscous forces, fluid is ejected from the meniscus. A balance in flow rate and applied potential will then cause the meniscus to achieve different states, such as a dripping mode, the cone-jet mode, or the multiple jet modes (listed in order of increasing voltage, for a fixed flow rate).

The ability to eject charged particles is the real power of the MESH technique. Most biological fluids have free ions in solution and will respond to an externally applied voltage. The range of tubing and fluid properties typical in SFX experiments leads to currents at, and typically below, the μA range. Considering Faraday's constant (~96,485 C/mol), a fluid with 100 mM ionic strength operating in MESH conditions at 1 μA is ejecting 1 pmol of charged particles per second or consuming approximately 6 nL/min. Therefore, once a flow rate and voltage are found to stably create a cone-jet Taylor cone, there is minimal additional cost in sample consumption by the electrokinetics. The applied voltage focuses the meniscus and creates a thin liquid stream, but it minimally affects the fluid's flow rate, or the sample consumption.

The GDVN partially mitigates freezing in vacuum by always running a sheath gas, which accelerates the meniscus to make a jet and keeps the fluid at ambient temperature [13, 65]. To run the MESH in vacuum, the crystal slurry must be doped with a *cryoprotectant,* such as 40% glycerol [53]. The addition of glycerol interferes with the fluid's properties such that it will not evaporate and freeze or precipitate in vacuum. Typical—not all—additives tend to also increase the viscosity and alter other fluid properties, such as density and surface tension, which aid in reducing the flow speed as well as maintaining a homogeneous protein crystal suspension for longer.

5.4.1.1 Charging the System

To create a MESH injection from the free surface of a liquid drop at the end of a capillary, an electric field must be present across the meniscus. The free ions in the solution will begin to accumulate and will slowly begin to be repelled from one another, because of their like charges, until they can overcome the surface tension and viscous forces holding them together as a liquid meniscus. Eventually, the meniscus will go from a hanging drop to the more familiar Taylor cone shape. A stable cone-jet mode passing protein crystals from the reservoir, through the capillary to the thin jet filament, and finally probed by photons is essentially the MESH technique.

In order to create this electric field, the meniscus can be placed between two charged electrodes, or a charged and grounded electrode. In the case of the MESH injector, the fluid is charged at its reservoir. This is intentionally done to avoid unwanted voltage arcs to discharge from charged metal towards sensitive scientific equipment in the sample chamber (e.g., detectors or motors). The difference between charging at the tip and creating an external electric field, versus charging within the fluid and having the meniscus create the electric field, are subtle but very important to understanding the operation of the MESH technique and its shortcomings.

5.4.1.2 Creating the Driving Electric Field

The charged fluid in the capillary acts like a conductive wire, where the fluid is poorly conducting and the fused silica and polyamide coating of the outer capillary walls insulate the charge. In order for the MESH technique to work properly, all charge must be contained in the capillary and reservoir and allowed only to escape through the meniscus via the *leaky capacitor* of the cone-jet physics.

At the meniscus, it should be the charge buildup that ejects charge via the cone-jet physics. If charge is taken away from the meniscus by the local atmosphere, this will prevent sufficient charge to buildup and MESH jetting will not be observed. MESH operation has been observed in air at ambient pressure, but below ($<\sim$200 Torr) ceased to operate until the vacuum pressure was sufficiently low ($<5 \times 10^{-2}$ Torr), where it would begin to form a cone-jet again [53].

When the meniscus is the source of the electric field, any changes in the electrical properties of the meniscus create a varying electric field as non-conductive phases, such as the protein crystals, pass by. The ionizing X-rays can also affect the field if shot too deep in the meniscus. Even the vacuum dehydrating the meniscus can play a role in disrupting the local electrokinetics of the meniscus. A shortcoming of charging the reservoir is that any bubbles trapped in the sample capillary line, even small bubbles that do not fill the full diameter, severely hinder or completely stop the transfer of charge. The air bubble can act like an infinite resistor and will prevent the Taylor cone from forming, while the vacuum continues to draw in fluid as a large droplet; so crystal slurries should be bubble free for best injection.

5.4.1.3 A Poorly Conducting Wire

The fluid-filled capillary can be treated as a wire with a known resistance that increases linearly with length, and inversely with both the cross sectional area and the conductivity. For commonly used fluid conductivities, through a meter-long, 50 μm capillary, resistances [typically] vary from 0.1 to 10 GΩ. The flow of ions typically have the MESH operating in the 100s of nA regime. These values indicate that no more than a few hundred volts can be dropped across the length of the capillary; however, if higher viscosity fluids are used, which diminish the conductivity as charge cannot easily move, then higher driving voltages might be needed to account for the high resistance of the line. Currents much higher than this are indicative of a charge leak, which can severely diminish the functionality of the MESH injection.

Since the crystal solution is conductive, there can be no internal electric fields, as the free ions will quickly move to shield any unbalanced charge. The structures determined using MESH thus far have not shown any differences between their MESH and conventionally-determined structures. Voltage-sensitive structures would potentially be affected by the electrokinetics of the MESH; recent [unpublished] data has suggested further studies are necessary. To date, there are no apparent adverse effects on structures by applying voltages typical of the MESH technique.

To properly MESH, a non-conductive tubing, such as fused silica capillary or polymer tubing, and a counter electrode is needed. The droplet exiting the capillary will wet upwards along the outside of the capillary if the closest grounding sources are the polymer fittings and mounting material holding the capillary into vacuum. The capillary should protrude at least 1 mm away from these to minimize external wetting. Note that if the capillary is protruding too much, the capillary starts to oscillate from the downstream spinning process. Everything in the MESH geometry should be a polymer or a non-conducting material except for the charging material. Any excess buildup of liquid, or fluids that wet backwards towards the sleeves and fittings can disrupt the field or cause a current leak.

The next sections will discuss counter electrode designs that will draw the charged meniscus in the correct direction for the experiment.

5.4.1.4 Counter Electrode Design

Counter electrode (CE) design and function are very important to the MESH injection. Aside from the need to dope the fluid for vacuum injection, the counter electrode design considerations differentiate MESH injection from standard electrospray and electrospin techniques. A counter electrode close to the meniscus will give a high driving electric field, but it might interfere with the diffraction data. The shape should promote a homogeneous electric field like those between two parallel plates, yet precipitates from the sample can buildup and interfere with the electric field and ultimately grow back into the interaction region.

Typically, MESH injection charges the fluid and leaves the counter electrode at earth ground. However, the counter electrode can be charged and the fluid left grounded and the familiar Taylor cone will form (likely at different voltages). The ions in the fluid will collect at the meniscus and mirror the charges at the counter electrode. Eventually they will build up enough charge to begin to repel themselves and begin the Taylor cone phenomenon again. However, it is strongly recommended that the counter electrode be simply connected to earth ground rather than charged via a bias voltage because it is more dangerous to electrify the counter electrode the higher density of "free" electrons in metals. Copper, assuming one free electron per copper atom, has a [charge] density of over 13,000 C/mL whereas a fluid with 100 mM ionic strength has barely 10 C of ions per mL. Consequently, the metal in vacuum can readily discharge three orders of magnitude more charges than a typical MESH fluid of the same volume but this is not a concern when the counter electrode is properly grounded instead.

5.4.1.5 Counter Electrode Geometries

An ideal geometry for the counter electrode is one that creates straight electric field lines from the meniscus to the counter electrode, much like those between two plates in a parallel plate capacitor, but is impractical to implement. Instead, a counter electrode with a hole to allow the ejected, non-ionized mass through with proper spacing to create a sufficient electric field while not shadowing the diffraction detector is necessary. The distance of the counter electrode for the MESH injection is typically at least 5 mm, depending on the radius of the counter electrode. The counter electrode should be the closest conductive object to the capillary tip. A smaller hole is ideal for straight field lines from the meniscus (ideally a point source of charge), but the small opening can quickly build up debris and strengthen the field towards mass buildup, leading to injection failure. Nominally, the opening size in the CE is of the same order as the spacing from the capillary tip, typically 5–10 mm. Figure 5.12 shows different iterations of CE geometries used in atmospheric and vacuum conditions.

5.4.1.6 Sample Reservoirs

The preferred reservoir for MESH injection is a pressure-driven vessel. This method was described in Sierra et al. [53] and can be seen in Fig. 5.13. Here an open centrifuge vial is loaded with the 0.1–1 mL of sample and into an acrylic pressure vessel. The vessel is pressurized with an inert gas ($<\sim$100 psig), such as compressed air or nitrogen, while the capillary and a charging wire are submerged into the fluids like straws. The advantage of this technique is the simplicity of the sample loading as it uses common centrifuge vials. The plastic acrylic housing is adapted from an electrospray aerosol generator (Model 3480, TSI) and easily screws open to hold the reservoir and interface the capillary and wire. A steel nut on the lid of the acrylic housing has a platinum wire soldered to the bottom that is long enough to reach the bottom of the vial; the high voltage is connected to the outside of the nut, which in turn charges the submerged wire and the fluid. With the capillary submerged to the bottom of the vial, there is less concern with settling and clogging, as the sample is drawn in, as opposed to being forced in with other reservoirs. In this setup, there is a possibility for the sample to settle and become so concentrated that the fluid may continue to enter the capillary while the crystals aggregate and form a filter-like structure that prevent further crystals from entering.

It's important to recall that many fundamental fluid mechanics equations assume a homogenous fluid or dilute solutions. In the case of SFX sample delivery, homogeneity is not always a luxury. Growing the crystals in large batches rarely produces identical crystals. Even if the crystals are identical in size and morphology they are suspended in the solution as opposed to solvated in a solution. This contrast from a solid object to a liquid media makes a difference in injection behavior. The solid objects passing through a meniscus provide enough disruption that the Rayleigh breakup of the liquid jet comes sooner than the breakup of a pure liquid jet.

Fig. 5.12 Different counter electrode geometries used for MESH injections. (clockwise) conical with flat and large hole (capillary visible above the hole); flat surface with hole (stainless steel optical post; capillary above hole); 2 mm pointed stainless steel pin (protein debris deposited on electrode after injection); thin-walled conical stainless steel with hole (deposited polymer and protein debris buildup visible); flat stainless steel with hole (MESH-on-a-stick geometry); copper wire loop. For scale, capillaries are 360 μm outer diameter, standard PEEK fittings are for 1/16" OD tubing, optical post hole is tapped 1/4"–20 (in top middle image)

It is important to recognize these potential differences and understand that testing simply the mother liquor without suspended crystals is not enough to guarantee a jetting success as the bulk fluid slurry will behave like a more viscous media and need typically higher parameters to create an equivalent jet of pure fluid. Homogeneity is also a problem with viscous injection methods where insufficient mixing to create the media can create pockets of low and high viscosity causing running instabilities and erratic sample consumption. Many times the difference between a jet running mother liquor and a slurry carrying protein crystals can be evident from optical imaging; however, the jet behavior can be more telling. If the jet's behavior at the interaction region begins to stabilize and improve it is likely too good to be true: the crystals are no longer flowing and obstructing the meniscus. A later section discusses the effects these suspended solids had on the design of the MESH injection method.

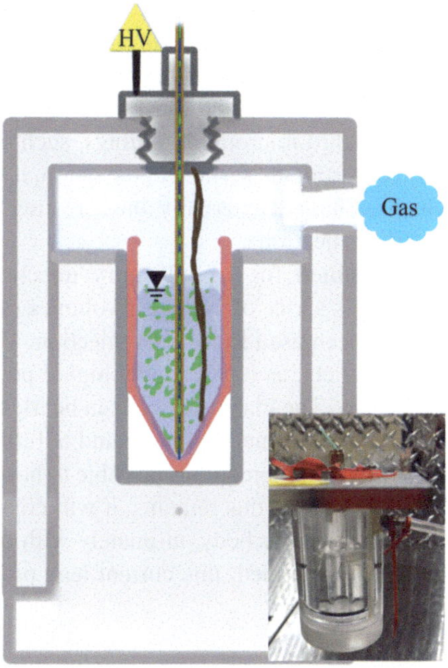

Fig. 5.13 A schematic of the modified pressure reservoir used for typical MESH injections, with actual picture inset. The pink centrifuge vial holds the sample and is held in place by the bottom of the acrylic housing. A Swagelok adapter with appropriate polymer tubing sleeve (green in photo) allows a capillary to be submerged into the vial, while maintaining pressure. A platinum wire electrically connected to the Swagelok fitting is also submerged into the vial

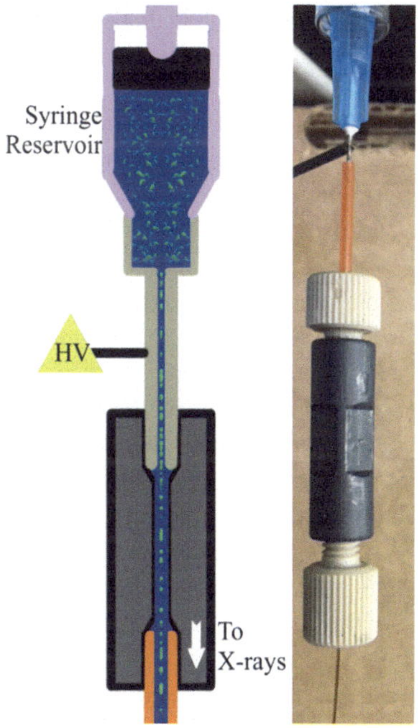

Fig. 5.14 A common syringe with a stainless steel needle as a conduit between the fluid and the applied high voltage (HV). The outer diameter of the needle is fitted with standard polymer sleeves and microfluidic fittings to interface with the capillary of the MESH injection

The second preferred reservoir for MESH injection is to use a syringe and syringe pump as shown in Fig. 5.14. This setup is most amenable to quick and modular deployments of MESH injection; it is especially ideal if compressible gases are not immediately available or an appropriate pressure vessel is not available. Any variety of plastic or glass syringes can be used as a sample reservoir. The needle must be conductive and must have an appropriate outer diameter that can interface with standard microfluidic fittings, such as 1/32" OD gaged needles. As the slurry advances from the syringe to the union, it wets the inner metal surface of the needle, which is charged externally, thus creating the voltage connection needed to perform a MESH injection.

Best suited for crystal slurry injection are glass gas-tight syringes, such as Hamilton's series of syringes. Volumes of 0.050, 0.100, 0.250, 0.500, 1, and 5 mL have all been used for MESH injection. The preferred syringe end is the removable needle variety, as it can handle higher pushing forces while still maintaining some modularity. The glass syringes can be tricky to keep electrically isolated. Most glass syringes have a metal plunger and a Teflon head to create the seal. When loading the syringe it is sometimes possible to have fluid get trapped between the Teflon and the glass wall. If this remains, it will create an electrical contact with the fluid and the metal plunger body, ultimately with the syringe pump. Although most devices are safely grounded, this current leak prevents the MESH injection from properly working.

5.4.1.7 The Effects of Low Reynold's Number Flow

Capillary Selection Fused silica capillaries are commonly used for sample injection in SFX. The appropriate diameter size and the electrical insulation of the fused silica and polyamide coating make it an ideal tubing material for MESH injection. It is worth mentioning that other tubing materials are compatible with the MESH, such as borosilicate glass, quartz, and polymer tubing, such as PEEK.

One of the most important parameters in capillary selection is the inner diameter. As previously discussed, pipe flow is still applicable in μm-sized capillary flows and a useful metric is the hydrodynamic resistance of the system. Much like an element in an electronic circuit has some resistance to the flow of current from an applied voltage, a capillary of fluid will restrict volumetric flow with the application of a volumetric potential, that is, the pressure. The capillary diameter, length and fluid viscosity dictate the resistance to flow and can then be used to estimate the effect of changing these parameters.

Another restriction on the capillary geometry selection is from the sample itself. Again, the fluid properties from a crystal's mother liquor might dictate the viscosity and thus cannot be readily changed; however, the size of the protein crystals themselves begin to limit the geometries suitable for the SFX experiment. In most fluid mechanics calculations applicable to SFX, the fluid is treated as a continuum and species inside of the fluid are treated as dilute, solvated species not causing changes in the bulk properties of the fluid. This assumption cannot be maintained when dealing with a suspension of solid protein crystals on the order of 100s of nanometers to dozens of micrometers in dimension.

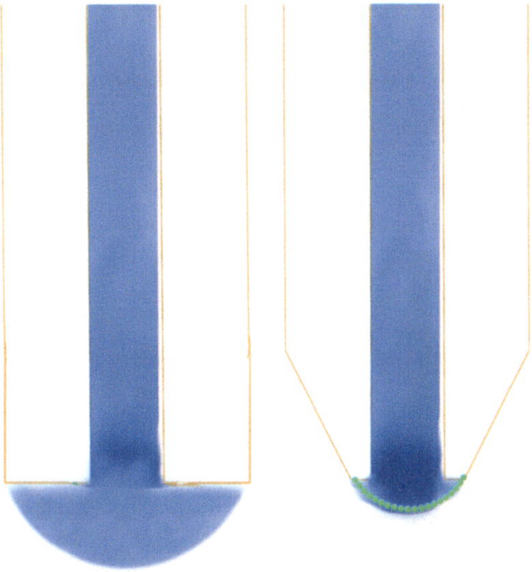

Fig. 5.15 Schematic of a non-tapered capillary (left) with meniscus wetting the exposed flat surface out towards the outer diameter. By comparison, a tapered capillary (right) with the same outer diameter allows for a smaller meniscus and therefore smaller Taylor cone geometry

Tapering or polishing the capillary tip is not a necessity for MESH injections; however, it creates a smaller meniscus and subsequently smaller Taylor cones, thin filaments and progeny droplets. Figure 5.15 shows a schematic of the tapered versus non-tapered geometry. Crystal dimensions on the order of the meniscus size can cause jetting instabilities; for example, 25 μm crystals passing through a 75 μm tapered capillary face will be much more unstable than the same crystals through a non-tapered capillary with a 360 μm capillary face.

The MESH injector flow rates vary from the 100s of nL/min for 50 μm capillaries to almost 3 μL/min for a 100 μm capillary—assuming moderately viscous fluid additives (e.g., 5 mPa·s; water is ~1 mPa·s). Figure 5.16 shows a plot of flow rates at different inner diameters for a MESH-compatible fluid. If the doped fluid has a viscosity different than ~5 mPa·s, then the flow rate will scale inversely; for example, doubling the viscosity will halve the flow rate.

Typical Fluid Properties for MESH In crystallography, the fluid selection is typically governed by the protein crystallization conditions. Changing the properties of the mother liquor once the crystal is grown can lead from mild to severe degradation of the crystals [71] if not done properly; sometimes it is necessary for the experiment to succeed [72] and it can also improve the crystal quality [73]. As mentioned before, to work in vacuum, the MESH must contain some kind of cryoprotectant-like additive that allows it to stay liquid despite evaporation once in vacuum.

Ionic Strength The most basic of requirements for creating MESH injection is the free ions in solution. The free ions build up at the meniscus until a Taylor cone is formed (see previous sections; Gañán-Calvo and Montanero [74] and references within). A typical ionic strength for the fluid is on the order of 100 mM with

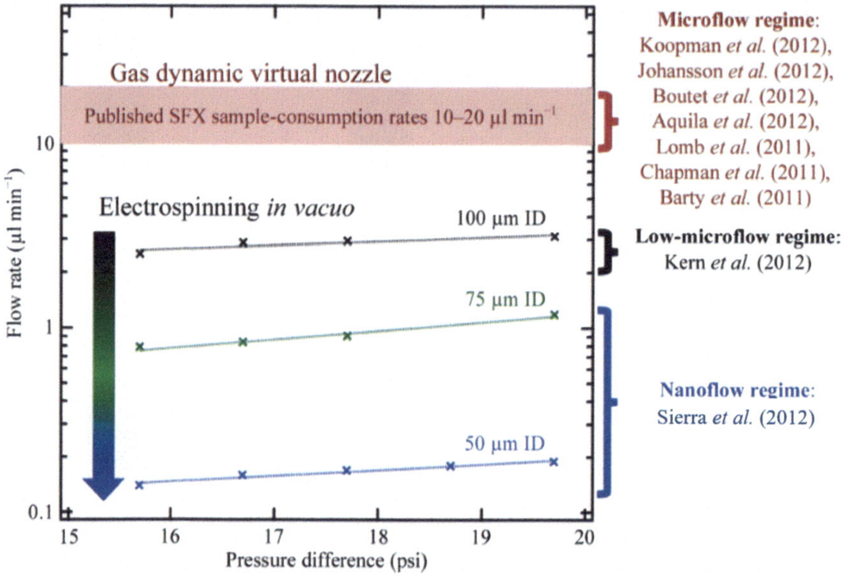

Fig. 5.16 Reproduced with permission of the International Union of Crystallography [53] The figure shows the average flow rate versus overall pressure difference (sample reservoir pressure to nominally 0 psi in vacuum) for different capillary inner diameters on the MESH. The rate was measured by weighing the volume of sample before and after an elapsed injection time. The fluid is a crystal-free solution composed of a common mother liquor: 30% w/v glycerol, 10% w/v PEG2000, 100 mM MES pH 6.5 and 5 mM $CaCl_2$. The solution viscosity is approximately 5 cP. The compared flow rates are from the cited references [3, 22, 55, 66–70]

conductivities approximately 1 mS/cm; however, these values can vary above or below this point.

pH Buffer The free ions can be easily solvated ions, such as the divalent, cationic magnesium ions and the monovalent, anionic chloride ions of a magnesium chloride solution. The free ions can also come from a weak acid, such as anionic acetate ions and cationic sodium ions from a properly titrated acetate buffer solution. Either of these sets of free ions will create a MESH. The flow of ions is irreversible as the fluid leaves the meniscus towards the counter electrode and will thus change the electrochemistry. Without a proper pH buffer the solution can change pH and can adversely affect the protein crystal.

Cryoprotectant Some additives that have successfully created a MESH injection in vacuum have been: 30% glycerol + 10% PEG2000, 40% glycerol, 40% ethylene glycol, 1.4 M sucrose, 40% PEG400, 40% ethanol, 1:1 DMF:THF, 40% MPD. Ionic strengths (pH buffer and salts) of 100 mM are good starting points, with values ranging lower and higher; for example, 15 mM $MnCl_2$ solution has been the only free ions in a successful in-vacuum MESH injection for soft X-ray emission studies. Higher salt values are not prohibitive for the electrokinetics but can be problematic in vacuum; for example, 2.5 M ammonium sulfate injects with the MESH in atmospheric conditions but does not survive the vacuum injection well due to precipitation. A benefit to the cryoprotectant additives is that they can potentially increase the fluid's viscosity, thus reducing the flow rate. It is possible that either the viscosity and/or density can also change favorably to help prevent any sedimentation of the suspended crystals.

Capillary Length At the CXI instrument of LCLS [75], nozzle rods over 1 m long set a minimum capillary length. Some additional length is typically necessary to get from the sample reservoirs to the interaction region. Capillary lengths of 1.5–2 m are therefore typically sufficient for MESH injection. If the fluid is too viscous it is likely that the meniscus will not focus unless higher voltages are applied, the diameter is increased or the capillary is shortened, that is, the effective conductivity is increased.

The MESH injection typically does not use sample shakers to prevent settling. In most cases, the increased viscosity aids in deterring the rate of sedimentation, or as discussed in the reservoirs section, the submerged capillary in the pressure-driven vial reservoir is less subject to the adverse effects of sedimentation. The other reservoir, the syringe-pump-driven approach, is more difficult to rock back and forth. Although early versions of sample shakers for SFX essentially rotated a syringe pump, and smaller syringe pumps [20, 21] have been mounted on oscillation motors to mitigate sedimentation during MESH injection, the settling problem is still prevalent and ultimately not worth the additional complication introduced to the system. In the case of the MESH, many times larger crystals are used since the MESH can accommodate larger inner diameter capillaries, which lead to much faster sedimentation times. If this is inevitable, finding more viscous or dense media might be necessary. Also, priming the capillary with sample slurry as it is freshly suspended will buy some time before the sedimentation in the reservoir prevents the sample from entering the sample capillary. A useful method to monitor settling during the experiment is to bring a 1 mL vial of sample, load 500 μL into the reservoir and keep the other 500 μL in the vial, vertical on a bench in the lab. Once the crystals in the vial have settled a few millimeters below the top fluid surface, it might have settled more than the half of the syringe reservoir's diameter and it is likely time to intervene in order to increase the data collection rate.

5.4.2 Concentric MESH

In order to find a way to increase the effectiveness of the MESH, a concentric design was developed that leverages the laminar flow properties of fluids at that scale and allows a protective outer fluid to flow around and shield the sample-carrying inner fluid from the adverse effects of the vacuum. The idea of concentric fluid flows is not novel, even in the field of SFX—the GDVN uses an outer fluid of helium gas to focus the inner sample-containing fluid. The next logical step is to use another liquid instead of a gas to flow around an inner fluid. The application of this idea to the GDVN is presented in Sect. 5.5. At these low Reynold's numbers the interface between the fluids will be subject mostly to viscous, diffusive forces, rather than inertial, convective forces and mixing on short length scales can be neglected. The diffusion time is low, that is, the fluids remain separate and continue in their typical laminar flow, where the outer fluid only applies a shear rate to the internal meniscus while protecting it from any ambient environment.

MESH is limited to the subset of protein crystal slurries that could tolerate the doping with ∼40% cryoprotectant. Not all crystals can handle this change in their mother liquor and thus the need for a new method was developed the coMESH [21]. In the case of the ribosome and Photosystem II (PSII) structures [21], the MESH and coMESH were ideal to limit the sample consumption; however, the large unit cell of the ribosome made it very sensitive to changes to the mother liquor and necessitated the coMESH. For PSII, the coMESH aided the data collection by protecting the mother liquor from the dehydration of the meniscus [21, 57].

5.4.2.1 coMESH Geometries

Figure 5.17 shows a schematic of the coMESH setup. Commonly, a 100 × 160 μm inner sample line approximately 2 m in length will run uninterrupted from the sample reservoir to the interaction region. The capillary is thin enough that it will pass through common microfluidic tees, such as those from IDEX Health Science (P-727-01) or Labsmith (seen in Fig. 5.17). The inner diameter of the tees is typically 500 or 250 μm, respectively, allowing for fluid to pass around the outside of the inner sample capillary. At the reservoir end of the tee, an appropriate sleeve (e.g., IDEX F-246 for 150 and 160 μm OD tubing and F-240 blue for 200 μm OD) will seal off the outer diameter of the sample capillary. The parallel port on the tee will have the concentric outer capillary and will seal with the appropriate sleeve (IDEX F-242 for 360 μm OD tubing). The length of this tubing is dictated by the experimental geometry. In some cases, the experimental geometry can permit a larger injector mounting surface and thus allowed the Labsmith piece to be near the interaction region [21]. This afforded a shorter concentric capillary approximately five centimeters in length; the short distance, resulted in a centered inner capillary. Since, a MESH-on-a stick [63] geometry has been employed (seen in Fig. 5.12), requiring the off-the-shelf tees to sit outside of vacuum, meaning that the concentric line must now be approximately 1.2 m in length in order to safely sit outside of the vacuum chamber, while sufficiently long to connect the reservoir to the interaction region.

The outer flowing liquid is chosen similarly to the MESH, where a 40% additive helps prevent freezing and can inadvertently increase the viscosity of the fluid. This should be carefully considered and tested prior to the experiment. The inner line carrying the mother liquor and crystals can have much less fluidic resistance than a 10–50 μm annulus, meaning that the fluid that should not be exposed to vacuum

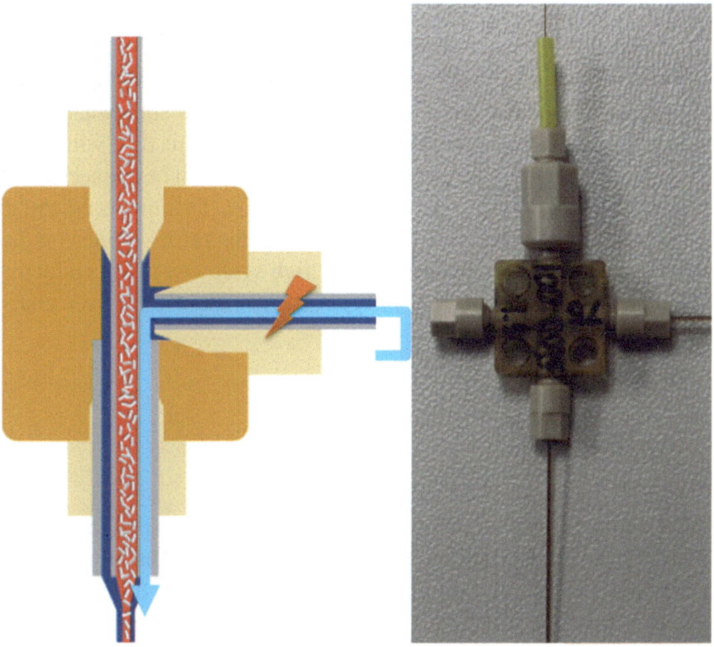

Fig. 5.17 A picture and schematic of the concentric assembly enabling the coMESH. The Labsmith cross has a 100 × 160 μm fused capillary passing through uninterrupted (top channel through bottom); this is the sample line carrying the crystals in mother liquor (red). A concentric line 180, 200, or 250 × 360 μm is connected to the bottom channel of the cross. The length depends on the overall experiments geometry; the inner and outer lines are typically left coterminal. The right channel is of arbitrary geometry sufficient to connect the accompanying sister liquor reservoir and charge conduit to the concentric annulus (blues). Appropriate Labsmith fittings and polymer sleeves are used to interface the capillaries to the cross

to avoid freezing can readily get to the interaction region with potentially no *sister liquor* there to protect it.

This is not to say that the other annulus geometries could not work, rather it suggests that they will require higher driving pressures to ensure sufficient fluid delivery. As discussed in prior sections, the reservoir of the MESH needs to be electrically isolated, which results in the use of plastic pressurized vessels or plastic or glass syringes pushed by a syringe pump. These methods typically are not high pressure applications, with plastic pressure reservoirs only capable of safely holding hundreds of psi (<70 bar) and common syringe pumps having linear forces of approximately 89 N (it is possible to have a syringe pump with a higher linear force). If a commonly-used 1 mL plastic syringe (bore diameter of ~4.5 mm) is used to drive the 160, 180 combination of capillaries it would struggle to achieve the 55 bar (800 psi) required to drive the flow. The syringe plunger might also fail before the syringe pump fails and cause charged liquid to leak behind the plunger and undesired current leaks. This can get worse if the inner fluid is low viscosity and flows faster than a more viscous outer fluid.

The third and final line, attached to the perpendicular section of the tee junction (right of the tee schematic and picture of Fig. 5.17), is somewhat arbitrary, and should likely be a large bore polymer tube or capillary. This tube will interface back to a syringe filled with the appropriate sister liquor and will be connected to allow for charging, as explained in the previous MESH section. This line is essentially acting as a wire connecting the charge to the system. See the prior sections on discussions about making the tube sufficiently short for high viscosity, poor conducting media. A 1-m tube with an ID >250 μm is typically sufficient to deliver charge to the concentric portion.

5.4.3 Conclusion

The MESH and coMESH are two electrokinetic techniques that are used to deliver fluids with a wide range of viscosity to SFX experiments. The power of the applied voltage allows the sample consumption to remain low, ranging from hundreds of nanoliters per minute in a 50 μm ID capillary through 3 μL/min in a larger 100 μm capillary, compared to pressure-driven flows. The MESH and coMESH techniques can tolerate a wide range of flow viscosities, from volatile solvents like ethanol, dimethylformamide (DMF), and tetrahydrofuran (THF) [60–62] to high molecular weight PEGs and possibly more [21, 53, 55, 57, 59, 63]. The electrokinetic injection methods fill a large gap between the other liquid injection methods like the GDVN and LCP (or high viscosity extrusion) methods, which work well for water-like slurries and highly viscous media, respectively.

5.5 Double Flow Focusing and Mixing Nozzles

5.5.1 Challenges in Liquid Jet Sample Injection

Despite tremendous progress in sample delivery for serial crystallography [76] in the past years, there are still many obstacles to overcome in order to make SFX as efficient as traditional macromolecular crystallography. This subchapter will focus on liquid jet technologies and their challenges and possible improvements, whereas other methods to introduce sample into the X-ray focus are discussed in the rest of this chapter. As discussed in Sect. 5.2, the first SFX experiments [3, 68, 77] at LCLS [78] were conducted using GDVN [13] for sample delivery because of the reduced risk of nozzle clogging and the ability to produce jets with flow rates and diameters compatible with the initial serial crystallography experiments. In these initial experiments detector geometries and challenges in data analysis were the limiting factors. With these challenges addressed it became more and more clear that there are certain disadvantages [21, 23, 79] of the GDVN mode of injection:

(a) The low probability of an X-ray pulse interacting with a crystal
(b) Very high sample consumption compared to traditional crystallography
(c) Short runtime of a nozzle
(d) Danger of debris and ice formation (in vacuum) on nozzle tip
(e) Danger of debris and ice growth (in vacuum) from the sample catcher back to the nozzle
(f) The low reproducibility of the GDVN; each GDVN is one of a kind.

Given all these disadvantages that may prevent the success of a serial crystallography experiment, one would expect a shift away from GDVN, or liquid jet injection in general, towards other methods of sample delivery. However there are a number of advantages of liquid jet sample delivery that inspired the community to address the challenges rather than to abandon the technique. Liquid jet injection is ideally suited for fast sample replenishment, the X-ray background is small because most of the time crystals are of similar size to the jet diameter or are even larger, which is apparent in the missing “water-ring” in many diffraction images, and liquid jets can easily be used in vacuum. Most importantly, protein crystals can be injected in their native buffer and the injection method is compatible with light and/or oxygen-sensitive samples. Moreover liquid jet injection can relatively easily be combined relatively easily with all kinds of triggering methods for time-resolved crystallography, including light activation of native chromophores [80–83], mixing [84, 85] and light activation of photo-caged compounds. Double-flow focusing nozzles (DFFN) [23] tackle almost all the abovementioned disadvantages of the GDVN while retaining most, if not all, of the advantages.

5.5.2 Double-Flow Focusing Nozzles

5.5.2.1 Principle of DFFN

A double-flow focusing nozzle (DFFN) is a GDVN (see Fig. 5.18a) with an additional capillary inserted into the liquid capillary. The sample—in most cases a slurry of protein microcrystals—is flowing in the additional innermost capillary and is injected into the liquid stream from the manifold capillary, into which the innermost capillary is inserted, as shown in Fig. 5.18b. This combined liquid jet in turn is focused by helium gas, just as in a normal GDVN, which results in a liquid cone with the sample at its core (see Fig. 5.18b), as described by Gañán-Calvo et al. [86]. The flow conditions to form a jet [87–89] at a certain nozzle geometry [89] only need to be obeyed by the outer liquid and thus (a) the sample flow rate can be changed almost without interfering with the properties of the jet, which means that (b) one can change between samples on the innermost line without the need to interrupt the jet. Both points are very important, since the amount of

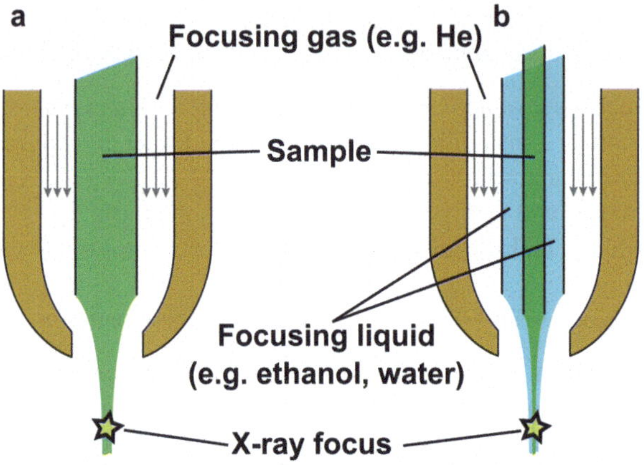

Fig. 5.18 Comparison of GDVN (**a**) with a DFFN (**b**). The sample (usually suspensions of protein nanocrystals or microcrystals in crystallization buffer) is shown in green, the focusing/sheath liquid in blue. In a DFFN (**b**) the sample is injected into an accelerated focusing liquid jet. To achieve this the innermost capillary that delivers the sample protrudes out of the capillary that delivers the focusing liquid. (Figure by Dominik Oberthuer and licensed under a Creative Commons Attribution 4.0 International License)

sample is often limited and one strives to reduce the flow rate as much as possible under the counteracting constraint to retain a stable jet. In a normal GDVN, stable jetting conditions [87–89] are dependent on the flow rate of the sample at the given viscosity and surface tension, and are further affected by of the high content of solids within the samples (crystals). Therefore the sample flow rate cannot be changed independently of the sample composition when using a GDVN. Moreover, changes between samples or between sample and buffer/water always require a short interruption of the jet. Jet interruptions can be highly undesirable, since it increases the risk of nozzle clogging in general but especially in vacuum. Using a DFFN instead of a GDVN thus results in a decreased clogging risk and a decrease in sample consumption, both making experiments more efficient.

It should be noted though that for both GDVN and DFFN, air bubbles in a not properly loaded sample can cause severe problems. These air bubbles are compressed in the pressurized sample delivery system and expand once the pressure drops at the outlet of the nozzle, resulting in jet interruption and splash-back of sample to the nozzle tip. The sheath liquid of the DFFN cannot prevent this from happening and it is thus of extreme importance to load the sample properly and as gas free as possible, even when using a DFFN. The outer/sheath liquid of a DFFN can be almost freely chosen, as long as it is miscible with the sample buffer. Liquids immiscible with the sample buffer lead to severe jet instabilities and thus impair successful serial crystallography experiments [23]. For all liquids miscible with the aqueous buffers used for crystallization of biological macromolecules, one has to consider their viscosity and surface tension, since—at a given nozzle geometry—these two parameters are most important for determining jetting conditions [87–89].

The focusing liquids should be non-toxic, contain no impurities and be available at low cost and large quantities. For X-ray background considerations they should consist of mostly low-Z atoms. Ideal in this respect are short alcohols, such as ethanol or 2-propanol. In some cases interference of the sheath liquid with the buffer upon slow diffusive mixing of the two co-flowing liquids poses a problem. This has been seen recently in the case of high concentrations (>500 mM) of sulfate-anions in the buffer where the immediate formation of salt crystals at the contact area to the alcohol sheath leads to clogging of the nozzle. In those cases other sheath liquids such as ultrapure water can be used. Sheath flow rates for a stable jet are slightly higher for pure water than for ethanol or 2-propanol, due to different viscosities and surface tensions, thus using water as a sheath liquid results in higher X-ray background due to a larger jet diameter. At the same sheath liquid flow rates, ethanol yields a slightly higher hit-fraction (fraction of X-ray pulses that interact with a protein crystal) than water [23]. Moreover for the case of injection into vacuum, the lower freezing point of the short alcohols compared to water effectively prevents the freezing and icing problems—both from the nozzle tip and from the sample catcher. Since the formation of ice in vicinity of the X-ray interaction point poses a big risk for the detectors (death by Bragg-reflection), the use of a DFFN with sheath liquids, primarily alcohols, can increase the lifetime of the very expensive detectors. Furthermore, the use of DFFN essentially eliminates ice or solidified desiccated sample growing from the catcher or the shroud used to contain the sample, which in most cases requires venting of the chamber, leading to interruptions of the experiment and less efficient use of the beamtime available.

5.5.2.2 Development of DFFN

Double flow focusing in combination with gas dynamic virtual nozzles (GDVN) was first reported by Gañán-Calvo et al. [86] in 2007. In this paper they investigated double flow focusing using simple (non-biological) liquids, in which the sheath liquid was immiscible with the central liquid. The goal of their research was the development of ultra-thin (sub-micron diameter) jets for biotechnology and medical applications. They achieved double flow focusing by placing a capillary within a manifold capillary and positioned the inner capillary such that its tip extended out of the manifold capillary by about one capillary diameter. The sheath flow was gas focused through a round orifice within a plate [86]. This nozzle design cannot be used for serial crystallography, since the aperture plate would be too close to the X-ray interaction point and it would block the scattered X-rays on the way to the detector. In a second approach, Wang et al. [33] adapted this design and integrated it within a GDVN design typically used for serial crystallography experiments to create a nozzle for mix-and-diffuse [90] time-resolved experiments. To achieve mixing they made a nozzle with a retracted inner capillary in order to inject the sample into an outer flow containing a ligand. Due to the retracted inner nozzle, the sample is injected into the unfocused outer flow. The sample solution is focused to a smaller diameter by the sheath liquid, which flows at a higher speed. The focused

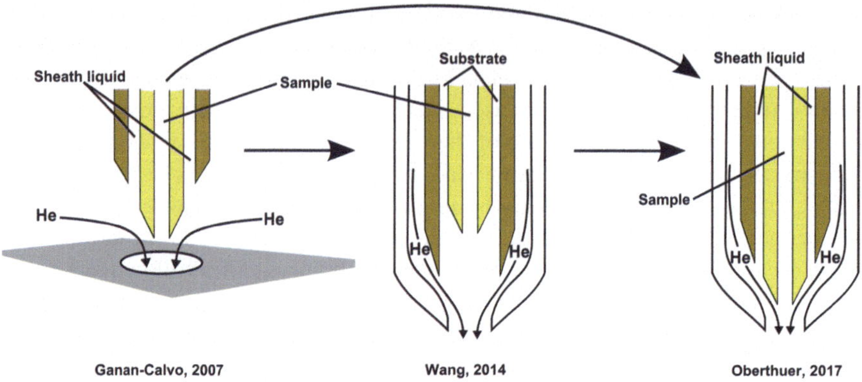

Fig. 5.19 Evolution of double flow focusing nozzles. The first double flow focusing GDVN, presented by Gañán-Calvo et al. in 2007, had no outer glass capillary for the focusing gas. In 2014 Wang et al. repurposed the double flow focusing idea to make a microfluidic mixer for time-resolved crystallography. The DFFN successfully tested and used for serial femtosecond crystallography experiments at LCLS [23] combined the protruding innermost capillary introduced by Gañán-Calvo with the outer glass capillary from the design by Wang. (Figure by Dominik Oberthuer and licensed under a Creative Commons Attribution 4.0 International License)

sample and sheath liquid flow co-axially and mix through laminar-flow diffusive mixing. The homogenized mixture of sample and ligand would then be focused and accelerated together into a single jet, with crystals ideally being aligned in the center, to get a large hit fraction and a homogenous time delay. It has been shown with fluorescence imaging that mixing can be achieved with such a nozzle. However the parameters for successful mixing (ratio of flow $\text{rate}_{\text{ligand}}$–flow $\text{rate}_{\text{sample}}$) again impaired [91] its use for serial crystallography experiments. For the DFFN, both approaches have been combined (see Fig. 5.19 for a comparison) where a similar gas-focusing layout is used as in Wang et al. [33] and a capillary arrangement with the inner capillary protruding out of the outer capillary as in Gañán-Calvo et al. [86]. This design avoids diffusive mixing and creates a stable double focused jet. Initially sheath liquids immiscible with the aqueous buffers were used. With sheath liquids such as n-decane, n-hexane, and ethyl acetate, double flow focusing at low flow rates could be achieved and detected with fluorescence imaging.

5.5.2.3 Use of DFFN at LCLS

A challenge for any experiment at X-ray FELs is the interface of the sample delivery device with the experimental chamber (if) under vacuum. For liquid jets this can be facilitated by a load lock through which nozzles that are attached to a long pipe (nozzle rod) can be exchanged without the need to vent the vacuum chamber (see Sect. 5.2). There have been approaches to decouple nozzle exchange from the use of the nozzle rod, but this requires hardware to be installed within the experimental chamber [92] and is challenging for short beamtimes such as Protein Crystal Screening (PCS) experiments. The biggest disadvantage of the nozzle rod is its narrow inner diameter. All liquid and gas lines have to go through this rod and have to be connected in such a way that it is compatible with the vacuum environment within the experimental chamber. For normal GDVN this is not a problem: liquid line and gas connection are put in place in a small metal tube using epoxy-glue and sealed with standard HPLC-fittings [79] to a connector to the nozzle rod. For the double flow focusing nozzle, the central sample capillary must be embedded into the sheath liquid line. In the proof-of concept experiments, embedding of the capillary was achieved using a standard HPLC T-connector. The diameter of this connector however is too large for the typically used nozzle rod. Therefore, an injector was designed that is compatible with the experimental setup. The injector connects directly to the lower end of the nozzle rod and all fluid lines are arranged to form a DFFN. The design was mostly restrained by the diameter of the differential pumping tube in use at the time, through which the nozzle rod is inserted into the load lock and vacuum chamber. The first custom built DFFN injector used at LCLS can be seen in Fig. 5.20a, b. It was subsequently improved, slightly miniaturized and its production automated through metal 3D-printing (sintering) techniques (Fig. 5.20c). These DFFN were successfully tested at CXI/LCLS and it could be demonstrated that (a) much lower sample flow rates are possible (5 vs 30 μL/min), (b) the jet is very stable and the X-ray background is lower than for jets from a GDVN, (c) sample change is possible without stopping the jet, (d) no ice formation/debris formed on the nozzle tip, (e) no ice/debris grows back from sample shroud/catcher and (f) the nozzles run a long time without clogging. Almost all of the challenges described above for GDVNs could be addressed with the use of a DFFN, and it was now possible for the first time to obtain enough diffraction data to solve the room temperature structures of RNA polymerase II and a dioxygenase [23].

5.5.3 From Art to Science: 3D Printed Nozzles

One challenge not tackled by the DFFN design described above was the low reproducibility of the handmade nozzles, that is, the quality of the assembly of capillaries and outer glass piece strongly depends on technical experience and craftsmanship. Ideally the nozzles would be mass-produced and all nozzles would have the same jetting properties. Moreover, there should be a certain flexibility of design to accommodate the diversity of possible experiments and sample buffers. Both requirements are met by rapid-prototyping technologies, and early attempts used PDMS-based soft lithography [93]. However, even with the use of microfluidic device fabrication technology, the nozzle quality still strongly depends on the capabilities of the technician and again not every nozzle of one batch is the same. Recently 3D-printing technology [94] with submicron-accuracy was introduced, and such accuracy enables the design, rapid prototyping and production of nozzles

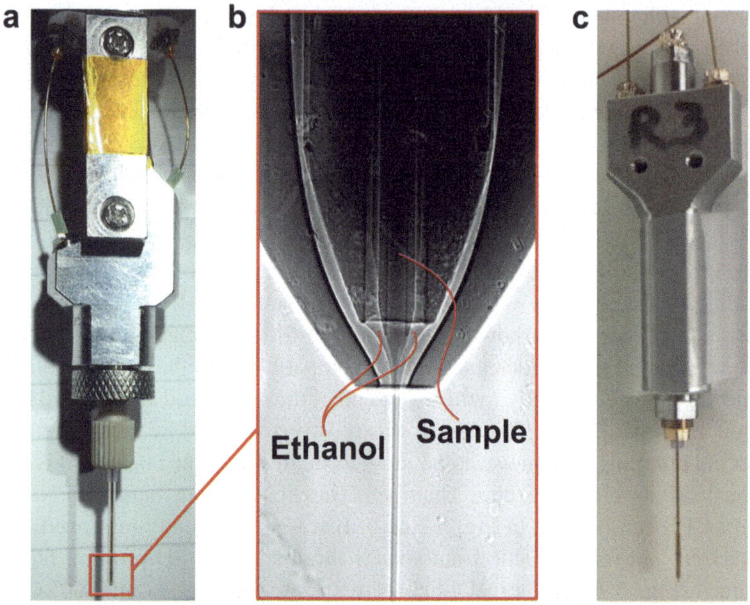

Fig. 5.20 Double flow focusing nozzles used at LCLS. During the first successful experiments with DFFN at LCLS a nozzle body as shown in (**a**) was used. It was machined from pieces of bronze and steel. In (**b**) an X-ray radiograph (beamline P05/PETRAIII, Deutsches Elektronen-Synchrotron, DESY and Helmholtz-Zentrum Geesthacht, HZG) reveals a detailed view of a DFFN. 4 M potassium iodine solution ("sample") is focused by ethanol, which is being focused to a jet by helium gas. A second generation DFFN, produced by metal 3D-sintering is shown in (**c**). All injectors consist of inner line for sample, sheath liquid line, and focusing gas line. They can be mounted to the standard "nozzle rod" at the CXI endstation of LCLS using an adaptor piece or easily in other atmospheric pressure instruments at existing X-ray FELs. The second generation can be operated without this adaptor piece when mounted to other instrumentation such as a goniometer. The position of the inner capillary relative to the gas orifice of the nozzle and the end of the sheath-liquid capillary can be adjusted by a screw, to optimize jetting conditions. (Figure was adapted from Oberthuer, D. et al. Double-flow focused liquid injector for efficient serial femtosecond crystallography. *Sci Rep* 7, 44628 (2017) and is licensed under a Creative Commons Attribution 4.0 International License)

for serial crystallography. Nelson et al. reported the first 3D-printed GDVN and its use in serial crystallography experiments at LCLS [32]. The nozzles were designed using a CAD-program and printed with a Photonic Professional GT (Nanoscribe, Germany). One disadvantage of this technology is that the high resolution of printing comes at the price that only small volumes can be printed within a reasonable time frame. While the actual nozzle tip and the fluidic channels are small enough, the fluid connections to the sample reservoir and the lines for the focusing gas pose a challenge. In the future there might be possibilities to directly print the nozzle tip onto connecting lines. However Nelson et al. used a different approach in their work [32]. They printed the nozzle wide enough such that a sharpened fused glass capillary could be glued to the sample channel and the whole nozzle-capillary assembly was then glued to a larger glass capillary, which in turn forms the connection to the gas line. This still requires a certain set of manufacturing skills, but the critical parts for jetting, the geometry of the nozzle tip (gas orifice diameter, distance of capillary tip to orifice, angle of outer capillary) are 3D-printed and thus independent of the manufacturer. Each nozzle design can now be printed, processed using developed protocols and tested. The feedback from testing is in turn used to evolutionarily optimize the nozzle design.

5.5.4 Development of Mixing Injectors for Time-Resolved Studies

Liquid jet injection can be combined with a variety of triggering methods for time-resolved serial crystallography [95–98] (see Chaps. 11 and 12). For light-activated processes [66, 82, 83, 99], GDVN and DFFN nozzles can be used as is, or slightly modified with an opaque coating on the liquid lines to avoid pre-activation through stray light. For the study of chemically activated processes (e.g., enzyme-catalyzed reactions) the nozzle needs to be combined with a mixing device (mix-and-diffuse approach [90]). In the mixer the sample under investigation is mixed with a chemical compound that binds to the sample and triggers a response in the sample. Mixing should occur fast and in a controlled way throughout the whole crystal, which requires crystals smaller than 10 μm, since reduced size decreases the diffusion time of the substrate into the crystal. Over the years various microfluidic mixing technologies have been developed for studies with proteins in solution [100–104]. For serial crystallography these need to be adapted to the special case of a thick microcrystalline slurry at high salt content and at times high viscosity. Wang et al. [33] reported a first mixing nozzle (see Fig. 5.19) for serial crystallography that was not suited for actual serial crystallography experiments due to the required high dilution factor of the sample, which would lead to very low hit rates. Another approach by Trebbin et al. [93] is based on PDMS soft lithography and has the mixer integrated with the nozzle in a double-flow architecture similar to that of Wang et al. A clear advantage is the design flexibility of PDMS in comparison to the limitations of a manifold of fused silica fibers. Thus it should be possible to achieve sufficiently short time delays for many enzyme reactions at acceptable hit-fraction using PDMS based mixing nozzles. In the first two successful mix-and-diffuse experiments at LCLS [84, 85] longer time delays were probed and thus a simple T-mixer introduced into the sample line more than 1 m upstream of the X-ray interaction point could be used. Calvey et al. [91] presented a more sophisticated mixing nozzle for shorter and more defined mixing time points (see Fig. 5.21). The glass-based nozzle with integrated mixer is inspired by the double-flow design of Wang et al. [33] and is improved such that fast mixing can be achieved at a better ratio of ligand–sample (from 2000:1 to 30:1) [91]. This dilution of sample can still be tolerated in respect of the expected hit-fraction during an experiment at LCLS or SACLA. The resulting

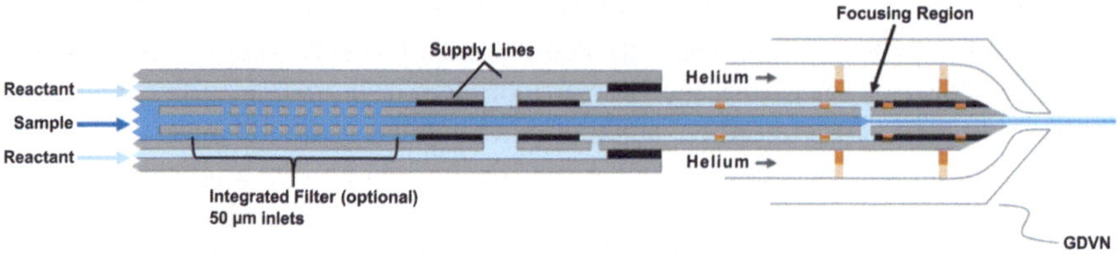

Fig. 5.21 Advanced flow focusing mixing nozzle. In the focusing region sample and reactant mix through laminar flow diffusive mixing. The reacting mixture is then accelerated by Helium, just as in a normal GDVN. The distance between the focusing region and the acceleration region and the flow rates of sample and reactant define the time-delay to be probed. (Figure was adapted from Calvey, G.D., Katz, A.M., Schaffer, C.B. & Pollack, L. Mixing injector enables time-resolved crystallography with high hit rate at X-ray free electron lasers. *Structural Dynamics* 3, 054301 (2016) and is licensed under a Creative Commons Attribution 4.0 International License)

reduction in hit-fraction is even less than 1/30, if it is considered that the design leads to a focusing of the sample stream and that most crystals (if sufficiently small) can be expected to align with the flow, in a narrow channel near the middle of the jet.

5.5.5 Outlook

Many of the challenges in liquid jet injection have been addressed in the past years and a clear trend has emerged in moving away from manual assembly of nozzles towards automation. The inherent precision, accuracy, reproducibility and three-dimensional flexibility of modern 2-photon polymerization direct laser writing (2PP-DLW) techniques is—at least for now—the superior way to make GDVN. And there is no reason why the same techniques should not be employed to make DFFN and mixing nozzles. A combination of a 3D-printed DFFN with a mixer printed close to the nozzle tip should yield an excellent injection device for long sought time-resolved crystallography experiments at high spatial and temporal resolution of enzymatic reactions. These nozzles would be highly reproducible, they could be produced in large numbers in an automated fashion and they could easily be adapted to different experimental parameters. A combination of experimental evolutionary rapid prototyping and fluid dynamic simulations (as already described for the double-flow focusing nozzles [23] and recently for GDVN [105]) would yield the ideal nozzle for every sample and ligand-solution composition and desired time delays. Fluid-dynamic simulations, fluorescence imaging and UV-VIS-spectroscopy could be used to investigate the properties of the mixers and to optimize the mixing process, building up on years of experience with microfluidic mixers in other fields of science. The next step is now to develop methods for automated sample exchange to further reduce sample consumption and to integrate the sample reservoir close to the nozzle. A possible future operation mode of LCLS-II for hard X-rays at repetition rates of $\gg$1000 Hz, and the recently operational European XFEL, will enable the collection of full data-sets (one static structure, one time point) in about a minute, in principle. Mass produced 3D-printed nozzles and automated sample exchange will need to be able to cope with this data-collection speed and will enable high-throughput and high-output structural biology of fascinating targets.

5.6 Highly Viscous Carrier Media and Droplet Injector

We describe two methods, a matrix carrier technique and droplet injection, for serial sample delivery at SACLA using X-ray free-electron lasers. The first is a high viscosity micro-extrusion injector, extruding viscous samples such as lipidic cubic phase (LCP), which has been installed at SACLA. The micro-extrusion technique was also employed with other highly viscous matrices as a crystal carrier, such as grease and oil-free hydrogel medium, for serial sample delivery, which allows the structure determination of a wide variety of proteins with low sample consumption, typically less than 1 mg. In the second method, a piezo-driven droplet injector that delivers single drops containing crystals to synchronize with individual X-ray pulses was developed.

5.6.1 Serial Sample Delivery at SACLA

Serial femtosecond crystallography (SFX) [3, 67, 76, 78, 106] using XFELs has the potential for structure determination by collecting diffraction signals up to a few angstroms in resolution, even from submicrometer-size crystals. This has greatly improved the understanding of the room temperature structures of proteins [22, 40, 46, 48, 77, 107–110], and offers new opportunities for time-resolved studies of light-driven structural changes and chemical dynamics [80, 83, 111–114]. A single-pulse X-ray exposure will completely destroy small individual crystals; therefore, fresh specimens must be supplied for subsequent X-ray pulses to continue data acquisition. Liquid jet injection of small protein crystals using a gas dynamic virtual nozzle (GDVN) is often exploited for serial sample delivery [13]. The GDVN can provide fundamental and crucial insights in double-focusing mixing jet (liquid mixing jet) experiments in SFX to study the real-time enzyme-substrate interactions as well as reactions [33, 85, 91]. In the SFX research at SACLA (the SPring-8 Angstrom Compact Free Electron Laser), a liquid jet injector was installed into the experimental instruments of the DAPHNIS (Diverse Application Platform for Hard X-ray Diffraction in SACLA) system, which basically consists of a sample chamber, fluid injectors and a two-dimensional detector [39]. Continuous flow at a relatively high speed ($\sim$10 m/s) and high flow rate results in 10–100 mg of protein used, that is, the amount of sample consumption for structure determination is significant. Moreover, due to the high speed of the jet, the liquid-jet injectors may not be ideal for X-ray FELs with low repetition rates (e.g., 30 or 60 Hz at SACLA), because most of the sample is wasted between X-ray pulses. In addition, the implementation is complicated by a variety of factors, including difficulties imposed by viscous solutions and unpredictable trajectories of drops that contain crystals of varied shapes and sizes. For membrane proteins, a lipidic cubic phase (LCP) injector with a low flow rate was developed by Uwe Weierstall and coworkers at Arizona State University [45]. However, a more universal method for serial sample delivery that is applicable to a wide variety of soluble and membrane proteins is essential to firmly establish the SFX method, because the LCP micro-extrusion technique is generally limited to membrane proteins crystallized in LCP or crystals that do not dissolve when mixed with LCP. We have since introduced mineral oil-based grease as a protein carrier in SFX [40]. We originally developed an LCP injector consisting of a hydraulic cylinder, a removable sample reservoir, and a nozzle using the same basic concept of the LCP injector as reported by Weierstall et al. Using the LCP injector, the sample injection method with viscous media is technically simple. Micro-extrusion of specimens using a viscous media such as LCP [45], grease [40, 115, 116], Vaseline (petroleum jelly) [41], agarose [42], hyaluronic acid [115], or hydroxyethyl cellulose [116] can maintain a stable stream at a low flow rate of 0.02–0.5 μL/min, which helps to reduce sample consumption to less than ~1 mg. At SACLA the sample injection method with viscous media is mainly used for SFX as well as LCP-SFX. In addition, a piezo-driven droplet injector, that delivers single drops containing crystals to individual X-ray pulses, was developed for viscous-medium sensitive proteins [109]. We describe these two approaches for serial sample delivery here.

5.6.2 *Highly Viscous Carrier Media*

5.6.2.1 Grease Matrix

In protein X-ray crystallography, mineral oil is used as a cryoprotectant [117]. The mineral-oil based grease [40] provides protection against cracking and dissolution of protein crystals. The grease-matrix based approach is applicable to structure determination for a wide range of proteins at room temperature requiring less than 1 mg of the sample. The sample preparation in this technique can be performed by simply mixing crystals with a matrix medium (details described below in the Sect. 5.6.2.3). We have also introduced Super Lube synthetic grease (synthetic grease) [115] and Super Lube nuclear grade approved grease (nuclear grease) [116] to SFX, to reduce background scattering commonly observed with mineral-oil based grease. Weaker background scattering was noted when using nuclear grease compared with other grease matrices. To date, the adaptability of grease matrices in SFX has been demonstrated at SACLA using a wide variety of soluble and membrane proteins: lysozyme, glucose isomerase, thaumatin, fatty acid-binding protein type 3 or proteinase K [40, 115, 116, 118], copper nitrite reductase [107], photosystem II (PSII) [114], luciferin-regenerating enzyme [119],

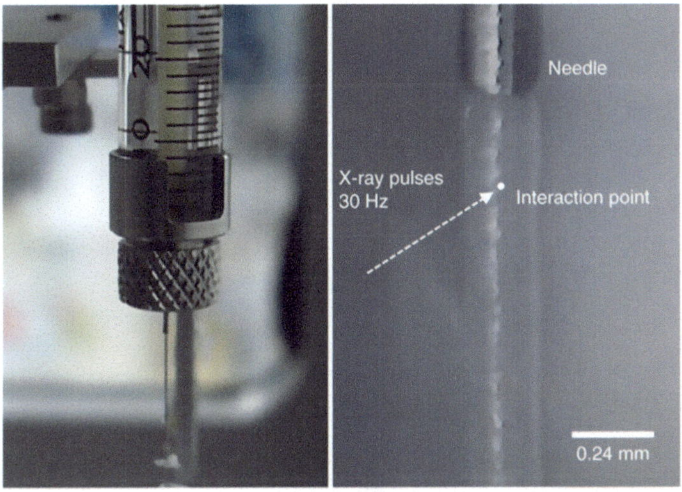

Fig. 5.22 (*left*) A 100 μL commercial glass syringe (Hamilton), is shown extruding samples mixed into a grease matrix. (*right*) A zoomed image of the interaction region with X-ray interaction region labeled. Modified with permissions from Sugahara et al. [40]

the photoswitchable fluorescent protein IrisFP [120], bacteriorhodopsin [121], and bacterial phytochrome [122]. These results suggest that grease has potential as a versatile matrix carrier. Figure 5.22 shows the simplicity of the extrusion, as it utilizes many commercial components and needs minimal customization.

5.6.2.2 Hydrogel Matrix

In spite of its versatility, dissolution of crystals in soluble and membrane protein samples in the grease matrix is occasionally observed. In addition, the grease media tends to produce stronger X-ray scattering in the resolution range of 4–5 Å, which increases the noise level. A new crystal carrier with low background scattering (e.g., agarose [42]) is essential to improve the signal-to-noise ratio. Recently, hyaluronic acid [115] and hydroxyethyl cellulose matrices [116] were introduced for SFX experiments at SACLA. The hydrogel matrices have lower background scattering compared to the diffuse scattering generated by a grease matrix in the resolution range of 4–5 Å. There are no significant differences of the background scattering between hyaluronic acid and hydroxyethyl cellulose matrices. The usability of hydrogel matrices was confirmed for all oil-sensitive crystals that were tested. The hyaluronic acid and hydroxyethyl cellulose matrices provide alternative choices for grease sensitive protein crystals. The hydroxyethyl cellulose matrix is less adhesive than the hyaluronic acid matrix and prevents clogging of the suction tube used as a sample catcher [41] and also prevents adhesion of the matrix to the injector nozzle surface. Using the hydroxyethyl cellulose matrix, the proteinase K crystal structure

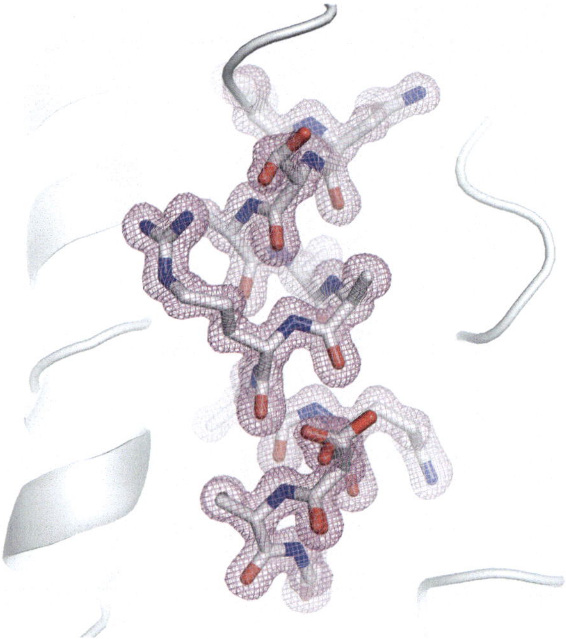

Fig. 5.23 A close-up view of the proteinase K from *Engyodontium album* at 1.20 Å resolution with a $2F_O–F_C$ electron-density map contoured at the 1.5σ level. Reproduced with permissions from Masuda et al. [123]

was determined at 1.20-Å resolution (Fig. 5.23) [123]. A total of ~82,000 indexed patterns were collected at 13 keV photon energy. This atomic resolution structure clearly allows for visualization of the hydrogen atoms forming hydrogen bonds in secondary structures.

5.6.2.3 Matrix Preparation

In the matrix technique using viscous media, the first step is to find a carrier for the protein crystals of interest that is suitable for data collection at room temperature. In SFX, a grease or hydrogel matrix may not always be useful, because some proteins are damaged while being mixed and soaked in these media. Viscous media tends to cause cracking and dissolution of protein crystals due to various physical or chemical events, such as osmotic shock arising from the properties of the viscous media. Grease has potential as a versatile matrix carrier, but the hydrogels, such as hydroxyethyl cellulose and hyaluronic acid matrices, would enable SFX experiments for grease-sensitive protein crystals or data collection with low background scattering. Grease and hydrogel crystal carriers are therefore complementary.

To maintain a continuous sample column, a grease matrix is typically created as a grease and crystal solution in a ratio of 9:1 (*v*/*v*). As an example, after a 100-μL sample of crystal solution (e.g., a crystal number density of ~10^7 crystals/mL) is centrifuged, a 90-μL aliquot of supernatant solution is removed. A 10-μL aliquot of the crystal solution is dispensed into 90 μl of grease on a glass slide and then mixed.

The hydrogel concentration is ideally ~10 to 20% (*w*/*v*). The hydrogel (hydroxyethyl cellulose or hyaluronic acid) solution is saturated with the crystal harvest solution, or the supernatant solution from the crystal suspension solution can be used for the protein crystals. As an example, after a 100-μL sample of crystal solution is centrifuged, a 90-μL aliquot of the supernatant solution is removed. A 10-μL aliquot of the crystal solution is dispensed into 90 μL of a ~16% (w/v) hydroxyethyl cellulose solution on a glass slide and then mixed with a spatula. In some cases, a hydrogel aqueous solution (without buffer) acts as a matrix. However, optimizing the hydrogel solution buffer is important to avoid potential osmotic shock to the crystals after mixing with the medium.

Dehydration of protein crystals can be induced during the sample preparation process of the water-free grease matrix. The unit-cell axes of the lysozyme crystals for the grease matrix are slightly shorter than those for the hyaluronic acid matrix [115]. In such cases, a water-based hydrogel medium can be helpful for preventing the contraction of the unit cell in SFX experiments. Although the medium mixing technique using a syringe coupler may prevent crystal dehydration [42, 124], the technique may cause mechanical damage to brittle crystals by the physical contact between crystals and the interior surface of the coupler, resulting in the deterioration of diffraction quality. Simple and quick mixing with a spatula on a glass slide [40] would be better for brittle crystals.

5.6.2.4 Matrix Extrusion

An increase in the hit rate can be accommodated with increased crystal number densities, which could, however, cause multiple sample hits in one shot and complicate the indexing procedures. A crystal number density of ~10^7 and ~10^8 crystals/mL is suitable for SFX data collection under reported experimental conditions using ~100- and ~50-μm-i.d. nozzles, respectively. For micro extrusion of the matrix, a sample column with a smaller diameter contributes to reducing sample consumption and background noise from the matrices. The conventional grease matrices (mineral-oil based grease [40] and synthetic grease [115]) extruded through a 110-μm-i.d. nozzle tended to produce a larger-diameter grease column (approximately ~210 μm) about the size of the outer diameter (o.d.) of the nozzle. On the other hand, the nuclear grease and the ~16% (w/v) hydrogel matrices are extruded as a continuous column with a diameter of ~100 μm through a 100-μm-i.d. nozzle.

From ~30 μL of protein crystals (size 5–30 μm, crystal number density ~10^7 crystals/mL) dispersed in a matrix, a total of ~20,000–30,000 indexed patterns were typically recorded (indexing success rate ~30%) within 1 h at a 30 Hz repetition rate and a flow rate of ~0.5 μL/min through a 100-μm-i.d. nozzle. When the matrix is passed smoothly through a 50-μm-i.d. nozzle a reduced flow rate as low as 0.1 μL min^{-1} is sufficient for the 30-Hz operation. We were able to substantially lower sample consumption, using only ~0.2 mg of the sample instead of ~1 mg with a 100-μm-i.d. nozzle.

5.6.2.5 De Novo Phasing

For the de novo phasing of macromolecules, experimental phasing has been mainly performed on heavy-atom derivatives of protein crystals [63, 107, 116, 119, 121, 125–127]. Using the grease matrix the phasing was successfully performed by Hg-SIRAS (single isomorphous replacement with anomalous scattering) [119], Cu-SAD (single-wavelength anomalous diffraction) [107] and I-SAD [121] at SACLA. Additionally, the de novo structure determination of proteinase K from Pr-derivatized crystals was demonstrated using hydroxyethyl cellulose matrix [116]. In total, 2000 images (native/derivative: 1000/1000) were sufficient for SIR and SIRAS phasing of proteinase K, while SAD phasing required 3000 images. For SFX, native sulfur SAD phasing was also successful [50, 118, 127]. Using the grease matrix, the structure of native lysozyme was determined with SAD by utilizing the anomalous signal of sulfur and chlorine (Fig. 5.24) [118]. Native SAD phasing was successfully demonstrated using femtosecond X-ray pulses at 1.77 Å wavelength using 150,000 indexed patterns to 2.1 Å resolution. These results demonstrate that de novo phasing can now be used for SFX. However, SFX requires a higher volume of crystals and many hours of beam time for a large degree of data redundancy to effectively utilize the relatively weak anomalous signal. One of the major challenges for phasing in SFX is to improve the signal-to-noise ratio. Techniques using matrices with low background scattering noise will contribute significantly to the measuring of weak anomalous signals for de novo phasing from SFX data.

5.6.2.6 Other Applications

Room-temperature measurements enable time-resolved studies and eliminate the need to find a suitable cryoprotectant; however, in time-resolved experiments, the sample consumption is significant. Matrix carriers with a stable sample flow and low sample consumption should be applicable for time-resolved studies using pump–probe techniques. The structural changes of photosystem II (PSII) induced by 2-flash (2F) illumination at a resolution of 2.35 Å have been demonstrated using the grease matrix techniques and femtosecond X-ray pulses from SACLA at BL3 [114]. Matrix carriers have various applications in SFX experiments, such as femtosecond to millisecond time-resolved studies of light-driven structural changes, and chemical dynamics using pump–probe techniques.

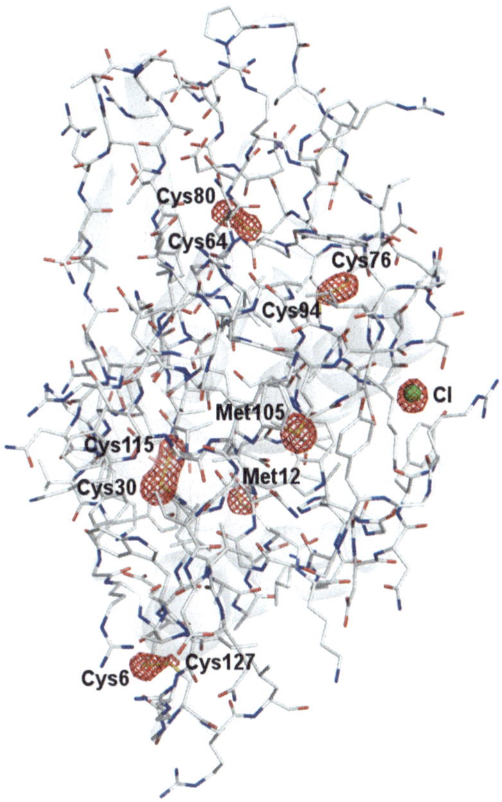

Fig. 5.24 Anomalous difference Fourier map (contoured at 6.0σ) calculated by ANODE [128], showing sulfur and chlorine atoms. Reproduced with permission of the International Union of Crystallography [118]

Synchrotron-based serial crystallography data collection at room temperature using viscous carrier media has been repeatedly demonstrated in recent years [41, 52, 129]. In the immediate future, sample delivery techniques using a viscous medium that help reduce sample consumption will become more important in serial millisecond crystallography using synchrotron radiation. It is valuable to provide a wide repertoire of carrier media for a wide variety of proteins, but a versatile carrier medium would be preferable.

5.6.3 Droplet Injector

5.6.3.1 Introduction

The microcrystal extrusion technique using highly viscous carrier media allows the collection of a complete dataset from less than 1 mg of protein. While the technique has an advantage for sample delivery, highly viscous carrier media are prone to breaking protein crystals due to the physical shock during mixing or by chemical damage. Some viscous media, such as grease and lipids used in LCP, produce intense X-ray scattering, resulting in high background noise that is likely to affect very weak signals such as resulting from small structural changes in a time-resolved SFX experiment.

To address these issues, another sample delivery method that utilizes pulsed liquid droplets was developed [109]. It has been noted that it is difficult to maintain synchronization of droplets loaded with microcrystals at the higher repetition rate of X-ray laser pulses such as LCLS, which fires at 120 Hz [130]. On the other hand, the X-ray FELs at SACLA fire at a maximum repetition rate of 60 Hz, so it is not necessary to continuously supply microcrystal samples to the intersection point, which allows for discontinuous sample delivery synchronized to the beam.

The experiment demonstrated protein microcrystals in a crystallization buffer ejected from a piezo-driven droplet nozzle as a pulsed liquid droplet, controlled in time and space, resulting in synchronization to XFEL pulses at 30 Hz. In this sample delivery method, microcrystals can be introduced in a pulsed manner, which reduces sample consumption compared to the continuous sample delivery. Since the method does not require additive materials such as lipids or oil, the background noise is lower than that from oil-based viscous media.

5.6.3.2 Droplet Injector Configuration

The droplet injector basically consists of a piezo-driven nozzle, a sample reservoir, a pressure controller (SF-100S, Microjet Inc.), and an electric pulse generator (IJK-200H, Microjet Inc.) (Fig. 5.25). The pressure controller is used to adjust the pressure on the loaded sample solution to keep the meniscus at the nozzle tip in an optimum position. The electric pulse generator provides pulsed voltages to a piezoelectric element in the nozzle at the same repetition rate as the X-ray FEL. The optimum amplitude and duration of the electric pulse (V and Δt in Fig. 5.25) changes depending on the samples used. Parameters of 80–100 V and 30 μs were used for an SFX experiment with lysozyme crystals, as described below. The timing of the electric pulse for droplet ejection is controlled by a delay generator (DG645, Stanford Research Systems Inc.) that is synchronized to the X-ray FEL pulses. Droplets from the nozzle are monitored by using a CCD camera (IPX-VGA120-LMCN, Imperx Inc.) with a strobe light.

5.6.3.3 Example of SFX Experiment Using Droplet Injector

The capabilities of the droplet injector were demonstrated using ca. 5-μm lysozyme crystals suspended in buffer solution (1 M sodium acetate buffer (pH 3.0) and 10% sodium chloride). The injector was installed into a chamber in DAPHNIS [39] and consists of a piezo-driven droplet nozzle with an 80-μm aperture (IJHDS-1000, Microjet Inc.). The droplet is basically spherical but at times slightly elongated with a diameter almost the same as that of the nozzle aperture (80 μm). Therefore,

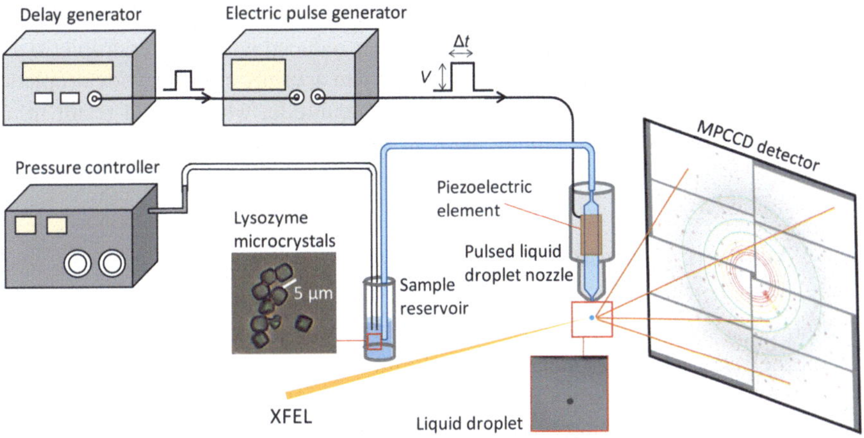

Fig. 5.25 Experimental setup of SFX using droplet injector: pulsed liquid droplets containing microcrystals are ejected from a piezo-driven nozzle, to which a pulsed electric voltage is applied with an electric pulse generator. The droplets are introduced into a DAPHNIS chamber (not shown here), the inside of which is filled with helium. The timing and the position of the pulsed-droplet nozzle is adjusted with a delay generator and a motorized stage. Synchronization with XFEL is monitored with a CCD camera. Reproduced with permission of the International Union of Crystallography [109]

the sample flow rate was about 30 μL/h when operated at 30 Hz. The sample solution in the reservoir was stirred vigorously to avoid any settling of the crystals due to gravity. The chamber was filled with helium gas to reduce air scattering. Also, the ambient conditions helped to prevent liquid droplets from freezing due to evaporative cooling at reduced pressure.

Fine tuning the sample delivery method in space and time is crucial for the droplets to overlap with the X-ray FEL pulses. The tuning was judged by the hit rate, which is defined as the number of images exhibiting more than 20 Bragg spots divided by the total number of images with the Cheetah software [131]. Using lysozyme crystals with 3.2×10^8 crystals/ml, the hit rate fluctuated greatly between 14 and 93% for the first 30 min, while fine tuning of the liquid droplets in space and time was fully achieved. After that, diffraction images from the crystals decreased to 3% of the hit rate since the crystals precipitated in the tube connecting the nozzle and reservoir, due to the very slow flow speed. This issue can be solved by optimizing the configuration of the nozzle and reservoir. Also, it is possible to prevent crystals from settling by increasing the viscosity of the buffer solution. A stable injection of 5-μm lysozyme crystals in a buffer solution with 10% polyethylene glycol 6000 was achieved, which is a relatively low concentration compared to a cryoprotectant. All combined, this allowed for structure determination of lysozyme at 2.3 Å from the SFX data obtained using the droplet injector.

Although the flow rate of 30 μL/h was approximately 20× lower than the typical flow rate of the liquid injector, it was still comparable with or higher than the flow rates of the LCP injector [45] and the electrospun microjet [21]. The sample consumption of the droplet injector can be further reduced with smaller droplet size using a finer nozzle. A droplet nozzle with a 30-μm aperture could provide droplets with a volume that is smaller by an order of magnitude, which could result in a decrease in the sample consumption, with an increased risk of clogging.

5.7 Acoustic Drop and Tape Drive

The previous section describes the generation of monodispersed droplet streams via piezoelectric triggers, synchronized with the SACLA FEL [109]. Roessler et al. [132] studied metalloenzymes with an Acoustic Droplet Ejector (ADE) where the drop generation was synchronized to the incident FEL pulses at 120 Hz using a pressure wave from an acoustic transducer to propel fluid away from a free liquid surface. The ADE allows the use of larger crystals (20–100 μm) with minimal clogging problems, since the crystals do not have to pass through a capillary with small inner diameter; however, sedimentation can be an issue, hence the inverted system seen in Fig. 5.26. The larger crystals sediment away from the free surface in the original modified commercially available Echo system from Labcyte, *up-shot*, configuration, whereas they sediment towards the free surface in the inverted, *down-shot*, configuration. The latter promotes more droplets generated with crystals inside.

While serial crystallography techniques continued to develop at X-ray FELs, many of the techniques began to return to bright synchrotron sources. Beyerlein et al. [133] performed a serial crystallography experiment at the P11 instrument of PETRA III in Hamburg, Germany. A brass chopper created 7.5 ms pulses at a frequency of 25 Hz, from the original 10^{13} photons/second continuous synchrotron source. Figure 5.27 shows a slurry of crystals was sheared away from their free surface via a polyamide tape which subsequently carried them towards the X-ray focus (4×8 μm). The slurry of lysozyme crystals were mixed with chitotriose with 2 s and 50 s delays by varying the mixing junction geometry and position, liquid flow rates, and tape speed. The results indicate that room temperature structures of longer time point intermediates can be solved at a synchrotron using serial crystallography.

A combination of the ADE method and a belt system can provide increased versatility for pump–probe measurements. This has been demonstrated by Fuller and Gul et al. [134] in a study of the Kok cycle of photosystem II [135] with simultaneous collection of diffraction data and emission spectra from the water-splitting Mn cluster. The droplets are acoustically ejected, and deposited on a polyamide ribbon, which acts as a conveyor belt system. As the droplets progress on the tape, they pass through three laser illumination points, appropriately spaced to match the belt speed and desired timing. Finally, the belt passes close enough to the interaction region that the protruding droplet is hit by the fourth and final laser, followed by the FEL pulse; the X-ray pulse does not pass through the polyamide tape, thus not directly damaging the tape, instead it passes through the droplet parallel to the tape surface. Roughly, half of the forward scattering passes through

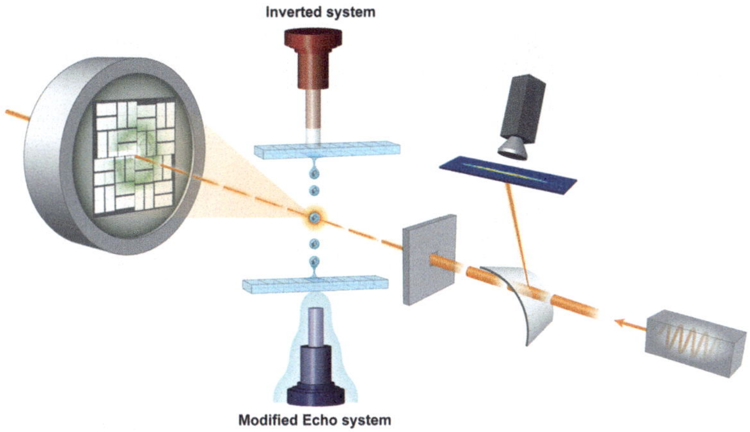

Fig. 5.26 An acoustic transducer is coupled via an agarose plug to a sample well full of the protein slurry. The sample is ejected from the meniscus and tracked with an imaging system to receive the synchronized FEL pulse at 120 Hz at the LCLS. Two configurations are shown. The original *up-shot* configuration, modifies an Echo 555 system (Labcyte) to shoot a droplet upwards towards the X-ray focus, ideally carrying a crystal. To help create more sample-containing droplets, the inverted *down-shot* configuration is shown. Here the crystals sediment towards the liquid's free surface and are more likely to eject with the liquid. The two configurations are not concurrent and are simply overlaid into one schematic. Modified from Roessler et al. [132]

the belt, making a correction of the Bragg intensities in this area necessary. Bragg diffraction and emission data can be simultaneously measured from the sample within the droplet.

The optical pump can be replaced by a *reaction region*, seen in Fig. 5.28b. The entire Drop-On-Tape (DOT) system is in a helium enclosure to minimize air scatter. The reaction region can therefore be a smaller chamber, in which instead of pumping optically, another gas is flushed, such as oxygen, in order to study aerobic transients. The electronic structure of ribonucleotide reductase R2 was reported in the work, with and without the presence of oxygen.

The DOT system is versatile and can run up to 120 Hz, lower collection rates reported in the work were limited by the data detection devices rather than the sample or DOT's capabilities. The DOT has sample consumptions of 0.8–6 nL/shot depending on the crystal size, and thus the required sample droplets. Assuming a 120 Hz data acquisition rate, the sample consumption rates can be nontrivial: 5.8–43 μL/min. These flow rates fall in line with common liquid injection rates used in the GDVN and DFFN.

An important distinction must be made about the droplet systems of this section when comparing them to the liquid jets of the prior sections: the synchronization is more efficient than the liquid delivery. The liquid jets travel at speeds of 10–100 m/s and thus have samples traverse the interaction region, but not necessarily when X-ray photons are present, whereas the synchronized systems have samples

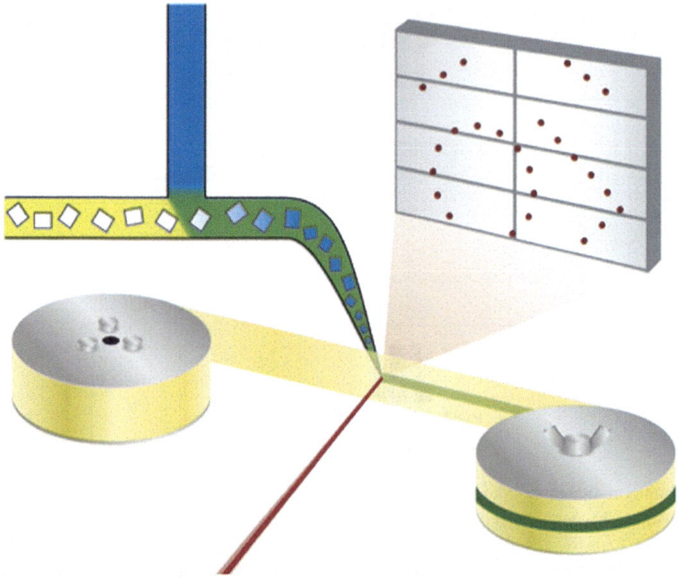

Fig. 5.27 A polyamide tape passes in front of a chopped synchrotron beam (P11, PETRA III, 10^{13} photons/s, 7.5 ms pulses every 40 ms, focused to 4×8 μm). The X-ray photons (red) pass through the tape and sample and continue onto a beamstop (not shown) and detector. A commercial microfluidic tee-junction is used to introduce a slurry of lysozyme crystals (white blocks in yellow media) sheared onto the tape (green streak). An additional fluid containing a competitive inhibitor, chitotriose, was introduced (blue) and mixed (green media with blue soaked blocks) with 2 s and 50 s delays. Modified with permission of the International Union of Crystallography [133]

rarely present when no photons are present. Mentioned in the first section of the chapter, a successful sample delivery system will have a new sample present and an old sample removed in between incident pulses. This efficiency can be broken down further into how often the sample delivery mechanism can replenish a volume and how often is a replenished volume containing a sample, the former dictated by the delivery method while the latter is dictated more by the sample concentration. Crystals passing through a liquid jet can destabilize the liquid jet and cause the stream to fluctuate in and out of the X-ray focus, this combined with the particles concentration can lead to low data collection rates. More simply, the sample is concentrated enough but the jet is too unstable, or the jet is stable but the sample is not in high enough concentration or some combination of the two. Even when the data collection rate is *good*, the liquid jets are still wasting un-shot samples between pulses.

The synchronized nature of these droplet and tape systems [109, 132–134] mean that there is more sample delivery efficiency; for example, most of the 2 mL of sample delivered by a synchronized system will go towards the final solved structure, whereas a lower percentage of that volume is used towards the solved structure when using a liquid jet. High viscosity extrusions, for example, are more

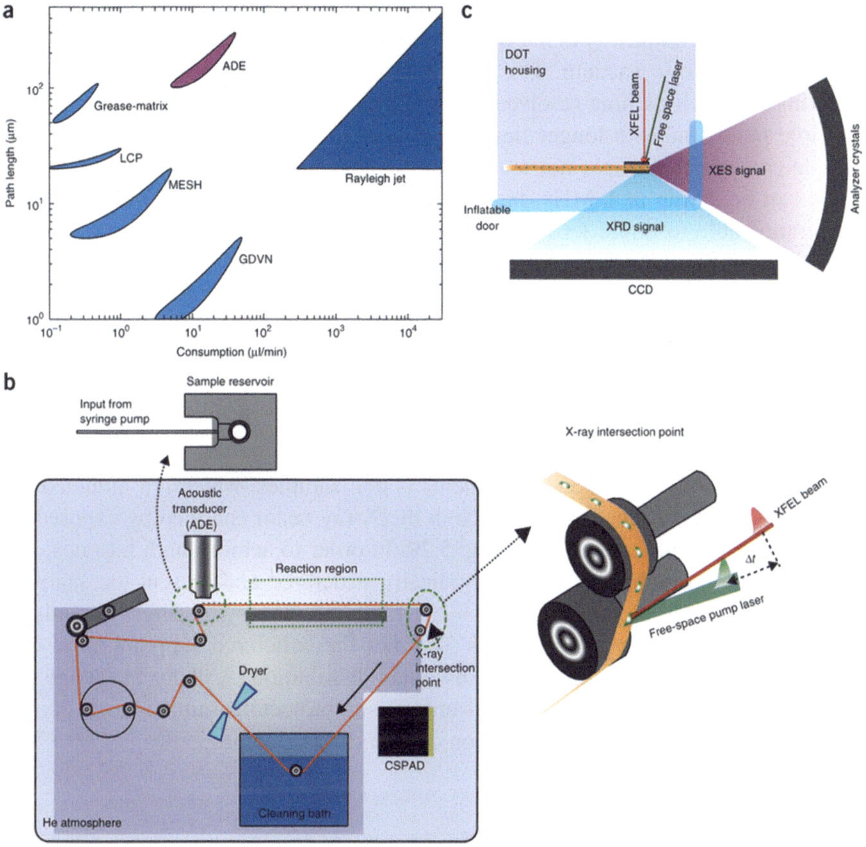

Fig. 5.28 (**a**) Schematic showing the path length the X-rays typically need to traverse within a sample for different sample injection methods vs the sample consumption. Longer path length means larger crystals can be used. In the case of an ADE, the crystals can be large without having to pass through a small 50 μm orifice, while maintaining the sample consumption of a GDVN. (**b**) Schematic of a tape drive system where the drops pass through a large bore capillary towards the acoustic droplet ejector (ADE) and are deposited onto a polyamide tape. They pass through the reaction region, which can contain optically pump lasers or a liquid mixing stage, followed by the X-ray interaction region. A cleaning bath removes debris from the polyamide tape, which is then dried to be reused again. (**c**) The open geometry allows for simultaneous X-ray diffraction and emission spectra to be collected. Reproduced with permissions from Fuller and Gul et al. [134]

efficient than GDVN and DFFN, but still wasting un-shot sample between pulses. The tape drives rarely wastes sample between shots and mostly need to worry about particle concentration when trying to optimize one crystal per droplet. More dense slurries with larger crystals suspended, can lead to droplet instabilities and reduce the repeatable nature of the droplet streams as well as potentially lead to multiple crystals per droplet, complicating the data analysis.

These synchronized systems have only been demonstrated out of vacuum, because of their complexity to interface within vacuum, as well as the exposed liquid that would freeze in vacuum. This, however, can open the way to more complex experiments, such as time-resolved mixing with fluids and optical pumping with multiple pumps of with longer time delays than can be achieved with a jet. The sample consumption might not scale up to higher repetition rates as a droplet per pulse can be taxing at 1 MHz, but at 120 Hz repetition rate the consumption is still comparable to liquid jet methods, with the added advantage of more sample delivery efficiency and higher path length for signal starved approaches, such as X-ray emission spectroscopy.

5.8 Fixed Target Sample Delivery for SFX

The main idea of fixed target experiments is that samples, which are mounted on solid supports, are raster-scanned through the X-ray beam and thereby exposed to the X-ray FEL pulses as shown in Fig. 5.29. In order to achieve high hit rates, the sample support should ideally be periodically structured and present the samples at known and well-defined positions—the highest hit rate for single crystals if the sample is randomly distributed is 37% but the structured supports allow the approach to exceed the limit of the Poissonian distribution. A major challenge for fixed target experiments of biological samples is to protect the samples from drying out, which would result in a degradation of their diffraction properties.

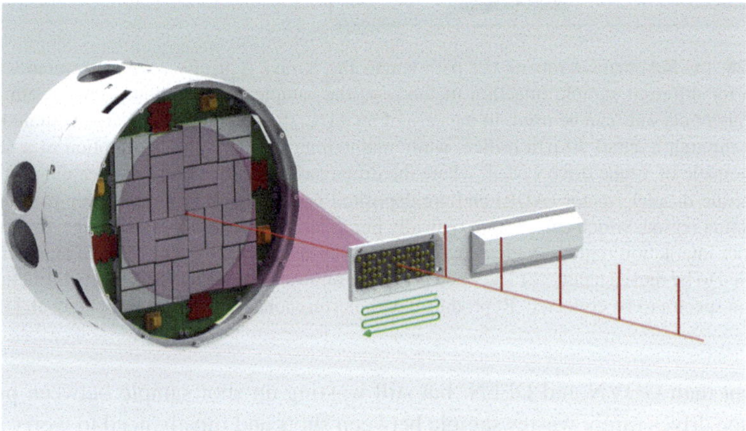

Fig. 5.29 Scanning approach for fixed target serial crystallography experiments: The fixed target sample holder is raster-scanned through the X-ray beam and samples are exposed to the X-ray pulses at predefined positions. The resulting diffraction patterns are recorded with an integrating X-ray detector located behind the sample. The red delta functions in the above image represent the individual X-ray FEL pulses

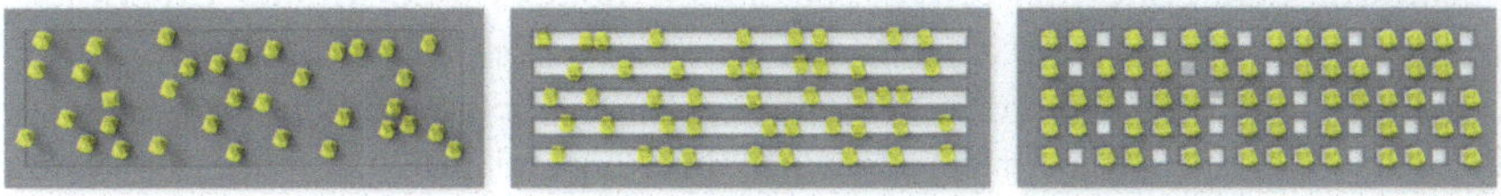

Fig. 5.30 Three different categories of fixed target sample delivery approaches for SFX: (*left*) Samples are arbitrarily distributed on a flat membrane. (*center*) Microcrystals arrange themselves in one dimension according to the geometry of the substrate. (*right*) Micro-crystals arrange themselves in a 2 dimensional periodic fashion according the pattern of the micro-structured substrate. See Sect. 5.8.2 for more information

5.8.1 Types of Fixed Targets for SFX

Fixed target sample holders can be classified in three different categories as shown in Fig. 5.30: In the first approach the samples are arbitrarily distributed for example on a flat membrane [136, 137]. In the second approach the samples are arranged in a 1-dimensional periodic fashion as shown in Fig. 5.30 (*center*) which should facilitate higher hit rates compared to the first approach [138]. In the third approach samples are arranged in a 2D periodic fashion on a micro-structured sample holder, as reported by [138–141]. This approach should allow achieving the highest hit rates, as the samples are ideally all located at predefined position.

Different materials have been used as substrate materials for the solid supports, such as silicon nitride [136, 137], single crystalline silicon [140–142], and polycarbonate and other plastics [138, 139, 143, 144] (Cohen 2016).

The first fixed target SFX experiments at room temperature used a REP-24 microcrystal suspension that was mixed with Paratone-N, a mixture of hydrocarbons used as a cryoprotectant for traditional crystallography, to avoid dehydration of the sample [136]. The emulsion was applied as a thin film on thin silicon nitride membranes that spanned lithographically etched silicon crystals. The diffraction experiments were performed in the vacuum chamber of the CXI instrument at LCLS and they achieved a hit rate of 38.2% at an average data acquisition rate of 3 Hz resulting in a hit rate of about 1 hit/s. Unfortunately, the Paratone-N caused a large scattering background and the emulsion was challenging to apply to the fixed targets without damaging the supports. A similar approach was followed by Kimura et al., who performed X-ray FEL imaging experiments of living cells sandwiched between two silicon nitride windows [137].

Sample holders structured 1-dimensionally, in which the crystals are located in channels with a width of 20 μm and a length of about 100 μm have been used as a support. In this case the windows were etched into a 15 μm thick photoresist layer located on top of a 150 nm silicon nitride membrane [138]. The solid supports are sealed on both sides with Kapton foil to prevent the crystals from drying out.

Sample holders structured 2-dimensionally were first used at an X-ray FEL by using larger crystals at cryogenic temperatures in a similar way to conventional synchrotron data collection [139]. The sample holders used in their experiments provided space for about 50 crystals and were made from a low-Z polymer.

Sample holders structured in 2D with very fine structures (micro-structured) on silicon can be the basis for serial data collection and provide space for several hundreds to hundreds of thousands of crystals. They were first introduced for synchrotron experiments [141]. In this design, crystals are located in individual compartments of a silicon mesh structure ('silicon chip') with both sides of the chip sealed with Kapton foil to prevent the crystals from drying out. The chip is designed for room temperature experiments and the well sizes for the crystals vary between 45 μm and 60 μm. An improved version of this chip design with well sizes ranging from 30 μm to 100 μm has been successfully used for data collection from Myoglobin crystals at the XPP instrument at LCLS [142]. The crystal sizes used for these experiments matched the well size.

A more compact silicon chip with a size of 4.5×2.5 mm^2 for microcrystals with dimensions of a few micrometers was developed and first employed for synchrotron experiments at cryogenic temperatures [140]. The design was later successfully used for SFX data collection at both cryogenic and at room temperature, again at the XPP instrument of LCLS [145]. Using this chip design in combination with a specially designed goniometer, called Roadrunner I, hit rates of up to 90% at a sample scan rate of 120 Hz were achieved. This approach allowed collecting complete data sets from CPV18 and BEV2 crystals in less than 10 min each, without post-refinement, similar to the time required for liquid-jet-based SFX under smooth running conditions.

5.8.2 Sample Loading

Sample loading should be easy and sample efficient, which means all sample material loaded onto the solid supports should be accessible to the X-rays and wasting material during loading should be avoided. Different approaches have been developed for loading samples onto fixed targets. Generally microcrystals of biological macromolecules have to be kept in a humid environment in order to prevent them from drying and thereby losing their diffraction properties. So the samples are ideally handled as suspensions using micropipettes. Depending on the chip size, typical amounts for loading samples on the chip are in the range of a few microliters.

In case of non-structured fixed target sample holders such as silicon nitride membranes, a small volume of the sample suspension is typically pipetted on the sample holder and then evenly distributed using a soft tool (e.g., a nylon loop) in order to avoid destroying the membrane [136, 137]. Structured sample holders aim to harvest the sample at predefined positions. One approach to achieve this is the use of a combination of hydrophobic and hydrophilic surface coatings [138] or the use of beads with a specific surface composition, which attract the crystals out of the suspension [141].

In another approach, the fixed target sample support is equipped with pores, which extend through the entire support. In this case, the sample solution is applied from one side, for example, by using a micropipette. By attaching a piece of filter paper from the other side, the solution is flowing through the pores and all particles larger than the pore diameter are retained and organize themselves according to the pore pattern as shown in Fig. 5.31 [140, 145]. Instead of wicking, it is also possible to achieve a liquid-flow through the pores by applying suction [140, 142].

A different approach has been recently reported by Opara et al. [146], where they have crystallized their samples directly on silicon nitride membranes that were sealed afterwards for the diffraction experiments. This approach has the advantage that it avoids the sample-loading step. This is especially useful in case of very sensitive crystals or large crystals, which are difficult to handle in a suspension due to sedimentation.

5.8.3 Preventing Samples from Drying Out

A major challenge for fixed target experiments with biological samples, in particular if they are to be carried out in vacuum, is preventing the samples from dehydration, which typically causes them to lose their diffraction properties. In general, three approaches can be followed to maintain the hydration of the samples for the diffraction experiment: (1) flash cooling to cryogenic temperatures, (2) keeping the micro-crystals in a sealed enclosure, or (3) providing a local humid environment for the samples.

The cryogenic approach is most commonly performed in conventional X-ray crystallography at synchrotron sources. Here the water contained in the protein crystals is vitrified in a solid state and thereby protects the crystals from drying out. So-called cryo-crystallography has the further advantage that radiation damage effects are reduced by more than two orders of magnitude at cryogenic temperatures compared to room temperature [147, 148]. Fixed target protein crystallography experiments at cryogenic temperatures have been carried out at LCLS and SACLA [139, 145, 149]. An advantage of fixed target SFX at cryogenic temperatures is that radiation damage effects to crystals in neighboring compartments, which are often pre-exposed by the tails of the X-ray beam, are reduced and higher sample densities with spacing down 10 μm between two samples can be realized. For fixed target experiments at room temperature, a spacing of at least 20 μm is required [145]. A certain risk of fixed target experiments at cryogenic temperatures is unwanted ice formation on the sample, as the resulting very strong Bragg reflection from the ice crystals can damage the detector.

In contrast to cryogenic temperatures, performing SFX experiments at room temperature allows studying protein dynamics and enzyme reactions [84, 85, 150]. For room temperature experiments it is required to permanently keep the crystals at a specific, well-defined humidity. The most common approach is to keep a small volume of the mother liquor close to the crystals and to seal the fixed target sample support from both sides with a material that is relatively transparent for X-rays

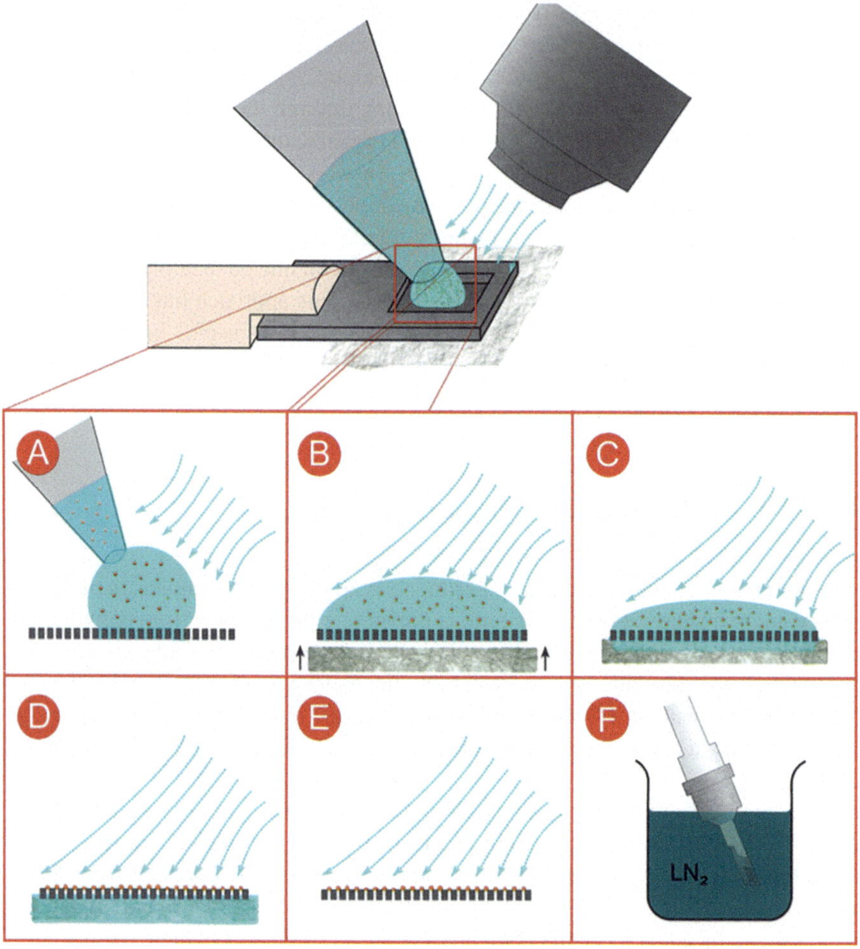

Fig. 5.31 Loading of microcrystals on a fixed target equipped with micro-pores for holding the crystals: (**a**) A few microliters of the microcrystal suspension are pipetted onto the chip while kept in a stream of humidified air. (**b–d**) Mother liquor is soaked-off through the micro-pores by wicking with a filter paper from the lower side and inducing a liquid flow through the pores. (**e**) The microcrystals are retained by the pores and arrange themselves in a periodic fashion for effective scanning with the X-ray beam. (**f**) The chips can be plunged into liquid nitrogen in case of cryogenic data collection. Reproduced with permissions from Roedig et al. [140]

such as Silicon nitride [136, 137, 151] or Kapton or Mylar foil [141, 142] or a combination of both [138]. Graphene has been proposed as an ideal sealing material as it is extremely thin and has a very low cross section for X-rays. Due to its high fragility its application as a sealing material is challenging. Freestanding graphene can cover only very small areas of typically a few micrometers and additional sealing with Mylar foil was required in a recent application [152].

By performing the fixed target measurements in an atmosphere of controlled humidity, it is possible to avoid any sealing material. This approach has been developed for conventional synchrotron crystallography at room temperature [153, 154] and recently adapted to fixed target crystallography experiments at synchrotrons and X-ray FELs [145, 155]. Biological samples typically require relative humidity (r.h.) levels above 80%, depending on solvent content and chemical composition, and most of them require r.h. close to 100%. Avoiding any sealing of fixed target sample holders has advantages for time-resolved experiments since the samples are more or less freely accessible on the support, which will facilitate soaking or mixing experiments. Unwanted scattering and reflections by pump laser light by the sealing material are also prevented using this approach.

5.8.4 Scattering Background

For all measurements and in particular for weakly scattering samples such as very small crystals of large unit cell systems or even aperiodic objects such as individual virus particles, it is essential to keep the background scattering levels as low as possible. This is best achieved by avoiding any non-sample material to interact with the X-ray beam. In contrast to other sample sample delivery methods, fixed target experiments are often accompanied by significantly higher background scattering levels caused by the substrate and sealing material of the targets compared to other methods such as liquid jets. In order to minimize scattering from the sample holder, the sample holder should ideally consist of very thin membranes, such as silicon and silicon nitride that can be manufactured with thicknesses of a few tens of nanometers [136] or even better graphene mono-layers [152], which can be of sub-nanometer thicknesses The use of a single crystalline and ideally defect free membrane material such as silicon is advantageous, since elastic scattering is limited to Bragg reflections, which occur only for specific orientations of the support material [140]. A certain limitation of this approach arises from risk of unwanted strong Bragg reflections of the support material, which can potentially damage the detector. In all cases, the thickness of the membranes should be kept thin in order to minimize inelastic scattering and X-ray fluorescence. Low-Z materials are generally preferable due to their weaker interaction with X-rays. For diffraction studies of larger objects such as microcrystals, the interaction of the support material with the X-rays can be further minimized by using a perforated supported material. Here the samples are located in micro-pores and the X-rays ideally only interact with the sample, which is larger than the pore diameter [140, 142, 145].

Further background scattering arises from interaction of the X-rays with air or gas. One approach is performing the fixed targets experiments in vacuum, which generally avoids air scattering. As a drawback this approach requires more sealing effort due to the pressure difference between the sample and the vacuum environment [136]. In addition, in-vacuum experiments provide limited access to the samples, for example, for performing time-resolved experiments. In another recent

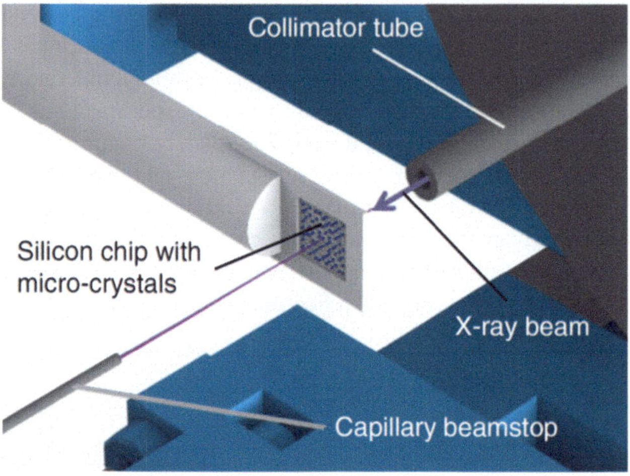

Fig. 5.32 A CAD model of the X-ray interaction region of a Roadrunner geometry. A collimator tube upstream of the sample chip is large enough to let the X-rays pass through unobstructed. Its material and thickness are such to absorb any X-rays scattered from the air or upstream equipment. The X-rays then pass through the sample and chip and produce the desired diffraction signal; however, the spent beam which did not interact with the sample continues to interact with air downstream, thus requiring a capillary beamstop to terminate the beam. Modified with permissions from Meents et al. [156]

approach it has been possible to reduce air scattering by more than a factor of one hundred compared to conventional X-ray crystallography experiments by replacing air with by helium gas (which provides a factor of ~25 lower cross section for X-rays) and by enclosing the direct beam shortly before and after the sample in capillary shields [156]. These upstream and downstream scatter guards, when placed close to the interaction region, can significantly reduce errant background scatter and improve the signal to noise, Fig. 5.32 shows more detail.

5.8.5 Scanning Approaches

For scanning of the fixed targets through the X-ray beam either 2D or 3D scanning devices based on stepper motor translation stages [136], piezo-driven motor driven translation stages [142, 145, 157], or goniometers offering an additional rotational degree of freedom have been employed so far [139, 145].

Most of the fixed target SFX experiments have been performed in step-scanning mode. Here the sample frame is accelerated, moving to the next position, decelerated, and then stopped at the predefined position where it is then exposed to the X-ray pulse [136–139, 157]. The maximum sample exchange rate using this approach is typically limited to a few Hz. Recently, step scanning fixed target experiments at a synchrotron were performed [158]. Faster scanning of fixed targets with sample exchange rates of the LCLS repetition rate of 120 Hz is achievable [145]. Instead of accelerating, moving, and decelerating the chip, a so-called fly-scan, where the fixed target is scanned at constant speed through the X-ray beam is applied. By inducing additional phase control to this scanning approach it was possible to expose the fixed target at the predefined samples positions. Using this approach, hit rates of up to 90% were achieved.

A certain complication in collecting complete datasets in fixed target experiments arises from preferred orientation of some crystal types on their supports. This is in particular severe for plate- and needle-shaped crystals mounted on flat membranes. The preferred orientation gives a range of reciprocal space that is not sampled for a fixed mounting angle. One approach to mitigate this problem is to use micro-structured sample holders which provide different well shapes in order to harvest the crystals in different orientations [140, 142]. Another approach is rotating the sample holder with respect to the incoming beam so that the crystals are exposed in different directions [139, 145].

5.8.6 Time-Resolved Experiments

So far only very few time-resolved experiments have been published using fixed target sample delivery methods [114], as this is accompanied with several challenges: for example, for pump probe experiments using laser excitation, the pre-exposure of crystals in the neighboring compartments with laser light. For other time-resolved experiments, such reaction initialization by micro-diffusion, sealing of the chips for example with Mylar foil to prevent the crystals from drying out, represents a major limitation, as the samples are not directly accessible, for example by spraying a substrate or ligand solution on the fixed target. For such experiments an approach with non-sealed chips kept in a humid atmosphere offers much more flexibility, as the samples are directly accessible to all different kind of spraying or microinjection approaches. Avoiding sealing of the chips has also advantages for laser excitation as the pump light has not to pass through any sealing material which in many cases results in scattering or even absorption of the light. In comparison to sample delivery with jets, where the surrounding liquid stream often acts as a lens for the laser light and thereby to undefined illumination conditions, fixed target pump probe experiments—at least in principle—should allow for a well-defined a reproducible illumination scheme.

5.8.7 Summary

Fixed target approaches for SFX offer an alternative means of delivering sample to fixed target experiments allow high hit rates of up to 90% at X-ray FEL's with data collection time of less than 10 min and require only very small amounts of sample, which can be as little as 4 μg for a complete SFX structure determination. The complicated machinery can be an obstacle to new users, but once standard in a facility's deployment, the barrier to entry for the crystallographer can be low. Custom chip substrates can be expensive for multiple datasets, but costs may diminish over time. Loading the sample onto the chip substrate can also require skill and practice and can vary from sample to sample. The high machine precision makes rapid scanning fixed target techniques an ideal candidate for high precision optical pump–probe experiments, where variations in flows can diminish high timing precision.

5.9 Outlook

The sample delivery methods presented in this chapter are plentiful and constantly evolving. The evolution of sample delivery methods continues to progress as SFX is attempted with new protein crystal systems, which introduce new complications. The serial crystallography sample delivery journey started from largely inefficient methods like aerosol lens stacks, Rayleigh jets and crude fixed target systems. Over the years, gas-focused approaches allowed the generation of smaller jets and increased the sample delivery efficiency with the GDVN, yet the sample consumption remained too high for certain systems. The electrokinetics of electro-spraying were leveraged to create a lower flow rate MESH and modified coMESH injection method. Concurrently, higher viscosity media were explored with reduced sample consumption; the improvement in detector technology and shorter X-ray FEL wavelengths allowed experiments to be performed at ambient conditions. The LCP injector and similar high viscosity extrusion devices have aided in the fight to reduce sample consumption while maintaining compatibility with the persnickety crystallization conditions. The DFFN improved the parameter space available to the GDVN and further improved the stability for troubling systems. Droplet injection systems at SACLA have been developed that match the droplet generation rate to the FEL repetition rate. Fixed targets have had a resurgence in an effort to get higher hit rates and reduced sample consumption, thus maximizing the sample delivery efficiency.

SFX experiments have matured to the point where sufficient preparation can lead to minimal surprises during an experiment and guarantee as successful an experiment as the crystal diffraction permits. Sample sedimentation continues to be an issue that can lead to clogging and diminished data collection. As 3D printing techniques improve, the reproducibility of complex injection nozzles, like the DFFN, GDVN, and mixing variants, are substantially improved, leading to better performance from nozzle to nozzle.

New techniques are being developed to help improve injection and sample preparation. Microfluidic chips [93] are being developed as injection and mixing nozzles and dielectrophoresis can be used to sort crystals based upon size [159].

As these injection methods have been developed for X-ray FELs, goniometer and fixed-target approaches from synchrotrons have been crossed between both former and current generation sources. Serial diffraction techniques have moved to synchrotrons and have proved successful in many instances, using lower flow rate HVE and MESH methods or scanned fixed target holders (Roadrunner), since millisecond exposure times are needed to obtain sufficient scattering signal from the available flux. Nanometer liquid sheets have also been developed for soft X-ray studies and may find a use in serial crystallography [160]. High speed injectors are currently being studied to maximize the efficiency of delivery at higher repetition rate sources [161].

The ability to obtain femtosecond snapshots from protein crystals will eventually lead to the realization of a longtime goal in structural biology and chemistry: to view structural intermediates. The next few years will see an increase in time-resolved studies beyond optical pump–probe. These mixing experiments will require further understanding of the underlying fluid mechanics and will likely be a significant cornerstone of a future text on sample delivery techniques of SFX.

References

1. Chapman, H. N., Barty, A., Bogan, M. J., Boutet, S., Frank, M., Hau-Riege, S. P., et al. (2006). Femtosecond diffractive imaging with a soft-X-ray free-electron laser. *Nature Physics, 2*(12), 839–843. https://doi.org/10.1038/nphys461.
2. Stan, C. A., Milathianaki, D., Laksmono, H., Sierra, R. G., McQueen, T. A., Messerschmidt, M., et al. (2016). Liquid explosions induced by X-ray laser pulses. *Nature Physics, 12*, 966. https://doi.org/10.1038/nphys3779.
3. Chapman, H. N., Fromme, P., Barty, A., White, T. A., Kirian, R. A., Aquila, A., et al. (2011). Femtosecond X-ray protein nanocrystallography. *Nature, 470*(7332), 73–77. https://doi.org/10.1038/nature09750.
4. Seibert, M. M., Ekeberg, T., Maia, F. R. N. C., Svenda, M., Andreasson, J., Jönsson, O., et al. (2011). Single mimivirus particles intercepted and imaged with an X-ray laser. *Nature, 470*(7332), 78–81. https://doi.org/10.1038/nature09748.
5. Bogan, M. J., Benner, W. H., Boutet, S., Rohner, U., Frank, M., Barty, A., et al. (2008). Single particle X-ray diffractive imaging. *Nano Letters, 8*(1), 310–316. https://doi.org/10.1021/nl072728k.
6. Awel, S., Kirian, R. A., Wiedorn, M. O., Beyerlein, K. R., Roth, N., Horke, D. A., et al. (2018). Femtosecond X-ray diffraction from an aerosolized beam of protein nanocrystals. *Journal of Applied Crystallography, 51*(1), 133–139. https://doi.org/10.1107/S1600576717018131.
7. Hantke, M. F., Hasse, D., Maia, F. R. N. C., Ekeberg, T., John, K., Svenda, M., et al. (2014). High-throughput imaging of heterogeneous cell organelles with an X-ray laser. *Nature Photonics, 8*(12), 943–949. https://doi.org/10.1038/nphoton.2014.270.
8. Munke, A., Andreasson, J., Aquila, A., Awel, S., Ayyer, K., Barty, A., et al. (2016). Coherent diffraction of single Rice Dwarf virus particles using hard X-rays at the Linac Coherent Light Source. *Scientific Data, 3*, 160064. https://doi.org/10.1038/sdata.2016.64.
9. Rayleigh, L. (1878). On the instability of jets. *Proceedings of the London Mathematical Society, 1*(1), 4 Retrieved from http://plms.oxfordjournals.org/content/s1-10/1/4.full.pdf.
10. Rayleigh, L. (1892). XVI. On the instability of a cylinder of viscous liquid under capillary force. *The London, Edinburgh, and Dublin Philosophical Magazine and Journal of Science, 34*(207), 145–154. https://doi.org/10.1080/14786449208620301.

11. Faubel, M., Schlemmer, S., & Toennies, J. (1988). A molecular beam study of the evaporation of water from a liquid jet. *Zeitschrift fur Physik D: Atoms, Molecules and Clusters, 10*(2), 269–277 Retrieved from http://www.springerlink.com/index/X6GXW8Q11QV3R085.pdf.
12. Weierstall, U., Doak, R., Spence, J. C. H., Starodub, D., Shapiro, D., Kennedy, P., et al. (2008). Droplet streams for serial crystallography of proteins. *Experiments in Fluids, 44*(5), 675–689 Retrieved from http://www.springerlink.com/index/Q7N2L3278806078U.pdf.
13. DePonte, D. P., Weierstall, U., Schmidt, K., Warner, J., Starodub, D., Spence, J. C. H., et al. (2008). Gas dynamic virtual nozzle for generation of microscopic droplet streams. *Journal of Physics D: Applied Physics, 41*(19), 195505. https://doi.org/10.1088/0022-3727/41/19/195505.
14. White, F., & Corfield, I. (2005). *Viscous fluid flow* (3rd ed.). New York: McGraw-Hill.
15. Eggers, J., & Villermaux, E. (2008). Physics of liquid jets. *Reports on Progress in Physics, 71*(1), 1–79. https://doi.org/10.1088/0034-4885/71/3/036601.
16. Gonzalez-Tello, P., Camacho, F., & Blazquez, G. (1994). Density and viscosity of concentrated aqueous solutions of polyethylene glycol. *Journal of Chemical & Engineering Data, 39*(3), 611–614. https://doi.org/10.1021/je00015a050.
17. Lee, R. J., & Teja, A. S. (1990). Viscosities of poly(ethylene glycols). *Journal of Chemical & Engineering Data, 35*(4), 385–387. https://doi.org/10.1021/je00062a003.
18. Einstein, A. (1905). On the motion of small particles suspended in a stationary liquid, as required by the molecular kinetic theory of heat. *Annalen der Physik, 322*, 549–560. https://doi.org/10.1002/andp.19053220806.
19. Probstein, R. F. (1994). *Physicochemical hydrodynamics*. New York: Wiley. https://doi.org/10.1002/0471725137.
20. Lomb, L., Steinbrener, J., Bari, S., Beisel, D., Berndt, D., Kieser, C., et al. (2012). An anti-settling sample delivery instrument for serial femtosecond crystallography. *Journal of Applied Crystallography, 45*(4), 1–5. https://doi.org/10.1107/S0021889812024557.
21. Sierra, R. G., Gati, C., Laksmono, H., Dao, E. H., Gul, S., Fuller, F., et al. (2015). Concentric-flow electrokinetic injector enables serial crystallography of ribosome and photosystem II. *Nature Methods, 13*(1), 59–62. https://doi.org/10.1038/nmeth.3667.
22. Johansson, L. C., Arnlund, D., White, T. A., Katona, G., Deponte, D. P., Weierstall, U., et al. (2012). Lipidic phase membrane protein serial femtosecond crystallography. *Nature Methods, 9*(3), 263–265. https://doi.org/10.1038/nmeth.1867.
23. Oberthuer, D., Knoška, J., Wiedorn, M. O., Beyerlein, K. R., Bushnell, D. A., Kovaleva, E. G., et al. (2017). Double-flow focused liquid injector for efficient serial femtosecond crystallography. *Scientific Reports, 7*, 44628. https://doi.org/10.1038/srep44628.
24. Taylor, G. (1953). Dispersion of soluble matter in solvent flowing slowly through a tube. *Proceedings of the Royal Society of London A: Mathematical, Physical and Engineering Sciences, 219*(1137), 186–203. https://doi.org/10.1098/rspa.1953.0139.
25. Taneda, S. (1979). Visualization of separating Stokes flows. *Journal of the Physical Society of Japan, 46*(6), 1935–1942. https://doi.org/10.1143/JPSJ.46.1935.
26. Purcell, E. (1976). Life at low Reynolds number. *AIP Conference Proceedings, 45*, 3–11.
27. Rayleigh, L. (1879). On the capillary phenomena of jets. *Proceedings of the Royal Society of London, 29*, 71–97.
28. Frohn, A., & Roth, N. (2000). *Dynamics of droplets*. Berlin, Germany: Springer Science & Business Media.
29. Gañán-Calvo, A. M. (1998). Generation of steady liquid microthreads and micron-sized monodisperse sprays in gas streams. *Physical Review Letters, 80*(2), 285.
30. Lomb, L., Steinbrener, J., Bari, S., Beisel, D., Berndt, D., Kieser, C., et al. (2012). An anti-settling sample delivery instrument for serial femtosecond crystallography. *Journal of Applied Crystallography, 45*(4), 674–678.
31. Weierstall, U., Spence, J. C. H., & Doak, R. B. (2012). Injector for scattering measurements on fully solvated biospecies. *The Review of Scientific Instruments, 83*(3), 035108.

32. Nelson, G., Kirian, R. A., Weierstall, U., Zatsepin, N. A., Faragó, T., Baumbach, T., et al. (2016). Three-dimensional-printed gas dynamic virtual nozzles for x-ray laser sample delivery. *Optics Express, 24*(11), 11515–11530.
33. Wang, D., Weierstall, U., Pollack, L., & Spence, J. (2014). Double-focusing mixing jet for XFEL study of chemical kinetics. *Journal of Synchrotron Radiation, 21*(6), 1364–1366.
34. Weierstall, U., Doak, R. B., & Spence, J. C. H. (2011). A pump-probe XFEL particle injector for hydrated samples. arXiv preprint arXiv:1105.2104.
35. Daurer, B. J., Okamoto, K., Bielecki, J., Maia, F. R. N. C., Muhlig, K., Seibert, M. M., et al. (2017). Experimental strategies for imaging bioparticles with femtosecond hard X-ray pulses. *IUCrJ, 4*(3), 251–262. https://doi.org/10.1107/S2052252517003591.
36. Perry, S. L., Guha, S., Pawate, A. S., Bhaskarla, A., Agarwal, V., Nair, S. K., et al. (2013). A microfluidic approach for protein structure determination at room temperature via on-chip anomalous diffraction. *Lab on a Chip, 13*(16), 3183–3187.
37. Zhu, L., Weierstall, U., Cherezov, V., & Liu, W. (2016). Serial femtosecond crystallography of membrane proteins. In I. Moraes (Ed.), *The next generation in membrane protein structure determination* (pp. 151–160). Cham, Switzerland: Springer.
38. Martin-Garcia, J. M., Conrad, C. E., Nelson, G., Stander, N., Zatsepin, N. A., Zook, J., et al. (2017). Serial millisecond crystallography of membrane and soluble protein microcrystals using synchrotron radiation. *IUCrJ, 4*, 439–454. https://doi.org/10.1107/S205225251700570X.
39. Tono, K., Nango, E., Sugahara, M., Song, C., Park, J., Tanaka, T., et al. (2015). Diverse application platform for hard X-ray diffraction in SACLA (DAPHNIS): Application to serial protein crystallography using an X-ray free-electron laser. *Journal of Synchrotron Radiation, 22*, 532–537. https://doi.org/10.1107/S1600577515004464.
40. Sugahara, M., Mizohata, E., Nango, E., Suzuki, M., Tanaka, T., Masuda, T., et al. (2015). Grease matrix as a versatile carrier of proteins for serial crystallography. *Nature Methods, 12*(1), 61–63.
41. Botha, S., Nass, K., Barends, T. R. M., Kabsch, W., Latz, B., Dworkowski, F., et al. (2015). Room-temperature serial crystallography at synchrotron X-ray sources using slowly flowing free-standing high-viscosity microstreams. *Acta Crystallographica. Section D, Biological Crystallography, 71*(2), 387.
42. Conrad, C. E., Basu, S., James, D., Wang, D., Schaffer, A., Roy-Chowdhury, S., et al. (2015). A novel inert crystal delivery medium for serial femtosecond crystallography. *IUCrJ, 2*(4), 421–430.
43. Sugahara, M., Song, C., Suzuki, M., Masuda, T., Inoue, S., Nakane, T., et al. (2016). Oil-free hyaluronic acid matrix for serial femtosecond crystallography. *Scientific Reports, 6*, 1–6.
44. Kovacsova, G., Grunbein, M. L., Kloos, M., Barends, T. R. M., Schlesinger, R., Heberle, J., et al. (2017). Viscous hydrophilic injection matrices for serial crystallography. *IUCrJ, 4*, 400–410. https://doi.org/10.1107/S2052252517005140.
45. Weierstall, U., James, D., Wang, C., White, T. A., Wang, D., Liu, W., et al. (2014). Lipidic cubic phase injector facilitates membrane protein serial femtosecond crystallography. *Nature Communications, 5*, 3309.
46. Zhang, H., Unal, H., Gati, C., Han, G. W., Liu, W., Zatsepin, N. A., et al. (2015). Structure of the angiotensin receptor revealed by serial femtosecond crystallography. *Cell, 161*(4), 833–844.
47. Fenalti, G., Zatsepin, N. A., Betti, C., Giguere, P., Han, G. W., Ishchenko, A., et al. (2015). Structural basis for bifunctional peptide recognition at human delta-opioid receptor. *Nature Structural & Molecular Biology, 22*(3), 265–268.
48. Kang, Y., Zhou, X. E., Gao, X., He, Y., Liu, W., Ishchenko, A., et al. (2015). Crystal structure of rhodopsin bound to arrestin by femtosecond X-ray laser. *Nature, 523*(7562), 561–567.
49. Nogly, P., Panneels, V., Nelson, G., Gati, C., Kimura, T., Milne, C., et al. (2016). Lipidic cubic phase injector is a viable crystal delivery system for time-resolved serial crystallography. *Nature Communications, 7*, 1–9.

50. Batyuk, A., Galli, L., Ishchenko, A., Han, G. W., Gati, C., Popov, P. A., et al. (2016). Native phasing of x-ray free-electron laser data for a G protein–coupled receptor. *Science Advances, 2*(9), e1600292.
51. Fromme, R., Ishchenko, A., Metz, M., Chowdhury, S. R., Basu, S., Boutet, S., et al. (2015). Serial femtosecond crystallography of soluble proteins in lipidic cubic phase. *IUCrJ, 2*(5), 545–551.
52. James, D., Wang, D., White, T. A., Zatsepin, N., Nelson, G., Liu, H., et al. (2015). Lipidic cubic phase serial millisecond crystallography using synchrotron radiation. *IUCrJ, 2*(2), 168–176.
53. Sierra, R. G., Laksmono, H., Kern, J., Tran, R., Hattne, J., Alonso-Mori, R., et al. (2012). Nanoflow electrospinning serial femtosecond crystallography. *Acta Crystallographica. Section D, Biological Crystallography, 68*(11), 1584–1587. https://doi.org/10.1107/S0907444912038152.
54. Hattne, J., Echols, N., Tran, R., Kern, J., Gildea, R. J., Brewster, A. S., et al. (2014). Accurate macromolecular structures using minimal measurements from X-ray free-electron lasers. *Nature Methods, 11*(5), 545–548. https://doi.org/10.1038/nmeth.2887.
55. Kern, J., Alonso-Mori, R., Hellmich, J., Tran, R., Hattne, J., Laksmono, H., et al. (2012). Room temperature femtosecond X-ray diffraction of photosystem II microcrystals. *Proceedings of the National Academy of Sciences of the United States of America, 109*(25), 9721–9726. https://doi.org/10.1073/pnas.1204598109.
56. Kern, J., Tran, R., Alonso-Mori, R., Koroidov, S., Echols, N., Hattne, J., et al. (2014). Taking snapshots of photosynthetic water oxidation using femtosecond X-ray diffraction and spectroscopy. *Nature Communications, 5*, 4371. https://doi.org/10.1038/ncomms5371.
57. Young, I. D., Ibrahim, M., Chatterjee, R., Gul, S., Fuller, F. D., Koroidov, S., et al. (2016). Structure of photosystem II and substrate binding at room temperature. *Nature, 540*(7633), 453–457. https://doi.org/10.1038/nature20161.
58. Alonso-Mori, R., Kern, J., Gildea, R. J., Sokaras, D., Weng, T.-C., Lassalle-Kaiser, B., et al. (2012). Energy-dispersive X-ray emission spectroscopy using an X-ray free-electron laser in a shot-by-shot mode. *Proceedings of the National Academy of Sciences of the United States of America, 109*(47), 19103. https://doi.org/10.1073/pnas.1211384109.
59. Kern, J., Alonso-Mori, R., Tran, R., Hattne, J., Gildea, R. J., Echols, N., et al. (2013). Simultaneous femtosecond X-ray spectroscopy and diffraction of photosystem II at room temperature. *Science, 340*(6131), 491–495. https://doi.org/10.1126/science.1234273.
60. Kroll, T., Kern, J., Kubin, M., Ratner, D., Gul, S., Fuller, F. D., et al. (2016). X-ray absorption spectroscopy using a self-seeded soft X-ray free-electron laser. *Optics Express, 24*(20), 22469. https://doi.org/10.1364/OE.24.022469.
61. Kubin, M., Kern, J., Gul, S., Kroll, T., Chatterjee, R., Löchel, H., et al. (2017). Soft x-ray absorption spectroscopy of metalloproteins and high-valent metal-complexes at room temperature using free-electron lasers. *Structural Dynamics, 4*(5), 054307. https://doi.org/10.1063/1.4986627.
62. Mitzner, R., Rehanek, J., Kern, J., Gul, S., Hattne, J., Taguchi, T., et al. (2013). L-edge X-ray absorption spectroscopy of dilute systems relevant to metalloproteins using an X-ray free-electron laser. *Journal of Physical Chemistry Letters, 4*, 3641–3647. https://doi.org/10.1021/jz401837f.
63. Colletier, J.-P., Sawaya, M. R., Gingery, M., Rodriguez, J. A., Cascio, D., Brewster, A. S., et al. (2016). De novo phasing with X-ray laser reveals mosquito larvicide BinAB structure. *Nature, 539*(7627), 43–47. https://doi.org/10.1038/nature19825.
64. Fernández de la Mora, J. (2007). The fluid dynamics of Taylor cones. *Annual Review of Fluid Mechanics, 39*(1), 217–243. https://doi.org/10.1146/annurev.fluid.39.050905.110159.
65. Gañán-Calvo, A. M., & Barrero, A. (1999). A novel pneumatic technique to generate steady capillary microjets. *Journal of Aerosol Science, 30*(1), 117–125. https://doi.org/10.1016/S0021-8502(98)00029-9.
66. Aquila, A., Hunter, M. S., Doak, R. B., Kirian, R. A., Fromme, P., White, T. A., et al. (2012). Time-resolved protein nanocrystallography using an X-ray free-electron laser. *Optics Express, 20*(3), 2706–2716 Retrieved from http://www.opticsinfobase.org/abstract.cfm?URI=oe-20-3-2706.

67. Barty, A., Caleman, C., Aquila, A., Timneanu, N., Lomb, L., White, T. A., et al. (2012). Self-terminating diffraction gates femtosecond X-ray nanocrystallography measurements. *Nature Photonics, 6*(1), 35–40. https://doi.org/10.1038/nphoton.2011.297.
68. Boutet, S., Lomb, L., Williams, G. J., Barends, T. R. M., Aquila, A., Doak, R. B., et al. (2012). High-resolution protein structure determination by serial femtosecond crystallography. *Science, 337*(6092), 362–364. https://doi.org/10.1126/science.1217737.
69. Koopmann, R., Cupelli, K., Redecke, L., Nass, K., DePonte, D. P., White, T. A., et al. (2012). In vivo protein crystallization opens new routes in structural biology. *Nature Methods, 9*(3), 259–262. https://doi.org/10.1038/nmeth.1859.
70. Lomb, L., Barends, T. R. M., Kassemeyer, S., Aquila, A., Epp, S., Erk, B., et al. (2011). Radiation damage in protein serial femtosecond crystallography using an x-ray free-electron laser. *Physical Review B, 84*(21), 1–6. https://doi.org/10.1103/PhysRevB.84.214111.
71. Garman, E. (1999). Cool data: Quantity AND quality. *Acta Crystallographica, Section D: Biological Crystallography, 55*(10), 1641–1653. https://doi.org/10.1107/S0907444999008653.
72. Garman, E. F., & Owen, R. L. (2006). Cryocooling and radiation damage in macromolecular crystallography. *Acta Crystallographica, Section D: Biological Crystallography, 62*(1), 32–47. https://doi.org/10.1107/S0907444905034207.
73. Ibrahim, M., Chatterjee, R., Hellmich, J., Tran, R., Bommer, M., Yachandra, V. K., et al. (2015). Improvements in serial femtosecond crystallography of photosystem II by optimizing crystal uniformity using microseeding procedures. *Structural Dynamics, 2*(4), 041705. https://doi.org/10.1063/1.4919741.
74. Gañán-Calvo, A. M., & Montanero, J. (2009). Revision of capillary cone-jet physics: Electrospray and flow focusing. *Physical Review E, 79*(6), 1–18. https://doi.org/10.1103/PhysRevE.79.066305.
75. Liang, M., Williams, G. J., Messerschmidt, M., Seibert, M. M., Montanez, P. A., Hayes, M., et al. (2015). The coherent X-ray imaging instrument at the Linac Coherent Light Source. *Journal of Synchrotron Radiation, 22*(3), 514–519. https://doi.org/10.1107/S160057751500449X.
76. Schlichting, I. (2015). Serial femtosecond crystallography: The first five years. *IUCrJ, 2*, 246.
77. Redecke, L., Nass, K., DePonte, D. P., White, T. A., Rehders, D., Barty, A., et al. (2013). Natively inhibited Trypanosoma brucei cathepsin B structure determined by using an X-ray laser. *Science, 339*, 227–230.
78. Emma, P., Akre, R., Arthur, J., Bionta, R., Bostedt, C., Bozek, J., et al. (2010). First lasing and operation of an ångstrom-wavelength free-electron laser. *Nature Photonics, 4*, 641–647.
79. Beyerlein, K. R., Adriano, L., Heymann, M., Kirian, R., Knoška, J., Wilde, F., et al. (2015). Ceramic micro-injection molded nozzles for serial femtosecond crystallography sample delivery. *The Review of Scientific Instruments, 86*, 125104.
80. Barends, T. R., Foucar, L., Ardevol, A., Nass, K., Aquila, A., Botha, S., et al. (2015). Direct observation of ultrafast collective motions in CO myoglobin upon ligand dissociation. *Science, 350*, 445–450.
81. Coquelle, N., Sliwa, M., Woodhouse, J., Schirò, G., Adam, V., Aquila, A., et al. (2018). Chromophore twisting in the excited state of a photoswitchable fluorescent protein captured by time-resolved serial femtosecond crystallography. *Nature Chemistry, 10*, 31–37.
82. Pande, K., Hutchison, C. D., Groenhof, G., Aquila, A., Robinson, J. S., Tenboer, J., et al. (2016). Femtosecond structural dynamics drives the trans/cis isomerization in photoactive yellow protein. *Science, 352*, 725–729.
83. Tenboer, J., Basu, S., Zatsepin, N., Pande, K., Milathianaki, D., Frank, M., et al. (2014). Time-resolved serial crystallography captures high-resolution intermediates of photoactive yellow protein. *Science, 346*, 1242–1246.
84. Kupitz, C., Olmos Jr., J. L., Holl, M., Tremblay, L., Pande, K., Pandey, S., et al. (2017). Structural enzymology using X-ray free electron lasers. *Structural Dynamics, 4*(4), 044003.
85. Stagno, J. R., Liu, Y., Bhandari, Y. R., Conrad, C. E., Panja, S., Swain, M., et al. (2017). Structures of riboswitch RNA reaction states by mix-and-inject XFEL serial crystallography. *Nature, 541*, 242–246.

86. Gañán-Calvo, A. M., González-Prieto, R., Riesco-Chueca, P., Herrada, M. A., & Flores-Mosquera, M. (2007). Focusing capillary jets close to the continuum limit. *Nature Physics, 3*, 737–742.
87. Acero, A. J., Ferrera, C., Montanero, J. M., & Gañán-Calvo, A. M. (2012). Focusing liquid microjets with nozzles. *Journal of Micromechanics and Microengineering, 22*, 065011.
88. Montanero, J. M., Rebollo-Munoz, N., Herrada, M. A., & Gañán-Calvo, A. M. (2011). Global stability of the focusing effect of fluid jet flows. *Physical Review. E, Statistical, Nonlinear, and Soft Matter Physics, 83*, 036309.
89. Vega, E. J., Montanero, J. M., Herrada, M. A., & Gañán-Calvo, A. M. (2010). Global and local instability of flow focusing: The influence of the geometry. *Physics of Fluids, 22*, 064105.
90. Schmidt, M. (2013). Mix and inject: Reaction initiation by diffusion for time-resolved macromolecular crystallography. *Advances in Condensed Matter Physics, 2013*, 1–10.
91. Calvey, G. D., Katz, A. M., Schaffer, C. B., & Pollack, L. (2016). Mixing injector enables time-resolved crystallography with high hit rate at X-ray free electron lasers. *Structural Dynamics, 3*, 054301.
92. Chavas, L. M., Gumprecht, L., & Chapman, H. N. (2015). Possibilities for serial femtosecond crystallography sample delivery at future light sources. *Structural Dynamics, 2*, 041709.
93. Trebbin, M., Krüger, K., DePonte, D., Roth, S. V., Chapman, H. N., & Förster, S. (2014). Microfluidic liquid jet system with compatibility for atmospheric and high-vacuum conditions. *Lab on a Chip, 14*, 1733–1745.
94. Au, A. K., Huynh, W., Horowitz, L. F., & Folch, A. (2016). 3D-printed microfluidics. *Angewandte Chemie (International Ed. in English), 55*, 3862–3881.
95. Moffat, K. (2014). Time-resolved crystallography and protein design: Signalling photoreceptors and optogenetics. *Philosophical Transactions of the Royal Society of London. Series B, Biological Sciences, 369*, 20130568.
96. Neutze, R. (2014). Opportunities and challenges for time-resolved studies of protein structural dynamics at X-ray free-electron lasers. *Philosophical Transactions of the Royal Society of London. Series B, Biological Sciences, 369*, 20130318.
97. Neutze, R., & Moffat, K. (2012). Time-resolved structural studies at synchrotrons and X-ray free electron lasers: Opportunities and challenges. *Current Opinion in Structural Biology, 22*, 651–659.
98. Schlichting, I., & Goody, R. S. (1997). Triggering methods in crystallographic enzyme kinetics. *Methods in Enzymology, 277*, 467–490.
99. Barends, T., White, T. A., Barty, A., Foucar, L., Messerschmidt, M., Alonso-Mori, R., et al. (2015). Effects of self-seeding and crystal post-selection on the quality of Monte Carlo-integrated SFX data. *Journal of Synchrotron Radiation, 22*, 644.
100. Brennich, M. E., Nolting, J. F., Dammann, C., Nöding, B., Bauch, S., & Herrmann, H. (2011). Dynamics of intermediate filament assembly followed in micro-flow by small angle X-ray scattering. *Lab on a Chip, 11*, 708–716.
101. Knight, J., Vishwanath, A., Brody, J., & Austin, R. (1998). Hydrodynamic focusing on a silicon chip: Mixing nanoliters in microseconds. *Physical Review Letters, 80*, 3863–3866.
102. Park, H. Y., Qiu, X., Rhoades, E., Korlach, J., Kwok, L. W., & Zipfel, W. R. (2006). Achieving uniform mixing in a microfluidic device: Hydrodynamic focusing prior to mixing. *Analytical Chemistry, 78*, 4465–4473.
103. Pollack, L., & Doniach, S. (2009). Time-resolved X-ray scattering and RNA folding. *Methods in Enzymology, 469*, 253–268.
104. Pollack, L., Tate, M. W., Darnton, N. C., Knight, J. B., Gruner, S. M., Eaton, W. A., et al. (1999). Compactness of the denatured state of a fast-folding protein measured by submillisecond small-angle x-ray scattering. *Proceedings of the National Academy of Sciences, 96*, 10115–10117.
105. Zahoor, R., Belšak, G., Bajt, S., Weckert, E., & Hajdu, J. (2018). Simulation of liquid micro-jet in free expanding high-speed co-flowing gas streams. *Microfluidics and Nanofluidics, 22*, 87. https://doi.org/10.1007/s10404-018-2110-0.

106. Neutze, R., Wouts, R., van der Spoel, D., Weckert, E., & Hajdu, J. (2000). Potential for biomolecular imaging with femtosecond X-ray pulses. *Nature, 406*, 752–757.
107. Fukuda, Y., Tse, K. M., Nakane, T., Nakatsu, T., Suzuki, M., Sugahara, M., et al. (2016). Redox-coupled proton transfer mechanism in nitrite reductase revealed by femtosecond crystallography. *Proceedings of the National Academy of Sciences of the United States of America, 113*, 2928–2933.
108. Liu, W., Wacker, D., Gati, C., Han, G. W., James, D., Wang, D., et al. (2013). Serial femtosecond crystallography of G protein–coupled receptors. *Science, 342*, 1521–1524.
109. Mafuné, F., Miyajima, K., Tono, K., Takeda, Y., Kohno, J. Y., Miyauchi, N., et al. (2016). Microcrystal delivery by pulsed liquid droplet for serial femtosecond crystallography. *Acta Crystallographica, Section D: Biological Crystallography, 72*(Pt 4), 520–523.
110. Zhou, Q., Lai, Y., Bacaj, T., Zhao, M., Lyubimov, A. Y., Uervirojnangkoorn, M., et al. (2015). Architecture of the synaptotagmin–SNARE machinery for neuronal exocytosis. *Nature, 525*, 62–67.
111. Kupitz, C., Basu, S., Grotjohann, I., Fromme, R., Zatsepin, N. A., Rendek, K. N., et al. (2014). Serial time-resolved crystallography of photosystem II using a femtosecond X-ray laser. *Nature, 513*, 261–265.
112. Nango, E., Royant, A., Kubo, M., Nakane, T., Wickstrand, C., Kimura, T., et al. (2016). A three dimensional movie of structural changes in bacteriorhodopsin. *Science, 354*, 1552–1557.
113. Nogly, P., Panneels, V., Nelson, G., Gati, C., Kimura, T., Milne, C., et al. (2016). Lipidic cubic phase injector is a viable crystal delivery system for time-resolved serial crystallography. *Nature Communications, 7*, 12314.
114. Suga, M., Akita, F., Sugahara, M., Kubo, M., Nakajima, Y., Nakane, T., et al. (2017). Light-induced structural changes and the site of O=O bond formation in PSII caught by XFEL. *Nature, 543*, 131–135.
115. Sugahara, M., Song, C., Suzuki, M., Masuda, T., Inoue, S., Nakane, T., et al. (2016). Oil-free hyaluronic acid matrix for serial femtosecond crystallography. *Scientific Reports, 6*, 24484.
116. Sugahara, M., Nakane, T., Masuda, T., Suzuki, M., Inoue, S., Song, C., et al. (2017). Hydroxyethyl cellulose matrix applied to serial crystallography. *Scientific Reports, 7*, 703.
117. Hope, H. (1988). *Acta Crystallographica. Section B, 44*, 22–26.
118. Nakane, T., Song, C., Suzuki, M., Nango, E., Kobayashi, J., Masuda, T., et al. (2015). Native sulfur/chlorine SAD phasing for serial femtosecond crystallography. *Acta Crystallographica Section D: Structural Biology, 71*, 2519–2525.
119. Yamashita, K., Pan, D., Okuda, T., Sugahara, M., Kodan, A., Yamaguchi, T., et al. (2015). An isomorphous replacement method for efficient de novo phasing for serial femtosecond crystallography. *Scientific Reports, 5*, 14017.
120. Colletier, J. P., Sliwa, M., Gallat, F. X., Sugahara, M., Guillon, V., Schirò, G., et al. (2016). Serial femtosecond crystallography and ultrafast absorption spectroscopy of the photoswitchable fluorescent protein IrisFP. *Journal of Physical Chemistry Letters, 7*, 882–887.
121. Nakane, T., Hanashima, S., Suzuki, M., Saiki, H., Hayashi, T., Kakinouchi, K., et al. (2016). Membrane protein structure determination by SAD, SIR or SIRAS phasing in serial femtosecond crystallography using a novel iododetergent. *Proceedings of the National Academy of Sciences of the United States of America, 113*, 13039–13044.
122. Edlund, P., Takala, H., Claesson, E., Henry, L., Dods, R., Lehtivuori, H., et al. (2016). The room temperature crystal structure of a bacterial phytochrome determined by serial femtosecond crystallography. *Scientific Reports, 6*, 35279.
123. Masuda, T., Suzuki, M., Inoue, S., Song, C., Nakane, T., Nango, E., et al. (2017). Atomic resolution structure of serine protease proteinase K at ambient temperature. *Scientific Reports, 7*, 45604.
124. Cheng, A., Hummel, B., Qiu, H., & Caffrey, M. (1998). A simple mechanical mixer for small viscous lipid-containing samples. *Chemistry and Physics of Lipids, 95*, 11–21.
125. Barends, T. R. M., Foucar, L., Botha, S., Doak, R. B., Shoeman, R. L., Nass, K., et al. (2014). De novo protein crystal structure determination from X-ray free-electron laser data. *Nature, 505*, 244–247.

126. Hunter, M. S., Yoon, C. H., DeMirci, H., Sierra, R. G., Dao, E. H., Ahmadi, R., et al. (2016). Selenium single-wavelength anomalous diffraction de novo phasing using an X-ray-free electron laser. *Nature Communications, 7*, 13388.
127. Nass, K., Meinhart, A., Barends, T. R., Foucar, L., Gorel, A., Aquila, A., et al. (2016). Protein structure determination by single-wavelength anomalous diffraction phasing of X-ray free-electron laser data. *IUCrJ, 3*, 180–191.
128. Thorn, A., & Sheldrick, G. M. (2011). ANODE: Anomalous and heavy-atom density calculation. *Journal of Applied Crystallography, 44*(6), 1285–1287.
129. Stellato, F., Oberthür, D., Liang, M., Bean, R., Gati, C., Yefanov, O., et al. (2014). Room-temperature macromolecular serial crystallography using synchrotron radiation. *IUCrJ, 1*, 204–212.
130. Weierstall, U. (2014). Liquid sample delivery techniques for serial femtosecond crystallography. *Philosophical Transactions of the Royal Society of London. Series B, Biological Sciences, 369*, 20130337.
131. Nakane, T., Joti, Y., Tono, K., Yabashi, M., Nango, E., Iwata, S., et al. (2016). Data processing pipeline for serial femtosecond crystallography at SACLA. *Journal of Applied Crystallography, 49*, 1035–1041.
132. Roessler, C. G., Agarwal, R., Allaire, M., Alonso-Mori, R., Andi, B., Bachega, J. F. R., et al. (2016). Acoustic injectors for drop-on-demand serial femtosecond crystallography. *Structure, 24*, 631–640. https://doi.org/10.1016/j.str.2016.02.007.
133. Beyerlein, K. R., Dierksmeyer, D., Mariani, V., Kuhn, M., Sarrou, I., Ottaviano, A., et al. (2017). Mix-and-diffuse serial synchrotron crystallography. *IUCrJ, 4*, 769–777. https://doi.org/10.1107/S2052252517013124.
134. Fuller, F. D., Gul, S., Chatterjee, R., Burgie, E. S., Young, I. D., Lebrette, H., et al. (2017). Drop-on-demand sample delivery for studying biocatalysts in action at X-ray free-electron lasers. *Nature Methods, 14*(4), 443–449.
135. Kok, B., Forbush, B., & Mcgloin, M. (1970). Cooperation of charges in photosynthetic O_2 evolution-I. A linear four step mechanism. *Photochemistry and Photobiology, 11*(6), 457–475. https://doi.org/10.1111/j.1751-1097.1970.tb06017.x.
136. Hunter, M. S., Segelke, B., Messerschmidt, M., Williams, G. J., Zatsepin, N. A., Barty, A., et al. (2014). Fixed-target protein serial microcrystallography with an x-ray free electron laser. *Scientific Reports, 4*, 6026.
137. Kimura, T., Joti, Y., Shibuya, A., Song, C., Kim, S., Tono, K., et al. (2014). Imaging live cell in micro-liquid enclosure by X-ray laser diffraction. *Nature Communications, 5*, 3052.
138. Murray, T. D., Lyubimov, A. Y., Ogata, C. M., Vo, H., Uervirojnangkoorn, M., Brunger, A. T., et al. (2015). A high-transparency, micro-patternable chip for X-ray diffraction analysis of microcrystals under native growth conditions. *Acta Crystallographica Section D: Biological Crystallography, 71*, 1987–1997.
139. Cohen, A. E., Soltis, S. M., González, A., Aguila, L., Alonso-Mori, R., Barnes, C. O., et al. (2014). Goniometer-based femtosecond crystallography with X-ray free electron lasers. *Proceedings of the National Academy of Sciences of the United States of America, 111*, 17122–17127.
140. Roedig, P., Vartiainen, I., Duman, R., Panneerselvam, S., Stübe, N., Lorbeer, O., et al. (2015). A micro-patterned silicon chip as sample holder for macromolecular crystallography experiments with minimal background scattering. *Scientific Reports, 5*, 10451.
141. Zarrine-Afsar, A., Barends, T. R. M., Müller, C., Fuchs, M. R., Lomb, L., Schlichting, I., et al. (2012). Crystallography on a chip. *Acta Crystallographica Section D: Biological Crystallography, 68*, 321–323.
142. Mueller, C., Marx, A., Epp, S. W., Zhong, Y., Kuo, A., Balo, A. R., et al. (2015). Fixed target matrix for femtosecond time-resolved and in situ serial micro-crystallography. *Structural Dynamics, 2*, 054302.

143. Feld, G. K., Heymann, M., Benner, W. H., Pardini, T., Tsai, C. J., Boutet, S., et al. (2015). Low-Z polymer sample supports for fixed-target serial femtosecond X-ray crystallography. *Journal of Applied Crystallography, 48*(4), 1072–1079.
144. Lyubimov, A. Y., Murray, T. D., Koehl, A., Araci, I. E., Uervirojnangkoorn, M., Zeldin, O. B., et al. (2015). Capture and X-ray diffraction studies of protein microcrystals in a microfluidic trap array. *Acta Crystallographica Section D, 71*(4), 928–940.
145. Roedig, P., Ginn, H. M., Pakendorf, T., Sutton, G., Harlos, K., Walter, T. S., et al. (2017). High-speed fixed-target serial virus crystallography. *Nature Methods, 14*, 805. https://doi.org/10.1038/nmeth.4335.
146. Opara, N., Martiel, I., Arnold, S. A., Braun, T., Stahlberg, H., Makita, M., et al. (2017). Direct protein crystallization on ultrathin membranes for diffraction measurements at X-ray free-electron lasers. *Journal of Applied Crystallography, 50*, 909–918.
147. Meents, A., Gutmann, S., Wagner, A., & Schulze-Briese, C. (2009). Origin and temperature dependence of radiation damage in biological samples at cryogenic temperatures. *Proceedings of the National Academy of Sciences, 107*(3), 1094–1099. https://doi.org/10.1073/pnas.0905481107.
148. Owen, R. L., Rudiño-Piñera, E., & Garman, E. F. (2006). Experimental determination of the radiation dose limit for cryocooled protein crystals. *Proc Natl Acad Sci USA, 103*(13), 4912–4917.
149. Suga, M., Akita, F., Hirata, K., Ueno, G., Murakami, H., Nakajima, Y., et al. (2015). Native structure of photosystem II at 1.95 Å resolution viewed by femtosecond X-ray pulses. *Nature, 517*(7532), 99–103. https://doi.org/10.1038/nature13991.
150. Fraser, J. S., van den Bedemb, H., Samelsona, A. J., Langa, P. T., Holton, J. M., et al. (2011). *Proceedings of the National Academy of Sciences of the United States of America, 108*(39), 16247–16252. https://doi.org/10.1073/pnas.1111325108.
151. Coquelle, N., Brewster, A. S., Kapp, U., Shilova, A., Weinhausen, B., Burghammer, M., et al. (2015). Raster-scanning serial protein crystallography using micro- and nano-focused synchrotron beams. *Acta Crystallographica Section D: Biological Crystallography, 71*(Pt 5), 1184–1196. https://doi.org/10.1107/S1399004715004514 Epub 2015 Apr 25.
152. Sui, S., Wang, Y., Kolewe, K. W., Srajer, V., Henning, R., Schiffman, J. D., et al. (2016). Graphene-based microfluidics for serial crystallography. *Lab on a Chip, 16*(16), 3082–3096. https://doi.org/10.1039/c6lc00451b.
153. Kiefersauer, R., Than, M. E., Dobbek, H., Gremer, L., Melero, M., Strobl, S., et al. (2000). *Journal of Applied Crystallography, 33*, 1223–1230.
154. Sanchez Weatherby, J., Bowler, M. W., Huet, J., Gobbo, A., Felisaz, F., Lavault, B., et al. (2009). Improving diffraction by humidity control: A novel device compatible with X-ray beamlines. *Acta Crystallographica. Section D, Biological Crystallography, 65*, 1237–1246.
155. Roedig, P., Duman, R., Sanchez-Weatherby, J., Vartiainen, I., Burkhardt, A., Warmer, M., et al. (2016). Room-temperature macromolecular crystallography using a micro-patterned silicon chip with minimal background scattering. *Journal of Applied Crystallography, 49*, 968–975.
156. Meents, A., Wiedorn, M. O., Srajer, V., Henning, R., Sarrou, I., Bergtholdt, J., et al. (2017). Pink beam serial crystallography. *Nature Communications, 8*, 1281. https://doi.org/10.1038/s41467-017-01417-3.
157. Sherrell, D. A., Foster, A. J., Hudson, L., Nutter, B., O'Hea, J., Nelson, S., et al. (2015). A modular and compact portable mini-endstation for high-precision, high-speed fixed target serial crystallography at FEL and synchrotron sources. *Journal of Synchrotron Radiation, 22*, 1372–1378.
158. Owen, R. L., Axford, D., Sherrell, D. A., Kuo, A., Ernst, O. P., Schulz, E. C., et al. (2017). Low-dose fixed-target serial synchrotron crystallography. *Acta Crystallographica Section D: Biological Crystallography, 73*, 373–378.
159. Abdallah, B. G., Zatsepin, N. a., Roy-Chowdhury, S., Coe, J., Conrad, C. E., Dörner, K., et al. (2015). Microfluidic sorting of protein nanocrystals by size for X-ray free-electron laser diffraction. *Structural Dynamics, 2*, 041719. https://doi.org/10.1063/1.4928688.

160. Koralek, J. D., Kim, J. B., Brůža, P., Curry, C. B., Chen, Z., Bechtel, H. A., et al. (2018). Generation and characterization of ultrathin free-flowing liquid sheets. *Nature Communications, 9*(1), 1–8. https://doi.org/10.1038/s41467-018-03696-w.
161. Wiedorn, M. O., Awel, S., Morgan, A. J., Ayyer, K., Gevorkov, Y., Fleckenstein, H., et al. (2018). Rapid sample delivery for megahertz serial crystallography at X-ray FELs. *IUCrJ, 5*(5), 574–584.

Chapter 6
When Diffraction Stops and Destruction Begins

Carl Caleman and Andrew V. Martin

6.1 Introduction

It is now possible to solve protein structures with femtosecond X-ray free-electron laser (XFEL) pulses that were previously inaccessible to continuous synchrotron sources due to radiation damage [1, 2]. The key to this success is that diffraction probes the protein structure on femtosecond timescales, whereas nuclear motion takes tens to hundreds of femtoseconds to have a significant effect on the crystal structure. This is the essential idea behind the diffraction-before-destruction principle that underlies serial femtosecond crystallography (SFX) with XFELs [1]. In practice, the principle works well enough to determine protein structures of comparable resolution to synchrotron protein crystallography [2], which has led to the many successes of XFEL crystallography to date.

The reality is, however, that radiation damage begins from the first femtosecond that the pulse interacts with the sample and begins to affect diffraction through electron motion. The signal adds incoherently during the exposure and the measured pattern is the accumulated diffraction from an initially undamaged crystal as it gradually degrades and becomes disordered [3]. A striking feature of crystal samples is that the induced motion changes the nature of the diffraction during the pulse

C. Caleman (✉)
Department of Physics and Astronomy, Uppsala University, Uppsala, Sweden

Center for Free-Electron Laser Science, Deutsches Elektronen-Synchrotron, Hamburg, Germany
e-mail: carl.caleman@physics.uu.se

A. V. Martin
School of Science, RMIT University, Melbourne, VIC, Australia

ARC Centre of Excellence for Advanced Molecular Imaging, University of Melbourne, Melbourne, VIC, Australia
e-mail: andrew.martin@rmit.edu.au

S. Boutet et al. (eds.), *X-ray Free Electron Lasers*,
https://doi.org/10.1007/978-3-030-00551-1_6

from sharp Bragg peaks to continuous diffuse scattering. The crystalline Bragg diffraction may stop completely before the end of the pulse, but the measured pattern will still contain Bragg peaks so long as they are larger than the background noise generated by the diffuse scattering integrated over the whole pulse duration. This has been dubbed the "self-gating" effect [3], as crystal diffraction appears to have been generated by shorter pulse of the same intensity. The self-gating effect is resolution dependent as Bragg diffraction will only stop at a particular scattering angle when the damage has spread to the corresponding length-scale in the sample.

Damage processes that affect all atoms approximately equally independent of atom species or position within the sample are commonly referred to as "global damage". These processes include the average ionization level throughout the sample and ion motion due to the rising average temperature during the exposure. The effect of global damage on the diffraction can be modelled as an attenuation factor that varies with scattering angle (i.e. resolution). This is a similar effect to the temperature factors used in synchrotron crystallography, except the dependence on scattering angle can be different (non-Gaussian) in the XFEL case and sensitive to experimental conditions. We will review the physical origins of the attenuation of diffraction by global damage in this chapter.

Local damage is any damage process that does not affect all atoms in the sample equally, i.e. is element specific, depends on local environment or depends on the position of an atom. The effect most likely to be encountered is a relative change in the scattering strength of different elements caused by differences in ionization rates. This is expected when heavy elements are present and there are large differences in atomic number between elements in the sample [4, 5]. In some cases, this is an advantage because it can increase multiple anomalous absorption signal that can be used for ab initio phasing [6]. Variable ionization rates between elements also have subtler effects on the coherence of the diffracted X-rays [7].

While most of the damage effects encountered in XFEL protein crystallography are captured by the concepts of global and local damage, there have also been observations of exotic, cooperative structural changes in inorganic crystals. These are dynamical changes in highly ionized crystals on femtosecond timescales that modify the observed crystal structure. A striking example of electron dynamics is the cooperative motion that was observed in C_{60} fullerene crystals [8], effectively changing crystal structure as seen by the X-rays. It has also been shown using a two-colour X-ray pump-probe experiment on crystalline xenon clusters that when widespread ionization drives the sample into a plasma-like state, the lattice spacing of a crystal can contract on sub-100 fs timescales [9]. While these exotic, cooperative effects have not been observed in protein crystals, and may even be unlikely, they do serve as a warning that nothing can be taken for granted in XFEL experiments.

In synchrotron crystallography, a key measure of damage tolerance in crystals is dose, which is the absorbed energy in the sample per unit mass. For cryocooled macromolecules, an estimated limit is 30 MGy [10]. In these terms, XFELs massively exceed this limit by solving protein structures with doses of tens of giga-Gray [11]. However, for the femtosecond timescale of XFEL pulses, dose is not the best predictor of radiation damage effects. Induced sample dynamics are more sensitive to the rate at which energy is absorbed, which could be expressed as a dose rate, but more commonly is quantified by the incident beam intensity, which facilitates the specification of experimental parameters.

While the concept of diffraction-before-destruction serves as a good working principle to understand why XFEL crystallography works, there is a complex competition between damage and diffraction that ultimately determines suitable beam conditions for XFEL crystallography. Our aim in this chapter is to review the background theory of damage and diffraction in XFEL crystallography and to discuss what is known about radiation damage from XFEL crystallography experiments so far.

6.2 Damage Processes and Modelling

6.2.1 Ionization Processes

X-rays interact with matter through scattering and absorption. Photon scattering can either be elastic, where the photon energy is conserved, or inelastic, where some of the photon energy is transferred to the atom. Generally, elastic scattering contributes to the recordable information in the diffraction pattern, while inelastic scattering does not carry structural information that is easily decipherable. In photoabsorption, the energy of the photon is transferred to an electron, which is then ejected from the atom, leaving a positively charged ion behind. Figure 6.1 shows the cross section

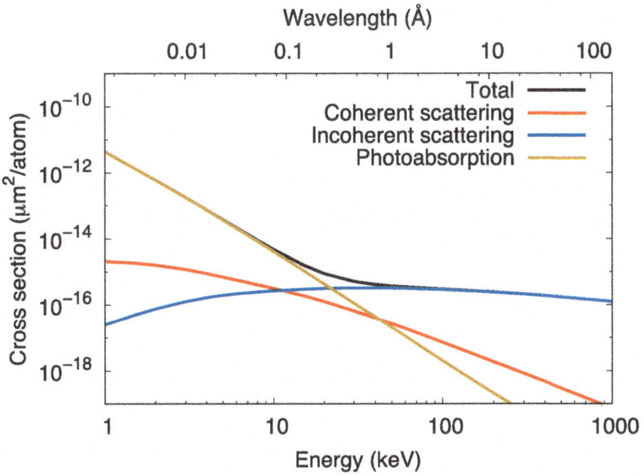

Fig. 6.1 Cross sections vs energy. Atomic cross sections of neutral carbon for photoabsorption, coherent scattering and incoherent (Compton) scattering

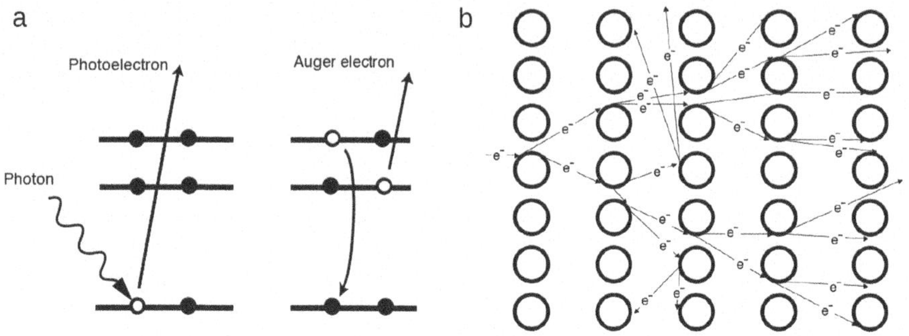

Fig. 6.2 Ionization processes. Diagrams illustrating the processes of (**a**) photoabsorption and Auger decay, and (**b**) the secondary electron cascade. The crystal approaches a plasma-like state as secondary ionization becomes widespread. Reproduced with permission from Caleman [18]

for the photon interaction with carbon at energies relevant for SFX. At 10 keV the probability for photoabsorption is around ten times higher than that for coherent scattering. Hence, this is the main process that leads to damage in the sample.

When the photoelectron ejected from a core level leaves behind a vacancy, an electron from a higher energy level may fall into the empty orbit, resulting in a release of energy. This energy may be emitted in the form of a photon (dominant process with high-Z elements), but it can also be transferred to an outer shell electron, which is then ejected from the atom in a process called Auger decay [12] as illustrated in Fig. 6.2a. The Auger electron carries the kinetic energy corresponding to the difference between the shell binding energy and the energy of the initial electronic transition. Compared to the photoelectron generated by an X-ray photon, the energy of the Auger electron is significantly lower, and it is ejected at a later time than the photoelectron. In biologically relevant material Auger electrons have energies between 250 eV and 2 keV [13], compared to the photoelectron carrying the energy of the incoming photon minus the shell binding energy (typically between 2 and 20 keV with X-rays). The physics of this decay is well understood [14]. The lifetime of the inner level vacancy caused by photoionization can be determined by measurements of the Auger line widths [15]. For atoms abundant in biological samples (such as C, N, O and S), the K-hole lifetimes are up to around 10 fs. During the photoionization process, the photoelectron may interact with valence electrons, leading to the so-called shake-up and shake-off effects [16]. For light elements of biological significance, electron emission from these effects is on the order of 10–30% of the events where a low energy electron (10–100 eV) is emitted [17].

Inelastic scattering is a relatively rare event at X-ray frequencies although it is the main source of energy deposition with hydrogen, and represents about 3% of all interactions between X-rays and a biological sample at 1 Å wavelength. During inelastic scattering, the incoming photon excites an electron to some virtual level and when the electron relaxes emitting a photon, it does not come back to the ground state. The photon emitted has therefore a different frequency from the photon absorbed, and it also has an altered phase.

As atoms are photo-ionized, the electrons ejected from the atom will interact with the surrounding sample. In a macroscopic sample, both the photoelectrons and the secondary Auger electrons become thermalized and trapped inside the sample. Thermalization is based on inelastic electron–electron interactions and, to a lesser degree, on electron–nuclear interactions. An electron scattering inelastically on an atom may cause a second ionization of an outer shell electron. This mechanism leads to an avalanche of electrons generated from one single photoionization event as shown in Fig. 6.2b. Thermalization produces a large number of such secondary ionization events, known as electron cascades or electron avalanche [19]. The number of these cascade electrons is roughly proportional to the energy of the impact electron triggering the cascade. These electrons are redistributed in the sample and can recombine with atoms. On the 100 fs timescale, considered relevant to damage formation in structural biology with XFELs, the recombination is low [20].

Collisions with atoms are highly relevant for understanding of X-ray induced damage. In a near neutral organic sample, where most atoms have their outer shell electrons bound, a single Auger electron from a carbon atom, carrying a kinetic energy of 278 eV, will generate up to 12 secondary electrons [21]. An 8 keV photon electron, on the other hand, might in the same sample cause up to 400 secondary ionization events, as illustrated in Fig. 6.3. The thermalization happens within tens of femtosecond, which means that 10–30 fs after a single photo-ionization event, more than 400 secondary ionization events have occurred in the sample [11].

The description above is valid for a neutral system, where most atoms carry all their electrons. In an SFX experiment, the photon bombardment is so intense

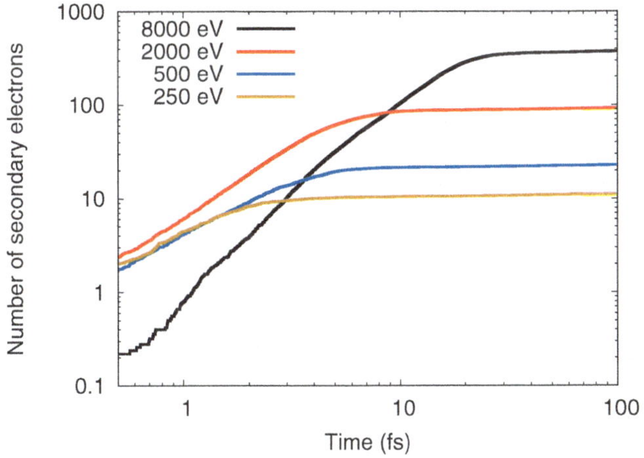

Fig. 6.3 Electron cascades at different energies. The cascade of electrons generated by a single electron of energy 250, 500, 2000 and 8000 eV in an neutral organic crystal, urea. The calculations were performed using a molecular dynamics code, as in Caleman et al. [22]. The number of secondary electrons generated is plotted as a function of time after the photon absorption event

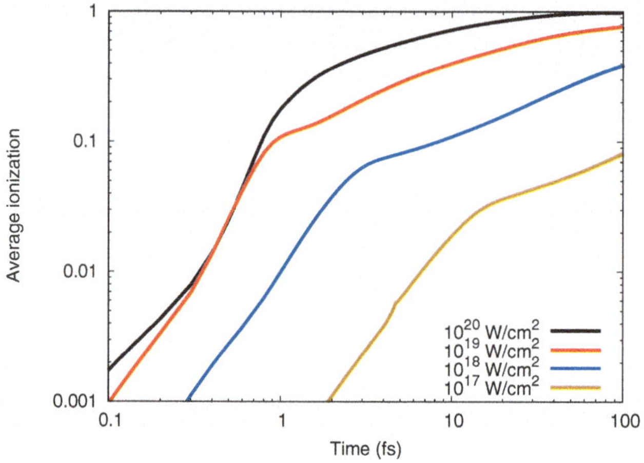

Fig. 6.4 Average ionization in a sample vs time for different intensities. Average ionization in a photosystem I protein crystal exposed to a 100 fs XFEL pulse with different pulse intensities. Photon energy was 6 keV and the simulations are performed using a continuum model [24, 25]. When the average ionization becomes constant in time, all the electrons are stripped from the atoms and the atoms become transparent to the X-rays

that the average ionization in a sample often rises well above one per carbon atom. This affects both the photoionization probability and the electron impact ionization. Photoionization at the relevant X-ray energies acts on the core s-electrons. If the vacancies caused by photoionization cannot be refilled by outer shell electrons, the sample in principle becomes transparent to the X-ray photons and cannot absorb more energy through photoionization. This has been described in early experiments at the linac coherent light source (LCLS) using Neon atoms as a target [23], but is equally relevant for biological samples. The electron–electron ionization probability is altered in a similar way. At high average ionization states the number of valence electrons is reduced and, hence, so is the inelastic free-electron valence–electron interaction. In Fig. 6.4 the average ionization of all atoms in a photosystem I protein crystal is shown and predicted by simulations using a continuum model [26], described below.

6.2.2 *Ion Motion*

The ionization caused by the X-ray bombardment causes the sample to heat up. The ions are put into motion due to the electron–ion collisions, ion–ion collisions and local electric fields. As the sample is heated up and ionized, it can be physically described as a plasma. In this description the ion diffusion coefficient D_i can be theoretically calculated, based on the ion temperature and the ion collision frequency ν_i [27]:

$$D_i(t) = \frac{k_B T_i(t)}{m_i \nu_i(t)} , \tag{6.1}$$

where m_i is the mass and k_B is the Boltzmann constant. This model assumes that each element i can be assigned a time-dependent temperature $T_i(t)$. The ion temperatures can be predicted using a quantum kinetic approach from warm dense plasma theory [28] using information about the trapped electrons that effectively heat the ions. From the ion diffusion coefficient it is further possible to calculate the root mean square displacement (RMSD) of the ions as

$$\sigma(t) = \sqrt{2ND_i(t)t} , \tag{6.2}$$

where N is the number of dimensions. The shape of the RMSD as a function of time is not strongly dependent on the intensity and could be represented by a scaling law;

$$\sigma(t) = Bt^n , \tag{6.3}$$

with $n \sim 1.5 \pm 0.4$ [25] and B is a fitting parameter that can be estimated from simulation. Figure 6.5 shows the RMSD for some examples relevant to SFX, estimated using the scaling law.

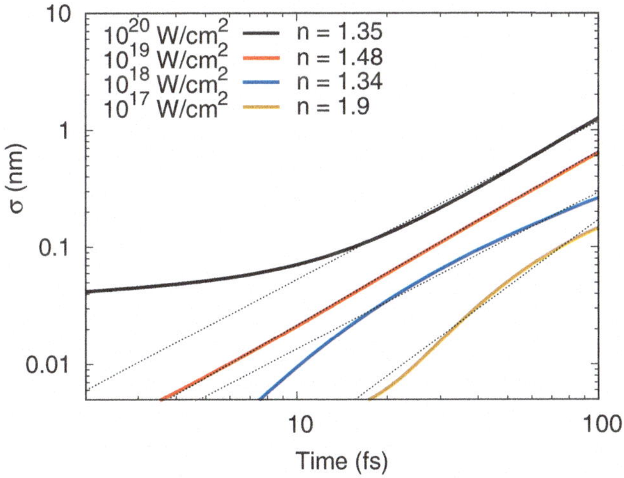

Fig. 6.5 Calculated ion displacement in a sample vs time at different intensities. Root mean squared displacement (σ) of the carbon ions in a photosystem I crystal exposed to a 100 fs XFEL pulse with photon energy of 6 keV, as simulated in Caleman et al. [25]. The dotted lines are the best fit using the scaling law described in the main text, with the B and n parameters estimated from continuum simulations

6.3 Damage Modelling

The potential in biological imaging using XFEL pulses was pointed out in a study by Neutze et al. [29] in the year 2000, well before any XFEL sources were available. In the following years several studies focusing on the optimal pulse parameters were published [30–32], all of them pointing out that the X-ray pulses needed to be shorter than the timescales of the destruction of the molecular structure of the sample. These studies and early experiments at the test facility at Stanford, SPPS [33, 34], were important to pave the way towards the investments and scientific efforts that lead to the building of the XFEL facilities we have today. The simulations, in particular the scientific case presented in Neutze et al. [29], were one of the major factors that made the scientific community realize the potential in biological imaging with an XFEL.

Simulations have been a natural part of the development of SFX. Early on they were used to explore the potential of using XFEL sources for structural biology, then later they informed the design of end stations and have been used to interpret the effects of radiation damage on experimental diffraction data. For large organic samples such as biomacromolecules and protein crystals, two major modelling approaches have been used. The study by Neutze et al. [29] applied a molecular dynamics (MD) approach. They used a well-established MD code that was adopted to include ionization and bond breaking. The models describing the physics in this early study were simplistic, but nevertheless the major conclusions drawn in the study still hold.

Since then several studies have used the MD approach and developed the model used in the early simulation. It is worth mentioning the so-called Cimarron Project [35], developed at Lawrence Livermore National Laboratory. The code is developed to simulate dense plasmas and has been applied to biological samples [36]. Another MD initiative developed to study FEL-matter interaction is the Xraypac [37] suite of programs that include both a MD part, XMDYN, and atomic ionization part, XATOM [38], that is based on nonrelativistic quantum electrodynamics and perturbation theory within the Hartree-Fock-Slater model. In general, molecular dynamics codes are rather computationally expensive to use, especially for large systems such as protein crystals. This necessitates the use of simplified descriptions of complex physical processes, such as bond breaking and ionization. However, MD codes have the advantage of giving direct information of the positions of the nuclei. Knowledge about the nuclei positions together with the ionization levels means that, in theory, the expected diffracted signal can be calculated directly from the simulations.

The second major approach to simulate the photon–matter interaction in an XFEL experiment has been by using rate equations to keep track of radiation transfer, ionization and temperatures. This description does not consider the time evolution of individual atoms, but rather describes a material as a continuum with specific properties. These models are often referred to as continuum models and are highly similar to models applied in other fields of research, like plasma and warm dense matter physics. A continuum model was first used for the purpose of understanding the dynamics in biological samples exposed to an FEL by Hau-Riege et al. in 2004 [30]. Since then this approach has been applied several times. Codes originally developed for warm dense matter physics applications, the so-called non-local thermal equilibrium codes (NLTE), have been used in several studies to describe the damage processes in SFX [3, 11, 25, 39]. The continuum model approach has the advantage over the MD approach that it is much less computationally expensive and does not scale with particle size. However, since everything is treated as ensembles it does not give any information about the individual atoms or ions. Any estimate of atomic displacement (and, in turn, decrease in diffracted signal) has to be calculated based on average ion temperatures. Changes in the diffracted signal caused by collective motion or local damage are not captured by this model, whereas such effects would be described, in principle, in a simulation considering the dynamics of the individual particles.

6.4 Diffraction and Damage

6.4.1 XFEL Diffraction Theory

X-rays are scattered by the electrons in the sample and the theory of XFEL crystal diffraction starts with a dynamical electron density for the sample, $\rho(q, t)$, and its Fourier transform, $f(q, t)$, called the atomic scatter factor. Diffraction at large scattering angles encodes high resolution structural information and is primarily caused by electrons that are still bound to atoms. The electrons that are ejected in ionization processes are delocalized and scatter X-rays diffusely to lower scattering angles, which is less important for crystallography experiments but contributes to the overall background.

There are two main impacts of XFEL damage on crystal diffraction: (1) the depletion of bound electrons and (2) the motion of ions due to Coulomb repulsion or diffusion. Both the depletion of bound electrons and ion motion break the translational symmetry of the crystal. This changes the nature of the diffraction from discrete to diffuse scattering. Hence, the characteristic effect of damage on the diffraction pattern is the decrease in Bragg diffraction relative to the diffuse background.

Due to the weak interaction between X-rays and matter, the X-ray diffraction pattern from a crystal can be expressed in terms of the crystal's time-dependent scattering factor $F(\mathbf{q}, t)$. The diffracted intensity $I(\mathbf{q})$ through a small solid angle $\Delta\Omega$ centred at scattering vector $\mathbf{q}$ is given by

$$I(\mathbf{q}) = r_e^2 \Delta\Omega \int I_0(t)|F(\mathbf{q}, t)|^2 dt \,, \tag{6.4}$$

where r_e is the classical electron radius, $I_0(t)$ is the incident intensity and t is time. For the simulation of an experiment, the intensity distribution measured on a detector can be estimated by setting $\Delta\Omega$ equal to the solid angle spanned by each pixel. The vector $\mathbf{q}$ indicates the point in reciprocal space that intersects the Ewald sphere. The crystal scattering factor is assumed to be a sum of the atomic scattering factors $f_i(\mathbf{q}, t)$ multiplied by a phase term that specifies the position of each atom in the crystal:

$$F(\mathbf{q}, t) = \sum_i f_i(\mathbf{q}, t) e^{-2\pi i \mathbf{q}\cdot\mathbf{r}_i(t)} , \tag{6.5}$$

where i ranges over the number of atoms in the crystal. The decomposition of the crystal scattering factor into a sum of atomic scattering factors is known as the isolated atom approximation, because it ignores bonding between atoms. The diffraction is impacted by ion diffusion through the time-dependence of the ion position $\mathbf{r}(t)$. The atomic scattering factors are parameterized into a $\mathbf{q}$-dependent factor $f_i^{(0)}(\mathbf{q})$ that depends on the electron density and wavelength dependent correction factors (f_i' and f_i''):

$$f_i(\mathbf{q}, t) = f_i^{(0)}(\mathbf{q}, t) + f_i' + i f_i'' . \tag{6.6}$$

The first term on the right-hand side, $f_i^{(0)}(\mathbf{q}, t)$, is proportional to Fourier transform of the atom's electron density $\rho(\mathbf{q}, t)$ and is thus sensitive to the number of electrons around an atom. Figure 6.6 shows how the scattering factor changes with ionization state for carbon. An important detail to note is that core-shell

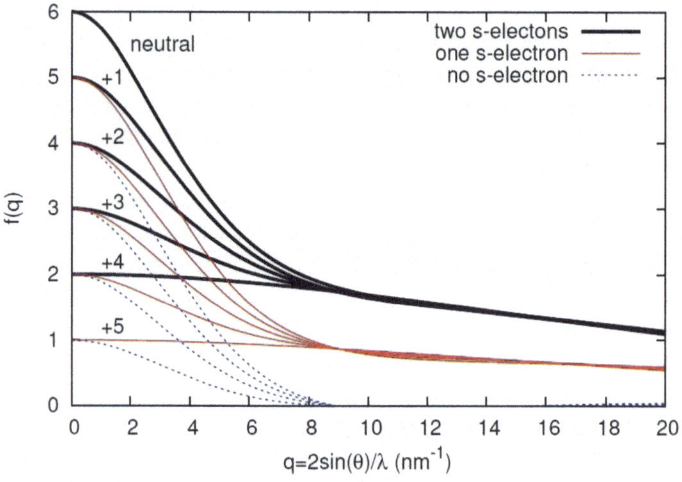

Fig. 6.6 Ionic scattering factors. The ionic scattering factors for carbon in different ionization states. Reproduced with permission from Caleman et al. [25]

electrons are almost the sole cause of scattering at resolutions greater than 3 Å. Hence, core-shell ionization and valence-shell ionization do not have equivalent effects on the diffraction pattern. As described in Sect. 6.2.1, core-shell and valence-shell ionization are also caused by different damage mechanisms. Valence-shell ionization, in particular, is more prevalent at later times during the exposure when the crystal is driven into a plasma-like state.

6.4.2 The Effect of Ionization on Crystal Diffraction

The difference between diffraction in synchrotron and XFEL crystallography is due to the extent to which scattering factors $f_i^{(0)}(\mathbf{q}, t)$ fluctuate during an exposure due to the widespread ionization. The effect of ionization is to reduce the number of bound electrons around an atom, which reduces the scattering strength $f_i^{(0)}(\mathbf{q}, t)$. This reduction of bound electrons occurs stochastically, which can have a significant effect on the diffracted intensities. For neutral atoms, scattering can be taken from parameterized tables [40]. For ionic scattering factors, the electron density is reconstructed from the densities of the remaining bound electrons. The densities of individual electron densities can be approximated with Slater orbitals [41]. Alternatively they can be calculated with quantum mechanically with code specifically designed for XFEL research, such as XATOM [38].

Both photoionization and secondary ionization are assumed to be random and not correlated between different unit cells or different atoms. The diffraction from atoms in different unit cells that are equivalent by lattice translation adds constructively at the Bragg condition. In the case of electronic damage, this effectively averages the scattering factor of an atom over the distribution of ionization states. Hence, the scattering factor for the crystal can be modelled by replacing each atomic scattering factor $f_i(\mathbf{q}, t)$ by its average over ionization states $\langle f_i(\mathbf{q}, t)\rangle$, as follows:

$$\langle F(\mathbf{q}, t)\rangle = \sum_i \langle f_i(\mathbf{q}, t)\rangle e^{-2\pi i \mathbf{q}\cdot\mathbf{r}_i(t)} . \tag{6.7}$$

Ignoring effects of the local structural environment on ionization leads to all atoms of the same species being modelled by the same average scattering factor. As a first approximation to model the global effects of damage, the diffracted intensity can be further averaged over different elements, which leads to the following approximation:

$$\langle |F(\mathbf{q}, t)|^2\rangle \approx \frac{\langle f^2(q, t)\rangle}{f_0^2(q, t)} |F(\mathbf{q}, t)|^2 + Nx(1 - x)\Delta f^2(t) , \tag{6.8}$$

where x is the fraction of atoms that have been ionized and N is the total number of atoms. The first term on the right-hand side of Eq. (6.8) is proportional to the undamaged scattering factor squared and represents coherent scattering, while the second term is the incoherent continuous background. We denote the global ionization scaling factor by

$$k(q,t) = \langle f^2(q,t)\rangle / f_0^2(q,t) \,. \tag{6.9}$$

We refer to this effect as global because it models the average ionization over all atoms. It has the same effect on each atom independent of element and atomic position.

Different elements will ionize at different rates and heavy elements have more electrons to lose than lighter elements. The change of relative strength of the different elements is a type of local damage that cannot be accounted for by an overall scaling of the diffraction. The difficulty with local damage is that it has the potential to change the interpretation of the structure. The bulk of atoms in protein crystals are light elements C,O and N, which will ionize at similar rates, and the effect of variable ionization is more likely to be observed in heavier elements relative to lighter elements. It is thus an important factor for methods of structure determination that explicitly use the scattering of heavy ions, such as anomalous dispersion or isomorphous replacement. It has even been proposed that in XFEL crystallography, higher ionization rates can increase the anomalous signal and be favourable [6].

6.4.3 The Effect of Ion Motion on Diffraction

To account for the effect of ion motion, the position of the ion can be written as a sum of its initial position plus a time-dependent displacement

$$\mathbf{r}_i(t) = \mathbf{R}_i + \boldsymbol{\epsilon}_i(t) \,. \tag{6.10}$$

Models of ion diffusion and temperature effects assume that $\boldsymbol{\epsilon}_i(t)$ is a random displacement with the statistics of a random walk. When the structure factor is averaged with respect to these statistics (and ionization is ignored) it takes the form

$$\langle |F(\mathbf{q},t)|^2\rangle = |F_0(\mathbf{q},t)|^2 e^{-4\pi^2 q^2 \sigma^2(t)} \,, \tag{6.11}$$

where

$$F_0(\mathbf{q},t) = \sum_i f(\mathbf{q}) e^{-2\pi i \mathbf{q}\cdot\mathbf{R}_i} \,. \tag{6.12}$$

The form of Eq. (6.11) is the same as the usual temperature factor in crystallography, except that the width is time-dependent to account for the effective change in temperature described in Sect. 6.2.2.

Ignoring ionization, the global correction factor for ion motion is constructed by integrating the time-dependent temperature factor:

$$g(q; T) = \int_0^T I_0(t) \exp^{-4\pi^2 q^2 \sigma^2(t)} dt . \tag{6.13}$$

The new expression for the diffracted intensity with the global correction for ion motion is

$$I(\mathbf{q}) = r_e^2 \Delta\Omega g(q; T) |F_0(\mathbf{q}, t)|^2 . \tag{6.14}$$

The effect of time-dependent ion diffusion on the diffraction is illustrated in Fig. 6.7. The crystal is initially highly ordered and produces Bragg diffraction. By 35 fs, ion motion has disrupted most of the high resolution structure of the crystal, leaving only periodicity at lower resolution, and by 70 fs all periodic structure is

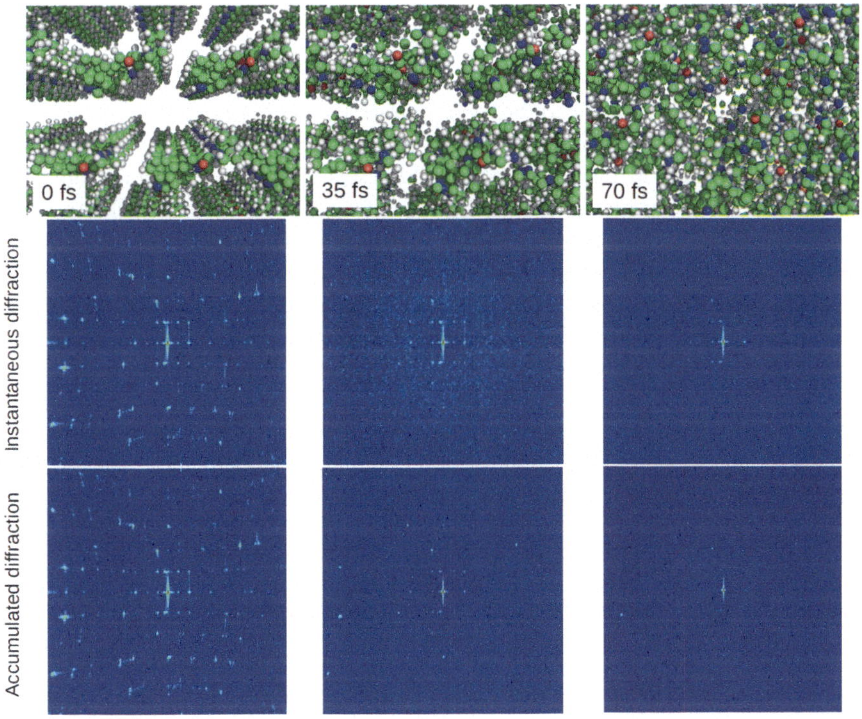

Fig. 6.7 The effect of ion diffusion on crystal diffraction. The top row shows the ion positions of a lysergic acid diethylamide crystal at different times during the exposure. The middle row shows the instantaneous diffraction pattern from each structure, which is the contribution to the measured intensity from that moment in time. The bottom row shows the accumulated diffraction which corresponds to the measured intensity

gone. This is reflected in the instantaneous diffraction patterns, which show only low resolution Bragg peaks at 35 fs and no Bragg peaks at 70 fs. However, the measured intensity is accumulated as the pulse traverses the sample and still shows Bragg peaks at 70 fs, because these were generated at earlier pulse times. Most importantly the Bragg peaks have greater magnitude than the noise generated by the increased diffuse background. Figure 6.8a shows how the accumulation of the Bragg signal stops at different moments during the pulse according to resolution. This leads to a relative resolution-dependent scaling of the Bragg peaks given by $g(q, T)$, which is plotted for different pulse times in Fig. 6.8b.

6.4.4 Global Correction Combining Ionization and Ion Motion

Combining the global corrections for ionization, Eq. (6.8), and for ion motion Eq. (6.11), we obtain a new expression for the diffracted intensity:

$$I(\mathbf{q}) = r_e^2 \Delta\Omega |F(\mathbf{q}, t)|^2 \int I_0(t) k(q, t) e^{-4\pi^2 q^2 \sigma^2(t)} dt \; . \tag{6.15}$$

The time integral on the right-hand side of Eq. (6.15) represents the global correction factor for both ion motion and ionization. As shown in Fig. 6.6, core-shell electrons make up a greater fraction of the scattering factor at high resolution. Hence, core-shell ionization is expected to reduce high resolution peaks relative to low resolution peaks, adding to the initial effects of ion motion. Interestingly, valence-shell ionization can have the opposite effect. Figure 6.6 shows that valence electrons contribute to the scattering factor at resolutions up to 3 Å and when valence ionization dominates low resolution peaks can be attenuated relative to high resolution peaks. This effect is predicted by the simulations shown in Fig. 6.9. At lower intensities ($10^{17} - 10^{18}$ W cm^{-2}) ion motion and core-shell ionization dominate and high resolution peaks are most affected, but at higher intensities ($10^{19} - 10^{20}$ W cm^{-2}) valence ionization dominates global damage and the low resolution peaks are more greatly affected.

6.4.5 Partial Coherence Effects in XFEL Crystallography

In addition to modifying the crystal scattering factor, XFEL radiation damage can also change the coherence properties of the diffracted X-rays. This effect has been predicted for single molecule studies [42], but was largely assumed to be absent for crystal diffraction. Although each unit cell is damaged differently within a single exposure, the scattered waves from all the units cells interfere constructively at the Bragg condition, motivating the assumption that diffraction can be modelled from the average crystal electron density. However, the time integral in Eq. (6.4)

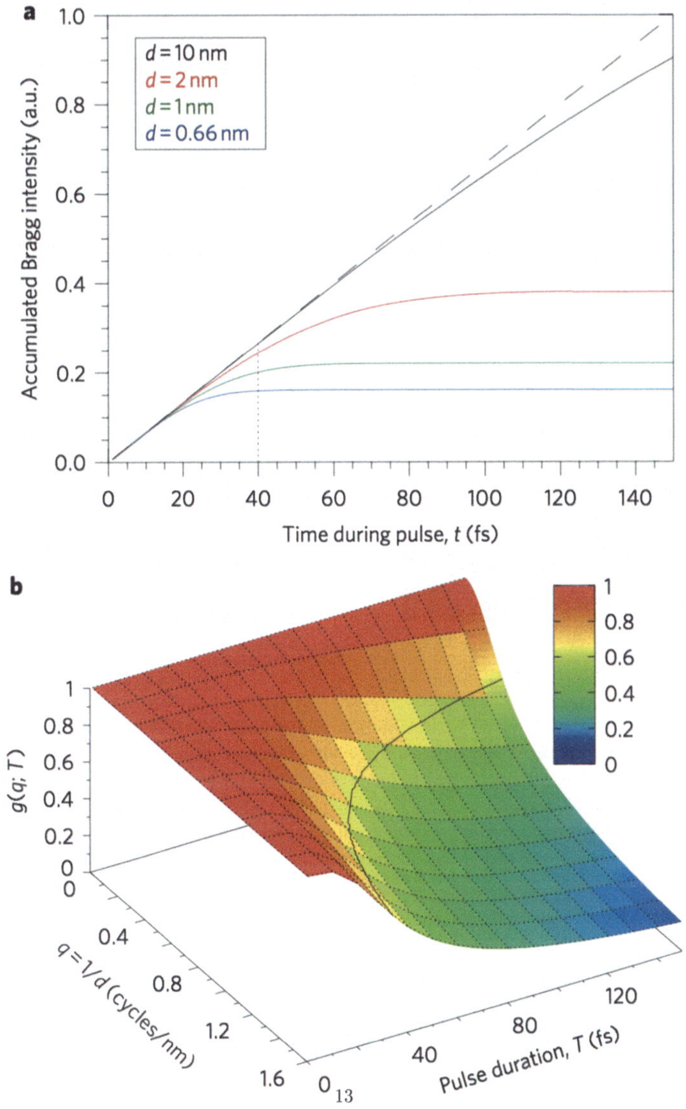

Fig. 6.8 Attenuation function due to ion motion. (**a**) The accumulation of the Bragg peak intensity at different resolutions as a function of the interaction time with the pulse predicted by the code CRETIN for a homogeneous protein sample in water. (**b**) The attenuation function due to ion motion as a function of pulse duration and resolution. Adapted from Ref. [3]

means that this is not entirely true. Fortunately, a more detailed analysis finds that for crystals, a full-coherence (single mode) approximation is valid, albeit with a slightly modified interpretation of the scattered wave [7]. With coherence effects included, the contribution of an ion to the scattered beam will not necessarily be

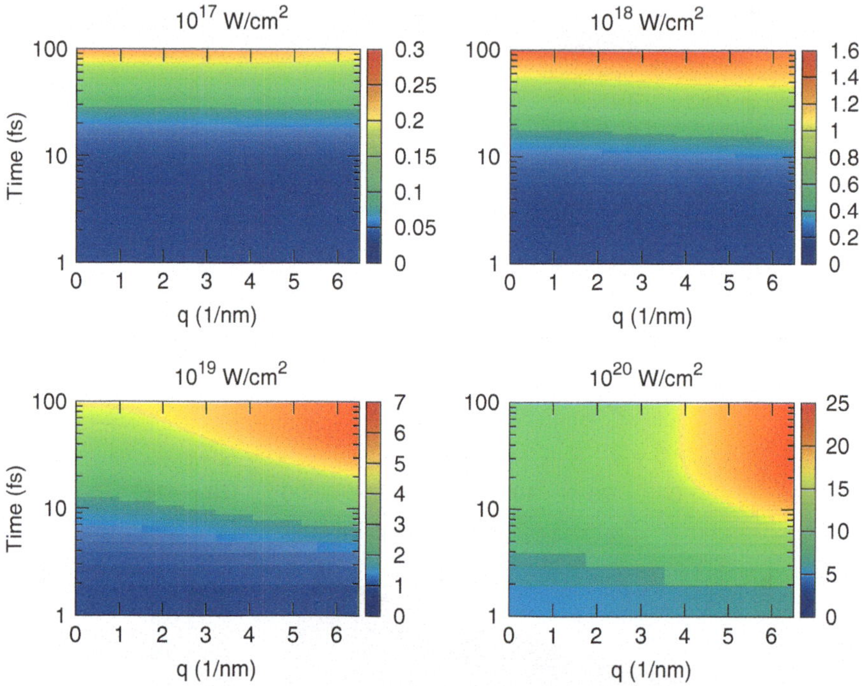

Fig. 6.9 Attenuation function due to ion motion and ionization. The attenuation function due to both ion motion and ionization as a function of pulse duration and resolution for different incident beam intensities. The relative attenuation of low and high angle scattering changes as intensities increase from low values (10^{17} W/cm^2–10^{18} W/cm^2) where diffusion dominates to higher values (10^{19} W/cm^2–10^{20} W/cm^2) where ionization effects dominate. The value for 6 keV and 10^{17} W/cm^2 is normalized to 1 to facilitate comparison with the different cases. Only ionization and displacement of carbon atoms are shown. Reproduced with permission from Caleman et al. [25]

equal to time-averaged scattering factor for each atom. This is only likely to affect crystals that have heavy elements and exposed to very high beam intensities, e.g. an effect of up to 20 % is predicted for sulphur exposed to high beam intensities.

6.5 SFX Damage Experiments

The first two experiments [3, 43] to study damage in SFX were performed at 2 keV at LCLS and spanned pulse durations from 70 to 400 fs. By studying long pulse durations, extensive radiation damage occurred and unique aspects of XFEL radiation damage were observed for the first time in serial crystallography samples. In particular, the experiment on photosystem I crystals [3] probed a region where ion motion effects dominated the global damage. This enabled the validation of a time-dependent temperature factor given in Sect. 6.4.3, as reasonably good agreement was achieved between the experimental data and modelling with CRETIN. This established the principle of damage effectively gating the pulse, so that the pulse appeared apparently shorter in duration. Excitingly, it gave explanation for why Bragg peaks were observed in SFX experiments even when the damage processes were expected to destroy the crystalline order before the end of the pulse. It is the diffraction from the early part of the pulse that generates the Bragg peaks, while the latter parts of the pulse produce a greater proportion of diffuse scattering. The experiment on lysozyme [43] under very similar experimental conditions was analysed with conventional time-independent temperature factor analysis and found inconsistencies between datasets at different pulse lengths, and between the XFEL data and synchrotron data of the same sample. The lack of agreement between datasets was not correctable with a global (time-dependent) factor for ion motion, leaving open the increased impact of global or local ionization effects on this sample.

The photosystem and lysozyme experiments formed part of the early development of SFX prior to the availability of shorter wavelength instruments. It was a positive development for SFX that hard X-ray pulses with photon energy 8–10 keV produce less radiation damage than the early experiments at 2 keV. Using crystals close to one micron in size with a comparable beam size, the intensity is sufficiently low to enable standard crystallographic analysis, without further modification for global or local XFEL damage. This was first confirmed in an experiment on lysozyme [2] and was soon confirmed with studies of other protein crystals. Structures had resolutions in the range of 2–3 Å sufficient for fitting atomic models. The validity of conventional analysis for a useful range of XFEL beam conditions has underpinned the rapid rise of this technique to date.

There are many SFX experiments that require higher intensities or smaller beam sizes for which radiation damage is critically important. In particular, recent efforts have focused on the potential for local damage in heavier ions such as sulphur or iron, as these elements are of biological importance and are also important for phasing based on anomalous scattering or isomorphous replacement. An important development for the SFX community was the demonstration of ab initio phasing by anomalous scattering of lysozyme soaked in an organometallic gadolinium complex [44], which showed that SFX data from an XFEL was of sufficient quality for this technique. This implies also that the extensive ionization does not eliminate heavy atom signal on which this technique is based. The extent of the ionization of gadolinium atoms in lysozyme was studied as a function for two different fluences (high and low) [45], and an average difference of almost nine electrons per Gd atom was found. The ionization was significantly lower than predicted by the XATOM code [38], and the possible explanations given included electron transport, recombination, ion motion or the unknown beam intensity profile.

Single anomalous dispersion has been demonstrated with sulphur [46, 47] and selenium [48] using XFELs, which is highly positive development for de novo phasing because sulphur and selenium are widely used for de novo phasing in synchrotron crystallography. This followed an earlier study of ionization in sulphur that reported an increase in the anomalous signal for XFEL sources compared to synchrotron sources [4].

An experiment designed to specifically target local damage effects was conducted on ferredoxin crystals with a beam intensity approaching the highest available at LCLS (1.8×10^{19} W cm^2) [5]. This protein contained two [4Fe–4S] clusters that displayed effects of element specific ionization rates that are a signature of local damage. Hence, the effect of damage on this sample cannot be corrected with global correction factors. Of even greater interest was the observation that the electron density of the two [4Fe–4S] clusters was different in the XFEL dataset but not in the synchrotron dataset. This is evidence that local structure and bonding can impact ionization rate or the distribution of local electron density in XFEL experiments. In the same XFEL experiment, it was unexpectedly observed that the ionization of Fe did not change significantly above and below an ionization edge, which was not expected from simulation. Identifying the physical explanations for the many observations of the ferrodoxin experiment is an outstanding challenge that motivates deeper research into the complex dynamics that can be induced by intense femtosecond X-ray pulses.

It should be noted that femtosecond timescales are sufficient for extensive electron motion or rearrangement. When single crystal xenon clusters are pumped into a plasma state with a femtosecond X-ray pulse, the lattice spacing has been observed to contract on sub-100 fs timescales [9]. Another example is the remarkable observation that the apparent crystalline C_{60} at the maximum available beam intensity at LCLS is different from the known fcc structure that is observed with lower beam intensities and with synchrotron sources [8]. The high-intensity data was shown to agree with a model in which bound electrons rearrange cooperatively across many unit cells to produce a density with lower translational symmetry. Ion motion is a less likely explanation due to the insufficient time for nuclear rearrangement. While such large-scale cooperative effects have not yet been observed in protein crystals, the xenon cluster and C_{60} observations are a reminder that nothing can be taken for granted with extreme intensities produced by X-ray free-electron lasers.

6.6 Damage and Diffraction of Single Molecules

A long-standing goal of XFEL research is to determine structures of single molecules and avoid the need for crystallization [29]. Single molecule diffraction is continuous and much weaker than crystal diffraction, producing as few as a thousand photons per pattern. Nevertheless, it is predicted by imaging analysis theory that sub-3 Å resolution could be achieved with the pulse intensities currently available assuming ideal damage-free, background-free data of reproducible single protein molecules [49]. This has provided motivation for XFEL single molecule imaging to be pursued by large collaborations, most notably the single particle imaging initiative [50]. Currently the best 3D images are of viruses at around ten nanometres resolution. Recovered images have shown the expected size and shape of virus particles at these resolutions [51], which suggests that damage is not significant at these length scales.

XFEL damage to single molecules has been studied with both molecular dynamics [29, 32, 37] and rate equations theory [30]. Theory suggests that there is a self-gating effect in single molecules for global damage processes [52], in the sense that the continuous diffraction retains a strong similarity to the diffraction of the undamaged molecule despite the extra background and noise generated by damage. However, a significant complication is the Coulomb explosion, which can be viewed as a local damage process that has a greater affect on the surface of the molecule [29, 30]. It has been predicted that the trapped electron redistributes to neutralize the core of the positively charged particle, exposing a positively charged shell on the surface of the particle, so that the explosion proceeds layer by layer [30]. The Coulomb explosion is seen as a critical limitation because imaging analysis algorithms that require a rigid structure or have limitations on their ability to handle heterogeneity. One proposal for mitigating the effect of the Coulomb explosion is to use pulses less than 10 fs to outrun the nuclear motion [53]. Another proposal is to use a tamper layer of water around the molecule [54, 55], although the water will generate a background scattering signal that may be problematic for small molecules like proteins.

Even with short pulses to outrun nuclear motion, electronic damage will occur and modify single molecule diffraction. This can viewed as a loss of coherence using the approaches described in Sect. 6.4.5. Encouragingly, a fully coherent model of the diffraction is predicted to adequately account electronic effects of photoionization [42] and secondary ionization [56] in single protein molecules, but would exhibit a modified electron density that differs from the average ion scattering factor. Knowledge of ionization rates would be needed as part of the imaging analysis to correct this effect or at stronger beam intensities where the fully coherent approximation breaks down. There has been some initial work on extending these ideas to the Coulomb explosion and sample heterogeneity [7].

Ultimately the issue of damage in single molecules will not be resolved until higher resolution data is available to test damage theory and to explore the impact of damage on imaging analysis. It remains an outstanding challenge for single molecule imaging development that will become increasingly prominent as single molecule imaging techniques improve in resolution.

6.7 Conclusion

It is a positive outcome for SFX that there are XFEL beam conditions for which conventional crystallography analysis can be applied, and the majority of experiments aimed at structure determination have exploited this knowledge so far.

It is also a positive result that there are sufficient anomalous signals to perform de novo phasing. The developments in understanding global damage corrections point the way to improve conventional crystallographic analysis for XFEL sources.

It has become clear that there are limits on the conditions when local damage can be ignored, which have been exceeded by current sources at their highest beam intensities. These local effects are concentrated on heavier elements that are often biologically important or are used for de novo phasing methods. We expect that the recent availability of two pulse modes at XFELs for X-ray pump-probe experiments will enable future experiments to probe local damage processes and more exotic damage processes observed in xenon clusters and C_{60} crystals in greater detail.

Finally, we note that damage will play an increasingly significant role in extensions of SFX to new phasing methods. Specific mention can be made of direct phasing methods that attempt to exploit scattering between the Bragg peaks or phasing of imperfect crystals that contain structural information in the diffuse continuous scattering. In both these cases, the effect of damage on the diffuse scattering plays a much bigger role that in standard SFX, and will drive the need for better understanding of XFEL damage processes.

References

1. Chapman, H. N., Fromme, P., Barty, A., White, T. A., Kirian, R. A., Aquila, A., et al. (2011). Femtosecond X-ray protein nanocrystallography. *Nature, 470*, 73–77.
2. Boutet, S., Lomb, L., Williams, G. J., Barends, T. R. M., Aquila, A., Doak, R. B., et al. (2012). High-resolution protein structure determination by serial femtosecond crystallography. *Science, 337*, 362–364.
3. Barty, A., Caleman, C., Aquila, A., Timneanu, N., Lomb, L., White, T. A., et al. (2012). Self-terminating diffraction gates femtosecond X-ray nanocrystallography measurements. *Nature Photonics, 6*, 35–40.
4. Galli, L., Son, S.-K., Klinge, M., Bajt, S., Barty, A., Bean, R., et al. (2015). Electronic damage in S atoms in a native protein crystal induced by an intense X-ray free-electron laser pulse. *Structure & Dynamics, 2*, 041703.
5. Nass, K., Foucar, L., Barends, T. R. M., Hartmann, E., Botha, S., Shoeman, R. L., et al. (2015). Indications of radiation damage in ferredoxin microcrystals using high-intensity X-FEL beams. *Journal of Synchrotron Radiation, 22*, 225–238.
6. Son, S.-K., Chapman, H. N., & Santra, R. (2011). Multiwavelength anomalous diffraction at high X-ray intensity. *Physical Review Letters, 107*, 218102.
7. Martin, A. V., & Quiney, H. M. (2016). Coherence loss by sample dynamics and heterogeneity in X-ray laser diffraction. *Journal of Physics B: Atomic, Molecular and Optical Physics, 49*, 244001.
8. Abbey, B., Dilanian, R. A., Darmanin, C., Ryan, R. A., Putkunz, C. T., Martin, A. V., et al. (2016). X-ray laser-induced electron dynamics observed by femtosecond diffraction from nanocrystals of buckminsterfullerene. *Science Advances, 2*, e1601186.
9. Ferguson, K. R., Bucher, M., Gorkhover, T., Boutet, S., Fukuzawa, H., Koglin, J. E., et al. (2016). Transient lattice contraction in the solid-to-plasma transition. *Science Advances, 2*(1), e1500837.

10. Owen, R. L., Rudiño-Piñera, E., & Garman, E. F. (2006). Experimental determination of the radiation dose limit for cryocooled protein crystals. *Proceedings of the National Academy of Sciences of the United States of America, 103*, 4912–4917.
11. Chapman, H. N., Caleman, C., & Timneanu, N. (2014). Diffraction before destruction. *Philosophical Transactions of the Royal Society of London. Series B, Biological Sciences, 369*, 20130313.
12. Auger, P., Ehrenfest, R., Maze, R., Daudin, J. & Fréon, R. A. (1939). Extensive Cosmic-Ray showers. *Reviews of Modern Physics, 11*, 288–291.
13. X-ray data booklet. revision. Technical report, 1986.
14. Hubbell, J. H., Veigele, W. J., Briggs, E. A., Brown, R. T., Cromer, D. T., & Howerton, R. J. (1975). Atomic form factors, incoherent scattering functions, and photon scattering cross sections. *Journal of Physical and Chemical Reference Data, 4*, 471–538.
15. Krause, M. O., & Oliver, J. H. (1979). Natural widths of atomic K and L levels, Kα X-ray lines and several KLL auger lines. *Journal of Physical and Chemical Reference Data, 8*, 329–338.
16. Siegbahn, K. (1970). *ESCA applied to free molecules*. Amsterdam: North-Holland.
17. Persson, P., Lunell, S., Szöke, A., Ziaja, B., & Hajdu, J. (2001). Shake-up and shake-off excitations with associated electron losses in X-ray studies of proteins. *Protein Science, 10*, 2480–2484.
18. Caleman, C. (2007). *Towards Single Molecule Imaging - Understanding Structural Transitions Using Ultrafast X-Ray Sources and Computer Simulations*. Digital Comprehensive Summaries of Uppsala Dissertations from the Faculty of Science and Technology.
19. Ziaja, B., London, R. A., & Hajdu, J. (2005). Unified model of secondary electron cascades in diamond. *Journal of Applied Physics, 97*, 064905.
20. Landau, L. D., & Lifshitz, E. M. (1981). *Quantum mechanics: Non-relativistic theory*. Amsterdam: Elsevier.
21. Caleman, C., Ortiz, C., Marklund, E., Bultmark, F., Gabrysch, M., Parak, F. G., et al. (2009). Radiation damage in biological material: Electronic properties and electron impact ionization in urea. *Europhysics Letters, 85*, 18005.
22. Caleman, C., Huldt, G., Maia, F. R. N. C., Ortiz, C., Parak, F. G., Hajdu, J., et al. (2011). On the feasibility of nanocrystal imaging using intense and ultrashort X-ray pulses. *ACS Nano, 5*, 139–146.
23. Young, L., Kanter, E. P., Krässig, B., Li, Y., March, A. M., Pratt, S. T., et al. (2010). Femtosecond electronic response of atoms to ultra-intense X-rays. *Nature, 466*, 56–61.
24. Caleman, C., Bergh, M., Scott, H. A., Spence, J. C. H., Chapman, H. N., & Tîmneanu, N. (2011). Simulations of radiation damage in biomolecular nanocrystals induced by femtosecond X-ray pulses. *Journal of Modern Optics, 58*, 1486–1497.
25. Caleman, C., Tîmneanu, N., Martin, A. V., Jönsson, H. O., Aquila, A., Barty, A., et al. (2015). Ultrafast self-gating Bragg diffraction of exploding nanocrystals in an X-ray laser. *Optics Express, 23*, 1213–1231.
26. Scott, H. A. (2001). Cretin—A radiative transfer capability for laboratory plasmas. *Journal of Quantitative Spectroscopy and Radiation Transfer, 71*, 689–701.
27. Book, D. L., & Naval Research Laboratory (U.S.). (1987). *NRL plasma formulary*.
28. Gericke, D. O., Murillo, M. S., & Schlanges, M. (2002). Dense plasma temperature equilibration in the binary collision approximation. *Physical Review E, 65*, 036418.
29. Neutze, R., Wouts, R., van der Spoel, D., Weckert, E., & Hajdu, J. (2000). Potential for biomolecular imaging with femtosecond X-ray pulses. *Nature, 406*, 752–757.
30. Hau-Riege, S. P., London, R. A., & Szoke, A. (2004). Dynamics of biological molecules irradiated by short X-ray pulses. *Physical Review. E, Statistical, Nonlinear, and Soft Matter Physics, 69*, 051906.
31. Bergh, M., Tîmneanu, N., & van der Spoel, D. (2004). Model for the dynamics of a water cluster in an X-ray free electron laser beam. *Physical Review. E, Statistical, Nonlinear, and Soft Matter Physics, 70*, 051904.

32. Jurek, Z., Oszlányi, G., & Faigel, G. (2004). Imaging atom clusters by hard X-ray free-electron lasers. *Europhysics Letters, 65*, 491–497.
33. Krejcik, P., Decker, F.-J., Emma, P., Hacker, K., Hendrickson, L., O'Connell, C. L., et al. (2003). Commissioning of the SPPS linac bunch compressor. In: *Proceedings of the 2003 particle accelerator conference*.
34. Lindenberg, A. M., Larsson, J., Sokolowski-Tinten, K., Gaffney, K. J., Blome, C., Synnergren, O., et al. (2005). Atomic-scale visualization of inertial dynamics. *Science, 308*(5720), 392–395.
35. Graziani, F. R., Batista, V. S., Benedict, L. X., Castor, J. I., Chen, H., Chen, S. N., et al. (2012). Large-scale molecular dynamics simulations of dense plasmas: The Cimarron Project. *High Energy Density Physics, 8*, 105–131.
36. Hau-Riege, S. P., & Bennion, B. J. (2015). Reproducible radiation-damage processes in proteins irradiated by intense X-ray pulses. *Physical Review E, 91*, 022705.
37. Jurek, Z., Son, S.-K., Ziaja, B., & Santra, R. (2016). XMDYN and XATOM: Versatile simulation tools for quantitative modeling of X-ray free-electron laser induced dynamics of matter. *Journal of Applied Crystallography, 49*, 1048–1056.
38. Son, S.-K., Young, L., & Santra, R. (2011). Impact of hollow-atom formation on coherent X-ray scattering at high intensity. *Physical Review A, 83*, 033402.
39. Jönsson, H. O., Tîmneanu, N., Östlin, C., Scott, H. A., & Caleman, C. (2015). Simulations of radiation damage as a function of the temporal pulse profile in femtosecond X-ray protein crystallography. *Journal of Synchrotron Radiation, 22*, 256–266.
40. Waasmaier, D., & Kirfel, A. (1995). New analytical scattering-factor functions for free atoms and ions. *Acta Crystallographica. Section A, 51*, 416–431.
41. Slater, J. C. (1930). Atomic shielding constants. *Physical Review, 36*, 57–64.
42. Quiney, H. M., & Nugent, K. A. (2011). Biomolecular imaging and electronic damage using X-ray free-electron lasers. *Nature Physics, 7*, 142–146.
43. Lomb, L., Barends, T. R. M., Kassemeyer, S., Aquila, A., Epp, S. W., Erk, B., et al. (2011). Radiation damage in protein serial femtosecond crystallography using an X-ray free-electron laser. *Physical Review B: Condensed Matter and Materials Physics, 84*, 214111.
44. Barends, T. R. M., Foucar, L., Botha, S., Doak, R. B., Shoeman, R. L., Nass, K., et al. (2014). De novo protein crystal structure determination from X-ray free-electron laser data. *Nature, 505*, 244–247.
45. Galli, L., Barends, T. R. M., Son, S.-K., White, T. A., Barty, A., Botha, S., et al. (2015). Crystal structure of gadolinium derivative of HEWL solved using Free-Electron laser radiation. *IUCrJ, 2*, 627–634.
46. Nakane, T., Song, C., Suzuki, M., Nango, E., Kobayashi, J., Masuda, T., et al. Native sulfur/chlorine SAD phasing for serial femtosecond crystallography. *Acta Crystallographica. Section D, Biological Crystallography, 71*, 2519–2525 (2015)
47. Nass, K., Meinhart, A., Barends, T. R. M., Foucar, L., Gorel, A., Aquila, A., et al. (2016). Protein structure determination by single-wavelength anomalous diffraction phasing of X-ray free-electron laser data. *IUCrJ, 3*, 180–191.
48. Hunter, M. S., Yoon, C. H., DeMirci, H., Sierra, R. G., Dao, E. H., Ahmadi, R., et al. (2016). Selenium single-wavelength anomalous diffraction de novo phasing using an X-ray-free electron laser. *Nature Communications, 7*, 13388.
49. Bortel, G., Faigel, G., & Tegze, M. (2009). Classification and averaging of random orientation single macromolecular diffraction patterns at atomic resolution. *Journal of Structural Biology, 166*, 226–233.
50. Aquila, A., Barty, A., Bostedt, C., Boutet, S., Carini, G., dePonte, D., et al. (2015). The linac coherent light source single particle imaging road map. *Structural Dynamics, 2*(4), 041701.
51. Ekeberg, T., Svenda, M., Abergel, C., Maia, F. R. N. C., Seltzer, V., Claverie, J.-M., et al. (2015, March). Three-dimensional reconstruction of the giant mimivirus particle with an X-ray free-electron laser. *Physical Review Letters, 114*, 098102.

52. Martin, A. V., Corso, J. K., Caleman, C., Timneanu, N., & Quiney, H. M. (2015). Single-molecule imaging with longer X-ray laser pulses. *IUCrJ, 2*, 661–674.
53. Hau-Riege, S., London, R., Huldt, G., & Chapman, H. (2005). Pulse requirements for X-ray diffraction imaging of single biological molecules. *Physical Review E, 71*, 061919.
54. Hau-Riege, S. P., London, R. A., Chapman, H. N., Szoke, A., & Timneanu, N. (2007). Encapsulation and diffraction-pattern-correction methods to reduce the effect of damage in X-ray diffraction imaging of single biological molecules. *Physical Review Letters, 98*, 198302.
55. Hau-Riege, S. P., Boutet, S., Barty, A., Bajt, S., Bogan, M. J., Frank, M., et al. (2010). Sacrificial tamper slows down sample explosion in flash diffraction experiments. *Physical Review Letters, 104*, 064801.
56. Lorenz, U., Kabachnik, N. M., Weckert, E.., & Vartanyants, I. A. (2012). Impact of ultrafast electronic damage in single-particle X-ray imaging experiments. *Physical Review B, 86*, 051911.

Chapter 7
Climbing the Data Mountain: Processing of SFX Data

Chun Hong Yoon and Thomas A. White

7.1 The Data Mountain

7.1.1 Why Does Serial Femtosecond Crystallography Produce So Much Data?

Serial crystallography represents a paradigm shift in macromolecular crystallography from the rotation method for data collection. It brings many benefits as described elsewhere in this book; however, it also brings a steep increase in the data volume. The main reason for this is simple statistics. Systematic rotation of a single crystal allows all the Bragg peaks, required for structure determination, to be swept through and recorded. Serial collection is a rather inefficient way of measuring all these Bragg peak intensities because each snapshot is from a randomly oriented crystal, and there are no systematic relationships between successive crystal orientations. In this chapter, we will elaborate on the quantities of data required, and how one can climb this data mountain to yield valuable and meaningful results.

Consider a game of picking a card from a deck of all 52 cards until all the cards in the deck have been seen. The rotation method could be considered as analogous to picking a card from the top of the deck, looking at it and then throwing it away before picking the next, i.e., sampling without replacement. In this analogy, the faces of the cards represent crystal orientations or Bragg reflections. Only 52

C. H. Yoon
Linac Coherent Light Source, SLAC National Accelerator Laboratory, Menlo Park, CA, USA
e-mail: yoon82@stanford.edu

T. A. White (✉)
Center for Free-Electron Laser Science, German Electron Synchrotron DESY, Hamburg, Germany
e-mail: taw@physics.org

S. Boutet et al. (eds.), *X-ray Free Electron Lasers*,
https://doi.org/10.1007/978-3-030-00551-1_7

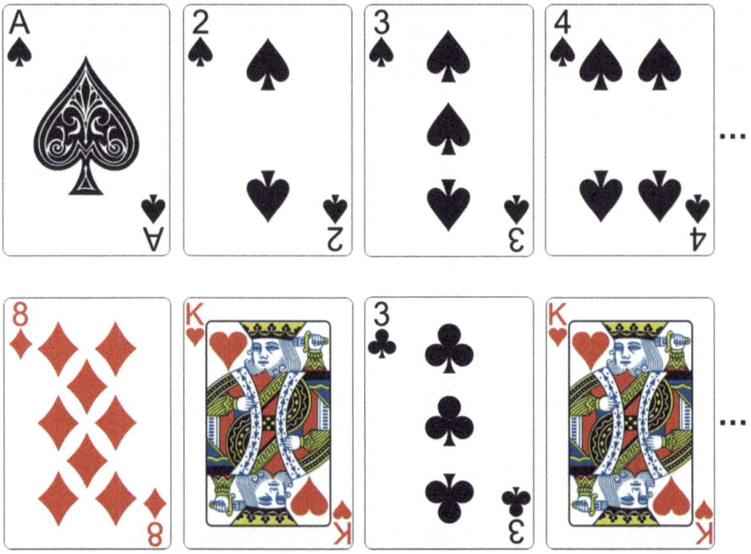

Fig. 7.1 Sampling without replacement (top) vs sampling with replacement (bottom) for a deck of cards. The expected number of turns to observe all cards by sampling one card at a time is 52 and 236, respectively.

turns are required to see all the cards in this case (Fig. 7.1 top), or analogously to acquire a complete dataset where all the symmetrically unique reflections (up to some resolution limit) have been measured. Serial collection is akin to randomly picking a card and then putting the card back in the deck before choosing the next card, i.e., sampling with replacement (Fig. 7.1 bottom). How many cards are needed to be drawn before all 52 have been seen? Intuitively, we can see that there is no guarantee that all cards will ever be observed. However, statistically speaking, the expected number of turns to complete the task, c, is given by:

$$c = n \sum_{k=1}^{n} \frac{1}{k},$$

where n is the total number of cards. For large n, c converges to $n \log(n)$. That is, for $n = 52$, it can reasonably be expected that all 52 cards will be observed only after about 236 turns! The problem is further exacerbated because a fraction of the images obtained in an SFX experiment will be blank because the X-ray pulse did not hit a crystal. This fraction varies depending on the sample preparation and delivery methods (see Chaps. 3–5), but is often higher than 60%. The random orientation of crystals and the random picking of this orientation on every measurement represent the primary reasons why SFX data volumes are inherently larger than rotation series data.

The second reason why SFX data volumes are so high is the high variability of many experimental parameters. The self-amplified spontaneous emission (SASE) process, by which the X-ray pulses are generated in the FEL, essentially amounts to amplifying a random fluctuation in the electron bunch by many orders of magnitude. This randomness becomes imprinted on the X-ray pulses, such that each one of them has a different intensity profile and photon energy spectrum. There may also be a wide variability in the crystals: their size, shape, crystalline order, and even their crystal structure. In effect, each frame in an SFX experiment is from a completely separate experiment to the others. As described later in this chapter, great progress has been made in compensating for this variation, but the main method for mitigating it is still to average intensity measurements from a large number of crystals.

Over the years, FEL facilities have been built or upgraded with higher repetition rates and larger detectors which help reduce the data collection time, but do nothing to reduce the data quantity required.

7.1.2 Facilities, Data Rates, and Detectors

In 2005, a series of proof-of-principle studies were performed at the Free electron LASer in Hamburg (FLASH) facility in Hamburg, Germany, demonstrating "diffraction-before-destruction" [2, 6, 15], the concept of side-stepping classical radiation damage limits by using X-ray pulses shorter than the damage processes (See Chap. 6). These results motivated high resolution experiments using shorter wavelength X-rays. In 2009, the Linac Coherent Light Source (LCLS) in Stanford, USA, started producing X-rays in the sub-nanometer range, and the first SFX experiments on protein crystals were performed [16], starting at a rate of 30 X-ray pulses per second. Shortly after, the LCLS repetition rate was increased to 120 pulses per second with the data acquisition infrastructure capable of reading out 5 GB/s per instrument. Table 7.1 shows the repetition rates of the currently operating (and soon to be operating) X-ray FEL facilities. In 2011, SPring-8 Angstrom Compact free electron LAser (SACLA) in Japan soon followed as a compact X-ray FEL facility that can be operated below 1 Å wavelength. In 2017, the Pohang Accelerator Laboratory XFEL (PAL-XFEL) in Korea and the European XFEL in Germany started user operation. Superconducting accelerator technology has led to much higher repetition rates that exceed the detector read-out rates. The European XFEL produces bursts of 2,700 pulses with only a few hundred nanoseconds between them, ten times a second, for a total of 27,000 pulses per second. LCLS-II, which is still a few years away, will produce a constant rate of X-ray pulses at a rate approaching one million pulses per second. Chapter 16 will present an outlook of X-ray FELs and discuss these new machines further.

Data rates will also increase with improvements in detector technology. Detectors will become larger and be able to read out data at higher rates. The short

Table 7.1 Photon pulses per second at facilities around the world

Facility	Pulses per second
FLASH (EUV and soft X-rays)	8000
LCLS	120
SACLA	60
FERMI (UV and soft X-rays)	50
PAL-XFEL	60
SwissFEL (from 2018)	100
European XFEL	27,000
LCLS-II (from 2020)	1,000,000

femtosecond pulses from X-ray FELs require the use of integrating detectors. Single-photon-counting detectors, such as the PILATUS [39] and the EIGER [14] detectors used at synchrotrons, offer many excellent features such as essentially zero background noise, but cannot be used with femtosecond pulses. The reason for this is that the ability to count photons requires that the electronic pulse from each photon be distinguishable from the others. When all the photons arrive within a few femtoseconds, a counting detector can only count zero or one photon. Instead, integrating detectors are used, in which the total charge created by the photons in the detector sensor is summed and read out after the pulse. Detectors used for crystallography are typically of multi-megapixel scale, which makes for formidable data rates at the repetition rates achievable for X-ray FELs. Detectors for SFX experiments need to have low noise and high dynamic range capable of capturing the Bragg peak intensities. The first detector specifically developed for LCLS was a 2.3-megapixel Cornell-SLAC Pixel Array Detector (CSPAD) which has a dynamic range of >2500 photons at 8 keV in low gain mode (Fig. 7.2) [10, 13, 50]. Even with the high dynamic range, the Bragg spots produced from the ultra-intense FEL pulses can saturate the detector pixels and Fourier amplitude of the protein structure cannot be determined in this case. The Rayonix (MX225-HS) detector at PAL-XFEL uses scintillators coupled to fiber optics that indirectly transmit signal to a light-sensitive camera to boost the dynamic range up to 45,000 photons at 12 keV. The European XFEL is equipped with a smarter detector that can adjust the gain depending on the charge deposited on the pixel, called Adaptive Gain Integrating Pixel Detector (AGIPD), giving it both single-photon sensitivity and a dynamic range of more than 10,000 photons at 12.4 keV [1]. Similarly, SwissFEL will have the adJUstiNg Gain detector FoR the Aramis User station (JUNGFRAU) which also automatically adjusts the gain depending on signal [45].

Instead of using a beam stop placed in front of the detector, X-ray FEL detectors typically allow the beam to go through a central gap. This allows the beam to be stopped far downstream of the detector, where due to its typical divergence it will be much larger and cause less damage. It further allows the "spent" beam to be reused for a second experiment in a downstream position, as shown in Fig. 7.3. The CXI beamline at LCLS operates in this configuration to double the number

Fig. 7.2 The Cornell-SLAC Pixel Array Detector (CSPAD) is an example of a PAD designed to operate at the conditions of a hard X-ray FEL. Photon-counting PADs are not usable for femtosecond pulses unless the expected signal is ≤ 1 photon per shot. The CSPAD has a dynamic range of >2500 photons at 8 keV in low gain mode with a total of 2.3 megapixels

of experiments [11, 32]. A similar serial setup is planned for the SPB/SFX beamline at the European XFEL [43]. There are also other ways that make more efficient use of the X-ray beam, such as splitting the beam using a monochromator into "monochromatic" and "rejected" components, both of which can be used for SFX or other types of experiments [67]. Since an SFX experiment is not very sensitive to pauses in the data acquisition, the pauses leading only to wasted sample as it flows past, the beam can be switched between an SFX experiment on a timescale of seconds. A common application of these multiplexing techniques is to "screen" sample batches in parallel to data collection on a primary sample. Of course, all this acts to further increase the amount of SFX data.

In an SFX experiment, the meeting of a crystal with an X-ray pulse is a chance event. As a result, there will be blank shots in an SFX dataset. The fraction of detector readouts that correspond to a crystal interacting with the pulse and producing a diffraction pattern is known as the hit rate and can be controlled by altering parameters such as the concentration of crystals or the thickness of the liquid jet, for example. To ease the data processing, we normally aim to maximize the number of "single hits," where only one crystal is hit by the X-ray pulse. The theoretical maximum for this rate is 37%, given by Poisson statistics as shown in Fig. 7.4, at which 23% of the hits will contain more than one overlapping diffraction pattern to produce an overall hit rate of 60% [31]. In most SFX experiments performed to date, the actual hit rate has been much lower, with a large fraction

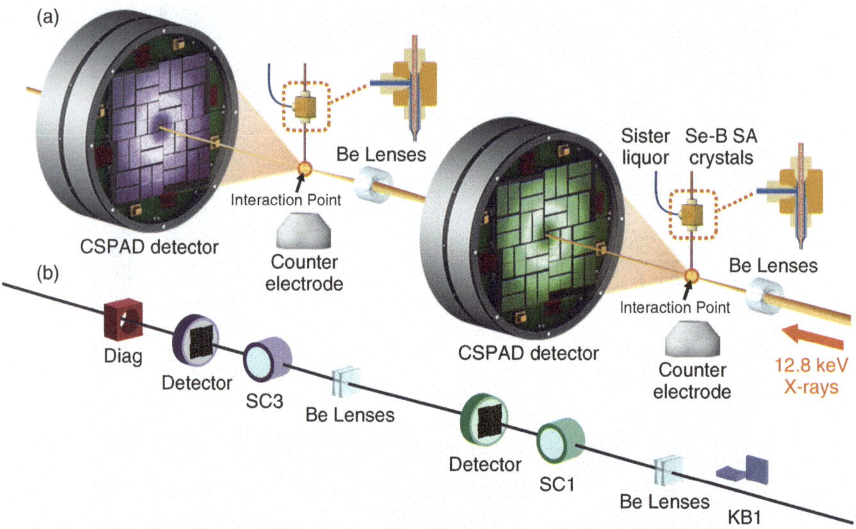

Fig. 7.3 Overview of the serial SFX setup at the CXI beamline, LCLS. (**a**) Data are collected simultaneously in two sample chambers using a serial SFX setup. (**b**) The X-rays are focused only using Be lenses with the Kirkpatrick–Baez mirrors (KB1) moved out. The X-rays enter the first sample chamber (SC1) and scatter from the protein crystal. The unscattered X-rays exit through the central hole in the CSPAD detector, and are then refocused for another scattering experiment in the downstream sample chamber (SC3) followed by the diagnostic (Diag). Reproduced from Yoon [65]

of blank shots. The first stage of data processing, described in the next section, is to identify the hits so that the blank frames can be ignored in the later stages of processing. In the high-throughput regime of future X-ray FELs, it may become necessary to perform hit finding prior to writing it to persistent storage, i.e., immediately and permanently deleting the blank frames.

7.2 Reducing the Mountain to a Hill: Hit Finding

7.2.1 Realities of Experimental Data

The aim of hit finding in SFX is to determine whether the snapshot contains Bragg spots or not. All the later processing stages are based on Bragg spots, and so frames which do not contain any of them are useless, at least as far as crystallographic data processing is concerned. Conceptually, hit finding seems trivial. However, in practice it can be challenging.

In an ideal case shown in Fig. 7.5a, the peaks are intense and there is no background noise. In this case, even a simple thresholding algorithm can locate

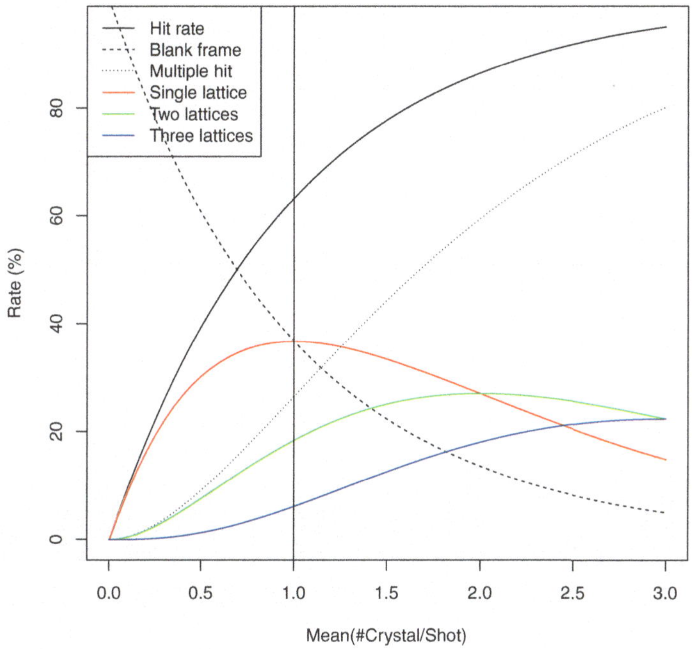

Fig. 7.4 Optimum hit rate for most single hits assuming Poisson statistics [31]. Image ©Takanori Nakane, reproduced by kind permission

the peaks. Unfortunately, real life is not so simple, and there are many additional features on the detector. The medium in which the crystals are embedded, necessary for them to stay hydrated, leads to coherent scattering. X-ray fluorescence from the sample can be significant for some elements and some photon energies, as can parasitic scattering from X-ray apertures and focusing optics. These effects all combine to give an overall background as shown in cyan in Fig. 7.5b. The detector will inevitably contain a small number of defective pixels, producing either very high or very low readouts.

Fluctuations due to beam fluence, photon energy and crystal size affect both the signal and background which makes it hard to set the right peak-finding parameters that will work for all the images. Given that each event is unique, peak-finding parameters will also be unique for each event. Indeed, a grid search over the peak-finding parameter space for each individual image can yield better results than a one-size-fits-all approach [41]. Shadowing due to particular experimental setup, such as upstream apertures or diagnostics including viewing camera optics close to the beam, can add abrupt nonuniform changes to the background, as shown in green in Fig. 7.5b. Multiple crystals in the beam can introduce extra Bragg spots that influence the local noise estimation (Fig. 7.5b). Some of these artifacts can be reduced by a series of calibration steps, as described below.

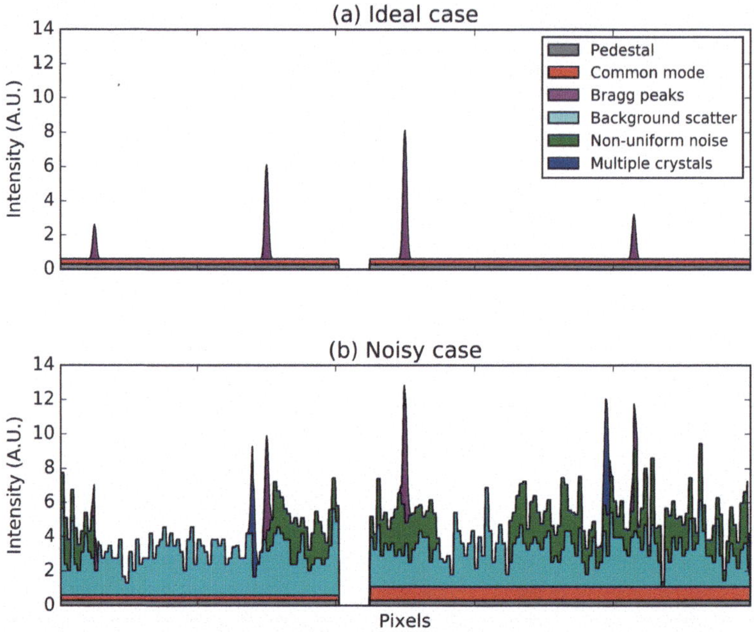

Fig. 7.5 Illustration of challenges associated with peak finding. (**a**) Ideal case, the Bragg spots (magenta) have no background noise apart from the detector noise due to pedestal (gray) and common mode (red). (**b**) Background scattering (cyan) from optics/crystal buffer/fluorescence can be significant. Bad pixels can often be brighter than the Bragg spots. Shadowing due to the chamber setup such as upstream aperture/camera wires add abrupt changes, nonuniform noise to the background (green). Multiple crystals in the beam can introduce extra Bragg spots that influence the local noise estimation (blue). Figure ©The Author, licensed under CC-BY-4.0

7.2.2 Correcting Detector Artifacts and Removing Background

Raw pixel values are read out by the detector for each event and calibrated in the following order: pedestal correction, common-mode correction, and gain correction. Pedestal refers to a large additive noise due to the electronic readout that is fairly constant over time (which slowly drifts during the course of an experiment); pedestals are independent of signal being measured. Before and during an experiment, the so-called dark images are collected (detector on, X-ray off) to measure the average pedestal. Dark runs also provide information about bad pixels.

Common mode is another form of additive noise that arises when a detector tile is bombarded by many photons producing a "common" offset for all pixels in the tile (Fig. 7.5a shown as red). Since this is signal-dependent noise, each tile has to be corrected per event basis by evaluating the offset experienced by the pixels with no photons. If all the pixels are illuminated, the common mode cannot be evaluated. A way to avoid this problem was devised in the CSPAD detector where unbonded pixels are strategically placed to read out the offset. Unbonded pixels are pixels where the X-ray sensor material is not connected to the electronics of the pixel and therefore cannot detect photons but still experience the common offset. It is important to carefully mask out the unbonded pixels during analysis.

Gain correction normalizes the response of the individual pixel. This is multiplicative noise that is proportional to the photons seen by the pixels.

For an ith event, the calibration step can be expressed as:

$$C_{\mathrm{i}} = \frac{R_{\mathrm{i}} - P - CM_{\mathrm{i}}}{G}, \tag{7.1}$$

where C_i is the ith calibrated image, R_i is the ith raw image, P is the pedestal, CM_i is the common mode due to R_i, and G is the gain factor.

The calibrated images are (to first approximation) devoid of electronic noise. However, the diffraction patterns still contain other forms of noise, such as the solvent scattering from crystal buffer, secondary scattering from experimental components around the interaction region, fluorescence and jet streaks from the injector that all contribute to the background noise making peak finding difficult. The term background noise used here is defined as any photons on the detector that are non-Bragg reflections. In severe cases, background subtraction algorithms are used to suppress slowly varying noise. There are largely two approaches: radial background subtraction and median background subtraction. Given that the dominant background comes from solvent scattering, the isotropic component of the background noise can be removed by subtracting the radial average of all pixels that are equidistant from the detector center. This algorithm requires a priori knowledge of the beam center and detector pixel positions which may not be available. Median background subtraction takes a different approach where a 2D sliding window of pixels is used to calculate the median intensity inside this window and subtracted from the pixel value at the center of the window. This algorithm suppresses both the isotropic and non-isotropic components of the background noise, and also does not require any geometric information.

7.2.3 Finding Peaks and Identifying Hits

There are several software packages that can perform peak finding at FEL facilities, most notably (in alphabetical order) CASS [21], cctbx.xfel [29], Cheetah [7, 46], and psocake [57]. All these packages implement their own unique algorithms that share common image processing ideas:

1. Locate all local maxima in a detector tile as potential Bragg spots. Brightest connected pixels of the local maxima are counted as signal and surrounding pixels as the estimation of local background noise.

2. Calculate peak properties, such as the signal-to-noise ratio (SNR), area of the spot, and sum of pixel values. The SNR is often evaluated to determine whether the peak found is significant enough to be a Bragg spot or just spurious noise:

$$\mathrm{SNR} = I/\sigma(I) = \mu_s/\sigma_n,$$

where μ_s is the mean of the signal S and σ_n is the standard deviation of noise estimate N. This SNR should not be confused with $\mathrm{I}/\sigma\,(\mathrm{I})$ used in crystallography tables for data collection and refinement statistics. Poisson statistics of photon arrival dictates the noise associated with the measurement of a Bragg spot intensity (I):

$$I = var(I) = \sigma^2(\mathrm{I}).$$

3. Return centroid of the peaks as peak positions if the peak properties meet the user-defined criteria.

A diffraction pattern is considered to be a hit if it has more than some significant number of peaks, typically around 15. A much lower number of reflections, as low as 2, may in principle be sufficient to determine the orientation of the crystal. However, 15 is a heuristically determined practical minimum number. Accurate peak finding is important; erroneous peak detection with many false negatives (i.e., unable to find Bragg peaks that are present) obviously leads to less hits being found ultimately leading to a poorer protein structure. On the other hand, peak detection with many false positives (i.e., finding peaks that are not Bragg peaks) artificially increases the number of hits found, giving an illusion of having collected more crystal data.

In Fig. 7.6, the peaks found are indicated by cyan squares. Note that some Bragg spots may have intensities close to zero, because the Bragg spot intensities are modulated by the molecular transform of the protein, which have zero intensities, and also systematic absences due to a screw axis of the unit cell can make certain Bragg spots disappear. Predicted lattice points are also shown in magenta rings which will be discussed later (Sect. 7.3.1). Identification of hits is an active area of research required for rapid structure solution during beamtime and may one day be performed more accurately and efficiently with machine learning techniques.

The peak-finding software packages mentioned above interface with various FEL facilities to access the diffraction patterns and save the crystal hits in a predefined data format that is easily accessible by the user regardless of where the data was collected.

7.2.4 Facility Frameworks and Data Formats

The large data volumes and high rates of acquisition calls for dedicated systems for recording, storing, and accessing the data. Instead of simply providing the data

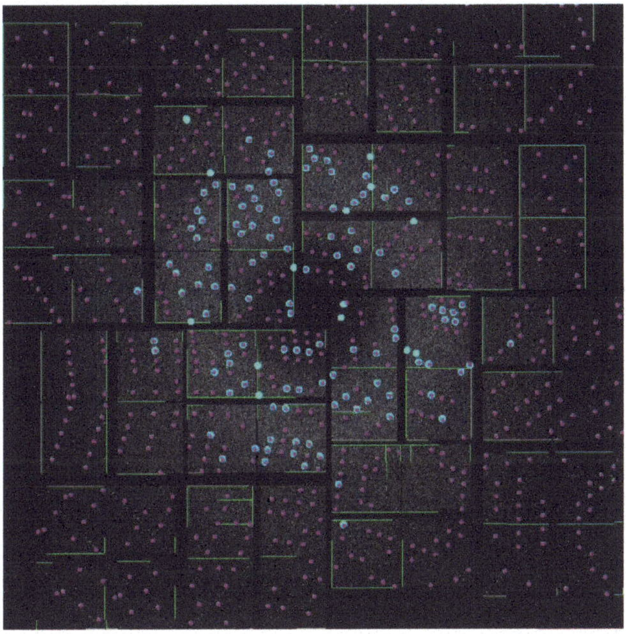

Fig. 7.6 Representative peak-finding (cyan) and indexing results (magenta) from a selenobiotinyl streptavidin crystal [65] at the CXI instrument, LCLS. Figure ©The Author, licensed under CC-BY-4.0

on disk in a familiar file format, most X-ray FEL facilities provide a software framework for accessing the data. The access may be "online"—during the data acquisition process itself on shared memory with minimal delay—or "offline," sometime afterwards when the data arrives on disk. LCLS provides a graphical online monitoring tool called the Analysis Monitoring Interface (AMI) [57], and a software framework "psana" (Photon Science ANAlysis) [18] for online and offline analysis. Psana offers programming interfaces in both C++ and Python and allows access to all data generated by the data acquisition system. Software packages such as Cheetah [7], OnDA [44], IOTA [41], and cctbx.xfel [29] use these interfaces to access the data and perform tasks such as calculating "live" hit rates and Bragg spot saturation levels, online, or writing the hits out to separate files, offline. The European XFEL is the other facility that has developed a dedicated software framework, Karabo [30], for accessing and analyzing data. SACLA offers its users a crystallography data processing pipeline through the SACLA data acquisition application programming interface (API) [46].

For storing hits, the HDF5 (Hierarchical Data Format, version 5) file format is commonly used across all of FEL science. This is a "container format" allowing a large amount of flexibility in the layout and representation of the data structures. For scattering data, a predefined schema by CXIDB [42] or NEXUS [38] is often used.

7.3 Indexing, Integration, and Merging

Having brought the data volume under control by isolating the patterns containing apparently useful diffraction signal, the next steps of data processing are concerned with turning those diffraction patterns into structure factor estimates which can be used by established crystallography software. Several software packages are now available to perform the processing, notably CrystFEL [60, 61], cctbx.xfel [29], nXDS [35], and cppxfel [26]. These packages have differences in their implementation, for instance, in how they interface with the facility's data processing framework, but are alike in more ways than they are different. They all sequentially perform the essential processing steps of indexing, integration, merging, and finally evaluating the data quality. Each of these steps is described below.

7.3.1 Indexing

Indexing a diffraction pattern means assigning Miller indices to the Bragg peaks in the diffraction pattern. Implicitly, this involves determining the lattice parameters of the crystal and its orientation relative to a reference ("laboratory") frame. It also acts as a powerful filter of data, because it is very unlikely that a frame containing only spurious peaks will pass through this process successfully.

Several algorithms and pieces of software exist for indexing rotation series data in macromolecular crystallography [20, 34, 51]. It has been found that the same algorithms and software are able to index snapshot diffraction patterns, without the advantage of recording a three-dimensional wedge of reciprocal space [37].

In a serial crystallography experiment, we can usually assume that each crystal has very similar lattice parameters. However, the indexing algorithms usually determine the parameters ab initio for each diffraction pattern. Each indexing result must therefore be checked for consistency with a reference set of lattice parameters provided by the user. If the parameters are unknown, they can be determined by plotting histograms of each of the parameters (a, b, c, α, β, and γ) and finding the most common values for a representative part of the dataset. SFX data processing programs provide graphical tools to assist this process. The indexing process can then be repeated using the parameters so determined. Some indexing algorithms can make use of prior information about the lattice parameters to increase their success rate [24], or even require this information to work [28].

Once the lattice parameters and orientation of the crystal have been determined, they can be used to calculate the positions where Bragg peaks are expected to appear on the detector. Successful indexing relies on having an accurate description of the detector geometry, which means that the positions of the detector panels (see Sect. 7.1.2) must be known accurately and precisely. Provided that the initial geometry is accurate enough to index at least a few patterns, it can be refined by comparing the observed and calculated positions of peaks on the detector. Since the indexing solution uses information about spot positions from the entire detector, the calculated peak positions can be taken as a reference, and a mispositioned panel will show a systematic offset between the observed and calculated peak locations [29]. After correcting the panel locations, the indexing process can be repeated until the detector geometry is known with high precision [64]. This process is made easier by the use of a strongly diffracting and readily available calibration sample such as lysozyme or thermolysin.

If the distance between the sample and detector (the "camera length") is set incorrectly in the description of the detector geometry, this will manifest as a systematic offset in the spot positions [64]. Small offsets of the camera length can also be seen in the histograms of lattice parameters. The peaks will be sharpest for the correct values, and become wider or even bimodal as the camera length deviates further from the true value [48].

If the concentration of crystals in the delivery medium is high, there may be a significant number of frames that contain multiple diffraction patterns (see Sect. 7.1.2). Most indexing algorithms assume that all the peaks in the diffraction pattern come from one lattice, and so can fail when presented with two or more overlapping lattices. As a result of this, algorithms for indexing multiple lattices have recently been developed. Some of these are based on the "delete and retry" method. Here, the pattern is indexed assuming that it is a single lattice, and the peaks which could be accounted for by the resulting lattice removed from the peak list prior to making another indexing attempt using the remaining peaks. This algorithm has been available in cctbx.xfel since the earliest released versions [29] and has been extended to larger numbers of lattices [24, 28].

This "delete and retry" method has an advantage of simplicity, not requiring much extra code to be added to software; however, it relies on the first indexing attempt succeeding in finding one of the lattices despite the extra peaks from other lattices. Another algorithm, called "FELIX," has recently been developed which operates on completely different principles [9]. Instead of treating the image as if it contained only one lattice at the outset and finding subsequent lattices in sequence afterwards, the FELIX algorithm assumes that there are multiple lattices present and searches for them simultaneously. It can index large numbers of overlapping lattices from a single snapshot: ten or more depending on the resolution of the patterns and the unit cell parameters.

The process for determination of the lattice parameters, described above, is quite a coarse one. Populations of crystals with different parameters will be identified only if at least one of the parameters is different from the rest of the population by more than the width of the distribution of that parameter. A *Data Exploration Toolkit* has been written [66], which, among other operations, can perform hierarchical clustering according to the Andrews–Bernstein distance metric to compare lattice parameters. Clustering the crystals in this way can reveal populations of crystals with subtly different lattice parameter, and appeared to reduce the number of outlying intensity values.

Some crystal symmetry classes produce an extra complication with serial data acquisition. These are the classes which allow the crystal structure to be rotated, usually by 90 or 180 degrees, in such a way as to overlap the reciprocal lattice points with their original locations while the structure itself does not look the same. Tables of which symmetry classes are affected can be found elsewhere [60]. As well as exact overlaps, there can also be approximate overlaps for certain unit cell parameters [3]. In these cases, the indexing solution is ambiguous between two (or, rarely, more) possibilities. While indexing is usually a purely geometrical procedure, which uses only the positions of the reflections, indexing ambiguities can only be resolved by additionally comparing the intensities of the reflections. Because of the large amounts of noise in the individual measurements, this was a significant problem in the first SFX experiments. However, the Brehm–Diederichs algorithm was later developed, which applies a clustering scheme to the intensities and effectively resolves the ambiguities [12]. This algorithm, or a simplified variant of it [27, 61], has now been implemented in all SFX data processing software packages.

7.3.2 Integration and Merging

The main aim of data processing in this chapter is to measure the structure factor moduli, which manifest themselves in the intensities of the Bragg peaks. There are many confounding factors affecting the relationship between the two, some of which were mentioned in Sect. 7.1.1. Nevertheless, the process begins by measuring the intensity above the background level at a location determined using the reciprocal lattice basis vectors (lattice parameters and orientation) determined by the indexing algorithm. There are several methods for doing this, the simplest of which, and the most commonly used one in SFX, being to estimate the background level from an annular region around the peak location and the intensity from a circular region inside it, as discussed in Sect. 7.2.3 and shown in Fig. 7.7. The average background level is subtracted from each pixel value in the peak region, and their sum calculated. Several other techniques have been used. These include two-dimensional profile fitting, where the shape of each peak is fitted using either an average of the shapes of the strong peaks [52] or a shape calculated from the properties of the crystals [40], or even using the crystal parameters to calculate exactly which pixels should contain signal [29].

Since the structure factor moduli can have very small values as well as large ones, this integration procedure must be performed even when no obvious peak is present. This introduces some additional considerations. The diffraction model, which includes the indexing solution as well as estimates of crystal parameters such as mosaicity and crystal size, must be as accurate as possible to avoid missing the true reflection locations. It also needs to be accurate to avoid making a large number of false measurements, integrating reflections which are not truly excited in the diffraction pattern (regardless of their structure factors). Most current software therefore performs a refinement stage, where these parameters are refined before integrating the reflections [27, 55, 61].

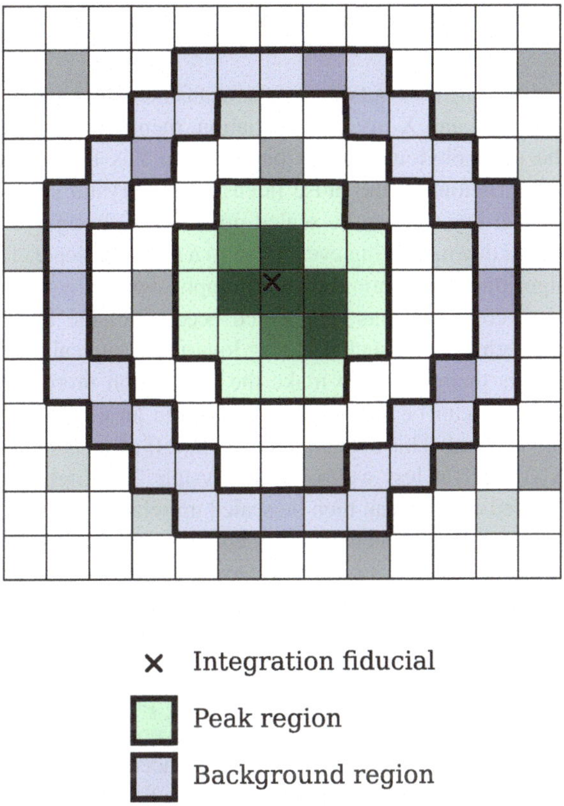

Fig. 7.7 Detailed view of a reflection integration "shoebox." Detector pixels are shown as squares in a grid, and pixel intensity values by the darkness of their shading. Note that the integration fiducial, which is the calculated location of the reflection, is not aligned with the pixel grid. Note further that it does not necessarily coincide with the centroid of the actual peak, because the detector geometry, cell parameters, crystal orientation, and other geometrical parameters may not be perfectly determined. Figure ©The Author, licensed under CC-BY-4.0

Compared with single-particle X-ray imaging, as described in Chap. 14, crystallography has the great advantage that the indices for each intensity measurement can be calculated geometrically, as described in the previous section. The intensities themselves are not important for indexing. The intensity measurements can be averaged together in "buckets" according to their indices (except for the indexing ambiguities previously discussed), and this averaging process is very powerful: apart from certain types of systematic effect, any confounding factor reducing the precision of the individual measurements can be overcome by using more measurements. For the initial experiments, hardly any attempt was made whatsoever to overcome any of these factors, and this approach was referred to as "Monte Carlo Integration" [37] because of its similarity to the numerical Monte Carlo integration procedures for calculating integrals. Since then, many techniques have been developed to compensate for the factors affecting the intensity of each peak, which are described in the next section.

The progress of the Monte Carlo procedure can be tracked by plotting a self-consistency figure of merit such as R_{split} (described in Sect. 7.3.5) against the number of crystals, n. R_{split} is proportional to $1/\sqrt{n}$, with the constant of proportionality depending on the dataset.

7.3.3 Scaling the Intensity Measurements

Getting the most out of the data means modeling and accounting for as many aspects of crystal and X-ray pulse variation, mentioned in Sect. 7.1.1, as possible. Perhaps the most obvious way to begin doing this is to scale the intensities to account for variations in the pulse intensity and crystal size. The weaker intensities from smaller crystals can be scaled up, and the stronger intensities from larger crystals scaled down, to bring everything to a common scale. This can be done using similar algorithms to rotation crystallography with a synchrotron or home source. Some extra considerations are needed because of the large crystal-to-crystal variations between frames, such as using logarithms of scaling factors instead of the scaling factors themselves, to make the calculation more numerically stable [35]. These methods can be extended to determine an effective Debye–Waller parameter for each crystal. This accounts for the falloff of intensity as resolution increases, which is greater for less well-ordered crystals. The high-resolution reflections from less ordered crystals can then be scaled up relative to others [54, 58, 61].

As was mentioned above, accurate values are needed for the parameters affecting which spots are to be integrated, to avoid "overprediction" or "underprediction." Automatically determining or refining these parameters for each crystal is another way to improve the modeling of the diffraction process, and has been found to improve the data quality [54, 58]. If the orientation of the crystal is not accurate, it is common to find that the prediction parameters, such as the spectral bandwidth of the X-ray pulse, are overestimated by the software in an attempt to fit the visible peaks despite the inaccurate orientation. These parameters have therefore been successfully used as a figure of merit in a grid search technique to determine the optimum processing parameters [41]. Combining geometrical refinement and parameter auto-determination techniques with scaling produces a significant combined improvement [61].

7.3.4 Partiality and Post-refinement

The idea of compensating for reflection *partiality* has seen a lot of discussion since the first X-ray FEL SFX experiments [16]. The "partiality" of a reflection in a given diffraction pattern is a quantity which describes the fact that not all regions of the crystal, nor all the X-rays in the beam, contribute simultaneously to the reflection. A partiality of 1 would mean that the entire crystal and all of the X-rays contribute to the reflection. The ideal situation, leading to the most straightforward data processing, would be for all reflections to have the same partiality. However, there are several reasons why they are not all equal.

Alongside the factors mentioned in Sect. 7.3.3, partiality is suspected to be a strong determiner of data quality in SFX. However, several schemes for modeling the diffraction conditions have been proposed, and do not yet appear to agree on

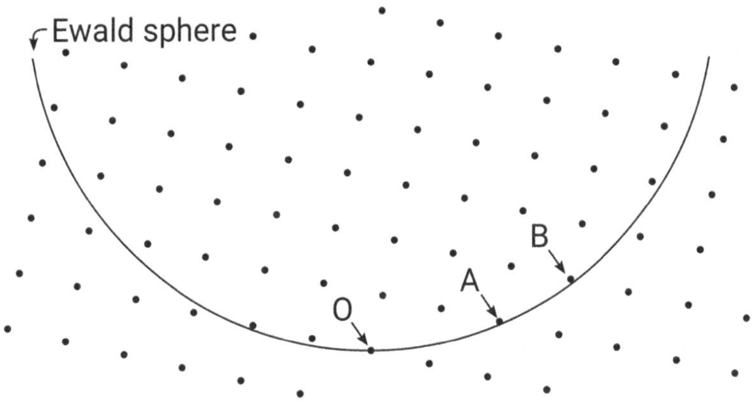

Fig. 7.8 Simple model of reflection partialities using a monochromatic X-ray beam. Point "O" is the origin of reciprocal space, and the small discs represent reciprocal lattice points. Point A is closer to the Ewald sphere than point B, therefore point A has a higher partiality than B. If the two reflections were to have the same structure factor, point A would produce a brighter reflection in the diffraction pattern. Figure © The Author, licensed under CC-BY-4.0. Reproduced from T. A. White, "Processing of XFEL data," in "Protein Crystallography," Wlodawer, Dauter and Jaskolski (eds), Methods in Molecular Biology 1607 (2017)

the most appropriate model. The simplest model is shown in Fig. 7.8, and has been used in the nXDS software [35] and cctbx.prime [58]. In this model, the X-ray pulses are assumed to be monochromatic, so the Ewald sphere is an infinitely thin shell. The partiality of a reflection is simply related to the distance between the reciprocal lattice point and the Ewald sphere. To compensate for the partiality, reflections which are further away from the Ewald sphere must be scaled up by a larger factor than ones closer to the exact Bragg condition. The model can be extended by including a description of the disorder of the crystal. Protein crystals are usually modeled as being made up of a large number of mosaic blocks, each perfectly ordered within itself, but with an orientational disorder between them. In this case, not all of the mosaic blocks need be in the Bragg condition at the same time. Small rotations of the crystal, as performed during rotation crystallography, would allow all the mosaic blocks to be sampled and therefore can record the intensity from the entire crystal. However, with X-ray pulses on the femtosecond timescale, there is no possibility to do this. Therefore, the partiality model includes an angular "smearing" of the reciprocal lattice points, which become larger with increasing distance from the origin of reciprocal space.

The X-ray beam used for SFX experiments is typically not monochromatic, but has a small bandwidth of around 0.1%. In this case, not all the wavelengths may satisfy the Bragg condition. The same applies if the incident X-ray beam is not completely collimated and therefore has a convergence or divergence angle, which is also usually the case due to the focusing optics. Furthermore, these effects may be convoluted with those mentioned above in a single snapshot. For example, a subset of mosaic blocks may be at the exact Bragg condition for one wavelength, while another subset of mosaic blocks may satisfy it for a different wavelength. A model based on a finite bandwidth X-ray beam has also been used successfully with X-ray FEL data [27].

Partialities are different for each reflection and each diffraction pattern. However, we can assume that the underlying structure factors should be the same among all patterns for each symmetrically unique reflection. The partialities are strongly affected by geometrical parameters such as the orientation of the crystal. This allows a refinement procedure to be performed where the geometrical parameters are optimized for each diffraction pattern in turn, aiming to maximize the fit between the underlying structure factor estimates derived from each pattern by modeling the partialities. This process is called post-refinement. Post-refinement has been implemented for rotation crystallography for many years [53]. For snapshot data, the first step was to determine whether the process could be performed stably with simulated data, which was found to be the case [59]. The first implementation of post-refinement aimed at experimental SFX data was by Sauter [54] in cctbx.xfel. This was followed by the software cctbx.prime, which interfaces with cctbx.xfel and is developed specifically for post-refinement and merging, including scaling steps as well. Partiality modeling and post-refinement has also been implemented in cppxfel [25]. Whereas cctbx.xfel and cctbx.prime assume, at least in the versions described by recent literature, that the X-ray pulses are monochromatic, cppxfel uses a different geometrical model of the diffraction process where the spectral bandwidth of the X-ray pulse is the dominant resolution-dependent effect.

Finally, comparing intensity values is not the only way to refine the factors affecting partiality. The EVAL15 software [56] instead compares simulated and experimental peak shapes and positions for each reflection, using a ray tracing method to model the diffraction geometry. This approach has been successfully tested on snapshot data from a laboratory source [40], and is now being investigated for SFX data.

7.3.5 Evaluating the Data Quality

Many of the conventional figures of merit for a crystallographic dataset also apply to SFX data. These include the intensities compared to their estimated errors ($I/\sigma(I)$) and the number of measurements per symmetrically unique reflection, for example. Due to the large errors in an individual intensity measurement, arising from all the causes mentioned above including SASE fluctuations, difficulties in accurately indexing snapshots, and un-modeled partialities, the minimum number of measurements per reflection considered acceptable in SFX is much higher than in rotation crystallography, where a single measurement can suffice.

The most widely used figures of merit for SFX data are those based on the self-consistency of the dataset. In the ideal case, all the factors introducing noise into the individual intensity measurements would be corrected for, thereby obtaining "perfect" data from only one measurement per symmetrically unique reflection. In this case, repeating the experiment and data processing under identical conditions would produce an identical set of intensity measurements. In practice, some amount of variation between separate measurements of the same reflection must be accepted. By quantifying this variation, we can estimate how much random error is contained in the data, which places a limitation on how well the model can fit the data. The variation can be quantified by splitting the experimental dataset into two halves, alternating patterns to avoid systematic variations between the start and end of the experiment, merging each one separately, then comparing the two half-datasets using a correlation coefficient, R-factor or other metric. $CC_{\frac{1}{2}}$, the name given to the correlation coefficient between two half-datasets, was introduced and is closely related to CC^* for conventional crystallography [36]. R_{split} expresses the same correlation as an R-factor, given by:

$$R_{\text{split}} = \frac{1}{\sqrt{2}} \frac{\sum |I_{\text{even}} - I_{\text{odd}}|}{\frac{1}{2} \sum (I_{\text{even}} + I_{\text{odd}})}. \tag{7.2}$$

This formula divides the total absolute difference between the intensity from the "odd" and "even" half-datasets by the total of their mean values. Dividing by $\sqrt{2}$ aims to adjust the figure of merit to account for the half-datasets containing only half the amount of data as the full dataset, to give an estimate of the error in the complete dataset. As has been described earlier, R_{split} usually decreases in inverse proportion to the square root of the number of diffraction patterns.

Several different methods have been proposed for estimating the standard error in each individual merged intensity. In CrystFEL, this is done by measuring the standard deviation of the individual measurements for one symmetrically unique reflection, then dividing this by $\sqrt{N}$, where N is the number of measurements for that reflection. This gives an estimate of the standard deviation of the mean value, which would hypothetically be found if the experiment were to be repeated many times under the same conditions and many mean values calculated. A problem with this method is that the standard deviation can only be calculated accurately when the number of measurements for the reflection is large. Unfortunately, this is precisely the case in which the merged intensity itself would be the most precise, and therefore when we *least* need a good estimate of the error!

The ease with which these metrics can be calculated means that they are very commonly used. However, it is much better to use figures of merit which measure the accuracy as well as the precision of the data. For these, the fit between the final model and the data (R_{work} and R_{free}) and the anomalous measurability [19] are useful.

7.3.6 Solving the Structure

Once the final set of reflection intensities has been produced by merging the individual measurements from all the diffraction patterns, solving the structure can follow the same procedure as for conventional rotation crystallography. As for conventional macromolecular crystallography, molecular replacement is the most commonly used structure solution method. This method relies on a "search model" which is similar in structure to the macromolecule under investigation. Because the search model provides a large amount of information itself, needing only small modifications to arrive at the final structure, this method has relatively low requirements on the data quality.

Experimental phasing techniques such as single anomalous diffraction (SAD) have now been applied successfully to SFX data. These methods have much more stringent requirements on the data quality, because they are based on differences between intensities that are small fractions of the absolute intensities, and therefore can provide a useful independent validation of the data quality in SFX. In SAD, these are the differences between intensities of reflections with inverse indices (hkl and $\overline{hkl}$). These differences can be made larger by incorporating atoms of a strongly scattering species into the structure, and recording data using an X-ray wavelength close to the resonance condition for that species. Improvements in SFX data acquisition and processing can be traced by following the progression of SAD phasing results, starting with the observation of an anomalous signal from sulfur atoms [5] progressing to phasing using a very strong anomalous signal from gadolinium atoms [4], phasing of a well-ordered protein using the weak anomalous signal from sulfur and chlorine atoms natively present in the protein [47], and then arriving at the native phasing of a membrane protein [8]. Along the way, it has been found that single isomorphous replacement with anomalous scattering (SIRAS), in which two sets of data are compared, with and without the additional heavy atom, gives a large reduction in the number of patterns required for successful phasing [63]. It has also been found that the number of patterns required can be significantly reduced by careful calibration of the detector geometry [48].

The most commonly used strongly scattering species for SAD phasing in conventional crystallography is selenium. It would be very useful to establish conditions for Se-SAD phasing at X-ray FEL facilities, because techniques for incorporating selenium atoms into macromolecular structures are widely understood. The resonance condition for selenium is at a wavelength of just less than 1 Å, which is currently difficult to achieve at LCLS, but routinely available at SACLA. However, Se-SAD phasing was recently demonstrated using LCLS [32] and SACLA [62], and is expected to be possible at future XFEL facilities including LCLS-II, PAL-XFEL, and the European XFEL.

The radiation damage mechanisms are quite different for FEL data as for synchrotron data (see Chap. 6). In FEL data, electronic damage processes such as "bleaching" of atoms dominate over disintegration of the crystal due to processes such as free radical diffusion, which occur on a timescale much longer than the FEL pulse length. The degree of ionization of atoms is expected to depend on the X-ray pulse fluence, with more bleaching expected at high intensities. This principle may lead to a new experimental phasing method similar to radiation-induced phasing (RIP). The heaviest atoms should be more ionized and hence scatter less strongly at high intensity than at low intensity. A pair of diffraction datasets at low and high intensities could therefore be used similarly to the datasets in single isomorphous replacement (SIR). These differences have been observed experimentally for heavy atoms [22] and natively occurring atoms [23].

Other aspects of FEL data, such as Fourier truncation fringes arising from the use of very small crystals and coherent X-rays, may give additional information as discussed in Chap. 8.

7.4 Conclusion

Data processing for SFX, as well as other related techniques using X-ray FELs and synchrotrons, has become an active research field in its own right. There are several themes to this research, including addressing the technical challenges of storing and archiving data, understanding the fundamental physics underlying the diffraction and radiation damage processes, and the software engineering challenges of creating processing software that meet these challenges while at the same time being accessible to a rapidly growing number of nonexpert users.

The amount of X-ray FEL beamtime available worldwide is oversubscribed several times over. The amount of data required per experiment has been reduced over the years, using the techniques described in this chapter. However, this is unlikely to translate to a reduction in the size of the data mountain. Instead, more experiments will be performed, each using a smaller amount of measurement time. This has already been seen at LCLS, which since 2013 has offered "protein crystal screening" shifts of only six hours to allow crystals to be tested in the LCLS beam before applying for time for a more ambitious experiment. Although intended only for a quick check that the crystals produce sufficient diffraction signal for the more ambitious experiment, several of these shifts have led to protein structures being determined, for example [17, 33].

Another factor affecting the future size of the data mountain is the type of experiment. In the past years, determination of static protein structure has dominated. However, dynamic experiments with many time points are on the rise, involving as many as 14 individual data sets [3]. When looking for small intensity differences between the datasets, consistency of acquisition conditions is paramount, and so all the time points should be recorded during one block of experiment time. This can result in hundreds of thousands of diffraction patterns (corresponding to many million detector frames) being processed for a single experiment [49].

Finally, there are high hopes that the next generation of high repetition rate X-ray FELs will allow us to record an entire data set in a matter of seconds. At this speed, the experimenters themselves would become the limiting factor for the speed of the experiment, and automated systems for injecting a lineup of samples would be useful. This could produce a whole new data mountain, consisting not just of carefully acquired individual datasets but instead of systematic parameter space investigation, for example, of different crystallization or ligand-binding conditions. The data mountain climbers will not be able to return to base camp for a long time yet!

Acknowledgements TAW acknowledges the Helmholtz Association via Programme-Oriented Funds. Portions of this research were carried out at the Linac Coherent Light Source (LCLS) at the SLAC National Accelerator Laboratory. LCLS is an Office of Science User Facility operated for the US Department of Energy Office of Science by Stanford University. Use of the Linac Coherent Light Source (LCLS), SLAC National Accelerator Laboratory, is supported by the US Department of Energy, Office of Science, Office of Basic Energy Sciences under Contract No. DE-AC02-76SF00515.

References

1. Allahgholi, A., Becker, J., Bianco, L., Delfs, A., Dinapoli, R., Goettlicher, P., et al. (2015). AGIPD, a high dynamic range fast detector for the European XFEL. *Journal of Instrumentation, 10*, C01023.
2. Bajt, S., Chapman, H. N., Spiller, E. A., Alameda, J. B., Woods, B. W., Frank, M., et al. (2008). Camera for coherent diffractive imaging and holography with a soft-x-ray free-electron laser. *Applied Optics, 47*, 1673–1683.
3. Barends, T. R. M., Foucar, L., Ardevol, A., Nass, K., Aquila, A., Botha, S., et al. (2015). Direct observation of ultrafast collective motions in CO myoglobin upon ligand dissociation. *Science, 350*, 445–450.
4. Barends, T. R. M., Foucar. L., Botha, S., Doak, R. B., Shoeman, R. L., Nass, K., (2014). De novo protein crystal structure determination from X-ray free-electron laser data. *Nature, 505*, 244–247.
5. Barends, T. R. M., Foucar, L., Shoeman, R. L., Bari, S., Epp, S. W., Hartmann, R., et al. (2013). Anomalous signal from S atoms in protein crystallographic data from an X-ray free-electron laser. *Acta Crystallographica D, 69*, 838–842.
6. Barty, A., Boutet, S., Bogan, M. J., Hau-Riege. S., Marchesini, S., Sokolowski-Tinten, K., et al. (2008). Ultrafast single-shot diffraction imaging of nanoscale dynamics. *Nature Photonics, 2*, 415–419. http://dx.doi.org/10.1038/nphoton.2008.128
7. Barty, A., Kirian, R. A., Maia, F. R. N. C., Hantke, M., Yoon, C. H., White, T. A., et al. (2014). Cheetah: Software for high-throughput reduction and analysis of serial femtosecond X-ray diffraction data. *Journal of Applied Crystallography, 47*(3), 1118–1131.
8. Batyuk, A., Galli, L., Ishchenko, A., Han, G. W., Gati, C., Popov, P. A., et al. (2016). Native phasing of x-ray free-electron laser data for a G protein-coupled receptor. *Science Advances, 2*, e1600292.
9. Beyerlein, K., White, T. A., Yefanov, O., Gati, C., Kazantsev, I. G., Fog-Gade, N., et al. (2017). FELIX: An algorithm for indexing multiple crystallites in X-ray free-electron laser snapshot diffraction images. *Journal of Applied Crystallography, 50*, 1075–1083.
10. Blaj, G., Caragiulo, P., Carini, G., Carron, S., Dragone, A., Freytag, D., et al. (2015). X-ray detectors at the Linac Coherent Light Source. *Journal of Synchrotron Radiation, 22(3)*, 577–583. http://dx.doi.org/10.1107/S1600577515005317

11. Boutet, S., Foucar, L., Barends, T. R. M., Botha, S., Doak, R. B., Koglin, J. E., et al. (2015). Characterization and use of the spent beam for serial operation of LCLS. *Journal of Synchrotron Radiation, 22*, 634–643. https://doi.org/10.1107/S1600577515004002
12. Brehm, W., & Diederichs, K. (2014). Breaking the indexing ambiguity in serial crystallography. *Acta Crystallographica Section D, 70*, 101–109.
13. Carini, G. A., Boutet, S., Chollet, M., Dragone, A., Haller, G., Hart, P. A., et al. (2014). Experience with the CSPAD during dedicated detector runs at LCLS. *Journal of Physics Conference Series, 493*, 012011.
14. Casanas, A., Warshamanage, R., Finke, A. D., Panepucci, E., Olieric, V., Nöll, A., et al. (2016). EIGER detector: Application in macromolecular crystallography. *Acta Crystallographica. Section D, Structural Biology, 72*(9), 1036–1048. http://doi.org/10.1107/S2059798316012304
15. Chapman, H. N., Barty, A., Bogan, M. J., Boutet, S., Frank, M., Hau-Riege, S. P., et al. (2006). Femtosecond diffractive imaging with a soft-X-ray free-electron laser. *Nature Physics, 2*, 839. http://dx.doi.org/10.1038/nphys461
16. Chapman, H. N., Fromme, P., Barty, A., White, T. A., Kirian, R. A., Aquila, A., et al. (2011). Femtosecond x-ray protein nanocrystallography. *Nature, 470*, 73–77.
17. Conrad, C. E., Basu, S., James, D., Wang, D., Schaffer, A., Roy-Chowdhury, S., et al. (2015). A novel inert crystal delivery medium for serial femtosecond crystallography. *IUCrJ, 2*, 421–430.
18. Damiani, D., Dubrovin, M., Gaponenko, I., Kroeger, W., Lane, T. J., Mitra, A., et al. (2016). Linac Coherent Light Source data analysis using psana. *Journal of Applied Crystallography, 49*, 672–679.
19. Dauter, Z. (2006). Estimation of anomalous signal in diffraction data. *Acta Crystallographica Section D, 62*, 867–876.
20. Duisenberg, A. J. M. (1992). Indexing in single-crystal diffractometry with an obstinate list of reflections. *Journal of Applied Crystallography, 25*, 92–96.
21. Foucar, L. (2016). CFEL-ASG Software Suite (CASS): usage for free-electron laser experiments with biological focus. *Journal of Applied Crystallography, 49*(4), 1336–1346.
22. Galli, L., Son, S. K., Barends, T. R. M., White, T. A., Barty, A., Botha, S., et al. (2015). Towards phasing using high X-ray intensity. *IUCrJ, 2*, 627–634.
23. Galli, L., Son, S. K., Klinge, M., Bajt, S., Barty, A., Bean, R., et al. (2015). Electronic damage in S atoms in a native protein crystal induced by an intense X-ray free-electron laser pulse. *Structural Dynamics, 2*, 041703.
24. Gildea, R. J., Waterman, D. G., Parkhurst, J. M., Axford, D., Sutton, G., Stuart, D. I., et al. (2014). New methods for indexing multi-lattice diffraction data. *Acta Crystallographica Section D, 70*, 2652–2666.
25. Ginn, H. M., Brewster, A. S., Hattne, J., Evans, G., Wagner, A., Grimes, J. M., et al. (2015). A revised partiality model and post-refinement algorithm for X-ray free-electron laser data. *Acta Crystallographica Section D, 71*, 1400–1410.
26. Ginn, H. M., Evans, G., Sauter, N. K., & Stuart, D. I. (2016). On the release of cppxfel for processing X-ray free-electron laser images. *Journal of Applied Crystallography, 49*, 1065–1072.
27. Ginn, H. M., Messerschmidt, M., Ji, X., Zhang, H., Axford, D., Gildea, R. J., (2015) Structure of CPV17 polyhedrin determined by the improved analysis of serial femtosecond crystallographic data. *Nature Communications 6*, 6435.
28. Ginn, H. M., Roedig, P., Kuo, A., Evans, G., Sauter, N. K., Ernst, O., et al. (2016). TakeTwo: An indexing algorithm suited to still images with known crystal parameters. *Acta Crystallographica Section D, 72*, 956–965.
29. Hattne, J., Echols, N., Tran, R., Kern, J., Gildea, R. J., Brewster, A. S., et al. (2014). Accurate macromolecular structures using minimal measurements from X-ray free-electron lasers. *Nature Methods, 11*, 545–548.
30. Heisen, B. C., Boukhelef, D., Esenov, S., Hauf, S., Kozlova, I., Maia, L., et al. (2013). Karabo: An integrated software framework combining control, data management, and scientific computing tasks. In *Proceedings of ICALEPCS*, San Francisco.

31. Hunter, M. S., Segelke, B., Messerschmidt, M., Williams, G. J., Zatsepin, N. A., Barty, A., et al. (2014). Fixed-target protein serial microcrystallography with an x-ray free electron laser. *Scientific Reports, 4*, 6026. http://dx.doi.org/10.1038/srep06026
32. Hunter, M. S., Yoon, C. H., DeMirci, H., Sierra, R. G., Dao, E. H., Ahmadi, R., et al. (2016). Selenium single-wavelength anomalous diffraction de novo phasing using an X-ray-free electron laser. *Nature Communications, 7*, 13388.
33. Hutchison, C. D. M., Cordon-Preciado, V., Morgan, R. M. L., Nakane, T., Ferreira, J., Dorlhiac, G., et al. (2017). X-ray free electron laser determination of crystal structures of dark and light states of a reversibly photoswitching fluorescent protein at room temperature. *International Journal of Molecular Sciences, 18* (1918). https://doi.org/10.3390/ijms18091918
34. Kabsch, W. (1988). Evaluation of single-crystal x-ray diffraction data from a position-sensitive detector. *Journal of Applied Crystallography, 21*, 916–924.
35. Kabsch, W. (2014). Processing of X-ray snapshots from crystals in random orientations. *Acta Crystallographica Section D, 70*, 2204–2216.
36. Karplus, P. A., & Diederichs, K. (2012). Linking crystallographic model and data quality. *Science, 336*, 1030–1033.
37. Kirian, R. A., Wang, X., Weierstall, U., Schmidt, K. E., Spence, J. C. H., et al. (2010). Femtosecond x-ray protein nanocrystallography — data analysis methods. *Optics Express, 18*, 5713–5723.
38. Könnecke, M., Akeroyd, F. A., Bernstein, H. J., Brewster, A. S., Campbell, S. I., Clausen, B., et al. (2015). The NeXus data format. *Journal of Applied Crystallography 48*, 301–305. http://dx.doi.org/10.1107/S1600576714027575
39. Kraft, P., Bergamaschi, A., Broennimann, C., Dinapoli, R., Eikenberry, E. F., Henrich, B., et al. (2009). Performance of single-photon-counting PILATUS detector modules. *Journal of Synchrotron Radiation 16*(3), 368–375. http://doi.org/10.1107/S0909049509009911
40. Kroon-Batenburg, L. M. J., Schreurs, A. M. M., Ravelli, R. B. G., & Gros, P. (2015). Accounting for partiality in serial crystallography using ray-tracing principles. *Acta Crystallographica Section D, 71*, 1799–1811.
41. Lyubimov, A. Y., Uervirojnangkoorn, M., Zeldin, O. B., Brewster, A. S., Murray, T. D., Sauter, N. K., et al. (2016). IOTA: Integration optimization, triage and analysis tool for the processing of XFEL diffraction images. *Journal of Applied Crystallography, 49*, 1057–1064.
42. Maia, F. R. N. C. (2012). The coherent x-ray imaging data bank. *Nature Methods, 9*(9), 854–855. http://dx.doi.org/10.1038/nmeth.2110
43. Mancuso, A. P., Aquila, A., Borchers, G., Giewekemeyer, K., & Reimers, N. (2013). Technical design report: scientific instrument single particles, clusters, and biomolecules (SPB). https://doi.org/10.3204/XFEL.EU/TR-2013-004
44. Mariani, V., Morgan, A., Yoon, C. H., Lane, T. J., White, T. A., O'Grady, C. P., et al. (2016) OnDA: Online data analysis and feedback for serial X-ray imaging. *Journal of Applied Crystallography, 49*(3), 1073–1080.
45. Mozzanica, A., Bergamaschi, A., Cartier, S., Dinapoli, R., Greiffenberg, D., Johnson, I., et al. (2014) Prototype characterization of the JUNGFRAU pixel detector for SwissFEL. *Journal of Instrumentation, 9*, C05010.
46. Nakane, T., Joti, Y., Tono, K., Yabashi, M., Nango, E., Iwata, S., et al. (2016). Data processing pipeline for serial femtosecond crystallography at SACLA. *Journal of Applied Crystallography, 49*, 1035–1041.
47. Nakane, T., Song, C., Suzuki, M., Nango, E., Kobayashi, J., Masuda, T., et al. (2015). Native sulfur/chlorine SAD phasing for serial femtosecond crystallography. *Acta Crystallographica Section D, 71*, 2519–2525.
48. Nass, K., Meinhart, A., Barends, T. R. M., Fourcar, L., Gorel, A., Aquila, A., et al. (2016). Protein structure determination by single-wavelength anomalous diffraction phasing of X-ray free-electron laser data. *IUCrJ, 3*, 180–191.
49. Pande, K., Hutchison, C. D. M., Groenhof, G., Aquila, A., Robinson, J. S., Tenboer, J., et al. (2016). Femtosecond structural dynamics drives the trans/cis isomerization in photoactive yellow protein. *Science, 352*, 725–729.

50. Pixel array detectors. http://bigbro.biophys.cornell.edu/research/pad. Accessed 20.11.2017.
51. Powell, H. R. (1999). The Rossmann Fourier autoindexing algorithm in *MOSFLM*. *Acta Crystallographica Section D, 55*(10), 1690–1695. https://doi.org/10.1107/S0907444999009506
52. Rossmann, M. G. (1979). Processing oscillation diffraction data for very large unit cells with an automatic convolution technique and profile fitting. *Journal of Applied Crystallography, 12*, 225–238.
53. Rossmann, M. G., Leslie, A. G. W., Abdel-Meguid, S. S., & Tsukihara, T. (1979). Processing and post-refinement of oscillation camera data. *Journal of Applied Crystallography, 12*, 570–581.
54. Sauter, N. K. (2015). XFEL diffraction: Developing processing methods to optimize data quality. *Journal of Synchrotron Radiation, 22*, 239–248.
55. Sauter, N. K., Hattne, J., Brewster, A. S., Echols, N., Zwart, P. H., & Adams, P. D. (2014). Improved crystal orientation and physical properties from single-shot XFEL stills. *Acta Crystallographica Section D, 70*, 3299–3309.
56. Schreurs, A. M. M., Xian, X., & Kroon-Batenburg, L. M. J. (2010). EVAL15: A diffraction data integration method based on ab initio predicted profiles. *Journal of Applied Crystallography, 43*, 70–82.
57. Thayer, J., Damiani, D., Ford, C., Dubrovin, M., Gaponenko, I., O'Grady, C. P., et al. (2017). Data systems for the Linac coherent light source. *Advances in Structural Chemical Imaging, 3*(1), 3. http://dx.doi.org/10.1186/s40679-016-0037-7
58. Uervirojnangkoorn, M., Zeldin, O. B., Lyubimov, A. Y., Hattne, J., Brewster, A. S., Sauter, N. K., et al. (2015). Enabling X-ray free electron laser crystallography for challenging biological systems from a limited number of crystals. *eLife, 4*, e05421.
59. White, T. A. (2014). Post-refinement method for snapshot serial crystallography. *Philosophical Transactions of the Royal Society B 369*, 20130330.
60. White, T. A., Barty, A., Stellato, F., Holton, J. M., Kirian, R. A., Zatsepin, N. A., et al. (2013). Crystallographic data processing for free-electron laser sources. *Acta Crystallographica D, 69*, 1231–1240.
61. White, T. A., Mariani, V., Brehm, W., Yefanov, O., Barty, A., Beyerlein, K. R., et al. (2016). Recent developments in CrystFEL. *Journal of Applied Crystallography, 49*, 680–689.
62. Yamashita, K., Kuwabara, N., Nakane, T., Murai, T., Mizohata, E., Sugahara, M., et al. (2017). Experimental phase determination with selenome-thionine or mercury-derivatization in serial femtosecond crystallography. *IUCrJ, 4*, 639–647.
63. Yamashita, K., Pan, D., Okuda, T., Sugahara, M., Kodan, A., Yamaguchi, T., et al. (2015). An isomorphous replacement method for efficient de novo phasing for serial femtosecond crystallography. *Scientific Reports, 5*, 14017.
64. Yefanov, O., Mariani, V., Gati, C., White, T. A., Chapman, H. N., Barty, A. (2015). Accurate determination of segmented X-ray detector geometry. *Optics Express, 23*, 28459.
65. Yoon, C. H., DeMirci, H., Sierra, R. G., Dao, H. E., Ahmadi, R., Aksit, F., et al. (2017). Se-SAD serial femtosecond crystallography datasets from selenobiotinyl-streptavidin. *Scientific Data, 4*, 170055. http://dx.doi.org/10.1038/sdata.2017.55
66. Zeldin, O. B., Brewster, A. S., Hattne, J., Uervirojnangkoorn, M., Lyubimov, A. Y., Zhou, Q., et al. (2015). Data exploration toolkit for serial diffraction experiments. *Acta Crystallographica Section D, 71*, 352–356.
67. Zhu, D., Feng, Y., Stoupin, S., Terentyev, S. A., Lemke, H. T., Fritz, D. M., et al. (2014). Performance of a beam-multiplexing diamond crystal monochromator at the Linac Coherent Light Source. *Review of Scientific Instruments 85*(6), 063106.

Chapter 8
Phasing Serial Crystallography Data

Richard A. Kirian, Joe P. J. Chen, and John C. H. Spence

8.1 Introduction

The development of serial femtosecond crystallography (SFX) at X-ray free electron lasers (X-ray FELs) allows for the use of tiny protein crystals down to just a few unit cells along an edge, measured at physiological temperatures, and with a time resolution far better than can be achieved with synchrotrons or electron microscopes. The unique properties of the X-ray FEL source has furthermore resulted in the appearance of entirely new ideas for solving the crystallographic phase problem. At the same time, in combination with work on phasing single-particle data (with one bioparticle per shot), SFX has stimulated research into new phasing methods for serial crystallography (SC) at synchrotrons, and protein crystallography in general. In the sense that these new phasing methods depend on the application of constraints, they might be considered developments of traditional "direct methods" such as density modification approaches.

It was the injection of new ideas from the signal processing and optics communities in a formative review article by Millane [1] that initiated the modern revival of interest in numerical iterative phasing methods for both single particles and crystals. These ideas go back, at least, to the paper by Sayre [2], who first pointed out that if scattering could be detected between Bragg reflections it would assist phasing (since intensity zero-crossings could then be identified), and to work by Gerchberg and Saxton [3] on iterative phasing for non-periodic samples in electron microscopy. The first successful algorithm based on this approach added feedback to the Gerchberg and Saxton algorithm and was described as the Hybrid Input-Output (HIO) algorithm by Fienup [4]. These algorithms, which are reviewed

R. A. Kirian (✉) · J. P. J. Chen · J. C. H. Spence
Department of Physics, Arizona State University, Tempe, AZ, USA
e-mail: rkirian@asu.edu; jpchen1@mainex1.asu.edu; spence@asu.edu

S. Boutet et al. (eds.), *X-ray Free Electron Lasers*,
https://doi.org/10.1007/978-3-030-00551-1_8

further in Chaps. 9 and 14, iterate between real and reciprocal space while imposing known constraints at each step, such as the boundary of the molecule, the sign of the scattering medium, and the measured scattering intensity. Many variants of these "iterated projection" algorithms are the subject is reviewed by Marchesini [5], Millane and Lo [6] and Spence [7]. In most cases, scattering that was more finely sampled in angle than twice the Bragg angle ("oversampling") was required. However, it is now understood that for many high-solvent crystals, where the molecule fills only a portion of the unit cell, one has access to sufficient information to solve the phase problem using only Bragg intensities, as we describe below in more detail. The control of hydration was also the basis for some early phasing efforts [8] aimed at sampling the molecular transform at several points by changing the unit cell volume. In Chap. 9, the production of diffuse scattering between Bragg reflections is described, which can also be used in this way. As we detail below, a means of determining if a unique solution can be expected for both single particles and crystals has now emerged through the introduction of a metric called the constraint ratio Ω [9]. At the same time that new algorithms were being considered, it was natural to try existing phasing methods with SFX data.

Below we begin with a brief overview of the phase problem and how conventional crystallographic phasing methods have been applied to SFX data (see also [10]). We then describe novel phasing techniques enabled by the unique properties of X-ray FELs.

8.2 The Phase Problem

Under the Born approximation, the far-field diffraction of an arbitrary object by X-rays is related to the object's structure by the Fourier transform

$$F(\mathbf{q}) = \int_{-\infty}^{\infty} f(\mathbf{x})\, e^{i\mathbf{q}.\mathbf{x}}\, \mathbf{dx} \tag{8.1}$$

where $\mathbf{x}$ and $\mathbf{q}$ are coordinates in real and Fourier (reciprocal) space, respectively, i is the square root of -1, $f(\mathbf{x})$ is the complex scattering density of the object, and $F(\mathbf{q})$ is the complex amplitude of the diffracted wavefield. To a first approximation, the scattering density of the object is proportional to its electron density, with the exception being cases where the X-ray energy is close to an electronic resonance (i.e., transition energy.) Resonance leads to a significant imaginary component of the scattering density and may be treated accurately with an explicit atomic model in which the total scattering factor is composed of a summation over atomic scattering factors.

The term "complex amplitude" means that $F(\mathbf{q})$ is a complex function, having a magnitude, $|F(\mathbf{q})|$, and a phase, $\varphi(\mathbf{q})$, and is expressible as:

$$F(\mathbf{q}) = |F(\mathbf{q})|\, e^{i\varphi(\mathbf{q})} \tag{8.2}$$

Knowing $F(\mathbf{q})$ allows one to obtain the scattering density $f(\mathbf{x})$ through the inverse Fourier transform:

$$f\left(\mathbf{x}\right)=\int_{-\infty}^{\infty}F\left(\mathbf{q}\right)e^{-i\mathbf{x}.\mathbf{q}}\,\mathbf{dq} \tag{8.3}$$

The lack of suitable materials to act as a lens for efficiently focusing scattered X-rays to form an image means that what is accessible in a diffraction experiment is just the intensity of the diffracted wavefront, which is the square of the magnitude of $F(\mathbf{q})$, that is, $|F(\mathbf{q})|^2$, and hence the phase function $\varphi(\mathbf{q})$ is not measured. This constitutes the so-called phase problem. In addition, individual diffraction patterns record scattering, which is constrained by the elastic scattering kinematics to lie on the Ewald sphere, so that these separate recordings must be indexed (oriented with respect to the lab frame), merged and assembled into the three-dimensional diffraction volume before these equations can be applied.

In crystallography, one is faced with an even more severe restriction than the absence of measured phases. The periodic nature of a crystal means that the diffracted intensity of the molecule is modulated by a periodic multiplicative term that peaks at the reciprocal lattice points and is of much lower values elsewhere in reciprocal space. For fully coherent illumination, this multiplicative function is often called the shape transform (see Sect. 8.4.1) since it depends strongly on the overall shape of the crystal. When the number of unit-cell repetitions is large, the shape-transform peaks become sharp and point-like, giving rise to the familiar concept of Bragg peaks, and effectively results in the diffracted intensity being sampled at only the reciprocal lattice points, which we sometimes refer to as "Bragg sampling."

Without any phase information, it can be shown that this Bragg sampling of the diffracted intensity is below the minimum amount required by Shannon's sampling theorem to uniquely determine the autocorrelation of the unit cell, which indicates that the Bragg sampling is also not sufficient to uniquely determine the unit cell itself (see [1, 11] for a more in-depth discussion.) This is illustrated in Fig. 8.1, which compares three diffraction scenarios: (1) an isolated non-crystalline molecule, (2) a crystal of finite size under coherent FEL-like illumination, and (3) an infinite crystal under similar coherent illumination. The intensity undersampling problem may be understood by first recalling that the Fourier transform of the diffraction intensities is equal to the autocorrelation of the real-space scattering density, and secondly by noting that only half of the Hermitian centrosymmetric autocorrelation function contains independent values. For the case of an infinite crystal, the essential problem is that the periodic autocorrelation function that results from Bragg-sampled intensities yields only half as many independent measurements as there are independent scattering densities to be determined. To solve the "phase problem" in conventional crystallography, one must therefore overcome the "undersampled-intensity problem." The solution to this conundrum usually requires additional experimental measurements in which the X-ray wavelength is varied, or the atomic structure is modified in a controlled way, or by making some assumptions about the unknown structure (e.g., that it is similar to a molecule whose structure is already

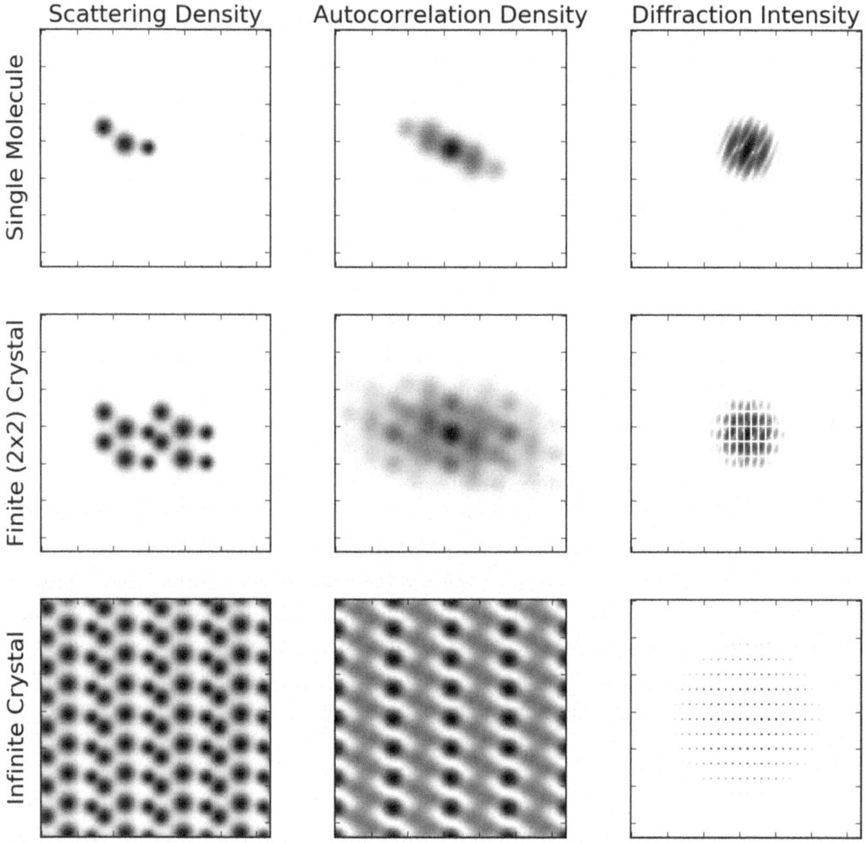

Fig. 8.1 Top row: scattering density of a single generic molecule (left), its autocorrelation function (middle), and diffraction intensities (right). The autocorrelation function is equal to the Fourier transform of the diffraction intensities, and hence is an equally valid representation of the data as the intensities, but more clearly reveals the number of independent measurements. Note the centrosymmetry of the autocorrelation function, in which only half of the non-zero area contain independent measurements. Middle row: the same molecule as in the top row, but arranged into a 2×2 crystal. In principle, a finite crystal may be treated much like any other object, if it were possible to measure appropriately sampled diffraction intensities at high signal-to-noise ratio. Bottom row: the same molecule arranged into an infinite crystal. In this case, the autocorrelation function has the same periodicity as the crystal itself, but due to the centrosymmetry there are only half as many independent measurements per unit cell as there are unknown scattering densities per unit cell; this is what gives rise to the crystallographic phase problem that suffers from a major deficit in the number of measurements

known). However, as to be discussed in Sect. 8.4, X-ray FELs offer some novel solutions to the phase problem in certain situations.

8.3 Conventional Methods

The standard crystallographic approaches to phasing are described in Rupp [12] where further details and references can be found. A brief summary of recent developments in de novo phasing in the context of X-ray FELs can be found in Schlichting [10]. Broadly speaking, all phasing techniques rely on increasing the number of unique measurements (beyond the Bragg-sampled intensities), and/or utilizing prior-known information about the sample. These include Single and Multiple Anomalous Diffraction (SAD and MAD), in which small differences due to anomalous scatterers are utilized; Single and Multiple Isomorphous Replacement (SIR and MIR), in which heavy atoms are placed in a protein crystal and thereby augment the diffraction intensities; Direct Methods, which use the likely zero sum of the phases around loops in reciprocal space; and Molecular Replacement (MR), which is based on modeling against macromolecules with similar sequence or fold in the Protein Data Bank (www.wwPDB.org). Of these methods, MR and SAD are the most popular de-novo technique over 70% of de novo structures deposited in the PDB in 2013 done by SAD [13], but the vast majority of all structures are now done by MR. The traditional Direct Methods require very high resolution data (atomic-resolution), and succeed typically only for small proteins.

It was not at all obvious that conventional phasing methods could be applied to "diffraction-before-destruction" SFX data, since the onset of radiation damage after a few tens of femtoseconds could mask the very small differences in structure factors that must be measured, for example, in the SAD method. In addition, the original Monte Carlo intensity merging method (without post-refinement procedures), described in detail in Chap. 7 of this book, has limited accuracy. Monte Carlo intensity merging requires a 100 times more data to obtain a single order of magnitude improvement in accuracy, as it averages partial reflection intensities over large numbers of crystals of different sizes. Shot-to-shot variations in the X-ray FEL beam intensity and wavelength are additional important stochastic variables to be accommodated in the Monte Carlo method. Despite issues with the stochastic nature of SFX measurements, it emerged that for micron-sized crystals, data collected using X-ray FEL pulses often showed higher resolution than synchrotron data from small crystals. The general trend seems to be that for microcrystals, radiation damage at synchrotrons results in lower resolution data than from X-ray FELs. For large crystals, the resolution achieved with synchrotrons may be better given sufficiently high-quality crystals (see [14] for a review of serial crystallography.)

The first report of SAD phasing with X-ray FEL data was carried out by Barends et al. [15], who recovered the structure of a heavy-atom derivative of lysozyme at 0.21 nm resolution. In this case, gadolinium atoms were used, which produced a relatively large anomalous difference signal of roughly 10%. Pulses of 50 fs duration, 2.6 mJ average power, and 8.5 keV photon energy were used to obtain 60,000 indexed diffraction patterns that yielded an anomalous correlation coefficient (CC_{ano}) of 0.48, with R_{split} about 5% at 0.3 nm resolution. The number of required patterns was later reduced to ~7000 upon improvements to the data processing, as discussed in Nass et al. [16]. A series of de novo phasing demonstrations have since been reported: isomorphous replacement was demonstrated on a mercury derivative [17] followed by the more recent application of direct SAD phasing to the same target [18]. The replacement of methionine residues by seleno-methionine, known as the "magic bullet" of structural biology, is a convenient means of SAD phasing (at 12.65 keV) and has now been demonstrated on SFX data [18, 19] using both the LCLS and SACLA facilities. The first previously unknown structure to be solved was the mosquito larvicide BinAB, which was solved using multiple isomorphous replacement with mercury, gadolinium, and iodine on in vivo grown nanocrystals [20].

Recent research has also focused on SAD phasing using native sulfur atoms, which constitute about 1% of non-hydrogen atoms in proteins. Anomalous signal from sulfur atoms was observed quite early in the development of SFX by Barends et al. [21], but at 7.3 keV the anomalous difference signal from sulfur is small and phase retrieval was not possible at that time, partly due to software limitations. This has since proven possible in more recent work by Nass et al. [16]. Native SAD phasing using sulfur and chlorine is also demonstrated in Nakane et al. [22] for lysozyme, and by Batyuk et al. [23] for the G-protein coupled receptor (GPCR) adenosine receptor A_{2a}. These studies, in which the anomalous difference signal is small (about 1%), demonstrate the increasing accuracy of SFX data analysis due to steady improvements in the SFX data analysis algorithms and detector metrology. In addition to algorithm developments, further improvements can be expected from the FEL sources; for example, the application of a new two-color mode of operation at X-ray FELs has very recently been exploited to gain further improvements in MAD phasing [24].

8.4 Novel Approaches

8.4.1 Finite Crystal Methods

From the observation of the first SFX diffraction patterns in 2009, it was clear that new data analysis methods would be needed. This was a consequence of the new experimental arrangements—protein crystals as small as a few dozen unit cells on a side (or bioparticles) sprayed across a pulsed beam, as in the first serial crystallography variants [25, 26]. Many of the patterns showed scattering in the form of interference fringes between Bragg reflections, which, as suggested by Sayre [2], could assist with phasing. The theoretical basis for describing protein nanocrystal diffraction in serial crystallography, including these "shape transform" effects, was provided by Kirian et al. [27], and the many developments of this theory and improved algorithms that have followed are described in Chaps. 7 and 9. For the purposes of phasing, the new opportunities offered by X-ray FEL data can best be understood through the history of the development of the field of Coherent Diffractive Imaging (CDI), which has been reviewed by Marchesini [5] and Spence [7]. In CDI, the electron density of an object is recovered from its Fourier intensities through iterative algorithms. These iterative solutions to the non-crystallographic phase problem work well for single-particle data, but, because they require scattering to be sampled at sub-Bragg intervals, were originally thought not to be useful for phasing scattering from crystals. It is now understood there are important exceptions to this conclusion, as we discuss here for the case of nanocrystals.

As alluded to in Sect. 8.2, measuring diffraction intensities between reciprocal lattice points can give enough information to uniquely determine the autocorrelation function of the crystal unit cell. Work by Perutz [28] aimed to determine the sufficiently sampled transform of the unit cell by modifying the solvent contents, and hence the unit cell size, of hemoglobin crystals. The term "sufficiently sampled" in this case means at least twice as fine as the Bragg-sampling. Perutz's work was met with some success but his particular technique fell out of favor with the invention of isomorphous replacement soon after.

The new experimental arrangements provided by X-ray FELs yielded diffraction patterns that showed measureable intensities between the Bragg reflections, as shown in Fig. 8.2. For an idealized nanocrystal immersed in a wide coherent beam, one finds (N-2) interference fringe maxima, for a crystal containing N planes normal to direction $\mathbf{g}$, running between Bragg reflections in the direction $\mathbf{g}$. This is akin to the (N-2) subsidiary maxima seen between the principle maxima in the optical transmission diffraction pattern from a grating of N slits. These fringes, running in several directions, therefore give the size of the crystal between facets. The interference fringes occur because the coherence width of the beam is larger than the microcrystals, meaning that the entire crystal is coherently illuminated. In contrast, conventional Bragg diffraction requires only that the coherence width exceed the size of the unit cell. The above situation can be modeled as follows. Write the electron density of the nth crystal, $f_n(\mathbf{x})$, as

$$f_n(\mathbf{x}) = \sum_{j=1}^{N_n} f\left(\mathbf{x} - \mathbf{r}_{nj}\right) \tag{8.4}$$

where $f(\mathbf{x})$ is the electron density of the unit cell, $\mathbf{r}_{nj}$ is the spatial shift for the jth unit cell required to construct the nth crystal which has N_n number of unit cells. Taking the Fourier transform, the far-field complex-valued diffraction amplitude of the crystal can be written as

$$F_n(\mathbf{q}) = F(\mathbf{q}) \sum_{j=1}^{N_n} e^{i\mathbf{q}\cdot\mathbf{r}_{nj}} \tag{8.5}$$

The quantity that is measured in the experiment is the intensity of the complex-valued diffracted amplitude, which is given by

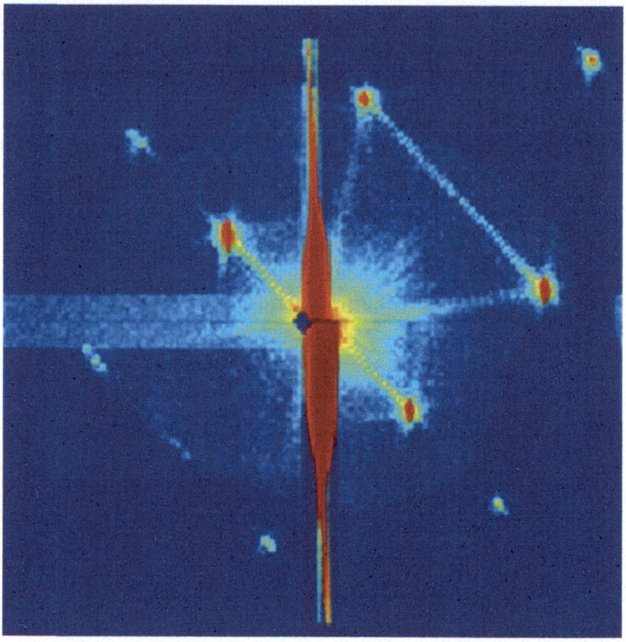

Fig. 8.2 A diffraction pattern obtained at LCLS from a submicrometer crystal of Photosystem I during the first serial femtosecond X-ray crystallography experiment [8]. The interference fringes between Bragg reflections provide the "oversampling" needed in principle to solve the phase problem. The red streak running vertically through the center of the pattern is the scattering of X-rays from the edge of the water jet that carried the crystals into the X-ray beam

$$I_n(\mathbf{q}) = |F(\mathbf{q})|^2 S_n(\mathbf{q}) \tag{8.6}$$

where

$$S_n(\mathbf{q}) = \left| \sum_{j=1}^{N_n} e^{i\mathbf{q}.\mathbf{r}_{nj}} \right|^2 \tag{8.7}$$

is the so-called "shape transform." Note that this simple product between the modulus-squared unit-cell transform and the shape transform results from the assumption that all unit cells are identical, which may not be the case, as discussed later.

Consider now the average over many diffraction patterns, measured from an ensemble of crystals of different sizes and shapes, which is the result of the data merging step in an SFX experiment. This averaged intensity is given by

$$\langle I_n(\mathbf{q})\rangle_n = |F(\mathbf{q})|^2 \langle S_n(\mathbf{q})\rangle_n \tag{8.8}$$

where $\langle . \rangle_n$ denotes the average over the entire ensemble of crystals. Spence et al. [30] suggested that the averaged shape transform $\langle S_n(\mathbf{q}) \rangle_n$ can be determined from $\langle I_n(\mathbf{q}) \rangle_n$ directly by averaging over all translations centered around the reciprocal lattice point $\mathbf{g_h}$ from the diffraction of all crystals. The vector $\mathbf{h}$ denotes a 3-tuple containing the Miller indices. By assuming that the molecular transform and the shape transform are uncorrelated, and that a sufficient number of reciprocal lattice points and crystals are used in forming this average, we can write

$$\langle S_n(\mathbf{q}) \rangle_n = \left\langle \langle I_n(\mathbf{q} - \mathbf{g_h}) \rangle_n \right\rangle_{\mathbf{h}} \tag{8.9}$$

In other words, the operation described by Eq. (8.9) is the average of the diffracted intensities over all Wigner–Seitz cells (i.e., the smallest primitive unit cell that can be constructed in reciprocal space); such an operation produces one period of the averaged shape transform. The averaged shape transform over all reciprocal space can then be obtained by replicating the averaged period throughout reciprocal space.

The molecular transform can in principle be obtained via a simple division, needing only the merged experimental intensity, that is,

$$|F(\mathbf{q})|^2 = \frac{\langle I_n(\mathbf{q}) \rangle_n}{\left\langle \langle I_n(\mathbf{q} - \mathbf{g_h}) \rangle_n \right\rangle_{\mathbf{h}}} \tag{8.10}$$

This resulting molecular transform is more finely sampled than the Bragg diffraction from conventional crystallography, thus compensating for the data deficiency impeding the solution of the phase problem from Bragg reflections alone. Having recovered a sufficiently sampled molecular transform, it can be phased in ways analogous to reconstructions in CDI by computational iterative methods, such as the HIO algorithm described in Sect. 8.1.

Simulations applying this method to ideal nanocrystals with P1 symmetry can be found in Spence et al. [30]. For all other space groups there exists more than one molecule in the unit cell, which needs to be taken into account. Unlike in conventional crystallography, the inter-Bragg intensities from finite crystals depend crucially on the electron density of the whole crystal rather than the electron density of the unit cell alone. Even for an idealized finite crystal, the way in which molecules terminate on the crystal surface determines which repeating unit cell the crystal is composed of. The usual point-group symmetries associated with Bragg reflections do not generally carry over to the inter-Bragg intensities, and the very notion of the unit cell breaks down because partial unit cells, with incomplete molecular occupancies, are very likely to occur at the crystal surface. With the above in mind, the diffraction intensity clearly cannot be written in terms of a simple product between a shape transform and a molecular transform. This problem, which requires important modifications to current phase-retrieval algorithms, is the subject of active research [31–34].

An experimental test of shape-transform phasing is shown in Fig. 8.3 (from [35]). In these experiments conducted at the FERMI facility at 32.5 nm wavelength, 2D crystals were formed from Pt islands deposited using a focused ion beam

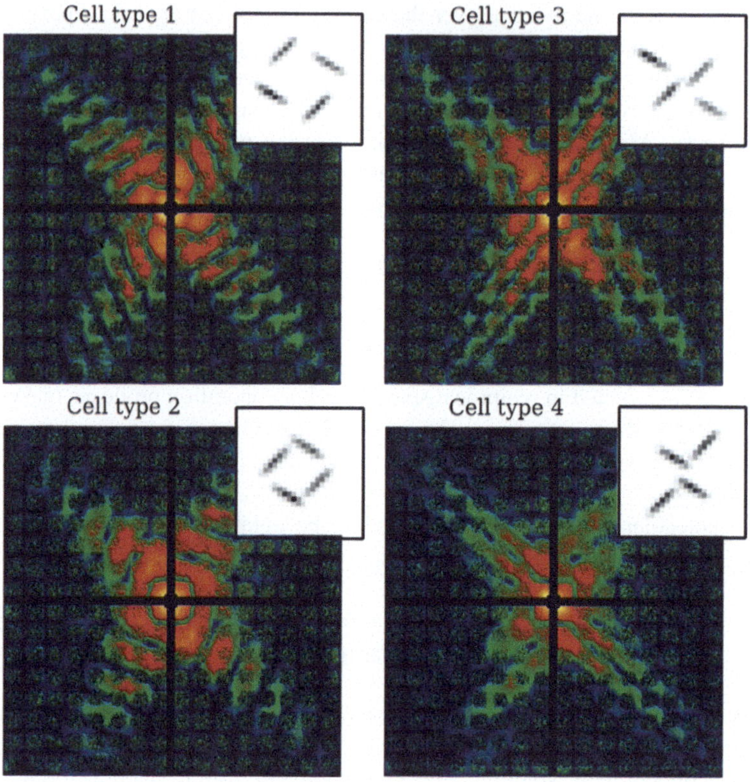

Fig. 8.3 Molecular transforms and real-space reconstructions (upper-right insets) corresponding to four different types of synthetic micro-crystals patterned with a focused ion beam. Each of the four crystal types differed only in their unit-cell configurations (i.e., only the surface truncations differed). The molecular transforms were recovered after averaging the diffraction intensities from multiple crystals with differing shapes and sizes, followed by a procedure to de-couple the crystal lattice transforms as described in the main text. Real-space reconstructions followed conventional CDI methods while utilizing intensities sampled between Bragg reflections [29]

(FIB) on a thin substrate. The same motif was used with four different edge terminations in order to clearly demonstrate the importance of the crystal surface. These experiments vindicate how the procedure of averaging over many shots (a couple dozen in this case), followed by division of the resulting average pattern by the periodically averaged Wigner–Seitz cell, reveals the underlying molecular transform of the crystal. It was also demonstrated that the X-ray beam need not be highly uniform in phase or amplitude over the entire crystal; it is sufficient if the X-ray beam is reasonably uniform over length scales corresponding to a few unit cells since the aim is to reconstruct the unit cell rather than the whole crystal. Some regions between Bragg reflections had very poor signal-to-noise ratio and were left unconstrained during the reconstruction, which is tolerable since, roughly speaking, it is only necessary to double the number of intensity measurements as compared to

Bragg sampling. The main limitation of this experiment is that the crystals had well-defined unit cells, whereas inter-Bragg diffraction observed from protein crystals in space groups other than P1, thus far, appear to arise from crystals that have surface truncations that are not consistent with a common unit cell throughout. As mentioned above, the surface terms remain an important challenge for practical implementations of shape-transform phasing.

For a diffraction-limited coherent beam of nanometer dimensions, the situation is analogous to that in the fully coherent scanning transmission electron microscope (STEM). If the beam divergence angle is larger than the Bragg angle, these coherent diffraction orders overlap at the detector, producing interference fringes that depend on the absolute position of the beam with respect to the crystal lattice, and may be analyzed according to the theory of ptychography for hard X-rays [36].

8.4.2 *Intensity Variation Methods*

Many phasing methods rely on an ability to record diffraction patterns before and after changing the scattering strength of just one species in a crystal, at a known site. In SIR/MIR, these are the heavy atom replacements; in MAD, the scattering factor is changed by varying the X-ray photon energy to alter the scattering potential. A related technique known as "radiation-induced-phasing" (RIP) [37] utilizes site-specific alterations to protein structures that may be induced by photoabsorption of UV light (for example). For the X-ray FEL, it is possible to create dramatic changes in scattering factors through the ionization of heavy atoms in proteins, which are virtually stationary during femtosecond pulses, and to use this "high-intensity MAD" effect for phasing. In this case, the X-ray diffraction is described by a time integral of the diffracted intensities throughout the duration of the exposure, during which many stochastic ionizations take place. The diffraction of a single shot takes the approximate form

$$I(\mathbf{q}) = \int dt\, I_0(t) \left| \sum\nolimits_n f_n(t) e^{i\mathbf{q}\cdot\mathbf{r}_n} \right|^2 \tag{8.11}$$

where $I_0(t)$ is the time-dependent incident intensity, and $f_n(t)$ is the time-dependent atomic scattering factor of atom n located at position $\mathbf{r}_n$. The single-shot diffraction must in turn be averaged probabilistically over all of the many configurations that are possible as a result of sequential ionizations, which leads to a unique form of partially coherent diffraction. A detailed theoretical treatment is presented by Son et al. [38], which provides a generalization of the Karle–Hendrickson equations for the high-intensity "diffraction-during-ionization" regime. Son et al. provide detailed calculations for the case in which a single atomic species (Fe) is independently ionized and the remainder of the atoms are assumed to diffract normally without ionization. The four coefficients of the generalized Karle–Hendrickson equations are shown as a function of photon energy and pulse fluence nearby the Fe absorption edge, which reveals how the contrast in these coefficients can be enhanced with increasing fluence.

Although high-intensity MAD phasing has not yet been demonstrated, a selective reduction in scattering power of sulfur atoms (with 2.47 keV K-edge) in protein was demonstrated by Galli et al. [39] by recording diffraction patterns with low fluence and high fluence (to fully ionize the sulfur and so reduce its scattering power.) This was done using 6 keV X-rays, allowing for high-resolution data to be collected, unlike a SAD study at 2.47 keV, where resolution may be limited by the 0.5 nm X-ray wavelength. Applications to a Gd derivative of lysozyme and to Cathepsin B, utilizing local electronic damage, are described in Galli et al. [40] and [41] respectively.

8.4.3 Constraint Ratios and High-Solvent Crystals

It was appreciated at an early stage from Pauling's work on the mineral bixbyite ($(Mn,Fe)_2O_3$) that the solution of the phase problem in crystallography is not unique, as for the class of homometric structures [42]. These are different crystal structures with the same Patterson (autocorrelation) function that therefore produce the same diffracted Bragg intensities. In practice, unique solutions can be expected if the number of independent Fourier equations relating measured Bragg intensities to crystal density is at least equal to the number of unknown phases [43–45]. This can be achieved by surrounding the molecule by an equal volume of known density. The subject is reviewed by Millane and Lo [6], who also include the effects of non-crystallographic symmetry. Elser and Millane [9] considered continuous diffraction from an isolated molecule and defined a constraint ratio Ω as the ratio of the number of independent diffraction intensity samples to the number of real space electron density samples. As alluded to earlier in Sect. 8.2, the number of independent diffracted intensity samples is limited by the Shannon sampling theorem, which gives the critical spacing between intensity samples in Fourier space. At larger spacings, the autocorrelation of the object cannot be uniquely recovered, yet finer spacings do not yield further information. Elser and Millane [9] showed that the constraint ratio can be defined as

$$\Omega = A/(2U) \tag{8.12}$$

where U is the volume of the support (the region in real space occupied by the molecule) and A is the volume of the autocorrelation function of the molecule (the 3D Fourier transform of the diffracted intensities), thus Ω depends on molecular shape, since the region occupied by the autocorrelation function depends on the shape of the molecule [46]. The factor of two in the denominator in the definition of Ω comes about because the autocorrelation function for a molecule is centrosymmetric, as described in Sect. 8.2. A unique solution to the phase problem requires $\Omega > 1$, being a necessary but not sufficient condition. The lower bound for Ω in

3D is 4, arising from support regions that are convex and centrosymmetric. Hence we see that the phase problem is highly over-constrained for continuous (diffuse) scattering that has been merged into a three-dimensional diffraction volume. These results can be related to the required angular sampling interval in diffraction experiments by recalling that, along one dimension, Shannon's theorem requires this continuous scattering to be sampled at intervals of $\Delta\theta = \lambda/L$ (in the small angle scattering approximation) for complete recovery of the density, where L, the width of the autocorrelation function, is twice the width of the molecule. Data is collected as 2D diffraction patterns, possibly affected by the curvature of the Ewald sphere so that they do not correspond to projections. However this data can be merged into a 3D volume to avoid issues associated with the curved sphere.

The analysis of the constraint ratio has been adapted to crystals by Millane and Arnal [47], where in the case of crystals, the autocorrelation function, about twice the size of the molecule in a unit cell, may overlap, so that A must be replaced by the volume V of the unit cell (the unique volume in the periodic Patterson map is half of this volume). The number of independent diffraction data and the number of electron density samples must also be reduced by the order of the space group. The previous equation then becomes

$$\Omega = R/(2f) \tag{8.13}$$

where $f = U/V$ is the fraction of the unit cell occupied by the molecule, V is the volume of the unit cell, and the crystal shows R-fold non-crystallographic symmetry. The constraint ratio for a crystal now depends only on the volume of the molecule, and not its shape. The phase problem may thus in principle be uniquely solved via Bragg intensities alone in cases where the unit cell contains more than 50% solvent in the absence of any non-crystallographic symmetry. In the presence of noise, values of Ω larger than unity are desirable.

Recent experiments using this method at 0.2 nm resolution suggest that a solvent content of greater than 65% is needed [48], however a lower fraction of solvent may be possible in crystals with larger values of R. By applying the Fienup HIO algorithm to crystals with high solvent contents, He and Su [49] have successfully phased a number of protein crystals. He et al. [50] have combined this approach with the Molecular Replacement method and applied it to three datasets.

8.4.4 Two-Dimensional Crystals

The weak scattering power of organic monolayers and two-dimensional crystals makes it extremely difficult to record diffraction patterns in transmission from these structures at synchrotrons, despite the importance of membranes in structural biology. They have been extensively studied by transmission electron diffraction, where imaging solves the phase problem. Here compact support along the beam direction can provide a useful constraint for iterative phasing [51]. Because a much higher dose can be applied without damage affecting the measured diffraction when using femtosecond X-ray FEL pulses, experimental patterns have now been published from 2D crystals by this method [52]. These patterns, from streptavidin and bacteriorhodopsin, were obtained without cryogenic cooling, and extend to about 0.8 nm resolution (although the data are measured to the edge of the detector). A full diffraction dataset requires the collection of a tilt series (crystal rotated about the normal to the beam), since reciprocal space for a 2D crystal consists of sharp rods running normal to the monolayer at each 2D lattice site. The phase problem in this case has been analyzed in terms of the constraint ratio formalism by Arnal and Millane [1] and Arnal et al. [53], who find that a smaller solvent content is needed than for 3D crystals in order to achieve uniqueness. They also discuss the helpful phasing effect of pores in a membrane, such as aquaporin 1 (AQP1) since the pores reduce the number of unknowns in real space while the volume of the autocorrelation of the unit cell is unchanged, keeping the amount of known Fourier space data constant. Once again, additional constraints, such as non-crystallographic symmetry and histogram matching of density map grey-levels, can assist.

8.4.5 Charge-Flipping and Atomic Resolution Data

A new iterative phasing algorithm for crystals appeared in 2004, in which the real-space operation consists of reversing the sign of the charge density at any pixel where it has a value less than some threshold δ (the only adjustable parameter in the algorithm) [54]. The algorithm starts with random phases satisfying Friedel's law. Experimental Fourier magnitudes are imposed in reciprocal space, and the phases are retained. It has been applied to many experimental datasets and modified for use with powder diffraction data, where it assists in resolving overlapping peaks and provides composition and space-group information [55]. The algorithm has been shown to be equivalent to an Output-Output algorithm in the Fienup scheme, with feedback parameter $\beta = -2$. For a review of the charge-flipping algorithm, see Oszlanyi and Suto [56].

The method requires atomic-resolution data so that, like Direct Methods, the space around atoms can be used as a support, since crystals consist mostly of vacuum. This once again relates to the idea that the phase problem is soluble from the Fourier intensities alone if the constraint ratio is greater than unity, which in this case arises from the diffraction data being of high enough resolution. As the quality and resolution of data from X-ray FELs continues to improve, we can expect this algorithm, and several others, which require atomic-resolution data, to become increasingly useful.

8.5 Conclusions

Imaging techniques based on X-ray FELs will continue to complement other techniques well into the future. X-ray FELs offer the advantage over synchrotrons and cryo-electron microscopy (cryo-EM) that samples can be studied under physiologically relevant temperatures without appreciable damage. These techniques may be extended to time-resolved "pump–probe" variants that enable biomolecular dynamics to be studied with an explicit time delay between the "pump" mechanism and the X-ray probe, with virtually no time-delay limit imposed by the X-ray source. This pump–probe method differs from cryo-EM studies of dynamics that are based on rapidly quenched equilibrium ensembles of molecules, whose images may subsequently be sorted by similarity, and which can provide an energy landscape [57] rather than time-resolved imaging. Recently we have seen the first such studies of gene expression from a virus by the single-particle X-ray FEL method [58].

Unlike the closely related technique of coherent diffractive imaging, conventional crystallographic imaging suffers from undersampled diffraction intensities. Conventional de novo phasing techniques have been applied to X-ray FEL data and the use of these techniques will continue to increase as analysis algorithms and software improve. Additionally, new opportunities for phasing X-ray diffraction data from both single particles and crystals have been enabled by the X-ray FEL. These new ideas, discussed above, show promise but most have not yet been applied to the solution of novel protein structures. This is partly because considerable experimental and theoretical obstacles still need to be overcome, but perhaps also because new techniques simply take time to catch on. With the increasing appearance of atomic-resolution XFEL data, the charge-flipping algorithm and its developments are expected to become more popular for de novo phasing since it does not require the use of a model or chemical modification to the sample. It will be interesting to see what kind of improvements may result from the use of large datasets from nanocrystals that are smaller than a single mosaic block, using coherent X-ray beams of a similar size.

Acknowledgments Supported by NSF STC BioXFEL award 1231306.

References

1. Millane, R. P. (1990). Phase retrieval in crystallography and optics. *Journal of the Optical Society of America A, 7*(3), 394–411.
2. Sayre, D. (1952). Some implications of a theorem due to Shannon. *Acta Crystallographica, 5*(6), 843–843.
3. Gerchberg, R. W., & Saxton, W. O. (1972). A practical algorithm for the determination of phase from image and diffraction plane pictures. *Optik (Stuttgart), 35*, 237–246.
4. Fienup, J. R. (1982). Phase retrieval algorithms: A comparison. *Applied Optics, 21*(15), 2758–2769.

5. Marchesini, S. (2007). A unified evaluation of iterative projection algorithms for phase retrieval. *The Review of Scientific Instruments, 78*, 011301.
6. Millane, R. P., & Lo, V. L. (2013). Iterative projection algorithms in protein crystallography. *Acta Crystallographica Section A: Foundations and Advances, A69*, 517.
7. Spence, J. C. H. (2017b). In P. Hawkes & J. C. H. Spence (Eds.), *Science of microscopy* (pp. 1196–1227). New York: Springer.
8. Bragg, L., & Perutz, M. F. (1952). The structure of haemoglobin. *Proceedings of the Royal Society of London A: Mathematical, Physical and Engineering Sciences, 213*(1115), 425–435 The Royal Society.
9. Elser, V., & Millane, R. P. (2008). Reconstruction of an object from its symmetry-averaged diffraction pattern. *Acta Crystallographica Section A: Foundations of Crystallography, 64*(2), 273–279.
10. Schlichting, I. (2017). Experimental phasing of serial femtosecond crystallography data. *IUCrJ, 4*, 517–517.
11. Thibault, P., & Elser, V. (2010). X-ray diffraction microscopy. *Annual Review of Condensed Matter Physics, 1*, 237–255.
12. Rupp, B. (2010). *Biomolecular crystallography: Principles, practice, and application to structural biology*. New York: Garland Science.
13. Hendrickson, W. A. (2014). Anomalous diffraction in crystallographic phase evaluation. *Quarterly Reviews of Biophysics, 47*(1), 49–93. https://doi.org/10.1017/S0033583514000018.
14. Spence, J. C. H. (2017). XFELs for structure and dynamics in biology. *IUCrJ, 4*, 322–339.
15. Barends, T. R., Foucar, L., Shoeman, R. L., Bari, S., Epp, S. W., Hartmann, R., et al. (2013a). Anomalous signal from S atoms in protein crystallographic data from an X-ray free-electron laser. *Acta Crystallographica Section A: Foundations and Advances, D69*, 838–842.
16. Nass, K., Meinhart, A., Barends, T. R., Foucar, L., Gorel, A., Aquila, A., et al. (2016). Protein structure determination by single-wavelength anomalous diffraction phasing of X-ray free-electron laser data. *IUCrJ, 3*(3), 180–191.
17. Yamashita, K., Pan, D., Okuda, T., Sugahara, M., Kodan, A., Yamaguchi, T., et al. (2015). An isomorphous replacement method for efficient de novo phasing for serial femtosecond crystallography. *Scientific Reports, 5*, 14017.
18. Yamashita, K., Kuwabara, N., Nakane, T., et al. (2017). Experimental phase determination with selenomethionine or mercury-derivatization in serial femtosecond crystallography. *IUCrJ, 4*, 639–647.
19. Hunter, M. S., Yoon, C. H., DeMirci, H., Sierra, R. G., Dao, E. H., Ahmadi, R., et al. (2016). Selenium single-wavelength anomalous diffraction de novo phasing using an X-ray-free electron laser. *Nature Communications, 7*, 13388.
20. Colletier, J.-P., Sawaya, M. R., Gingery, M., et al. (2016). De novo phasing with X-ray laser reveals mosquito larvicide BinAB structure. *Nature, 539*, 43–47. https://doi.org/10.1038/nature19825.
21. Barends, T. R. M., et al. (2013b). De novo protein crystal structure determination from X-ray free-electron laser data. *Nature, 505*, 244–247.
22. Nakane, T., Song, C., Suzuki, M., Nango, E., Kobayashi, J., Masuda, T., et al. (2015). Native sulfur/chlorine SAD phasing for serial femtosecond crystallography. *Acta Crystallographica Section D: Biological Crystallography, 71*(12), 2519–2525.
23. Batyuk, A., Galli, L., Ishchenko, A., Han, G. W., Gati, C., Popov, P. A., et al. (2016). Native phasing of x-ray free-electron laser data for a G protein–coupled receptor. *Science Advances, 2*(9), e1600292.
24. Gorel, A., Motomura, K., Fukuzawa, H., Doak, R. B., Grünbein, M. L., Hilpert, M., et al. (2017). Multi-wavelength anomalous diffraction de novo phasing using a two-colour X-ray free-electron laser with wide tunability. *Nature Communications, 8*, 1170. https://doi.org/10.1038/s41467-017-00754-7.
25. Shapiro, D. A., Chapman, H. N., DePonte, D., Doak, R. B., Fromme, P., Hembree, G., et al. (2008). Powder diffraction from a continuous microjet of submicrometer protein crystals. *Journal of Synchrotron Radiation, 15*(6), 593–599.

26. Spence, J. C., & Doak, R. B. (2004). Single molecule diffraction. *Physical Review Letters, 92*(19), 198102.
27. Kirian, R. A., Wang, X., Weierstall, U., et al. (2010). Femtosecond protein nanocrystallography - data analysis methods. *Optics Express, 18*(6), 5713–5723.
28. Perutz, M. F. (1954). The structure of haemoglobin. III. Direct determination of the molecular transform. *Proceedings of the Royal Society of London A: Mathematical, Physical and Engineering Sciences, 225*(1161), 264–286 The Royal Society.
29. Chapman, H. N., Fromme, P., Barty, A., et al. (2011). Femtosecond X-ray protein nanocrystallography. *Nature, 470*, 73–77.
30. Spence, J. C. H., Kirian, R. A., Wang, X. Y., Weierstall, U., Schmidt, K. E., White, T., et al. (2011). Phasing of coherent femtosecond X-ray diffraction from size-varying nanocrystals. *Optics Express, 19*(4), 2866–2873. https://doi.org/10.1364/Oe.19.002866.
31. Chen, J. P. J., Arnal, R. D., Morgan, A. J., Bean, R. J., Beyerlein, K. R., Chapman, H. N., et al. (2016). Reconstruction of an object from diffraction intensities averaged over multiple object clusters. *Journal of Optics, 18*(11), 114003.
32. Chen, J. P. J., & Millane, R. P. (2014). Diffraction by nanocrystals II. *Journal of the Optical Society of America A, 31*(8), 1730–1737.
33. Elser, V. (2013). Direct phasing of nanocrystal diffraction. *Acta Crystallographica Section A: Foundations of Crystallography, 69*(6), 559–569.
34. Kirian, R. A., Bean, R. J., Beyerlein, K. R., Yefanov, O. M., White, T. A., Barty, A., et al. (2014). Phasing coherently illuminated nanocrystals bounded by partial unit cells. *Philosophical Transactions of the Royal Society B, 369*(1647), 20130331.
35. Kirian, R. A., Bean, R. J., Beyerlein, K. R., Barthelmess, M., Yoon, C. H., Wang, F. L., et al. (2015). Direct phasing of finite crystals illuminated with a free-electron laser. *Physical Review X, 5*(1), 011015.
36. Spence, J. C. H., Zatsepin, N. A., & Li, C. (2014). Coherent convergent-beam time-resolved X-ray diffraction. *Philosophical Transactions of the Royal Society B: Biological Sciences, 369*(1647), 20130325.
37. Ravelli, R., Leiros, H., Pan, B. C., Caffrey, M., & McSweeney, S. (2003). Specific radiation damage can be used to solve macromolecular crystal structures. *Structure/Folding and Design, 11*(2), 217–224.
38. Son, S. K., Chapman, H. N., & Santra, R. (2011). Multiwavelength anomalous diffraction at high X-ray intensity. *Physical Review Letters, 107*(21), 218102.
39. Galli, L., Son, S. K., Klinge, M., Bajt, S., Barty, A., Bean, R., et al. (2015a). Electronic damage in S atoms in a native protein crystal induced by an intense X-ray free-electron laser pulse. *Structural Dynamics, 2*(4), 041703.
40. Galli, L., Son, S. K., Barends, T. R., White, T. A., Barty, A., Botha, S., et al. (2015b). Towards phasing using high X-ray intensity. *IUCrJ, 2*(6), 627–634.
41. Galli, L., Son, S. K., White, T. A., Santra, R., Chapman, H. N., & Nanao, M. H. (2015c). Towards RIP using free-electron laser SFX data. *Journal of Synchrotron Radiation, 22*(2), 249–255.
42. Hosemann, R., & Bagchi, S. N. (1954). On homometric structures. *Acta Cryst, 7*, 237–241.
43. Bricogne, G. (1974). Geometric sources of redundancy in intensity data and their use for phase determination. *Acta Crystallographica Section A: Foundations and Advances, A30*, 395–405.
44. Crowther, R. A. (1969). The use of non-crystallographic symmetry for phase determination. *Acta Cryst, B25*, 2571–2580.
45. Miao, J., Sayre, D., & Chapman, H. N. (1998). Phase retrieval from the magnitude of the Fourier transforms of nonperiodic objects. *Journal of the Optical Society of America A, 15*(6), 1662–1669.
46. Thibault, P. (2007). Algorithmic methods in diffraction microscopy. PhD Thesis, Cornell University.
47. Millane, R. P., & Arnal, R. D. (2015). Uniqueness of the macromolecular crystallographic phase problem. *Acta Crystallographica Section A: Foundations and Advances, A71*, 592–598.

48. Liu, Z.-C., Xu, R., & Dong, Y.-H. (2012). Phase retrieval in protein crystallography. *Acta Crystallographica Section A: Foundations of Crystallography, A68*, 256–265.
49. He, H., & Su, W. P. (2015). Direct phasing of protein crystals with high solvent content. *Acta Crystallographica Section A, Foundations and Advances, 71*(Pt 1), 92–98.
50. He, H., Fang, H., Miller, M. D., Phillips Jr., G. N., & Su, W. P. (2016). Improving the efficiency of molecular replacement by utilizing a new iterative transform phasing algorithm. *Acta Crystallographica Section A: Foundations and Advances, 72*(5), 539–547.
51. Spence, J. C. H., Weierstall, U., Fricke, T. T., Glaeser, R. M., & Downing, K. (2003). *Journal of Structural Biology, 144*, 209–218.
52. Frank, M., Carlson, D. B., Hunter, M. S., Williams, G. J., Messerschmidt, M., Zatsepin, N. A., et al. (2014). Femtosecond X-ray diffraction from two-dimensional protein crystals. *IUCrJ, 1*(2), 95–100.
53. Arnal, R. D., & Millane, R. P. (2017). The phase problem for two-dimensional crystals. I. Theory. *Acta Crystallographica Section A: Foundations and Advances, 73*, 438–448.
54. Oszlanyi, G., & Suto, A. (2004). Ab initio structure solution by charge flipping. *Acta Crystallographica Section A: Foundations and Advances, A60*, 134.
55. Wu, J., Leinenweber, K., Spence, J. C., & O'Keeffe, M. (2006). Ab initio phasing of X-ray powder diffraction patterns by charge flipping. *Nature Materials, 5*(8), 647–652.
56. Oszlanyi, G., & Suto, A. (2008). The charge flipping algorithm. *Acta Crystallographica Section A: Foundations of Crystallography, A64*, 123–134.
57. Dashti, A., Schwander, P., Langlois, R., Fung, R., Li, W., Hosseinizadeh, A., et al. (2014). Trajectories of the ribosome as a Brownian nanomachine. *PNAS, 111*(49), 17492–17497.
58. Hosseinizadeh, A., Mashayekhi, G., Copperman, J., et al. (2017). Conformational landscape of a virus by single-particle X-ray scattering. *Nature Methods, 14*, 877–881.

Chapter 9
Structure Determination by Continuous Diffraction from Imperfect Crystals

Kartik Ayyer, Oleksandr M. Yefanov, and Henry N. Chapman

9.1 Introduction

The far-field diffraction pattern of a finite and isolated object is continuous, unlike that of an ideal perfect crystal which consists of discrete spots, called Bragg peaks. The arrangement of macromolecules in a crystal lattice provides an enormous amplification of the diffraction signal in these peaks over that of a single molecule. This has been the key strategy for structure determination using X-rays from the earliest days, since it makes possible the measurement of a diffraction signal above background scattering within the very low exposure that can be tolerated before radiation damage modifies the structure under investigation. However, measurements confined only to Bragg peaks represent but a fraction of the information about a structure that could possibly be acquired in a diffraction experiment. This information loss, due to the fact that the Bragg peaks are too sparse to measure the entire content of the object's Fourier spectrum, usually prevents the possibility to derive the diffraction phases that are needed to synthesize the electron density from that spectrum. This so-called phase problem is the usual state of affairs in crystallography, requiring additional measurements such as multiple wavelength anomalous diffraction or isomorphous replacement to provide the needed missing information. The problem can also be overcome if extremely high resolution

K. Ayyer · O. M. Yefanov
Center for Free-Electron Laser Science, DESY, Hamburg, Germany
e-mail: kartik.ayyer@desy.de; oleksandr.yefanov@desy.de

H. N. Chapman (✉)
Center for Free-Electron Laser Science, DESY, Hamburg, Germany

Department of Physics, University of Hamburg, Hamburg, Germany

Centre for Ultrafast Imaging, University of Hamburg, Hamburg, Germany
e-mail: henry.chapman@desy.de

S. Boutet et al. (eds.), *X-ray Free Electron Lasers*,
https://doi.org/10.1007/978-3-030-00551-1_9

is available, such that the information content of the measurement exceeds the information needed to describe all atomic degrees of freedom (the positions of the atoms and their amplitudes of vibration). By contrast the measurement of the fully sampled continuous diffraction pattern of an isolated non-periodic object does not suffer from the phase problem since there are generally more independent measurements in the diffraction intensities than needed to describe the object, independent of resolution. For this reason, and because it is often difficult to produce highly ordered crystals of macromolecules, many approaches have been considered that may give access to the continuous diffraction.

Since there are more crystallographers than those who carry out coherent diffractive imaging of non-periodic objects, we have assumed that most who read this chapter are more familiar with the concepts of coherent diffraction as applied to crystalline systems. We therefore introduce, in Sect. 9.2, coherent diffraction generally, our nomenclature and conventions, and examine some insights about the information content of diffraction patterns and phase retrieval that have been developed in the field somewhat independently of crystallography. In Sect. 9.3, various situations are described where high-resolution continuous diffraction of macromolecules can be observed, many of which have long fascinated crystallographers and diffraction microscopists alike. Given the recent success of utilizing the continuous diffraction of translationally disordered photosystem II crystals for structure determination [1], summarized in Sect. 9.4.8, we devote Sect. 9.4 to such cases and show that the continuous diffraction of rigid units is remarkably resilient to different forms of disorder and correlation that might occur in such crystals, especially if one considers the structural information that can be extracted from the autocorrelation function (the generalized Patterson function [10], equal to the Fourier transform of the diffraction intensities). The exposition laid out in this section might inspire those working in coherent diffractive imaging to apply concepts such as partial-coherent diffraction analysis to macromolecular structure determination. Finally, in Sect. 9.5, we describe procedures and lessons learned to accurately measure macromolecular continuous diffraction, especially using X-ray free-electron lasers, which is somewhat more challenging to do than for Bragg peaks.

9.2 Coherent Diffraction of a General Object

Consider a diffraction experiment where a collimated quasi-monochromatic X-ray beam is elastically scattered by an object with a three-dimensional electron density distribution $\rho(\mathbf{r})$. The incident beam, of wavelength λ, can be described by a wave-vector $\mathbf{k_{in}}$ pointing in the direction of beam propagation with $|\mathbf{k_{in}}| = 1/\lambda$. The diffraction intensities are recorded in the far field on a pixelated detector. A particular detector pixel records radiation travelling from the object with a wave-vector $\mathbf{k_{out}}$ as shown in Fig. 9.1a. In the Born approximation, in which multiple scattering is neglected, the counts measured in this pixel are given by

$$I(\mathbf{q}) = I_0\, P\, \Omega_p\, r_e^2 \left| \int \rho(\mathbf{r}) \exp(2\pi i \mathbf{r} \cdot \mathbf{q})\, d\mathbf{r} \right|^2 , \tag{9.1}$$

where $\mathbf{q} = \mathbf{k}_{\text{out}} - \mathbf{k}_{\text{in}}$ is the photon momentum transfer vector, I_0 is the incident fluence (number of photons per unit area within the exposure time of the measurement), P the polarization factor, Ω_p the solid angle subtended by the pixel from the sample, and r_e the classical radius of the electron.

Equation (9.1) can be explained in terms of a ray description of scattering and considering the density $\rho(\mathbf{r})$ to be given by the sum of point scatterers of strength f_i located at positions $\mathbf{r}_i$: $\rho(\mathbf{r}) = \sum_i f_i\, \delta(\mathbf{r} - \mathbf{r}_i)$ (Fig. 9.1). A ray scattered in a direction $\mathbf{k}_{\text{out}}$ from atom 1 will acquire a path difference of $\ell_1 = (\mathbf{r}_1 \cdot \hat{\mathbf{k}}_{\text{out}} - \mathbf{r}_1 \cdot \hat{\mathbf{k}}_{\text{in}})$ relative to a ray scattering from the origin $\mathbf{O}$, where $\hat{\mathbf{k}}$ are unit vectors. This is the difference of the lengths of the thick lines in Fig. 9.1a. The accumulated phase, relative to this arbitrary origin, will therefore be $\phi_1 = (2\pi/\lambda)\ell_1 = 2\pi \mathbf{r}_1 \cdot \mathbf{q}$. The point scatterer itself may cause a modification to the scattered wave by the complex value f_1, giving a scattering amplitude $f_1 \exp(i\phi_1) = f_1 \exp(2\pi i \mathbf{r}_1 \cdot \mathbf{q})$. Equation (9.1) is simply the coherent summation of all scattered waves as obtained by integrating over all point scatterers in the object. The measured distribution of counts depends strongly on the phases ϕ of the scattered waves (and thus on the three-dimensional arrangement of scatterers), since these may lead to complete constructive or destructive interference, or something in between.

Equation (9.1) is further identified as the square modulus of the Fourier transform of the electron density of the object. The strength of the diffraction in a given direction $\mathbf{k}_{\text{out}}$ only depends on the Fourier component $\tilde{\rho}(\mathbf{q})$ where we use the definition of the Fourier transform, for any integrable function g, as

$$\tilde{g}(\mathbf{q}) \equiv \mathscr{F}_q\{g(\mathbf{x})\} \equiv \int g(\mathbf{x}) \exp(2\pi i \mathbf{x} \cdot \mathbf{q})\, d\mathbf{x}. \tag{9.2}$$

This component is a particular spatial frequency in the object, which may be thought of as a volume grating of a particular wavenumber $2\pi|\mathbf{q}|$ and direction $\hat{\mathbf{q}}$. From Fig. 9.1b, it is seen that the magnitude of $\mathbf{q}$ is given simply by

$$|\mathbf{q}| = 2|\mathbf{k}| \sin\theta = \frac{2}{\lambda} \sin\theta \tag{9.3}$$

for a scattering angle 2θ, and that due to the conservation of $|\mathbf{k}|$ (that is, elastic scattering) the vector $\mathbf{q}$ lies on the surface of a sphere (called the Ewald sphere). We see from the diagram in Fig. 9.1b that the scattered ray appears to reflect at an angle θ from a plane normal to $\mathbf{q}$. That is, the ray reflects from the volume grating which is tilted at the angle θ relative to the incoming wave-vector. The ray only reflects if the period of the volume grating, $d = 1/|\mathbf{q}|$, satisfies Eq. (9.3), which is to say $d = \lambda/(2\sin\theta)$ which is well recognized as Bragg's law.

For a given orientation of the object, there are only a subset of spatial frequencies (volume gratings) that can be observed by the diffraction measurement. These are

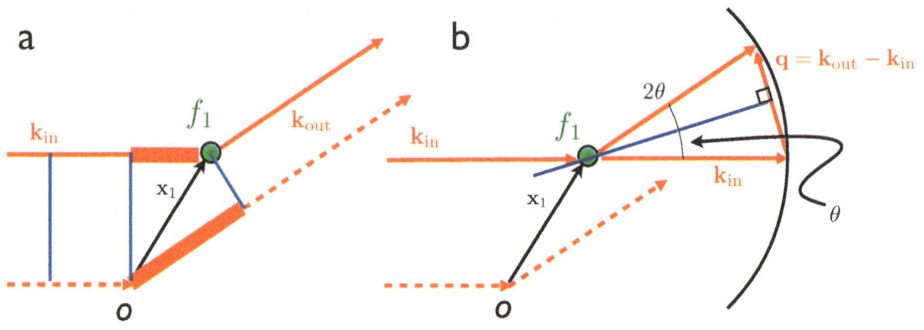

Fig. 9.1 (**a**) Far-field scattering can be described in terms of paths of rays scattered from atoms. (**b**) The Ewald-sphere construction

the frequencies that have the right periods d to obey the Bragg condition for given scattering angle 2θ and which happen to be in the right orientation for this to occur (by "reflection"). These are described exactly by the vectors $\mathbf{q}$ that lie on the Ewald sphere. In order to measure other spatial frequencies, not present on the Ewald sphere, the object must be rotated to bring these into the reflecting condition. It should be stressed that although Bragg's law and the Ewald sphere construction are well-known concepts in crystallography, there is no requirement of periodicity of the object in the derivation or application of these concepts. Diffraction from a crystal is a special case (a periodic object), as discussed next.

9.2.1 *Diffraction of a Periodic Object*

A crystal can be thought of as a special case of the general object (and thus crystallography as a special case of coherent diffractive imaging!). The electron density of an ideal finite crystal can be described as a sum of the unit cell contents convolved with a periodic lattice:

$$\rho(\mathbf{r}) = \sum_{b=1}^{N_b} \rho_b(R_b\mathbf{r} - \mathbf{a}_b) \otimes \sum_{c=1}^{N_c} \delta(\mathbf{r} - \mathbf{a}_c) \,, \tag{9.4}$$

where $\rho_b(\mathbf{r})$ is the asymmetric unit or rigid body which occurs N_b times in each unit cell in positions and orientations given by $\mathbf{a}_b$ and R_b, respectively (relative to an arbitrary cell origin), and $\mathbf{a}_c$ are the positions of all the N_c unit cells that make up the crystal.[1] These positions are usually periodic in all three dimensions. Below,

[1]Throughout this chapter we will use the indices abc to uniquely distinguish atom a of rigid body b in unit cell c. When it is not needed to report on which body or cell an atom is part of, we just use the index i.

we use the simplified notation of ρ_b representing each of the N_b differently oriented and translated asymmetric objects in the unit cell:

$$\rho(\mathbf{r}) = \sum_{b=1}^{N_b} \rho_b(\mathbf{r}) \otimes \sum_{c=1}^{N_c} \delta(\mathbf{r} - \mathbf{a}_c) \; . \tag{9.5}$$

The diffraction of a periodic object takes on a special form. Since the density is given by the convolution of the unit cell and the lattice, the Fourier transform $\tilde{\rho}(\mathbf{q})$ is given by the product of the Fourier transform of the unit cell with the Fourier transform of the lattice The diffraction pattern $|\tilde{\rho}(\mathbf{q})|^2$ is therefore also given by a product, where the diffraction intensities of the unit cell are modulated by Bragg peaks given by

$$L(\mathbf{q}) = \left| \mathscr{F}_q \left\{ \sum_{c=1}^{N_c} \delta(\mathbf{r} - \mathbf{a}_c) \right\} \right|^2 \tag{9.6}$$

$$= \left| \sum_{c=1}^{N_c} \exp(2\pi i \mathbf{a}_c \cdot \mathbf{q}) \right|^2 \; . \tag{9.7}$$

As the number of unit cells tends to infinity, $L(\mathbf{q})$ approaches a sum of delta functions, the reciprocal lattice, with spacing inversely proportional to the real-space lattice spacing. The existence of peaks can be easily explained by considering a crystal of symmetry P1 (i.e., with no additional symmetry) with unit cell dimensions in all directions of w. The electron density of this crystal is given by Eq. (9.4) with $N_b = 1$, $\rho(\mathbf{r}) = \sum_c \rho_b(\mathbf{r} - \mathbf{a}_c)$. The diffraction from each object in the crystal is coherently added, as described in Eq. (9.1). Since the phase of the diffracted wave of an object varies with its displacement $\mathbf{a}_c$ as $2\pi \mathbf{a}_c \cdot \mathbf{q}$, the diffraction from the various cells c in the crystal will usually destructively interfere, since their relative phases would tend to uniformly take on values from 0 to 2π. The exception is in directions where constructive interference occurs, which is when $\mathbf{a}_c \cdot \mathbf{q}$ forms whole numbers, or at values of q spaced by $\Delta q_B = 1/w$ for a unit cell spacing w. For a crystalline sample, one can therefore only make measurements of the unit cell transform at this minimum spacing of Δq_B. As discussed below, this limits the information content of the diffraction pattern.

The idealized situation of an infinite crystal and delta function Bragg peaks is not realized in practice. For a coherently illuminated finite crystal (such as discussed in Sect. 9.3.3) Bragg peaks have a width given by the transform of the shape of the crystal, with widths inversely proportional to the crystal width and a total integrated value proportional to the number of cells in the crystal. More usually, such as for measurements at synchrotron radiation beamlines, the transverse coherence length of the X-ray beam is smaller than the crystal size, and is what determines the width of peaks. The partially coherent diffraction pattern in this case can be well approximated by the convolution of the coherent diffraction $I(\mathbf{q})$ of Eq. (9.1) with the angular extent of the source (as seen by the sample). At such beamlines this angular extent is usually governed by the divergence of the beam, and not surprisingly this has the effect of matching the width of Bragg peaks to this angular extent. There also may be a length scale of the crystal over which strict periodicity persists, discussed in Sect. 9.4, again giving rise to a convolution of intensities. The convolution of $I(\mathbf{q})$ with a correlation function or coherence function of the incident radiation, $\Gamma(\mathbf{q})$, is to modulate the autocorrelation function of the object (described below in Sect. 9.2.4) with the Fourier transform $\tilde{\Gamma}$. As will be seen in this chapter, knowledge of the precise form of the correlation function over length scales of many unit cells is not necessarily required to determine the structures of the molecular constituents of crystals.

9.2.2 *Diffraction from an Atomic Object*

At the high spatial resolution that can be accessed by X-ray wavelengths, the electron density can be described in terms of the constituent atoms of the sample. Since the electron density is highest around the nuclei rather than the bonds, this density can be accurately modelled as a sum over all atoms as

$$\rho(\mathbf{r}) = \sum_{i=1}^{N} \zeta_i(\mathbf{r} - \mathbf{r}_i)\,, \tag{9.8}$$

where $\zeta_i(\mathbf{r})$ is the density of the ith atom, and there are N atoms in the entire object. The coherent diffracted intensity from this collection of atoms is therefore

$$I(\mathbf{q}) = \left|\sum_i f_i(\mathbf{q}) \exp(2\pi i \mathbf{r}_i \cdot \mathbf{q})\right|^2 = \sum_{ii'} f_i(\mathbf{q}) f_{i'}^*(\mathbf{q}) \exp(2\pi i (\mathbf{r}_i - \mathbf{r}_{i'}) \cdot \mathbf{q})\,, \tag{9.9}$$

where $*$ is the complex conjugate and we have dropped the pre-factors in Eq. (9.1). The structure factors $f(\mathbf{q})$, equal to the Fourier transform of the atom density $\tilde{\zeta}(\mathbf{q})$ of each atom, can be modelled as a sum of Gaussian functions [54] but are often considered to be constant (due to point-like atoms).

9.2.3 *Information Content of Diffraction Data*

The three-dimensional (3D) map of the electron density $\rho(\mathbf{r})$ of any general object can be synthesized from its Fourier amplitudes through an inverse Fourier transform. This is simply a coherent sum of all the volume gratings in real space that combine to make up the object,

$$m(\mathbf{r}) = \mathscr{F}_r^{-1}\{\tilde{\rho}(\mathbf{q})\} \equiv \int \tilde{\rho}(\mathbf{q}) \exp(-2\pi i \mathbf{q} \cdot \mathbf{r})\, d\mathbf{q}\,. \tag{9.10}$$

Each period and orientation of these gratings must be summed not only with the correct strength (or amplitude $|\tilde{\rho}(\mathbf{q})|$) but also with the correct shift (or phase $\arg\{\tilde{\rho}(\mathbf{q})\}$) with respect to other frequencies. While the modulus of the Fourier amplitudes can be obtained from the square root of the measured diffraction intensities, $\sqrt{I}(\mathbf{q})$, the phases are missing.

This so-called phase problem is perhaps one of the most studied inverse problems in science, and can be generally overcome from complete measurements of the independent diffraction intensities (except in some pathological cases) in two and three dimensions [2] (see Sect. 9.2.4). Unfortunately, the arrangement of objects in crystal lattices does not allow the required complete measurements to be made. For the simple example of a single molecule in a crystal of P1 symmetry, the Bragg peaks only provide half of the possible independent diffraction measurements that can be made in each direction, or an under-representation by a factor of 8 for a three-dimensional object, as is described below. This spells the difference between the feasibility or infeasibility of recovering the electron density directly from intensity measurements when no other information about the object is available.

9.2.4 Shannon Sampling and the Constraint Ratio

The diffraction pattern of a coherently illuminated finite object is "band limited," which is to say that the modulation of the diffraction intensity as a function of scattering angle θ or momentum transfer $\mathbf{q}$ has a certain minimum modulation period. This smallest period is inversely proportional to the width of the object. This is true even for diffraction of crystals, where the finest features in the pattern are the Bragg peaks themselves. As mentioned above, the width of a Bragg peak is inversely proportional to the width of the entire crystal, or at least the width that is coherently illuminated.

The frequency content of a diffraction pattern can be examined through Fourier analysis, by taking the Fourier transform of the diffraction intensities $I(\mathbf{q})$. That is

$$\begin{aligned} \tilde{I}(\mathbf{u}) \propto \mathscr{F}_u^{-1}\left\{|\tilde{\rho}(\mathbf{q})|^2\right\} &= \rho(\mathbf{r}) \otimes_u \rho^*(-\mathbf{r}) \\ &= \int \rho(\mathbf{r})\rho^*(\mathbf{r}-\mathbf{u})\, d\mathbf{r} \\ &\equiv A_\rho(\mathbf{u}). \end{aligned} \tag{9.11}$$

This is the autocorrelation function of the object, a 3D map of all pair correlations of points within the object, and is a function of the real-space difference vector $\mathbf{u}$. This function is zero for all $\mathbf{u}$ that are larger than the maximum separation of any two points in the object. For an object that has a largest width w, for example, the autocorrelation function extends from the origin by w, as well as by $-w$. Its extent is thus $2w$. Since the diffraction pattern $I(\mathbf{q})$ is a Fourier transform of the autocorrelation function, we see that the pattern is band limited with a minimum period equal to $1/w$. In essence, this means that the pattern is smooth at that reciprocal length scale. This can be verified from the Fourier transform of two delta functions spaced apart by $2w$:

$$\mathscr{F}_q^{-1}\{\delta(r-w)+\delta(r+w)\} = \exp(-2\pi i w q) + \exp(2\pi i w q) = 2\cos(2\pi w q)\,. \tag{9.12}$$

Shannon's theorem [43] states that a band-limited function can be completely specified from discrete samples of that function as long as there are at least two samples per smallest period. Thus, the diffraction pattern discussed above can be completely measured with samples spaced no more than $\Delta q = 1/(2w)$ apart. Measuring samples more finely than this does not increase our knowledge of the diffraction pattern, and so this defines the information content of the pattern. It specifies the quantity of independent measurements that can be made of the diffraction intensities. (In practice, a finer sampling than this may help overcome an effective decrease in coherence due to the finite pixel width, or to effectively increase dynamic range and signal to noise.) If the diffraction pattern is measured to a maximum resolution $q_{\max}$, or a range from $-q_{\max}$ to $q_{\max}$, then $N_S = 2q_{\max}/\Delta q = 4wq_{\max}$ samples across the diffraction pattern completely define it. Expressing the resolution as $q_{\max} = 1/d$, we find $N_S = 4w/d$. In two dimensions, there are thus N_S^2 independent measurements possible for an object whose extent fits in a square of width w, and N_S^3 for the corresponding case in three dimensions. Here we have assumed a complex-valued object, which gives a diffraction pattern that has no symmetry. If the object is real valued, then the diffraction pattern is centrosymmetric and the number of independent measurements is reduced by half.

How much information is needed to fully specify the complex-valued electron density $\rho(\mathbf{r})$ to a specific resolution? This time the minimum period to be considered is that of the finest volume grating that makes up the map $m(\mathbf{r})$ at resolution $q_{\max} = 1/d$, which is d. This modulation must be sampled at a spacing no larger than $d/2$. Since the object has an extent of w, then at least $2w/d$ samples are required in each dimension. The number of "unknowns" in the object is two per independent sample, considering that the density is complex-valued, and so the total is given by $2\,(2w/d)^n$ for an n-dimensional object, compared with $(4w/d)^n$ possible independent diffraction intensity measurements. Thus, for such an object (fitting within a cube of width w), the potential excess of measurements to unknowns is given by the ratio $\Omega = 2^{n-1}$ [16, 30]. That is, for a one-dimensional object there are an equal number of measurements to unknowns, two times as many in two dimensions, and a four times excess in three dimensions.

Now coming back to the crystal of symmetry P1 with unit cell dimensions in all directions of w, measurements of the intensity can only be made at the Bragg peaks which are spaced apart by $\Delta q_B = 1/w$ for a unit cell spacing w. These are twice as far apart as the minimum Shannon spacing $\Delta q_S = 1/(2w)$ required to completely measure the intensities from one of the objects on its own. That is, in this case of P1 symmetry, the Bragg peaks undersample the single-object diffraction pattern by a factor of two in each dimension, leading to $\Omega = 2^{n-1}/2^n = 1/2$.

The ratio Ω of the excess of measured intensities to those required to describe the object has been termed the *constraint ratio* [16]. Obviously, the recovery of the structure of the object from the measurements alone requires $\Omega > 1$. This condition is sufficient for reconstruction in the case of zero noise [2, 5], except in certain pathological cases where several different structures give rise to the same diffraction pattern. These cases are so-called homometric structures, defined as those with the same sets of interatomic distances (the same autocorrelation function, and hence the same diffraction pattern). Trivial examples are mirror images, but a homometric pair of objects can be constructed from the convolution of two non-centrosymmetric structures with one of the structures either being inverted or in its original position. The handedness of alpha helices in proteins means that such cases will not exist for macromolecules. Experience shows on the whole that the larger Ω, the easier it is to directly recover the electron density map.

Elser and Millane [16] have pointed out that since the diffraction intensities are equally represented by their Fourier transform, Ω is equal to the number of independent coefficients in the autocorrelation function divided by the number of independent object coefficients. Since the autocorrelation function of any complex-valued object ρ is Hermitian with $A_\rho^*(-\mathbf{u}) = A_\rho(\mathbf{u})$, the number of independent coefficients is equal to half the non-zero area (or volume) of the autocorrelation function divided by the area (or volume) of a resolution element in two (or three) dimensions. Since the resolution element is the same for the autocorrelation function and the object, Ω is equal to the ratio of areas or volumes of the autocorrelation and object, divided by two. For shapes other than cuboid discussed above, this may differ from 2^{n-1}. For example, triangular objects have $\Omega = 3$, making these structures potentially easier to determine directly from the diffraction observations. Crystal diffraction need not only result in $\Omega = 1/2$ as described above. If the volume of the asymmetric unit in the crystal is less than half the volume of the unit cell, there may indeed be sufficient measurements to determine the structure directly from Bragg intensities [22, 32].

The utilization of the excess of measured intensities to determine the structure presumes a shape of the constraint region to be used. This need not exactly conform to the actual boundary of the object, but must fully contain the object. The constraint that is applied in the process of phase retrieval is that the density is known outside this constraint volume (for example, it may be uniform or zero), consistent with the premise that the information required to describe the object is finite. Allowing this constraint volume to exceed the actual extent of the object reduces Ω, but may avoid applying an incorrect constraint. Prior knowledge about the shape of the object may therefore be helpful, which may indeed be available from microscopy or solution scattering. However, it is possible in many cases to determine the shape of an object from the shape of its autocorrelation function [12]—that is from the diffraction intensities themselves. Another strategy is to gradually improve the estimate of the constraint volume (known as the "support" of the object) based on the image obtained by phase retrieval based on a previous larger estimate [29]. This approach, known as "shrinkwrap," has been extremely successful because a more constraining tighter support produces an improved estimate of the image, which itself provides the means, by a simple threshold and blurring, to obtain an improved support constraint.

9.2.5 Iterative Phasing Algorithms

The feasibility of phasing sufficiently sampled diffraction data, as discussed in the previous section, has led to a vibrant field of research in applied mathematics to create phasing algorithms. This situation and activity has been pursued quite separate to developments in crystallography, where refinement of models constrained by the rules of chemistry or the use of anomalous diffraction are common approaches to solve structures. For continuous diffraction, the measured diffraction and the support constraint alone are sufficient to determine a 3D map of the electron density, without the need for a chemical model. Additional constraints that can be added, such as the positivity of electron density (if appropriate) or a presumed histogram of the electron density (generally known for protein structures for particular spatial resolutions), will improve the ability to phase the diffraction data and may make the solution more robust in the presence of noise. Recently, some of the ideas from phase retrieval of continuous diffraction have been successfully applied to crystal diffraction, including charge flipping [38] and the hybrid input–output algorithm [22, 23].

Much analysis of iterative phasing algorithms has been carried out in the context of images or maps of electron density $m(\mathbf{r})$ as finite-dimensional vectors. A particular map is represented by a point in an N-dimensional space, with the value along each coordinate given by the complex value at each of the N voxels. Out of all possible maps that can be formed, only a particular volume of the vector space will contain maps that obey a particular constraint, such as all maps that have Fourier amplitudes equal to the square root of the measured diffraction intensities. A different volume contains all maps that have zero density outside the support. The intersection of these volumes gives the solution—a map that obeys both sets of constraints. One possible strategy would be to exhaustively calculate images in an N-dimensional sphere of the vector space whose radius is limited by the maximum total intensity of the map, and test if they are in one or both constraint sets. One would need only compute maps within the volume of the support constraint (those that are zero outside the support boundary) and test if the Fourier amplitudes of the map $|\tilde{m}(\mathbf{r})|$ agree with the square root of the measured diffraction intensities, $\sqrt{I(\mathbf{r})}$. Such an approach is obviously too computationally expensive for maps with more than a few voxels. A tractable approach would be, starting from a trial point in the vector space, to calculate the next map in a direction that minimizes the error $\epsilon_M = \| |\tilde{m}| - \sqrt{I} \|$, where $\| \cdot \|$ is the Euclidean distance equal to the square root of the sum of the squares of the vector components. This can be easily achieved simply by setting the magnitudes $|\tilde{m}|$ equal to $\sqrt{I}$ at each reciprocal voxel $\mathbf{q}$. However, such a step will tend to move m out of the support constraint, increasing the corresponding error ϵ_S, so a correction will be needed to place the map back in that constraint space with the consequence of increasing ϵ_M from zero.

The error ϵ_M is the distance of the map m to the modulus constraint set. The error ϵ_S, equal to the intensity outside the support, is also the distance of the map (a point in the vector space) to the support constraint set. By iterating the steps indicated above of bringing the point to first the modulus constraint set and then to the support constraint set, it may be possible to eventually converge to the intersection point that we seek. This indeed would be the case if the sets were convex volumes and if the point m was always brought to its closest point in each constraint set. The latter condition is accomplished using a projection operator. That is, an updated point m_{i+1} is obtained from the current estimate m_i as $m_{i+1} = P_S P_M m_i$, where P_S is the projection that brings a point in the vector space onto the support constraint set, and P_M is the operator that brings a point onto the set of images that obey the Fourier modulus constraint. The repeated application of these operations approaches a fixed point $m^* \rightarrow (P_S P_M)^n m$. This fixed point will be the global minimum of the distances between the sets if the former condition of the sets being convex is satisfied. However, the modulus constraint set is decidedly not convex, and so the procedure may become trapped in a local minimum. Nevertheless, this formalism of projections has proven valuable in developing robust algorithms that can recover the phases even when given noisy measured Fourier intensities.

The projection operators can be easily constructed for the constraints mentioned above. The support projector P_S simply sets the values of all voxels outside the support to zero, which is the closest point in the vector space that satisfies the constraint. For the modulus constraint, $\tilde{m}$ is brought into agreement with $\sqrt{I}$ by rescaling the modulus of the complex value at each reciprocal voxel to equal $\sqrt{I}$, leaving the phase unchanged. In the complex plane, for a particular reciprocal voxel $\mathbf{q}$, this is the closest point to $\tilde{m}(\mathbf{q})$ on the circle of radius $\sqrt{I(\mathbf{q})}$. The modulus projection includes performing the Fourier transformation of the real-space map, and the inverse after rescaling:

$$P_M\, m(\mathbf{r}) = \mathscr{F}_r^{-1} \left\{ \sqrt{\frac{I(\mathbf{q})}{|\tilde{m}(\mathbf{q})|^2}}\, \tilde{m}(\mathbf{q}) \right\} . \tag{9.13}$$

The constraint errors can be seen to be equal to $\epsilon_S = \| P_S m - m \|$ and $\epsilon_M = \| P_M m - m \|$. In the latter case the error ϵ_M is invariant to the Fourier transform (through Parseval's theorem).

The algorithm $m_n = (P_S P_M)^n m_0$ was introduced by Feinup [18] as a generalization of the first iterative phasing approach of Gerchberg and Saxton [21].

They considered the related problem of recovering the complex-valued image from the measured transmission image and measured diffraction pattern in an electron microscope. Fienup's introduction of the support constraint brought the possibility of phasing diffraction data alone (without the need of a microscope). He called it the error reduction algorithm since the errors ϵ are non-increasing on each iteration step due to its equivalence to a steepest descent minimization [18]. However, due to the non-convexity of the constraint sets, it does not necessarily achieve the global minimum—often this simple algorithm gets trapped in a local minimum. This algorithm can be compared with density modification and solvent flattening that are used in crystal structure refinement, albeit without any structural model guiding the density in the volume inside the support. Perhaps a better analogy for the crystallographer is that this is an omit map where the entire molecule is omitted!

Since the error reduction algorithm often stagnates, Fienup introduced concepts from control theory to design algorithms with "feedback" that improved their convergence properties. Elser expanded on these ideas with his difference map algorithm [14] which actively explores space away from the constraint sets in order to avoid stagnation. He first noted that an algorithm constructed as

$$m_n = (\mathbf{I} + \beta\Delta)^n\, m_0 \tag{9.14}$$

converges to a fixed point given by $\Delta\, m^* = 0$ for any constant β and operator Δ. To ensure that this point is in the intersection of the two sets, Δ must take the form

$$\Delta\, m = (P_M f_S - P_S f_M)\, m \tag{9.15}$$

where f_S and f_M are any linear combination of operators for the support and modulus constraint, respectively. The key here is that the last operation in each term of Eq. (9.15) is P_M and P_S, respectively, taking m to the surface of one or the other constraint set, and giving zero when these intersect. The other operators f_S and f_M can be designed to give optimum convergence properties, which Elser finds to be

$$f_S\, m = [I + \alpha_S(P_S - I)]\, m \tag{9.16}$$

$$f_M\, m = [I + \alpha_M(P_M - I)]\, m\,. \tag{9.17}$$

The real-valued parameter α tunes these operators from the identity at $\alpha = 0$ to a projection with $\alpha = 1$ and a reflector (such as used in charge flipping) when $\alpha = 2$. Some particular choices are $\alpha_S = 0$ with $\alpha_M = 2$ [44], or $\alpha_S = 0$ with $\alpha_M = 1 + 1/\beta$. The latter is Fienup's hybrid input–output (HIO) algorithm [17]. The difference map algorithm of Eq. (9.14), like the HIO algorithm, tends to escape from local minima, sometimes by moving in a direction along the line of shortest approach between the two sets at the local minimum [28]. The solution is not the fixed point m^* to which the algorithm converges, but rather the nearest point on the constraint set, $P_M\, m^*$.

9.2.6 Phasing Twinned Data

As will be seen below, continuous diffraction often arises from ensembles of objects that are situated in several discrete orientations, without correlation of the relative positions of those objects. Examples are molecules aligned in a laser field (which may be oriented parallel or antiparallel to the polarization axis of the laser with equal probability, Sect. 9.3.1) or the four orientations of molecules in a crystal with $P2_12_12_1$ symmetry and exhibiting displacement disorder (Sect. 9.3.7). This results in an incoherent sum of the diffraction intensities in N_b various orientations, $I(\mathbf{q}) = \sum_{b=1}^{N_b} |\tilde{\rho}_b(\mathbf{q})|^2$, analogous to the diffraction from a twinned crystal. Such an incoherent sum cannot be represented by the square modulus of the Fourier transform of any single object (including the average over all orientations) and thus the application of the phasing algorithms described above will not succeed. Assuming that each of the ρ_b are differently oriented versions of the same rigid object as described in Eq. (9.4), the modulus constraint P_M of Eq. (9.13) must be modified simply as [16]

$$P_M\, m_1(\mathbf{r}) = \mathscr{F}_r^{-1}\left\{\sqrt{\frac{I(\mathbf{q})}{\sum_b |\tilde{m}_b(\mathbf{q})|^2}}\,\tilde{m}_1(\mathbf{q})\right\}, \tag{9.18}$$

where m_1 is the iterate of the single object reconstruction and m_b are the rotated versions, $m_b(\mathbf{r}) = m_1(\mathbf{R}_b\mathbf{r})$.

As emphasized by Elser and Millane [16], phasing the twinned data is feasible as long as the constraint ratio $\Omega > 1$. Twinning will always reduce Ω from the ratio Ω_1 for the single object, but it will only be as small as Ω_1/N_b for N_b orientations if the support is invariant under the rotation operations $\mathbf{R}_b$. Generally it will lay between these bounds as can be understood by constructing the union of the rotated supports of the object and determining the volume of the unique region as related by the symmetry operations. A molecule contained within a square prism support, for example, gives rise to a support autocorrelation that is also a square prism. If the length of this is greater than the width of the square face and is not oriented along a symmetry axis of the crystal itself, then the symmetrized support autocorrelation has some non-overlapping regions, with the possibility that Ω is considerably greater than unity.

9.3 Observations of Continuous Diffraction

Given that the lack of measured phases does not pose an obstacle to reconstructing an image of the electron density from a set of sufficiently sampled diffraction intensities, it is worth reviewing under which conditions the required continuous diffraction can be measured for the purpose of iterative phasing.

9.3.1 Single Object and a Gas of Aligned Identical Objects

The most obvious case for iterative phasing is that of a single non-periodic object illuminated with a coherent beam. The first demonstration of a fully sampled coherent diffraction pattern measured in the X-ray regime was by Miao et al. [31] after many years of effort led by David Sayre [41]. That the pattern was sufficiently sampled was proven by the fact that its Fourier transform gives an autocorrelation of limited extent. The real test of sufficiency, however, was that it could indeed be phased, by using an iterative phasing algorithm. The object was two-dimensional, fabricated by electron-beam lithography, and diffraction in only one view was needed to obtain the two-dimensional image. Such phasing is much more robust in three dimensions, for which diffraction must be measured also in three dimensions by rotating the sample. This is similar to data collection from a crystal, although still diffraction data frames are recorded as the object is rotated in steps. Each data frame, a measurement of $I(\mathbf{q})$ on the Ewald sphere, is then interpolated into a three-dimensional array, as illustrated in Fig. 9.2. In this example [7], the object consisted of an indent in a silicon nitride membrane that contained a number of colloidal gold particles. The silicon nitride was practically invisible to the X-rays due to its low scattering power, giving a compact object whose diffraction could be phased directly using the Shrinkwrap algorithm [29].

Diffractive imaging of single objects requires a high dose to the sample, which may exceed the tolerable dose to avoid damage at a particular resolution [25]. All methods to observe continuous diffraction at near atomic resolution from biological material therefore require measuring diffraction from many identical objects. One way to work around the dose limitation is by using femtosecond pulses from free-electron lasers, as described in the previous chapters in this book. Since the pulses are destructive, most schemes only give a two-dimensional snapshot diffraction pattern and so a supply of reproducible objects is still required to obtain a complete 3D diffraction dataset, as discussed in Chap. 14. Continuous diffraction can be

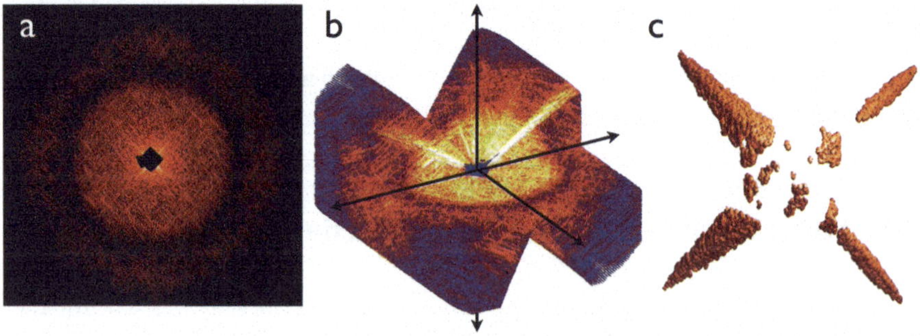

Fig. 9.2 Diffraction data collected from a 3D test object, showing (**a**) a diffraction pattern recorded at a single orientation, (**b**) 3D diffraction intensities collected at orientations from −70° to +70°, and (**c**) the reconstructed volume image. Adapted from [7]

readily combined from many objects if they share a common orientation. The diffraction will be the incoherent sum of the individual objects, equal to that of a single object multiplied by the number of objects, if the diffraction from these is mutually incoherent or the positions of the objects are random (as will be discussed below). For example, a gas of laser-aligned molecules [24] would give a diffraction pattern proportional to the single molecule, as would a long exposure made of a stream of aligned objects that sequentially pass across the beam [48]. Without alignment, enough signal is required per pattern to determine relative orientations of the particles, as detailed in Chap. 14.

9.3.2 Single Layers and Fibrils

Objects often come into alignment when placed in contact with each other. The most spectacular self-assembly of this kind is of course crystallization, but other examples include liquid crystals in the nematic phase where constituents are aligned in one of their dimensions. Such arrangements might give useful information about their cylindrically averaged density, for example. An early example of applying iterative phasing to such partially oriented systems was to biological membranes containing proteins. The arrangement of the proteins was disordered within the plane and with their orientations fixed only in the direction normal to the plane, but this gives a well-defined density thickness profile of the membrane. Such membranes can be layered, giving rise to a single column of Bragg peaks in the ordered direction, or a continuous rod of diffraction intensity, depending on the regularity of the spacings of the layers. In the latter case the Fourier transform of the intensity rod gives the entire autocorrelation of the thickness profile of the membrane. Stroud and Agard introduced the idea to phase this with a compact support [50] although later understanding as discussed in Sect. 9.2.4 showed that the density profile is not uniquely specified by 1D Fourier magnitudes without additional information [2]. Spence et al. [47] showed this could be overcome in 2D crystals where the diffraction is in the form of a lattice of intensity rods. An analogous case is a one-dimensional crystal, such as a single fibril that is periodic along its axis. The diffraction from this consists of two-dimensional planes of continuous diffraction separated by the reciprocal lattice spacing. The information content of the diffraction is certainly higher than for two-dimensional crystal, although Millane showed that the phase problem is not generally unique without an additional constraint such as the positivity of the electron density [33].

9.3.3 Finite Crystals and Finitely Illuminated Crystals

Deviations from an ideal infinite crystal can in principle give access to information additional to that restricted to the Bragg peaks. The diffraction of a finite crystal was considered by Laue [53] who derived the result that the diffracted wavefield is equal to the convolution of the Fourier transform of the shape of the coherently illuminated crystal with the delta-function Bragg peaks of the infinite crystal. That is, the 3D crystal "shape transform" is laid down on each lattice point. For a crystal with flat facets, this transform consists of continuous truncation rods in directions normal to the facets, giving an opportunity to measure the underlying molecular transform at locations away from the Bragg peaks. Elser suggested that this could be used to obtain, in addition to intensities at Bragg peaks $I(\mathbf{q}_{hkl})$, measurements of the gradient of the intensities, $\nabla I(\mathbf{q}_{hkl})$. The extra three independent values per Bragg peak increase the constraint ratio Ω and were shown to be enough to solve the structure by iterative phasing [15]. Spence et al. [49] extended this idea to an ensemble of crystals measured by serial crystallography, as described in Chap. 8, by dividing out the shape transform from the 3D diffraction intensity map, allowing the electron density of the unit cell to be recovered by iterative phasing. A similar effect can be obtained when a small focused beam illuminates the crystals. If the relative position of the beam and crystal is known on each shot, the dataset can be interpreted by the method of ptychography [27]. However, if the diffraction intensities are summed without regard to this relative displacement, the result is merely a convolution of the diffraction pattern that would usually be observed (with a collimated beam) with the angular distribution of the focused probe. This convolution operation can also be carried mathematically on data recorded with a collimated beam, or by using a detector with large pixels, and thus does not bring any new information. The sum described above also simulates the situation of illumination of a crystal with a beam of limited spatial coherence, showing that in that case no new information is revealed either, despite hope that this might mask the periodicity of the crystal and hence provide continuous diffraction of a single unit cell [52].

9.3.4 *Crystal Swelling*

In the early days of protein crystallography it was noted that the unit cells of some crystals expand or contract in different states of hydration. Bernal et al. suggested that measurements of such crystals could be used to map out the molecular transform with a fine sampling [3]. This is easy to imagine for a crystal of P1 symmetry, where a change in a unit cell dimension made just by changing the distance between molecules will cause Bragg peaks to move over the transform of the single molecule. The measurement of diffraction from crystals in several states would allow mapping the molecular transform in steps along a direction set by the swelling (e.g., in the 111 direction for a crystal that uniformly swells in all directions). For the P1 crystal, this "one dimensional" fine sampling should be enough to provide complete information for phase retrieval, at least to the resolution that the molecules remain identical (and in the same orientation) in the different crystal forms.

The situation is not so straightforward when the arrangements of molecules change upon swelling in crystals of other symmetries. Bragg and Perutz carried out measurements of a set of hemoglobin crystals with differing amounts of salt content [4]. In that case the crystals underwent a shear through a change in the angle β, without any significant change in the unit cell lengths. This could happen if whole layers of molecules in ab planes would slip relative to each other in the direction of the b axis. This then allowed for fine sampling of layer lines of the diffraction pattern in the c^* direction. However, only in the $00l$ direction could the ensemble of measurements be easily interpreted. The $0kl$ central plane of the diffraction pattern is the Fourier transform of the projection of the crystal structure down the a axis, and since molecules were slipping only in the b direction, this projection would be unchanged along c^*. Even in simpler cases of expansion without shear, the change of displacements of molecules *within* a unit cell means that even though fine samples can be made from different crystal forms, each of these samples possess a different unit cell transform. So far, a method to apply iterative phasing to such a dataset has not been found, but it is clear that the information content is twice that of a single crystal, which should allow de novo phasing. Structure refinement from multiple crystal forms can be carried out using density modification techniques, as in the programs *phenix.multi_crystal_average* and *DMMulti* [11].

9.3.5 Crystals with Large Solvent Fraction

As discussed in Sect. 9.2.4, iterative phasing should be feasible when the constraint ratio Ω exceeds unity, regardless of whether the measurements are from a continuous pattern or from Bragg spots. This ratio is equal to the number of independent coefficients in the autocorrelation function divided by the number of independent coefficients describing the object's density. For a P1 crystal (without any disorder) this will be half, since the autocorrelation function has the same periodicity as the crystal and it is centrosymmetric. However, if the object only actually accounts for less than half the volume of the unit cell, and the rest consists of solvent of uniform average density, then we will have $\Omega > 1$. Furthermore, if there are two or more identical objects in the unit cell that are not related by crystallographic symmetry, Ω can exceed unity even for solvent fractions smaller than 50% [34]. Iterative phasing can indeed succeed in this case, using only the Bragg intensities, as was recently demonstrated by He and Su [22] using Fienup's hybrid input–output algorithm ($\alpha_S = 0$ with $\alpha_M = 1 + 1/\beta$ in Eqs. (9.16) and (9.17)).

9.3.6 Crystals with Substitutional Disorder

Bragg peaks are a consequence of translational symmetry. Any deviation from that symmetry will disturb the constructive interference responsible for the peaks, reducing their intensities, and will also prevent the full cancellation of intensity between the peaks. One way to break this symmetry and still maintain molecular orientation is through substitutional disorder in the crystal. That is, a random occupation of lattice sites by a molecule or, more likely, sites randomly occupied by one or the other of two forms of a molecule can be described as the sum of a purely periodic density (given by the average structure $\bar{\rho}(\mathbf{r})$) and the difference $\Delta\rho(\mathbf{r})$ [10].

As an example consider a time-resolved experiment where an optical excitation pulse is set to a fluence where only half the molecules are isomerized. This will occur randomly throughout the crystal volume, and so the crystal can be considered as randomly occupied by two molecular structures, one in the ground state with a structure $\rho_1(\mathbf{r})$ and one with a structure $\rho_2(\mathbf{r})$. Considering for simplicity a P1 symmetry with just one molecule per unit cell, the density of this imperfect crystal can be described by a modification of Eq. (9.4) as

$$\begin{aligned}\rho(\mathbf{r}) &= \bar{\rho}(\mathbf{r}) + \Delta\rho(\mathbf{r})\\ &= \left[\bar{\rho}_b(\mathbf{r}) \otimes \sum_c \delta(\mathbf{r}-\mathbf{r}_c)\right] + \left[\Delta\rho_b(\mathbf{r}) \otimes \sum_c p_c\, \delta(\mathbf{r}-\mathbf{r}_c)\right], \qquad (9.19)\end{aligned}$$

where p_c is equal to either $+1$ or -1 in the case of 50% excitation. The diffraction intensity is equal to the Fourier transform of the autocorrelation of the density $\rho(\mathbf{r})$ as shown in Eq. (9.11). The autocorrelation of the sum Eq. (9.19) gives rise to four terms, given by the autocorrelation of $\bar{\rho}(\mathbf{r})$, the autocorrelation of $\Delta\rho(\mathbf{r})$, and two cross correlation terms, $\bar{\rho}(\mathbf{r}) \otimes \Delta\rho(-\mathbf{r}) + \bar{\rho}(-\mathbf{r}) \otimes \Delta\rho(\mathbf{r})$. Since $\bar{\rho}$ is periodic, each of these cross correlations will essentially carry out a sum of $\Delta\rho$ over all unit cells which will be equal to zero since (by definition) $\langle\Delta\rho\rangle = 0$. The autocorrelation is therefore a sum of just the autocorrelations of $\bar{\rho}(\mathbf{r})$ and $\Delta\rho(\mathbf{r})$, showing that the diffraction pattern is the incoherent sum $\left|\tilde{\bar{\rho}}(\mathbf{q})\right|^2 + |\Delta\tilde{\rho}(\mathbf{q})|^2$, where $\tilde{\bar{\rho}}$ is the Fourier transform of $\bar{\rho}$ and $\Delta\tilde{\rho}$ is the Fourier transform of $\Delta\rho$. Since $\bar{\rho}(\mathbf{r})$ is strictly periodic, the first term will give rise to Bragg peaks. The residual density $\Delta\rho(\mathbf{r})$ consists of components that are at lattice positions but differ from each other in a random manner and hence are not periodic. Again, considering the autocorrelation of $\Delta\rho(\mathbf{r})$ (see Eq. (9.11)), for small differences $\mathbf{u}$ one obtains the sum of cross correlations of difference densities within common unit cells. If the occupancies p_c are uncorrelated, once $\mathbf{u}$ crosses unit cell boundaries the terms $p_c p_{c'}$ cancel out, leaving just the autocorrelation of the single "object" $\Delta\rho_b$. The Fourier transform of this correlation of limited extent is of course continuous. In general, for a fraction x of excited molecules in the crystal, the continuous diffraction will be weighted by $x(1-x)$ [10] giving

$$I(\mathbf{q}) = \left|\tilde{\bar{\rho}}_b(\mathbf{q}) \sum_c \exp(2\pi i \mathbf{r}_c \cdot \mathbf{q})\right|^2 + x(1-x)\,|\Delta\tilde{\rho}_b(\mathbf{q})|^2\,. \qquad (9.20)$$

For a crystal consisting of several objects per unit cell (following Eq. (9.5)), Eq. (9.20) generalizes to

$$I(\mathbf{q}) = \left| \sum_b \tilde{\bar{\rho}}_b(\mathbf{q}) L(\mathbf{q}) \right|^2 + x(1-x) \sum_b |\Delta\tilde{\rho}_b(\mathbf{q})|^2 . \qquad (9.21)$$

The second term can be separated from the first by filtering out Bragg peaks (see Sect. 9.5) and then iteratively phased using a finite support constraint, and the modulus constraint of either Eq. (9.13) or (9.18) depending on the number of orientations of molecules in the crystal [51]. Since $\Delta\rho$ can be negative it is not appropriate to apply a positivity constraint.

Substitutional disorder exists in crystals of tris-*t*-butyl-benzene tricarboxamide [45]. This molecule crystallizes into a so-called two-component crystal with random occupation of one or the other component. In this case the two components consist of the molecule in one of two different orientations. These actually occur in columns, parallel to the c axis, of molecules of the same orientation. Looking down this axis one observes columns in a hexagonal close-packed lattice either pointing away or towards the observer. Although the occupational fraction x is 50%, there is a correlation between the positions of "up" and "down" columns due to a preference of antialignment of neighbors but a frustration in achieving this in a triangular lattice. This is revealed in the observed continuous diffraction as a characteristic honeycomb shape. The form of this correlation has been of interest to understand the solid state of the molecule, and the correlation function could be obtained by dividing the effect of the molecular contribution $\Delta\rho$ from the pattern [42] in a process somewhat similar to described in Sect. 9.3.3. (Eq. (9.20) only considers uncorrelated occupation.) However, in a beautiful analysis, Simonov et al. did the opposite to extract $|\Delta\tilde{\rho}(\mathbf{q})|^2$ which they then phased to obtain an atomic resolution image of the molecule [46].

9.3.7 Crystals with Displacement Disorder

The least amount of change that needs to be made to a crystal to disrupt translational symmetry is to randomly displace its elements. As with substitutional disorder (Sect. 9.3.6), if the mean displacement of the objects in the crystal is zero, then the imperfect crystal can be described as a repeat of an average unit cell $\bar{\rho}$ with strict translational symmetry and a difference term. As before, the correlations between the differences and the average sum to zero (since, by definition the mean of the difference is zero), so that the general form of the autocorrelation (and hence the diffraction intensities) is an incoherent sum of the periodic part, which gives Bragg peaks, and the non-periodic difference which gives continuous diffraction. However, as compared with substitutional disorder, the average unit cell is blurred out. We can assume small normally distributed displacements, which leads to a blurring given by the convolution of the unit cell density with a Gaussian. The effect of this convolution is to modulate the Bragg peaks by the well-known Debye-Waller factor $\exp(-4\pi^2\sigma^2q^2)$, for a root-mean square displacement σ.

The form of the continuous (diffuse) scattering depends on the object undergoing the displacement, and the nature of correlations of those displacements over the volume of the illuminated crystal. Below we make a general derivation of the diffraction intensities for a crystal consisting of randomly displaced molecules which themselves have randomly displaced atoms, with different correlations between them. A particularly favorable condition is when whole molecules move as rigid units. Again, choosing for simplicity the case of P1 symmetry with just one molecule per unit cell the density of the imperfect crystal is given by

$$\rho(\mathbf{r}) = \rho_b(\mathbf{r}) \otimes \sum_{c=1}^{N_c} \delta(\mathbf{r} - \mathbf{r}_c - \boldsymbol{\delta}_c) , \tag{9.22}$$

where δ_c is the displacement of the molecule in the cth unit cell, with $\langle \boldsymbol{\delta}_c \rangle = 0$ and $\langle \delta_c^2 \rangle = \sigma^2$. The autocorrelation of Eq. (9.22) will be equal to the autocorrelation of $\rho_b(\mathbf{r})$ convolved with the autocorrelation of the displaced lattice. This is the cross correlation of two blurred lattices (each with RMS displacements σ^2). Since we assume here that there is no correlation between the displacements of different unit cells the displacements $\mathbf{u}$ between the cells will be spread by a distribution of variance equal to $2\sigma^2$, for all peaks in this autocorrelation except for the lattice point at the origin (which has perfect correlation). The result, derived below in Sect. 9.4.2, is

$$I(\mathbf{q}) = |\tilde{\rho}(\mathbf{q})|^2 \left[N_c(1 - e^{-4\pi^2\sigma^2q^2}) + e^{-4\pi^2\sigma^2q^2}\, L(\mathbf{q}) \right] . \tag{9.23}$$

The continuous part of the diffraction pattern consists of the single-object diffraction modulated by the so-called complementary Debye-Waller factor $(1-e^{-4\pi^2\sigma^2q^2})$ which is zero at $\mathbf{q} = 0$ and increases to 1 for resolution lengths $d = 1/q < \sigma/(2\pi)$. The continuous and Bragg portions of the diffraction pattern can be separated from each other as discussed in Sect. 9.5, but it is seen in Eq. (9.23) that both sample the same single-object diffraction pattern. This is only true for a P1 crystal however. In other cases, the continuous diffraction is twinned in a similar fashion to the case of substitutional disorder (see Sect. 9.4.4), requiring phasing to be carried out using the modulus constraint of Eq. (9.18). Phasing with the continuous diffraction alone misses low-resolution information due to the complementary Debye-Waller factor, as discussed in Sect. 9.4.8.

The complementarity of the Debye-Waller factors in the two terms within the square brackets of Eq. (9.23) shows that the strength of the continuous and Bragg diffraction is comparable. The integrated intensity of Bragg peaks (not including the molecular transform) scales as N_c, as does the continuous term. The difference between these cases is that the Bragg counts are concentrated into easily measurable peaks whereas the continuous diffraction is spread over all detector pixels. By energy conservation the total counts in the pattern will not change as σ is varied, although the resolution at which one or the other dominates will. At a resolution where the continuous diffraction dominates, the counts per pixel would be equivalent to the average Bragg counts per pixel at that resolution, had the crystal been perfectly ordered. In such a case, if Bragg peaks were spaced apart by 10 pixels on average, for example, then the average count is actually only 1% of the peak height. Such levels are usually below the background noise, as discussed in Sect. 9.5.

It is interesting to note that the loss of Bragg peaks with resolution, and the corresponding increase in the continuous diffraction, does not need large displacements. For example, RMS displacements of 1 Å give a significant loss of Bragg intensity at a resolution of about 6 Å as can be seen from the expression for the Debye-Waller factor and the definition of q in Eq. (9.3). This large discrepancy is due to the fact that Bragg peaks are formed through constructive interference and are thus a phase effect. At the example resolution of 6 Å, a displacement of any one object by $\mathbf{a}_c = 3$ Å would cause its contribution, as seen in Eq. (9.7), to be completely out of phase and adding destructively. Displacements of 1 Å will already give contributions out of phase enough to suppress the formation of Bragg peaks. This extreme sensitivity of Bragg peaks to displacements may explain to a large degree the difficulty of recording high-resolution diffraction from protein crystals, especially those where the molecules are not highly constrained through crystal contacts. However, we will see in the following section that this very sensitivity can expose the continuously sampled Fourier transform which opens up exciting possibilities for de novo structure determination from crystal diffraction using iterative phasing.

9.4 Diffraction from Crystals with Displacement Disorder

After the survey in Sect. 9.3 of the different types of crystal imperfections which give the opportunity to reveal additional structural information about the molecular components, not present in Bragg peaks, in this section we expand upon the analysis of crystals with displacement disorder. This topic has been discussed previously [35, 36, 39], but here we will focus on the connection with continuous diffraction of the objects that undergo the displacements. After introducing a general formalism, we consider different types of random displacements which may be correlated or uncorrelated. We examine the cases of translation of entire unit cells in Sect. 9.4.2, the case of displacements of molecules which themselves exhibit atomic disorder in Sect. 9.4.3, multiple types of rigid object that are independently displaced (Sect. 9.4.4), and then two models which include correlations in the displacements. These are the liquid-like motions model of distortions within rigid bodies examined in Sect. 9.4.5 and displacements of rigid bodies that are influenced by their neighbors in Sect. 9.4.6. Finally, we look at the effects of rotational

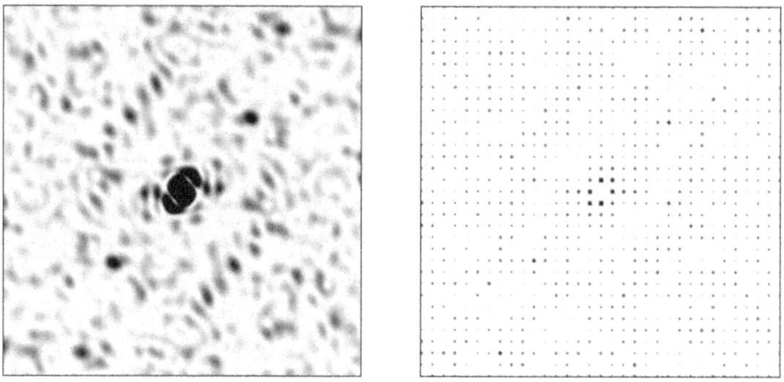

(a) (b)

Fig. 9.3 (**a**) Molecular transform intensities, $I_o(\mathbf{q})$, shown in a single planar slice passing through $q = 0$. (**b**) Ideal crystal intensities, $I_o(\mathbf{q}) \sum L_{cc'}(\mathbf{q})$ when there are no displacements. In both cases, for clarity, the atomic form factors have been set to constants (see Box 9.1)

rigid-body disorder in Sect. 9.4.7. Each of these cases is illustrated with a simulation of crystals of the lysozyme molecule as described in Box 9.1 and depicted in Fig. 9.3.

In order to give the most general formalism of a crystal with translational disorder we make use of the description of such an object as a collection of atoms, given by Eq. (9.8).

$$\rho(\mathbf{r}) = \sum_{i=1}^{N} \zeta_i (\mathbf{r} - \mathbf{r}_i) . \qquad \text{(9.8 revisited)}$$

We separate the position of each atom in the entire crystal into the sum of the position of the unit cell, the position of the atom within the unit cell, and the displacement from that ideal position,

$$\mathbf{r}_i = \mathbf{r}_c + \mathbf{r}_a + \boldsymbol{\delta}_{ac} .$$

Converting the sum of scattered waves from all atoms into a double sum over all unit cells and over all atoms within the unit cell, Eq. (9.9) becomes

$$\begin{aligned} I(\mathbf{q}) &= \sum_{cc'} e^{2\pi i (\mathbf{r}_c - \mathbf{r}_{c'})\cdot\mathbf{q}} \sum_{aa'} f_a(\mathbf{q}) f^*_{a'}(\mathbf{q}) e^{2\pi i (\mathbf{r}_a - \mathbf{r}_{a'})\cdot\mathbf{q}} e^{2\pi i (\boldsymbol{\delta}_{ac} - \boldsymbol{\delta}_{a'c'})\cdot\mathbf{q}} \\ &= \sum_{cc'} \mathrm{L}_{cc'}(\mathbf{q}) \sum_{aa'} \mathrm{I}_{aa'}(\mathbf{q}) e^{2\pi i (\boldsymbol{\delta}_{ac} - \boldsymbol{\delta}_{a'c'})\cdot\mathbf{q}} \end{aligned} \qquad (9.24)$$

where $\mathrm{L}_{cc'}$ is the lattice sum of the crystal which converges to a set of delta functions at the reciprocal lattice points [35, 36]. $\mathrm{I}_{aa'}$ is the contribution of each pair of atoms in their ideal positions (within one unit cell). The term $\sum_{aa'} \mathrm{I}_{aa'}(\mathbf{q}) =$

$|\sum_a f_a(\mathbf{q})|^2 = I_o(\mathbf{q})$ is commonly called the *molecular transform* of the molecule (Fig. 9.3a) since it is the Fourier transform of the molecular electron density function. Note that we have made no assumption as to the form of correlations between atoms, rigid bodies, or unit cells, just that they are statistically random.

Since the displacements are statistically random (though not necessarily uncorrelated), one is really interested in the average phase contribution from all the atoms. Thus, we can rewrite the expression for the intensity as follows:

$$I(\mathbf{q}) = \sum_{cc'} \mathrm{L}_{cc'}(\mathbf{q}) \sum_{aa'} \mathrm{I}_{aa'}(\mathbf{q}) \left\langle e^{2\pi i(\boldsymbol{\delta}_{ac}-\boldsymbol{\delta}_{a'c'})\cdot\mathbf{q}} \right\rangle . \tag{9.25}$$

Using the harmonic approximation for small displacements, the average over the exponentials can be simplified as

$$\left\langle e^{2\pi i(\boldsymbol{\delta}_{ac}-\boldsymbol{\delta}_{a'c'})\cdot\mathbf{q}} \right\rangle = e^{-2\pi^2\left\langle((\boldsymbol{\delta}_{ac}-\boldsymbol{\delta}_{a'c'})\cdot\mathbf{q})^2\right\rangle} .$$

With this simplification, we get the following general expression for the intensity distribution of a disordered crystal:

$$I(\mathbf{q}) = \sum_{cc'} \mathrm{L}_{cc'}(\mathbf{q}) \sum_{aa'} \mathrm{I}_{aa'}(\mathbf{q}) e^{-2\pi^2(\langle(\boldsymbol{\delta}_{ac}\cdot\mathbf{q})^2\rangle+\langle(\boldsymbol{\delta}_{a'c'}\cdot\mathbf{q})^2\rangle)} e^{4\pi^2\langle(\boldsymbol{\delta}_{ac}\cdot\mathbf{q})(\boldsymbol{\delta}_{a'c'}\cdot\mathbf{q})\rangle} . \tag{9.26}$$

The non-trivial behavior is now encoded in the last term which represents the covariance of the displacements among different atoms. This can be seen more clearly with the following:

$$\langle(\boldsymbol{\delta}_{ac}\cdot\mathbf{q})(\boldsymbol{\delta}_{a'c'}\cdot\mathbf{q})\rangle = \mathbf{q}^\mathsf{T} \left\langle \boldsymbol{\delta}_{ac}^\mathsf{T} \boldsymbol{\delta}_{a'c'} \right\rangle \mathbf{q} .$$

Here, the central term is the 3×3 covariance matrix of displacements between any pair of atoms, $\mathbf{C}_{ac\,a'c'}$.[2] One can also recognize the self terms $\left\langle(\boldsymbol{\delta}_{ac}\cdot\mathbf{q})^2\right\rangle$ as being the standard anisotropic B-factors (or Debye-Waller factors) for atom a, i.e.,

$$\left\langle(\boldsymbol{\delta}_{ac}\cdot\mathbf{q})^2\right\rangle = \mathbf{q}^\mathsf{T}\mathbf{U}_a\mathbf{q} .$$

Here we have made the assumption that the displacement distributions (not the actual displacements) for the same atom are identical among different unit cells. Putting it all together, we get

$$I(\mathbf{q}) = \sum_{cc'} \mathrm{L}_{cc'}(\mathbf{q}) \sum_{aa'} \mathrm{I}_{aa'}(\mathbf{q}) e^{-2\pi^2\mathbf{q}^\mathsf{T}(\mathbf{U}_a+\mathbf{U}_{a'})\mathbf{q}} e^{4\pi^2\mathbf{q}^\mathsf{T}\mathbf{C}_{ac\,a'c'}\mathbf{q}} . \tag{9.27}$$

---

[2] Although $\mathbf{C}_{ac\,a'c'}$ runs over four subscripts, this is really two dimensional, since any given atom in the crystal is specified by the indices ac (or $a'c'$) specifying which atom a in the molecule and which unit cell c in the crystal.

It is now possible to consider several kinds of disorder and examine how the (continuous) diffracted intensities relate to the molecular transform. The first, in Sect. 9.4.1, is the displacement of every atom in the crystal in an uncorrelated manner, such as may happen in a Coulomb explosion or simply due to thermal motion. This will be compared with the displacements of whole unit cells as single rigid units (Sect. 9.4.2), keeping displacements among different unit cells uncorrelated. For crystals with more than one asymmetric unit per cell, this choice is somewhat artificial since the choice of unit cell is arbitrary. Nevertheless, this will provide a simpler route to understanding what happens in the general case when there are multiple rigid units in the unit cell which are considered in Sect. 9.4.4 after first checking the effect of atomic disorder with rigid-body motion. Following that we investigate motions that are correlated with distance between atoms and units.

Box 9.1
The simulated diffraction images in this section were calculated using the lysozyme molecule (PDB: 4ET8). Each image represents a planar slice (not an Ewald sphere) through the 3D intensity distribution of the crystal. The resolution at the center edge is 2 Å. When showing Bragg peaks, the crystal unit cell was assumed to be 32 × 32 Å in the dimensions reciprocal to the displayed diffraction plane. This cell is too small to fit 4 molecules as demanded by the $P2_12_12_1$ space group simulated in Sect. 9.4.4. Nevertheless, the smaller unit cell leads to a larger Bragg peak spacing in reciprocal space, which results in a more esthetically pleasing image. In reality, the tetragonal lysozyme crystal has unit cell 79 × 79 × 38 Å (placing Bragg peaks closer together than simulated) with the space group $P4_32_12$.

Another simplification applied was to ignore the q-dependence of the atomic form factors $f_i(\mathbf{q})$ and consider them to be constant. This is equivalent to assuming point-like atoms. These simplifications from what one would see in a real experiment were made for the sake of clarity. Experimental details are discussed in Sect. 9.5.

9.4.1 Uncorrelated Random Disorder

For uncorrelated motions of atoms, the covariance matrix reduces to

$$\mathbf{C}_{aca'c'} = \left\langle \boldsymbol{\delta}_{ac}^{\mathsf{T}} \boldsymbol{\delta}_{a'c'} \right\rangle = \begin{cases} \mathbf{U}_a, & \text{if } c = c' \text{ and } a = a' \\ 0, & \text{otherwise.} \end{cases}$$

Separating the two cases of summing over identical or different atoms, from Eq. (9.27) we obtain

$$I(\mathbf{q}) = \sum_{\substack{cc' \\ c \neq c'}} \mathrm{L}_{cc'}(\mathbf{q}) \sum_{\substack{aa' \\ a \neq a'}} \mathrm{I}_{aa'}(\mathbf{q}) e^{-2\pi^2 \mathbf{q}^{\mathsf{T}}(\mathbf{U}_a + \mathbf{U}_{a'})\mathbf{q}} + \sum_{c} \mathrm{L}_{cc}(\mathbf{q}) \sum_{a} \mathrm{I}_{aa}(\mathbf{q})$$

where the two exponential terms cancel each other out in the second term. From Eq. (9.24), we can see that $\mathrm{L}_{cc}(\mathbf{q}) = 1$ and $\mathrm{I}_{aa}(\mathbf{q}) = |f_a(\mathbf{q})|^2$. Completing the sum in the first term, we obtain the familiar Debye-Waller suppression of Bragg intensities along with structureless diffuse scattering,

$$I(\mathbf{q}) = \sum_{cc'} \mathrm{L}_{cc'}(\mathbf{q}) \sum_{aa'} \mathrm{I}_{aa'}(\mathbf{q}) e^{-2\pi^2 \mathbf{q}^{\mathsf{T}}(\mathbf{U}_a + \mathbf{U}_{a'})\mathbf{q}} + N_c \sum_{a} |f_a(\mathbf{q})|^2 \left(1 - e^{-4\pi^2 \mathbf{q}^{\mathsf{T}} \mathbf{U}_a \mathbf{q}}\right). \tag{9.28}$$

In order to make the interpretation of this expression easier, let us assume isotropic displacement distributions. Thus, expressions of the form $\mathbf{q}^{\mathsf{T}}\mathbf{U}_a\mathbf{q}$ simplify to $\sigma_a^2 q^2$ where $q = |\mathbf{q}|$. One can always do the complete analysis with anisotropic distributions with some minor modifications. Applying this approximation and substituting in the full expression for $\mathrm{I}_{aa'}$ from Eq. (9.24),

$$I(\mathbf{q}) = \sum_{cc'} \mathrm{L}_{cc'}(\mathbf{q}) \left| \sum_{a} f_a(\mathbf{q}) e^{2\pi i \mathbf{r}_a \cdot \mathbf{q}} e^{-2\pi^2 \sigma_a^2 q^2} \right|^2 + N_c \sum_{a} |f_a(\mathbf{q})|^2 \left(1 - e^{-4\pi^2 \sigma_a^2 q^2}\right). \tag{9.29}$$

The two terms in this expression are conventionally called the Bragg term and the diffuse term, respectively. In the diffuse term, since each atomic form factor is spherically symmetric, the pattern as a whole is the same and just adds to the "background" similar to the solvent scatter. The Bragg term here is just the lattice sum multiplied by the Fourier transform of the B-factor-blurred molecule as discussed in Sect. 9.3.7. Thus, if one subtracts the background at each Bragg peak position and phases the integrated intensity, the electron density obtained can be thought of as replacing each atom by a Gaussian blob with a width σ_a. Note also that this is the average electron density over all unit cells. The total intensity is shown in Fig. 9.4 where one can see the reduced Bragg resolution as well as the structureless diffuse background.

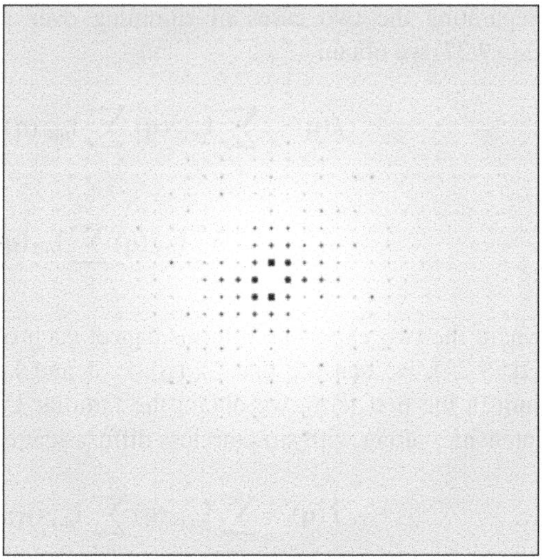

Fig. 9.4 Total intensity calculated in Eq. (9.28) when each atom is displaced independently. Here all atoms are displaced by an average of 0.8 Å ($B = 25.3$ Å^2) resulting in the Bragg peaks being suppressed at high-resolution and featureless diffuse scattering. For details see Box 9.1

Thus, we have seen that in the absence of correlated displacements, the Bragg peaks just represent the Fourier transform of the average unit cell electron density. In fact, this turns out to be generally true even in the case of strongly correlated motion. This is why that by applying the conventional analysis pipeline the crystallographer need not worry about correlated motion and can solve for the B-factor for each atom. In a sense, this has been a great boon for the field. One could even argue that the field may never have taken off as it did over the last 100 years if it were necessary to solve for the whole $C_{ac\,a'c'}$ matrix to solve the structure.

9.4.2 *Rigid-Body Translational Disorder of a Unit Cell*

We consider now a case in which the whole unit cell moves as a single rigid unit, with no correlations across unit cells. This is identical to what was discussed in Sect. 9.3.7, and we will obtain the same result by using the atomistic formalism. Note that this disorder model is only likely when there is a single asymmetric unit per unit cell. In other cases, the choice of the original asymmetric unit is arbitrary, leading to different possible choices of unit cells for the equivalent crystal. Such choices would, however, lead to different continuous diffraction if randomly displaced.

The analysis is very similar to the previous section. The covariance matrix reduces to

$$\mathbf{C}_{ac\,a'c'} = \left\langle \boldsymbol{\delta}_{ac}^{\mathsf{T}} \boldsymbol{\delta}_{a'c'} \right\rangle = \begin{cases} \mathbf{U}, & \text{if } c = c' \text{ for all } a \\ 0, & \text{otherwise.} \end{cases}$$

For an isotropic displacement distribution, $\mathbf{U} = \sigma^2 \mathbf{I}_3$ where $\mathbf{I}_3$ is the 3×3 identity matrix. Starting again from Eq. (9.27), separating the cases, and then completing the sum results in

$$I(\mathbf{q}) = e^{-4\pi^2\sigma^2 q^2} \sum_{cc'} \mathrm{L}_{cc'}(\mathbf{q}) \left| \sum_a f_a(\mathbf{q}) e^{2\pi i \mathbf{r}_a \cdot \mathbf{q}} \right|^2 + N_c \left(1 - e^{-4\pi^2\sigma^2 q^2}\right) \left| \sum_a f_a(\mathbf{q}) e^{2\pi i \mathbf{r}_a \cdot \mathbf{q}} \right|^2 . \tag{9.30}$$

Once again, the first term represents the Bragg intensities. The Bragg peaks sample the molecular transform, multiplied by the Debye-Waller factor. In fact, this term is the same as in Eq. (9.28), showing that the Bragg peak intensities are insensitive to whether displacements are correlated or not. However the diffuse term becomes something quite interesting. The same molecular transform term is present but now sampled everywhere. There is no lattice sum term reducing the sampling to just the reciprocal lattice points. In addition, there is the so-called complementary Debye-Waller term which grows with q. As q increases, the Bragg term vanishes and the intensity is just N_c times the intensity from a single unit cell. This can be seen by comparing Figs. 9.5 and 9.3a.

Since the diffuse term represents the continuously sampled Fourier transform intensities of the electron density of the molecule, $\rho(\mathbf{r})$, the constraint ratio Ω is greater than 1 and iterative phasing algorithms discussed in Sect. 9.2.5 can be used to recover the phases. It is as if there were N_c copies of perfectly aligned single molecules whose intensities were added on top of one another, as in Sect. 9.3.1.

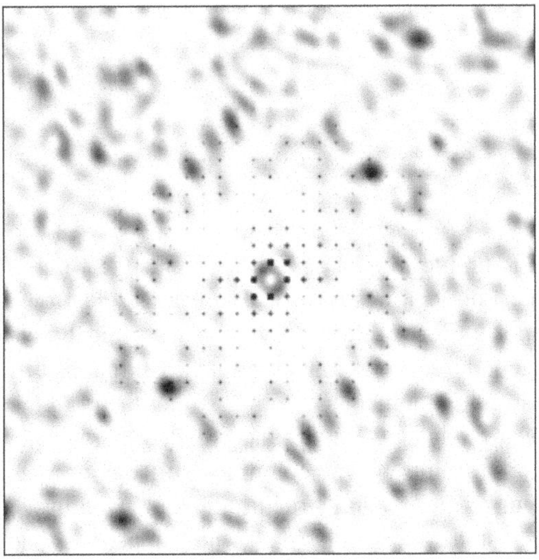

Fig. 9.5 Total intensity calculated in Eq. (9.30) when the entire unit cell is displaced as a rigid body. The Bragg peaks are identical to Fig. 9.4 but the diffuse scattering now represents the continuous diffraction of the molecule seen in Fig. 9.3a. For details see Box 9.1

9.4.3 Rigid-Body Translations Plus Uncorrelated Displacements

Now it is unlikely that the proteins are exactly rigid bodies with no internal motions. As a first approximation to address this, the atomic displacements can be modelled as a combination of rigid-body translation and uncorrelated displacements,

$$\boldsymbol{\delta}_a = \boldsymbol{\delta} + \boldsymbol{\eta}_a \,.$$

The corresponding B-factor matrices assuming isotropic displacement distributions are $\sigma^2\mathbf{I}_3$ and $\beta_a^2\mathbf{I}_3$ for rigid-body motion and the individual atomic B-factor "vibrations," respectively. Once again, we assume that there are no long range correlations across multiple unit cells. With these conditions, the covariance matrix can be written as

$$\mathbf{C}_{ac\,a'c'} = \left\langle \boldsymbol{\delta}_{ac}^{\mathsf{T}} \boldsymbol{\delta}_{a'c'} \right\rangle = \begin{cases} (\sigma^2 + \beta_a^2)\mathbf{I}_3, & \text{if } c = c' \text{ and } a = a' \\ \sigma^2\mathbf{I}_3, & \text{if } c = c' \text{ and } a \neq a' \\ 0, & \text{otherwise.} \end{cases}$$

The first case describes identical atoms (in the same unit cell), the second different atoms in the same unit cell, and the third are atoms in different unit cells. Since there are three cases, Eq. (9.27) must be separated into three terms. After completing the sum twice we obtain the following expression for the total scattered intensity:

$$\begin{aligned} I(\mathbf{q}) = & \sum_{cc'} \mathrm{L}_{cc'}(\mathbf{q}) \left| \sum_a f_a(\mathbf{q}) e^{2\pi i \mathbf{r}_a \cdot \mathbf{q}} e^{-2\pi^2(\sigma^2+\beta_a^2)q^2} \right|^2 \\ & + N_c \left(1 - e^{-4\pi^2\sigma^2 q^2}\right) \left| \sum_a f_a(\mathbf{q}) e^{2\pi i \mathbf{r}_a \cdot \mathbf{q}} e^{-2\pi^2\beta_a^2 q^2} \right|^2 \\ & + N_c \left(1 - e^{-4\pi^2\sigma^2 q^2}\right) \sum_a |f_a(\mathbf{q})|^2 \left(1 - e^{-4\pi^2\beta_a^2 q^2}\right) . \end{aligned} \quad (9.31)$$

The first term once again represents the Bragg intensities which are modulated by the Fourier transform of the average electron density. For this signal, each atom appears to be blurred by both the rigid body and the uncorrelated motion. The second term represents the same continuous diffraction as before modulated by the same complementary Debye-Waller factor of the rigid-body displacements, but now also modulated by the uncorrelated displacements which suppresses the signal at high resolution. The continuous diffraction sees an average molecule consisting of blurred atoms with structure factors given by $f_a(\mathbf{q})\exp(-2\pi^2\beta_a^2 q^2)$. The last term is just the remaining signal which appears as background containing no information about the structure, similar to Sect. 9.4.1. The calculated intensities are shown in Fig. 9.6.

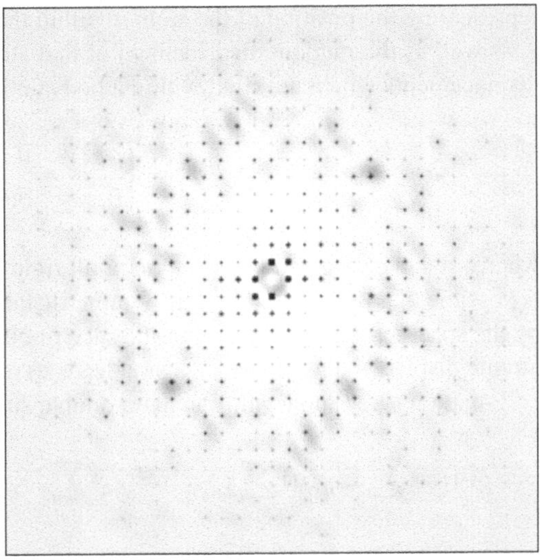

Fig. 9.6 Total intensity calculated in Eq. (9.31) when the total displacement of each atom has two components, rigid and uncorrelated. The total average displacement is still 0.8 Å, so the Bragg peaks are identical resulting in the Bragg peaks being suppressed at high resolution and featureless diffuse scattering. For details see Box 9.1

A striking feature of the expression of Eq. (9.31) is that the non-spherically symmetric part of the diffuse intensities can still be phased using iterative phasing techniques to give the average electron density of the rigid body. A method for separating the unstructured diffuse background from the structured continuous diffraction is described in Sect. 9.5. In addition, the q-dependence of these intensities is different from that of the average electron density of the unit cell probed by the Bragg peaks—the continuous diffraction can extend to higher resolution than the Bragg peaks depending on the magnitude of the rigid-body mean square displacement σ^2.

9.4.4 Multiple Rigid Bodies in a Unit Cell

As mentioned earlier, in any realistic situation with more than one asymmetric unit, the whole unit cell would not move as a rigid body since the unit cell is an arbitrary construction. This is the case for any space group other than P1. Instead it is conceivable that each asymmetric unit is displaced as a rigid body, or even smaller domains of molecules move as such. Different rigid bodies may be different orientations of the same molecule, as discussed in Sect. 9.3.7. Some units could consist of identical structures in like orientations, in which case their continuous diffraction contributions simply sum.

We consider the diffuse scattering intensity from a crystal with N_b rigid bodies per unit cell, each of which is displaced by an uncorrelated amount with respect to the other with a variance of $\langle \delta_r^2 \rangle = \sigma_r^2$. Each atom is assigned to a rigid body b and the position of any atom i can now be expressed as

$$\mathbf{r}_i = \mathbf{r}_c + \mathbf{r}_b + \mathbf{r}_a + \boldsymbol{\delta}_{abc} \ ,$$

representing the position of the atom a within the rigid body b that is located in cell c, as well as the random displacement of that atom δ_{abc}. The covariance matrix for displacements which are rigid within a body and uncorrelated otherwise is

$$\mathbf{C}_{abc\,a'b'c'} = \langle \boldsymbol{\delta}^{\mathsf{T}}_{abc} \boldsymbol{\delta}_{a'b'c'} \rangle = \begin{cases} \sigma_r^2 \mathbf{I}_3, & \text{if } c = c' \text{ and } b = b' \text{ for all } a \\ 0, & \text{otherwise,} \end{cases}$$

where we have assumed the same σ for all rigid bodies. This is not unusual since it is possible that the rigid bodies are symmetric units of the unit cell which are related by the space group symmetry. For the sake of clarity, we have ignored uncorrelated atomic displacements. Their inclusion leads to results analogous to Sect. 9.4.3.

Equation (9.27) now gains an extra double sum

$$I(\mathbf{q}) = \sum_{cc'} \mathrm{L}_{cc'}(\mathbf{q}) \sum_{bb'} e^{2\pi i (\mathbf{r}_b - \mathbf{r}_{b'})\cdot\mathbf{q}} \sum_{aa'} \mathrm{I}_{aa'}(\mathbf{q}) e^{-4\pi^2\sigma^2 q^2} e^{4\pi^2 \mathbf{q}^{\mathsf{T}} \mathbf{C}_{abc\,a'b'c'} \mathbf{q}} \,.$$

Separating into the different cases and then completing the sums as before, we obtain the following simple relation:

$$\begin{aligned} I(\mathbf{q}) =& e^{-4\pi^2\sigma^2 q^2} \sum_{cc'} \mathrm{L}_{cc'}(\mathbf{q}) \left| \sum_b e^{2\pi i \mathbf{r}_b\cdot\mathbf{q}} \sum_a f_a(\mathbf{q}) e^{2\pi i \mathbf{r}_a\cdot\mathbf{q}} \right|^2 \\ &+ N_c \left(1 - e^{-4\pi^2\sigma^2 q^2}\right) \sum_b \left| \sum_a f_a(\mathbf{q}) e^{2\pi i \mathbf{r}_a\cdot\mathbf{q}} \right|^2 . \end{aligned} \tag{9.32}$$

The first term is once again the familiar Bragg intensity which can be interpreted as the Fourier transform of the average unit cell, sampled at Bragg peaks. One sums over the atoms in all the rigid bodies with the positions relative to the origin of the unit cell $(\mathbf{r}_b + \mathbf{r}_a)$.

The second term of Eq. (9.32), however, is seen as the *incoherent* sum of the continuous diffraction from each average rigid body. The relative positions of the bodies $(\mathbf{r}_b)$ disappear. This equation is identical to that of the initial example of Sect. 9.4.2 with the scattering factors of individual atoms $f_a(\mathbf{q})$ of Eq. (9.29) replaced with the form factors of the individual rigid bodies, $\sum_a f_a(\mathbf{q}) \exp(2\pi i \mathbf{r}_a \cdot \mathbf{q})$. What was a background due to unstructured (almost point-like) atoms, becomes an incoherent sum of structured molecules or molecular units. The distinction between the coherent sum of units in the Bragg term of Eq. (9.32) and the incoherent sum of the second term is important. For rigid units of different orientation, the incoherent sum is analogous to diffraction of a twinned crystal (see Sect. 9.3.7). For iterative phasing this requires the modulus constraint of Eq. (9.18).

The calculated intensities are displayed in Fig. 9.7. Close inspection reveals another crucial difference between the "untwinned" Bragg diffraction and the "twinned" continuous diffraction is that even at reciprocal lattice points, the Bragg and

Fig. 9.7 Total intensity calculated in Eq. (9.32) when each rigid body in the unit cell is displaced. In this case, there are four rigid bodies, each of which is the asymmetric unit of the $P2_12_12_1$ space group. One can see the systematic absences in the Bragg peaks but since the diffuse scattering is the *incoherent* sum of the intensities, there are no cancellations due to the space group at the forbidden positions. For details see Box 9.1

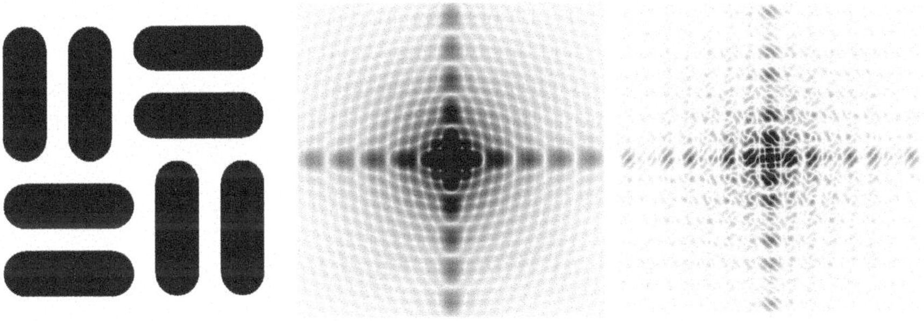

(a) (b) (c)

Fig. 9.8 The continuous and Bragg diffraction do not necessarily sample the same underlying transform, as illustrated in a 2D crystal of C4 symmetry consisting of eight rods which are all randomly displaced independently in the crystal (**a**). (**b**) The continuous diffraction consists of equal weightings of the single-rod diffraction in the two unique orientations, whereas the Bragg peaks sample the transform of the average unit cell (**c**). The patterns are markedly different and even display different symmetries

continuous terms do not sample the same underlying transform. Here is it seen that absences in the Bragg peaks do not necessarily correspond to minima of the continuous diffraction and in some places strong Bragg peaks overlay minima of the continuous diffraction. This situation is in stark contrast to the results of Sect. 9.4.3 where the unit cell transform could be factored out. In that case, the end result (Eq. (9.30)) is a more complicated lattice function than $L(\mathbf{q})$, as can also be seen within the square brackets of Eq. (9.23). The difference between incoherent and coherent addition can also be illustrated with a simple toy problem of diffraction of a two-dimensional array of rod-like structures, shown in Fig. 9.8. Here there are two orientations of these independently displaced rigid units. The continuous diffraction consists of the incoherent sum of rod diffraction in the two different orientations, Fig. 9.8b, whereas the Bragg peaks sample the underlying unit cell transform of Fig. 9.8c.

In Sect. 9.2.4, we saw how the feature size in Fourier space (the Shannon voxel or speckle) is inversely related to the size of coherently diffracting volume. If one combines two objects incoherently (by adding the intensities) however, the feature size does not change and contrast is reduced [8]. This means that the observed speckle size gives a measure of the size of the rigid bodies that contribute to the diffraction pattern. This can be examined directly by taking the Fourier transform of the measured 3D diffraction intensities, giving the autocorrelation function of the disordered crystal. If the Bragg peaks are first filtered from the diffraction intensities, the transform $\tilde{I}(\mathbf{u})$ of the second term of Eq. (9.32) is the sum of the complete autocorrelation functions of the rigid units (convolved with a narrow Gaussian given by the Fourier transform of the complementary Debye-Waller factor). More broadly, as expanded in the next section, the autocorrelation of the rigid units will be convolved with the correlation function of their displacements, making it possible to determine if the boundary between the rigid bodies is "soft," for example. When small domains of molecules are displaced, a case of great interest to understand protein dynamics and function, the resulting speckle size will be large, and the incoherent sum of the continuous diffraction from the many independent domains will be of low contrast, perhaps not distinguishable from uncorrelated atomic disorder.

9.4.5 Liquid-Like Motions Within a Rigid Body

A popular model of disorder is liquid-like motions [9, 39]. Here the covariance matrix between two atoms a and a' is a diagonal matrix which decays exponentially with the distance between them. This can be expressed as

$$\mathbf{C}_{aa'} = \left\langle \boldsymbol{\delta}_{ac}^{\mathsf{T}} \boldsymbol{\delta}_{a'c'} \right\rangle = \sigma^2 \exp(-r_{aa'}/\gamma)\, \mathbf{I}_3 \,. \tag{9.33}$$

This is termed liquid-like because it treats the molecules like an ideal fluid. There is no shearing motion which would lead to off-diagonal terms in the covariance matrix and the correlations are strictly non-negative. One can also have different values for σ and γ for atom pairs within the same molecule and in different molecules. To investigate how the scattered intensity is related to the molecular transform in this case, we consider a P1 crystal with the following covariance matrix:

$$C_{ac\,a'c'} = \left\langle \boldsymbol{\delta}_{ac}^{\mathsf{T}} \boldsymbol{\delta}_{a'c'} \right\rangle = \begin{cases} \sigma^2 \exp(-r_{aa'}/\gamma)\, \mathbf{I}_3, & \text{if } c = c' \\ 0, & \text{otherwise.} \end{cases}$$

This means that within the unit cell, correlations decay with distance but there are no correlations across unit cells. Using the now-familiar procedure, Eq. (9.27) becomes

$$I(\mathbf{q}) = \sum_{cc'} \mathrm{L}_{cc'}(\mathbf{q}) \sum_{aa'} \mathrm{I}_{aa'}(\mathbf{q}) e^{-4\pi^2\sigma^2 q^2}$$
$$+ N_c e^{-4\pi^2\sigma^2 q^2} \sum_{aa'} \mathrm{I}_{aa'}(\mathbf{q}) \left(e^{4\pi^2\sigma^2 q^2 \exp(-r_{aa'}/\gamma)} - 1 \right) .$$

As a sanity check, we see that as $\gamma \to \infty$, the expression becomes the same as in Eq. (9.30) with a B-factor suppressed Bragg term and continuous diffraction in the second term. To evaluate the second term for finite γ, it is helpful to expand the $I_{aa'}$ term

$$I_D(\mathbf{q}) = N_c e^{-4\pi^2\sigma^2 q^2} \sum_{aa'} f_a(\mathbf{q}) f_{a'}(\mathbf{q}) \left(e^{4\pi^2\sigma^2 q^2 \exp(-r_{aa'}/\gamma)} - 1 \right) e^{2\pi i \mathbf{r}_{aa'} \cdot \mathbf{q}}$$
$$= N_c e^{-4\pi^2\sigma^2 q^2} \int d^3\mathbf{u} \sum_{aa'} f_a(\mathbf{q}) f_{a'}(\mathbf{q}) \left(e^{4\pi^2\sigma^2 q^2 \exp(-u/\gamma)} - 1 \right) \delta(\mathbf{u} - \mathbf{r}_{aa'}) e^{2\pi i \mathbf{u} \cdot \mathbf{q}}$$
$$= N_c e^{-4\pi^2\sigma^2 q^2} \int d^3\mathbf{u}\, P(\mathbf{u}) e^{2\pi i \mathbf{u} \cdot \mathbf{q}}$$
$$= N_c e^{-4\pi^2\sigma^2 q^2} \mathscr{F}_q \left[P(\mathbf{u}) \right]$$

where $P(\mathbf{u})$ is the autocorrelation of the electron density weighted by the covariance terms and $\delta(\mathbf{u})$ is the 3D Dirac delta function. The advantage of this conversion to a continuous integral is that the Fourier transform is directly evident. The ideal autocorrelation function is the inverse Fourier transform of the molecular transform intensities,

$$P_o(\mathbf{u}) = \mathscr{F}_u^{-1} \left[I_o(\mathbf{q}) \right] .$$

The weighted autocorrelation can be written as the product of the ideal one with the liquid-like motion weighting

$$P(\mathbf{u}) = P_o(\mathbf{u}) \times \left(e^{4\pi^2\sigma^2 q^2 \exp(-u/\gamma)} - 1 \right) . \tag{9.34}$$

Using the fact that the Fourier transform maps products to convolutions and using a Taylor-series expansion, the diffuse term I_D can be written as

$$I_D(\mathbf{q}) = N_c e^{-4\pi^2\sigma^2 q^2} \mathscr{F}_q \left[P_o(\mathbf{u}) \times \left(e^{4\pi^2\sigma^2 q^2 \exp(-u/\gamma)} - 1 \right) \right]$$
$$= N_c e^{-4\pi^2\sigma^2 q^2} \sum_{n=1}^{\infty} \frac{(2\pi\sigma q)^{2n}}{n!} I_o(\mathbf{q}) \otimes \mathscr{F}_q \left[e^{-nu/\gamma} \right]$$
$$= N_c e^{-4\pi^2\sigma^2 q^2} \sum_{n=1}^{\infty} \frac{(2\pi\sigma q)^{2n}}{n!} I_o(\mathbf{q}) \otimes \left[\frac{8\pi n \gamma^3}{\left(n^2 + 4\pi^2\gamma^2 q^2\right)^2} \right] \tag{9.35}$$

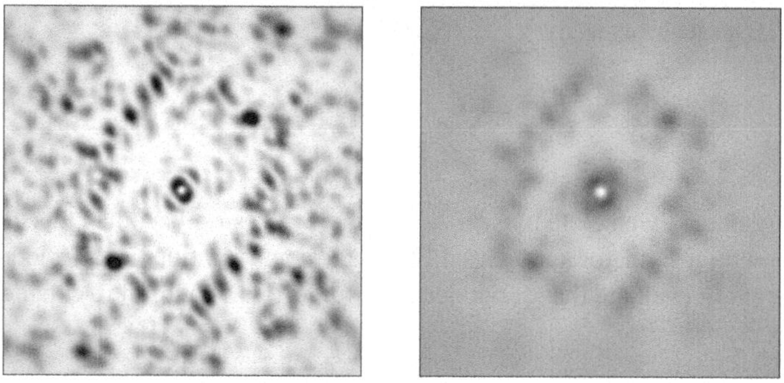

(a) (b)

Fig. 9.9 Diffuse scattering evaluated from the liquid-like motions model, Eq. (9.35), with the parameters $\sigma = 0.8$ Å in both cases and $\gamma = 100$ Å and 10 Å in (**a**) and (**b**), respectively. The pattern in (**a**) is fairly similar to Fig. 9.3a, except near the corners of the image. But if the correlation length is significantly smaller than the particle, which is around 40 Å wide, the blurring will make phase retrieval difficult. For details see Box 9.1

where the last term is the 3D Fourier transform of the spherically symmetric function $e^{-n|\mathbf{u}|/\gamma}$. Thus, the effect of liquid-like correlations within the rigid body results in a series of convolutions with incrementally broader kernels. In addition, the contribution of each successive term is weighted more at higher q. At very low resolution, the higher order terms can be neglected and the molecular transform is only slightly modified. At higher resolution, more terms have to be taken into account which blur the intensity more. This can be understood intuitively in that at low resolution (long length scales), the rigid body moves highly coherently since the deviations from rigid-body motion are small. At higher resolution, the effective correlation length decreases which broadens the features in Fourier space.

Figure 9.9 shows our standard example pattern modified for different values of the correlation length γ. In the first case γ is around 2.5 times the size of the rigid body so it can be approximated well as a rigid body, while in (b), the correlation length is only a quarter of the size. Thus, at higher resolution, the best approximation is that of uncorrelated displacements leading to structureless diffuse scattering.

The first line of Eq. (9.35) emphasizes the power of investigating the continuous diffraction (after filtering the Bragg peaks) in real space by Fourier transformation. After accounting for the Debye-Waller term, which could be obtained through a Wilson plot analysis [20], the transform of $I_D(\mathbf{q})$ yields the weighted autocorrelation function of Eq. (9.34). It may be possible to determine the form of the correlation function through inspection or modelling, or to solve for an approximation of it during the phasing process, similar to procedures of partially coherent diffraction imaging [19, 58].

9.4.6 Crystal Growth Model

One can consider the case where the bodies such as whole molecules are rigid but there is an exponentially decaying correlation between the displacements of those rigid bodies from their ideal positions. This situation can be surmised to be more likely than uncorrelated displacements of molecules, since molecules are in contact in the crystal and the position of one molecule is influenced by the positions of these neighbors. The so-called growth models capture this dependence of the position of one molecule on the positions of those that came before it [55], which we simplify here. Again, for simplicity, staying with the P1 model of the whole unit cell being a rigid body, we obtain the following covariance matrix in analogy to Eq. (9.33):

$$C_{aca'c'} = \left\langle \boldsymbol{\delta}_{ac}^{\mathsf{T}} \boldsymbol{\delta}_{a'c'} \right\rangle = \sigma^2 \exp(-r_{cc'}/\gamma)\, \mathbf{I}_3 \,.$$

Since the covariance only depends on the distance between the unit cell origins, this represents each unit cell moving as a rigid body. Now Eq. (9.27) does not separate into different terms, but can just be written as

$$\begin{aligned} I(\mathbf{q}) &= e^{-4\pi^2\sigma^2 q^2} \sum_{cc'} \mathrm{L}_{cc'}(\mathbf{q}) e^{4\pi^2\sigma^2 q^2 \exp(-r_{cc'}/\gamma)} \sum_{aa'} \mathrm{I}_{aa'}(\mathbf{q}) \\ &= e^{-4\pi^2\sigma^2 q^2} I_o(\mathbf{q}) \sum_{cc'} \mathrm{L}_{cc'}(\mathbf{q}) e^{4\pi^2\sigma^2 q^2 \exp(-r_{cc'}/\gamma)} \,. \end{aligned}$$

This is the first time we consider correlated displacements across unit cells, which has the effect of modifying the lattice sum itself. Now even for an infinite crystal, the diffraction is not expressed as an array of delta functions. The sum can, however, be expressed as a sum of convolutions as in Sect. 9.4.5, Eq. (9.35), but with $I_o(\mathbf{q})$ replaced by the ideal lattice sum, which is the array of delta functions (the Dirac comb).

$$I(\mathbf{q}) = e^{-4\pi^2\sigma^2 q^2} \sum_{n=1}^{\infty} \frac{(2\pi\sigma q)^{2n}}{n!} \left[\sum_{cc'} \mathrm{L}_{cc'}(\mathbf{q}) \right] \otimes \left[\frac{8\pi n \gamma^3}{\left(n^2 + 4\pi^2\gamma^2 q^2\right)^2} \right] . \quad (9.36)$$

The Bragg peak is broadened by an increasing amount as q increases until it disappears at a high enough resolution and one obtains the direct continuous diffraction term $I_o(\mathbf{q})$. This can again be understood intuitively in the following way. At low resolution (long length scales), a large chunk of the crystal is well-ordered resulting in sharp, strong Bragg peaks. At smaller length scales, only a small region of the crystal has the necessary degree of periodicity to interfere coherently to produce fat Bragg peaks, as discussed in Sect. 9.3.3. Two cases, with γ respectively bigger and smaller than the unit cell are shown in Fig. 9.10a, b. A recent investigation discerned correlations over lengths of many unit cells in the

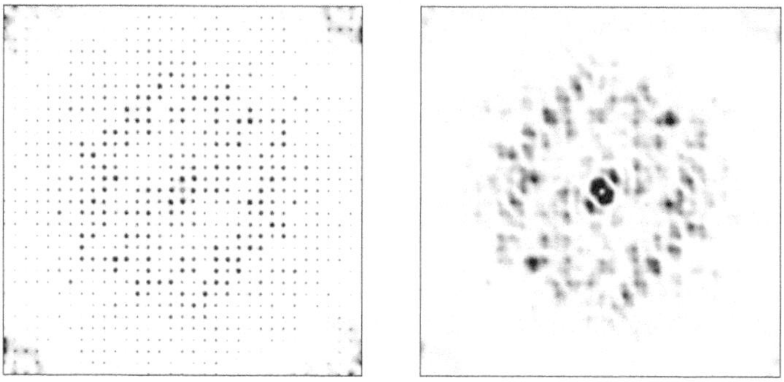

(a) (b)

Fig. 9.10 Scattered intensity evaluated from the growth model, Eq. (9.36), with the parameters $\sigma = 0.8$ Å and $\gamma = 100$ Å and 10 Å in (**a**) and (**b**), respectively. Even with a relatively high σ, since γ is so large in (**a**), Bragg peaks persist until the edge of the image since many unit cells move as a group. In (**b**), the situation is closer to rigid-body disorder of Sect. 9.4.2. For details see Box 9.1

merged experimental patterns of several protein crystals in which Bragg diffraction extended to high resolution [39]. The broadening and weakening of Bragg peaks with scattering angle is consistent with such growth models. As mentioned in Sect. 9.4.5 and summarized in Box 9.2, iterative algorithms to recover both the phases and the correlation function offer not only a route to obtain maps of electron density, but may also give new insights into the arrangements of molecules in crystals.

9.4.7 Rotational Rigid-Body Disorder

If the rigid body undergoes rotational oscillations (librations), the diffuse scattering is, in general, not a direct modulation of the continuous diffraction intensities. This is because rotations displace different parts of the molecule by different amounts. The atoms on the rotation axis for axial rotations do not displace at all. This effectively makes the rotationally disordered electron density fundamentally different from the rigid-body density. The situation is more tractable if translational disorder is also present and the displacement of all atoms due to rotations is small compared to translations. In that case, one observes a rotationally blurred copy of the continuous diffraction, which can still be phased under certain conditions using partial coherence methods. Figure 9.11 shows the intensity distributions when there is both translational and rotational disorder. The tolerance to rotation is similar to that of the Crowther condition for the angular step size of tomography [13], given by the resolution divided by the molecule width, and thus smaller rigid bodies can tolerate larger degrees of rotational disorder.

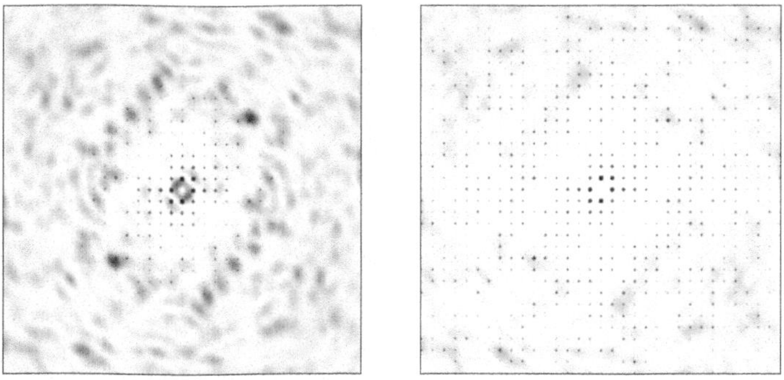

(a) (b)

Fig. 9.11 Intensity distribution from both rotational disorder about the center of mass of the molecule with standard deviation of 4° and rigid-body translational disorder with $\sigma = 0.8$ Å and 0.2 Å in (**a**) and (**b**), respectively. The pattern in (**a**) is recognizably a slightly rotationally blurred version of Fig. 9.3a. As the relative contribution of rotational disorder is increased, additional features appear and the diffuse scattering cannot be interpreted as simply transformed continuous diffraction. For details see Box 9.1

Box 9.2
In Sect. 9.4, we have seen the effect on the scattered intensity through various forms of crystalline disorder. In general, a few or more of these are likely to occur in all crystals to some extent. In conventional protein crystallography, uncorrelated disorder is assumed by default since Bragg intensities are insensitive to correlations in displacements. It is striking by comparing the figures in this section how persistent the structural information is in the continuous diffraction. This stems from the fact that the correlation functions are themselves relatively unstructured and their main effect is to broadly modulate the pair correlation (autocorrelation function) rather than to change the distribution of those pair correlations. As such the continuous diffraction, in combination with Bragg intensities and a suitable treatment of the correlation function, should provide robust information to obtain the molecular structure. This is akin to strategies of accounting for partial coherence, which have brought much success to the field of diffractive imaging [58].

9.4.8 An Example System: Photosystem II

Continuous diffraction from a disordered crystal was first identified as useful for structure determination in crystals of the membrane protein complex photosystem II (PS II) [1]. Membrane protein surfaces have both hydrophobic and hydrophilic parts. One way to generate a stable crystal in an aqueous environment is to use

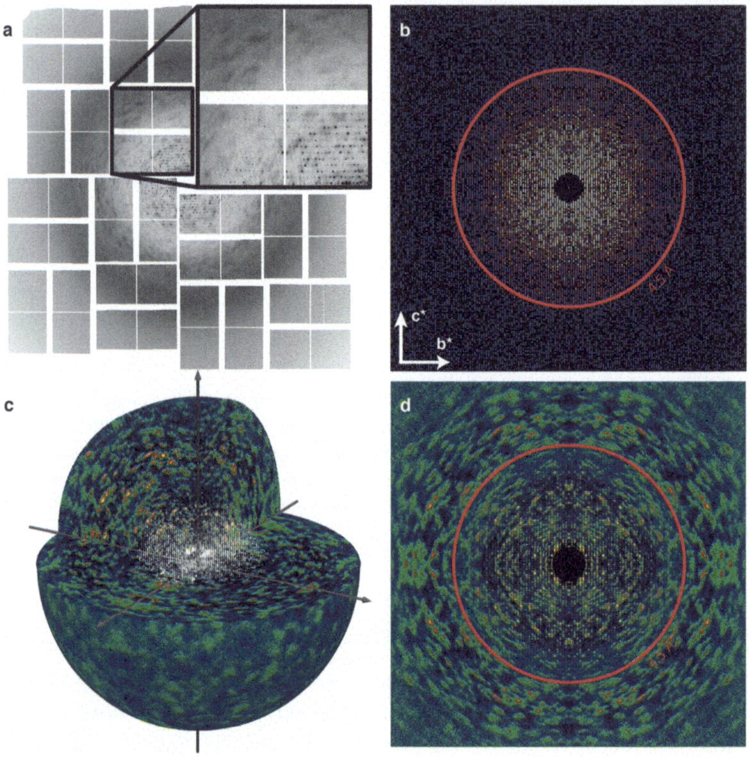

Fig. 9.12 (**a**) An X-ray FEL snapshot "still" diffraction pattern of a PSII microcrystal shows a weak speckle structure beyond the extent of Bragg peaks, which is enhanced in this figure by limiting the displayed pixel values. (**b**) Structure factors obtained from Bragg peak counts from 25,585 still patterns, displayed as a precession-style pattern of the [001] zone axis. (**c**) A rendering of the entire 3D diffraction volume assembled from the 2848 strongest patterns. (**d**) A central section of the diffraction volume in c normal to the [100] axis. Speckles are clearly observed beyond the 4.5-Å extent of Bragg diffraction (indicated by the white circles in **b** and **d**) to the edge of the detector. Caption and figure from Ref. [1]

detergents which form micelles around the hydrophobic surfaces. These micelles mediate the crystal contacts, but since they are flexible, the contacts are relatively soft. The crystals are therefore not perfectly ordered and, in this experiment, produced observable Bragg peaks to a resolution of only around 4.5 Å. However, one can observe weak continuous scattering at higher scattering angles all the way to the edge of the detector (Fig. 9.12). After subtracting the background and combining the strongest diffraction patterns as described in Sect. 9.5, one can observe a striking continuous signal. This continuous diffraction data is available as CXIDB Entry 59 (http://cxidb.org/id-59.html).

These molecules crystallize in the orthorhombic $P2_12_12_1$ space group, which means that there are four asymmetric units in each unit cell. Since the choice of which four units make up the unit cell is arbitrary, we do not consider that

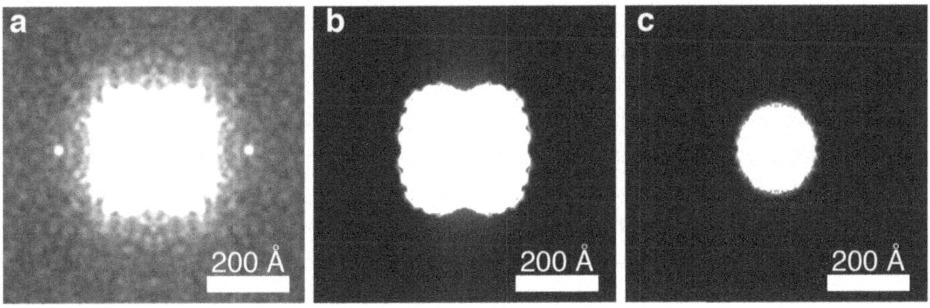

Fig. 9.13 (**a**) Electron density autocorrelation projected along the crystal c axis. (**b**) Point-group symmetrized autocorrelations calculated from the PS II dimer, and (**c**) the PS II monomer projected along the same axis. Reproduced from Ref. [1]

the whole unit cell moves as a single rigid body. In order to verify that this is indeed continuous diffraction from some rigid bodies, the sizes of the features (or speckles) in the pattern can be examined. This is done quantitatively by calculating the autocorrelation as described in Sect. 9.2.4 using small regions (masked with a Gaussian function) in the high-resolution parts of the 3D intensity distribution. It is clear from Fig. 9.13a that the autocorrelation is not only of finite extent (or support) but that it also has the same point-group symmetry as the crystal. From the previous sections, this suggests a strong possibility that rigid-body translations can explain the scattering and that each asymmetric unit is an independent rigid body. For PS II the asymmetric unit is itself a dimer, so it is possible that each monomer is independently displaced as a rigid body. However, the monomer is too small and the resulting symmetrized autocorrelation seen in Fig. 9.13c does not match the data. If the rigid body was smaller than the dimer, the features in the measured diffraction pattern would be bigger.

Using these parameters as justification, the authors proceeded to perform iterative phasing on these high-resolution diffuse intensities. As a proof of principle first step, the low-resolution data, where the Bragg peaks are visible, was replaced by the molecular transform obtained from the inverse Fourier transform of the phased Bragg peaks. Thus, the phase problem was reduced to just determining the high-resolution structure from the 222-twinned intensities. In addition, since the low-resolution model was assumed to be known, one could also obtain a good molecular envelope. The use of this envelope encodes the assumption that the protein is "compact," i.e., it is a connected volume and it does not consist of thin tendrils of electron density far from the bulk of the molecule. Both of these are reasonable for almost any protein, especially if the object moves as a rigid body. In Fourier space, the modulus projection P_M of Eq. (9.18) was applied to the high-resolution voxels while the entire complex Fourier amplitude was replaced by the Bragg model at low resolution. With this modified projection, the difference map algorithm was applied [14, 16] to obtain a higher resolution structure.

Further analysis confirmed that the continuous diffraction consisted of the sum of four independent rigid bodies [8]. This was achieved by examining the statistics of the diffraction intensities in shells of reciprocal space. As for Bragg peaks, the intensities of the continuous diffraction of a molecule closely follow a negative exponential distribution that would be expected from the coherent diffraction of a set of atoms in random locations [59]. The most likely intensity value in the pattern is zero, which is seen in the surroundings of the rarer maxima of speckles. The distribution changes markedly when summing the intensities of two different patterns. In this case it is unlikely that all the zeros coincide, and the distribution changes to a gamma distribution with reduced contrast and variance. Such distributions form the basis for analyzing the diffraction of twinned crystals [40], and the same can be applied here. Additionally, this was found to give a way to estimate the contribution of a spherically symmetric unstructured background (such as the third term in Eq. (9.31) which gives intensities that are approximately normally distributed), as described in Sect. 9.5. This analysis showed that the total counts in the continuous diffraction were about four times that of the Bragg peaks, and the background contained 100 times more photons. The improved analysis also gave a Pearson correlation between the modelled diffraction and the measured continuous diffraction of about 0.77, as shown in Fig. 9.14. This degree of correlation was achieved by assuming 1° of rotational disorder. Without this rotational blurring of the modelled patterns a correlation of 0.67 was obtained.

As we have discussed in the previous section, the real crystal probably consists of other kinds of disorder than just uncorrelated rigid-body displacements. There may be some amount of uncorrelated atomic disorder (Sect. 9.4.1), liquid-like motion within the rigid body (Sect. 9.4.5), correlations between rigid bodies (Sect. 9.4.6), and rotational rigid-body disorder (Sect. 9.4.7). In addition, there may be biologically relevant conformational motions reflecting the dynamic behavior of the protein in its native state. The exploration of all these possibilities to improve the retrieved structure is ongoing and will hopefully lead to reliable structure improvement using this continuous diffraction. Finally, since the continuous diffraction is also visible at lower resolutions, one could also envision fully de novo phasing using both the Bragg peaks and the inter-Bragg intensities.

9.5 Measuring and Processing Continuous Scattering

Although the technique of crystallography is well advanced and data collection at many beamlines is routine, the accurate measurement of continuous diffraction requires extra preparation and care, and the data analysis is different to conventional measurements. Diffraction patterns are recorded with a quasi-monochromatic collimated beam following the same source requirements as for monochromatic macromolecular crystallography. However, diffraction patterns are ideally recorded as "stills" or "snapshots" (no sample rotation during exposure) so that the pixelated diffraction pattern recordings can be mapped into voxels in a 3D array in a similar fashion to tomography or coherent diffractive imaging (see Fig. 9.2). The angular step size between measurements is set by the Crowther condition of tomography

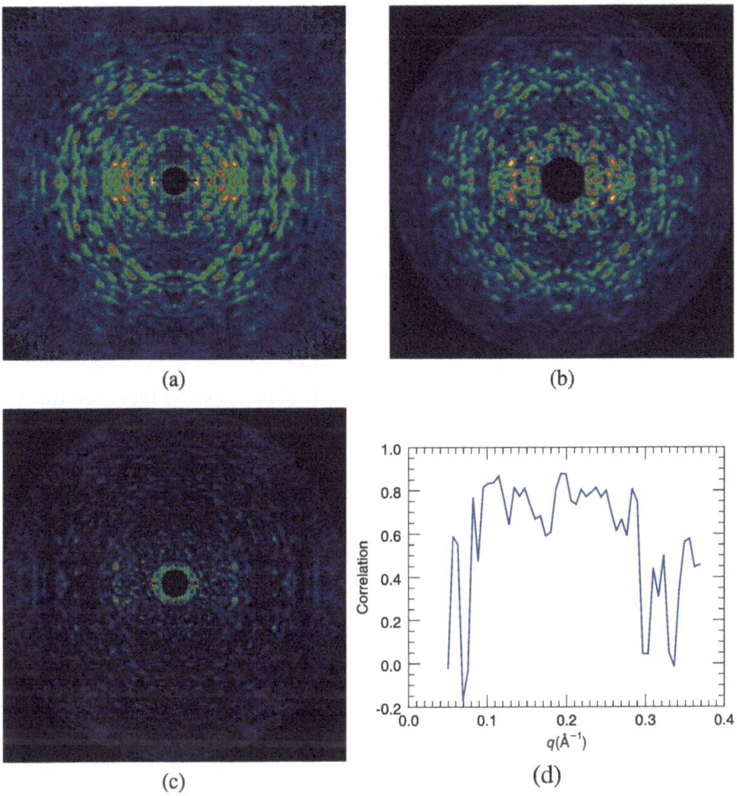

Fig. 9.14 (**a**) Central slices of the merged volume of experimental continuous diffraction intensities, normal to the (010) lattice vector, compared with (**b**) the same section of the simulated continuous diffraction assuming a rotational disorder of 1° RMS and a translational disorder of 2 Å RMS. (**c**) The difference of the experimental and simulated intensities, shown on the same color scale as (**a**) and (**b**). (**d**) Plot of the Pearson correlation coefficient in resolution shells between the experimental and simulated data, confirming that rigid-body displacements of the Photosystem II dimer account for the majority of the observed continuous diffraction. Reproduced from Ref. [8]

[13], equivalent to the Shannon sampling of speckles on subsequent patterns at the highest scattering angles q. This in turn depends on the size of the rigid unit, which can readily be estimated from the speckle size in any given diffraction pattern. The required coverage of reciprocal space depends on the point group of the crystal. For a P1 crystal, this corresponds to the half space (in fact a little more to compensate Ewald sphere curvature), unless anomalous diffraction is to be measured. Diffraction can be recorded from one or several crystals, using an X-ray tube, synchrotron radiation beamline, or free-electron laser, depending on the crystal size and experiment need. The latter source naturally gives snapshot patterns, usually of unknown orientation. As with serial crystallography, the orientations of patterns can be determined by indexing the observable Bragg peaks [56, 57]

as described in Chap. 7. Once relative orientations are known, patterns must be corrected for detector artifacts [61], background removed, and patterns scaled, before interpolating and summing them into a 3D array [60].

For the separate analysis of continuous diffraction from Bragg data, the Bragg peaks must be filtered from the data, which is best carried out prior to merging the patterns. There are several kinds of low-pass filters to remove sharp peaks and other features from the pattern, such as blurring with kernels of several pixels width or determining values that differ from the median within such a kernel. Since the locations of Bragg peaks are known, they can be further masked from the 3D array. Bragg peaks account for only a small fraction of all voxels, and one can quite aggressively mask out pixels at the peaks and surrounding them without losing information from the continuous diffraction. Indeed, when in doubt whether the peaks are influencing the continuous diffraction it is better to increase the mask of peaks rather than to phase with inappropriate data. The missing data can be allowed to float during the iterative phasing process, which will be heavily constrained by the support in any case.

The main difference between the measurement and treatment of Bragg and continuous diffraction of course stems from the diffuse nature of the continuous diffraction, which is much weaker per pixel than Bragg peaks since the counts are spread out over many more pixels. The continuous diffraction due to substitutional disorder in time-resolved measurements is presumably even more difficult to measure since it stems from smaller units than the entire molecule. Molecular continuous diffraction tends also to be much weaker than the structureless background that arises from air scatter, the medium containing the crystal, beamline optics, and atomic disorder in the crystal (possibly induced by the X-ray irradiation [6]). In the example of PS II (Sect. 9.4.8), the background was about 100 times higher than the diffuse scattering.

As compared with Bragg peaks, which can easily be distinguished and separated from any incoherent background, it is not so easy to separate continuous diffraction from it. For a crystal with a single rigid unit (P1 symmetry), it may be possible to use the local minima of the measured pattern or the merged intensities to estimate the level of the incoherent background. However, when the continuous diffraction consists of the incoherent sum of multiple rigid units the background estimation can no longer be guided by the local minima since intensity zeroes in the continuous diffraction are unlikely in this case. As long as the background is due solely to scattering from structureless components, such that it is spherically symmetric (after correction for the polarization of the incident beam), then the statistics of the intensities can be used to separate continuous diffraction of structured rigid objects from this unstructured background [8]. Briefly, as mentioned above, the distribution of intensities in shells of q of the continuous diffraction from a disordered crystal follows Wilson statistics. For a single rigid object per unit cell the distribution of intensities is the negative exponential distribution, but when the crystal consists of N_b rigid bodies (which incoherently sum as in Bragg diffraction from a twinned crystal) the intensities follow the Gamma distribution with a probability function

$$p(I) = \frac{N_b^{N_b} I^{N_b-1}}{\Sigma^{N_b}(N_b - 1)!} \exp\left(\frac{-N_b I}{\Sigma}\right), \quad I > 0, \tag{9.37}$$

with a mean of Σ and a variance of Σ^2/N_b. That is, in the absence of background, the number of rigid units can be determined by comparing the mean and variance of intensities in a q shell. As shown by Chapman et al. [8], the addition of an unstructured background with a normal distribution (or more accurately, Poisson when photon counts are low) alters the distribution of intensities and their corresponding moments. The mean of the structured diffraction (Σ), as well as the mean (μ_{back}) and variance (σ_{back}) of the background can be determined from the mean, variance, and skewness of the measured intensities. This allows the background μ_{back} to be estimated and subtracted in each q shell of every measured pattern, as well as the estimation of a scaling factor Σ that can be used when merging patterns into the 3D array. An improved variation of this method can be applied when photon counts are known [8].

This method of background estimation relies upon the fact that the background really is spherically symmetric. It is therefore very important to minimize any parasitic scattering in the experiment that is not symmetric, such as scatter from beamline optics and shadowing of that scattering by components downstream of the optics. The sensitivity of the analysis of continuous diffraction to this kind of artifact is much greater than for Bragg diffraction, so a beamline set up that is adequate for conventional crystallography is not necessarily suitable for continuous diffraction measurements. At synchrotron beamlines, such shadows can originate from a cryo or humidity jet nozzle, the beamstop holder, or the loop holder. The shadows are usually produced by the X-rays scattered upstream from the sample, for example from air. The beamstop and its holder must also be more carefully aligned than for conventional experiments to ensure that low-angle data is usable in subsequent analysis. If such shadowing is stable during the course of the measurement, then the effected regions of the detector should be masked and not used in further analysis. This may reduce the efficiency of data collection, or in the worst case may cause an incomplete measurement. A way to determine the detector mask to be used is to record long exposures with or without the sample or sum a large number of patterns together (a pixel-based sum in the detector frame of reference), and then subtract the rotationally averaged pattern after polarization correction. The resulting mask is then applied to the patterns to recompute the difference from the rotational average, and perhaps updated. This procedure is repeated several times to ensure the reliable detector pixels are identified.

Any non-symmetric background that varies over the set of measured patterns will not be correctable. This is most likely caused by the means to introduce the sample into the beam. Liquid jets used in X-ray FEL experiments (see Chap. 5) are particularly good since they tend only to give diffuse scattering from the liquid, although the tip of the nozzle may cause shadowing at high angles and misalignment of the jet to the focused X-ray beam causes streaks perpendicular to the liquid column at low resolution. Unstable jets can have different and unpredictable scattering from pattern to pattern. The recent double-flow focus jet is particularly stable [37]. "Fixed target" raster-scanning of crystals at X-ray FELs (Chap. 5) may give variable shadowing due to the movement of the sample support, and the support itself gives diffuse scattering. Chips made of Kapton can produce rather sharp rings, and single-crystal silicon chips can produce non-symmetric diffuse scattering due to strain or thermal disorder. For a tomographic series collected at a synchrotron radiation facility, crystals are often mounted in nylon loops which give different scattering depending on the angle of rotation of the loop. This can be avoided by mounting a large crystal sticking out of the loop or by measuring only in a limited angular range where the loop does not come close to occluding the beam. Other variable contributions to the dataset include diffraction from ice or salt, which produce strong Bragg peaks or Debye-Scherrer rings. These can be identified and removed plotting a radial average curve (intensity vs detector radius or q), smoothing it and analyzing the difference between the original and smoothed curves. This method works rather well because ice or salt rings are usually quite sharp.

Measurement of single crystals at synchrotron radiation beamlines must contend with radiation damage and care taken not to exceed tolerable doses. While a thorough study of the effects of radiation damage on continuous diffraction has not yet been made, global damage will have the effect of increasing background and reducing contrast of the continuous diffraction, as indicated by an increase in β_a in Eq. (9.31).

The final arrangement of data into a 3D volume is carried out by interpolating each diffraction pattern onto the voxels of that array that intersect with the corresponding Ewald sphere. Besides scaling by the mean signal, Σ, for each pattern, the voxel values must be divided by the number of observations in that particular voxel. This tends to be large for low-resolution voxels and reduces approximately as $1/q$ until the center edge of the detector is reached. For serial crystallography from randomly oriented crystals, this usually defines the boundary of accurately measured data since the detector corners give much lower coverage. In such experiments we find it is not usually the best strategy to include every indexed pattern in the 3D merge but to keep only the strongest with the best signal to noise ratio of the continuous diffraction. Adding a large number of weak patterns usually increases noise without the benefit of improved signal. Outlying patterns that have poor correlation with the constructed 3D volume can be excluded, although this process may introduce bias.

Box 9.3

Processing of the measured data for further analysis consists of several steps:

- Determination of the exact geometry of the experiment (relative position of the detector with respect to the sample) and masking out bad regions or pixels of the detector.
- Masking of statistically outlying regions in each pattern (such as ice or salt rings) and the removal of Bragg peaks.
- Correction of polarization and subtraction of symmetric background.
- Scaling of each pattern, either using the Bragg peaks (for example, from XDS output [26]) or by estimating continuous signal level from Wilson statistics [8].
- Merging of all diffraction patterns into a single 3D volume using geometrical information and beam parameters. For serial crystallography the orientation can be obtained by indexing [60].
- Final subtraction of the radially symmetrical background from the 3D merged data, based on Wilson statistics.
- If several datasets are merged, each 3D volume must be scaled, using again the procedure described in Chapman et al. [8].

Acknowledgements We acknowledge the Gottfried Wilhelm Leibniz Program of the DFG, and the European Research Council under the European Union's Seventh Framework Programme ERC Synergy Grant 609920 "AXSIS."

References

1. Ayyer, K., Yefanov, O. M., Oberthür, D., Roy-Chowdhury, S., Galli, L., Mariani, V., et al. (2016). Macromolecular diffractive imaging using imperfect crystals. *Nature, 530*, 202–206.
2. Bates, R. H. T. (1982). Fourier phase problems are uniquely solvable in more than one dimension: 1. Underlying theory. *Optik, 61*, 247–262.
3. Bernal, J. D., Fankuchen, I., & Perutz, M. (1938). An X-ray study of chymotrypsin and hæmoglobin. *Nature, 141*, 523–524.
4. Bragg, W. L., & Perutz, M. F. (1952). The structure of hæmoglobin. *Proceedings of the Royal Society of London, 213*, 425–435.
5. Bruck, Y., & Sodin, L. (1979). On the ambiguity of the image reconstruction problem. *Optics Communication, 30*, 304–308.
6. Caleman, C., Tîmneanu, N., Martin, A. V., Jönsson, H. O., Aquila, A., Barty, A., et al. (2015). Ultrafast self-gating Bragg diffraction of exploding nanocrystals in an X-ray laser. *Optics Express, 23*, 1213–1231.

7. Chapman, H. N., Barty, A., Marchesini, S., Noy, A., Hau-Riege, S. P., Cui, C., et al. (2006). High-resolution ab initio three-dimensional X-ray diffraction microscopy. *Journal of the Optical Society of America A, 23*, 1179–1200.
8. Chapman, H. N., Yefanov, O. M., Ayyer, K., White, T. A., Barty, A., Morgan, A., et al. (2017). Continuous diffraction of molecules and disordered molecular crystals. *Journal of Applied Crystallography, 50*, 1084–1103.
9. Clarage, J. B., Clarage, M. S., Phillips, W. C., Sweet, R. M., & Caspar, D. L. D. (1992). Correlations of atomic movements in lysozyme crystals. *Proteins: Structure, Function, and Bioinformatics, 12*(2), 145–157.
10. Cowley, J. M. (1981). *Diffraction physics*. Amsterdam: North-Holland.
11. Cowtan, K. (1998). Introduction to density modification. In *Direct methods for solving macromolecular structures*. Dordrecht: Springer.
12. Crimmins, T. R., Fienup, J., & Thelen, B. J. (1990). Improved bounds on object support from autocorrelation support and application to phase retrieval. *Journal of the Optical Society of America A, 7*, 3–13.
13. Crowther, R., DeRosier, D., & Klug, A. (1970). The reconstruction of a three-dimensional structure from its projections and its applications to electron microscopy. *Proceedings of the Royal Society of London, 317*, 319–340.
14. Elser, V. (2003). Phase retrieval by iterated projections. *Journal of the Optical Society of America A, 20*, 40–55.
15. Elser, V. (2013). Direct phasing of nanocrystal diffraction. *Acta Crystallographica Section A, 69*, 559–569.
16. Elser, V., & Millane, R. P. (2008). Reconstruction of an object from its symmetry-averaged diffraction pattern. *Acta Crystallographica Section A, 64*, 273–279.
17. Fienup, J. R. (1978). Reconstruction of an object from the modulus of its Fourier transform. *Optics Letters, 3*, 27–29.
18. Fienup, J. R. (1982). Phase retrieval algorithms: a comparison. *Applied Optics, 21*, 2758–2769.
19. Flewett, S., Quiney, H. M., Tran, C. Q., & Nugent, K. A. (2009). Extracting coherent modes from partially coherent wavefields. *Optics Letters, 34*, 2198–2200.
20. French, S., & Wilson, K. (1978). On the treatment of negative intensity observations. *Acta Crystallographica Section A, 34*, 517–525.
21. Gerchberg, R. W., & Saxton, O. (1972). Practical algorithm for determination of phase from image and diffraction plane pictures. *Optik, 35*, 237–246.
22. He, H., & Su, W.-P. (2015). Direct phasing of protein crystals with high solvent content. *Acta Crystallographica Section A, 71*, 92–98.
23. He, H., Fang, H., Miller, M. D., Phillips, G. N. Jr., & Su, W.-P. (2016). Improving the efficiency of molecular replacement by utilizing a new iterative transform phasing algorithm. *Acta Crystallographica Section A, 72*, 539–547.
24. Hensley, C. J., Yang, J., & Centurion, M. (2012). Imaging of isolated molecules with ultrafast electron pulses. *Physical Review Letters, 109*, 133, 202.
25. Howells, M. R., Beetz, T., Chapman, H. N., Cui, C., Holton, J. M., Jacobsen, C. J., et al. (2009). An assessment of the resolution limitation due to radiation-damage in X-ray diffraction microscopy. *Journal of Electron Spectroscopy and Related Phenomena, 170*, 4–12.
26. Kabsch, W. (2010). Integration, scaling, space-group assignment and post-refinement. *Acta Crystallographica Section D, 66*, 133–144.
27. Kewish, C. M., Thibault, P., Bunk, O., & Pfeiffer, F. (2010). The potential for two-dimensional crystallography of membrane proteins at future X-ray free-electron laser sources. *New Journal of Physics, 12*, 035,005.
28. Marchesini, S. (2007). A unified evaluation of iterative projection algorithms for phase retrieval. *The Review of Scientific Instruments, 78*, 011301.
29. Marchesini, S., He, H., Chapman, H. N., Hau-Riege, S. P., Noy, A., Howells, M. R., et al. (2003). X-ray image reconstruction from a diffraction pattern alone. *Physical Review B, 68*, 140,101.

30. Miao, J., Sayre, D., & Chapman, H. N. (1998). Phase retrieval from the magnitude of the Fourier transforms of nonperiodic objects. *Journal of the Optical Society of America A, 15*, 1662–1669.
31. Miao, J., Charalambous, P., Kirz, J., & Sayre, D. (1999). Extending the methodology of X-ray crystallography to allow imaging of micrometre-sized non-crystalline specimens. *Nature, 400*, 342–344.
32. Millane, R. P. (1990). Phase retrieval in crystallography and optics. *Journal of the Optical Society of America A, 7*, 394–411.
33. Millane, R. P. (2017). The phase problem for one-dimensional crystals. *Acta Crystallographica Section A, 73*, 140–150.
34. Millane, R. P., & Lo, V. L. (2013). Iterative projection algorithms in protein crystallography. I. Theory. *Acta Crystallographica Section A, 69*, 517–527.
35. Mizuguchi, K., Kidera, A., & Gō, N. (1994). Collective motions in proteins investigated by X-ray diffuse scattering. *Proteins: Structure, Function, and Bioinformatics, 18*, 34–48.
36. Moore, P. B. (2009). On the relationship between diffraction patterns and motions in macromolecular crystals. *Structure, 17*, 1307–1315.
37. Oberthuer, D., Knoška, J., Wiedorn, M. O., Beyerlein, K. R., Bushnell, D. A., Kovaleva, E. G., et al. (2017). Double-flow focused liquid injector for efficient serial femtosecond crystallography. *Scientific Reports, 7*, 44628.
38. Oszlanyi, G., & Suto, A. (2004). Ab initio structure solution by charge flipping. *Acta Crystallographica Section A, 60*, 134–141.
39. Peck, A., Poitevin, F., & Lane, T. J. (2018). Intermolecular correlations are necessary to explain diffuse scattering from protein crystals. *IUCrJ, 5*, 211–222.
40. Rees, D. C. (1980). The influence of twinning by merohedry on intensity statistics. *Acta Crystallographica Section A, 36*, 578–581.
41. Sayre, D. (2002). X-ray crystallography: The past and present of the phase problem. *Structural Chemistry, 13*, 81–96.
42. Schmidt, E., & Neder, R. B. (2017). Diffuse single-crystal scattering corrected for molecular form factor effects. *Acta Crystallographica Section A, 73*, 231–237.
43. Shannon, C. E. (1949). Communication in the presence of noise. *Proceedings of the IRE, 37*, 10–21.
44. Shapiro, D., Thibault, P., Beetz, T., Elser, V., Howells, M., Jacobsen, C., et al. (2005). Biological imaging by soft X-ray diffraction microscopy. *Proceedings of the National Academy of Sciences, 102*, 15343–15346.
45. Simonov, A., Weber, T., & Steurer, W. (2014). Experimental uncertainties of three-dimensional pair distribution function investigations exemplified on the diffuse scattering from a tris-*tert*-butyl-1,3,5-benzene tricarboxamide single crystal. *Journal of Applied Crystallography, 47*, 2011–2018.
46. Simonov, A., Weber, T., & Goodwin, A. (2017). Single crystal diffuse scattering—A solution to the phase problem? Acta Crystallographica Section A, 73, C1045.
47. Spence, J. C. H., Weierstall, U., Fricke, T. T., Glaeser, R. M., & Downing, K. H. (2003). Three-dimensional diffractive imaging for crystalline monolayers with one-dimensional compact support. *Journal of Structural Biology, 144*, 209–218.
48. Spence, J. C. H., & Doak, R. B. (2004). Single molecule diffraction. *Physical Review Letters, 92*, 198102.
49. Spence, J. C. H., Kirian, R. A., Wang, X., Weierstall, U., Schmidt, K. E., White, T. et al. (2011) Phasing of coherent femtosecond X-ray diffraction from size-varying nanocrystals. *Optics Express, 19*, 2866–2873.
50. Stroud, R. M., & Agard, D. A. (1979). Structure determination of asymmetric membrane profiles using an iterative Fourier method. *Biophysical Journal, 25*, 495–512.
51. Szoke, A. (1999). Time-resolved holographic diffraction at atomic resolution. *Chemical Physics Letters, 313*, 778–788.
52. Szoke, A. (2001). Diffraction of partially coherent X-rays and the crystallographic phase problem. *Acta Crystallographica Section A, 57*, 586–603.

53. von Laue, M. (1936). The external shape of crystals and its influence on interference phenomena in crystalline lattices. *Annales de Physique, 26*, 55–68.
54. Waasmaier, D., & Kirfel, A. (1995). New analytical scattering-factor functions for free atoms and ions. Acta Crystallographica Section A, 51, 416.
55. Welberry, T. R. (1985). Diffuse X-ray scattering models of disorder. *Reports on Progress in Physics, 48*, 1543–1593.
56. White, T. A., Kirian, R. A., Martin, A. V., Aquila, A., Nass, K., Barty, A., et al. (2012). *CrystFEL*: a software suite for snapshot serial crystallography. *Journal of Applied Crystallography, 45*, 335–341.
57. White, T. A., Mariani, V., Brehm, W., Yefanov, O., Barty, A., Beyerlein, K. R., et al. (2016). Recent developments in *CrystFEL*. *Journal of Applied Crystallography, 49*, 680–689.
58. Whitehead, L. W., Williams, G. J., Quiney, H. M., Vine, D. J., Dilanian, R. A., Flewett, S., et al. (2009). Diffractive imaging using partially coherent X rays. *Physical Review Letters, 103*, 243902.
59. Wilson, A. J. C. (1949). The probability distribution of X-ray intensities. *Acta Crystallographica, 2*, 318–321.
60. Yefanov, O., Gati, O., Bourenkov, G., Kirian, R. A., White, T. A., Spence, J. C. H., et al. (2014). Mapping the continuous reciprocal space intensity distribution of X-ray serial crystallography. *Philosophical Transactions of the Royal Society B, 369*, 1647.
61. Yefanov, O., Mariani, V., Gati, C., White, T. A., Chapman, H. N., & Barty, A. (2015). Accurate determination of segmented X-ray detector geometry. *Optics Express, 23*, 28459–28470.

Chapter 10
Advances in Structure Determination of G Protein-Coupled Receptors by SFX

Benjamin Stauch, Linda Johansson, Andrii Ishchenko, Gye Won Han, Alexander Batyuk, and Vadim Cherezov

10.1 Introduction

G protein-coupled receptors (GPCRs) are molecular gateways to the cell, allowing it to sense its environment and to communicate with other cells by decoding messages conveyed via diffusible signaling molecules. GPCRs are located on the cell surface and recognize a plethora of extracellular molecules, such as hormones, odorants, neurotransmitters, proteins, lipids, photons, and ions [1, 2], transmitting signals inside of the cell, primarily through coupling to heterotrimeric G proteins, but also through arrestins and other G protein-independent pathways [3–5]. This exposed cell surface location makes GPCRs ideal targets for therapeutic intervention. Indeed, about a third of current drugs approved by the FDA directly target GPCRs [6, 7]. However, out of over 800 human GPCRs, fewer than 110 currently have drugs designed for them [7, 8], while over 100 receptors remain orphans [9, 10], for which even endogenous ligands are unknown.

GPCRs undergo large-scale conformational changes between active and inactive conformations (Fig. 10.1a), with the active conformation making them amenable to interaction with G proteins and arrestins [3], initiating canonical signaling pathways and leading to intracellular responses. GPCR ligands can be classified based on their pharmacological efficacy [4] into agonists, (neutral) antagonists, and inverse agonists (Fig. 10.1a). While receptors are usually activated by agonist molecules, they can also display basal activity in the absence of any ligand. Molecules reducing

B. Stauch · L. Johansson · A. Ishchenko · G. W. Han · V. Cherezov (✉)
Department of Chemistry, Bridge Institute, University of Southern California, Los Angeles, CA, USA
e-mail: cherezov@usc.edu

A. Batyuk
Linac Coherent Light Source, Stanford Linear Accelerator Center (SLAC) National Accelerator Laboratory, Menlo Park, CA, USA

S. Boutet et al. (eds.), *X-ray Free Electron Lasers*,
https://doi.org/10.1007/978-3-030-00551-1_10

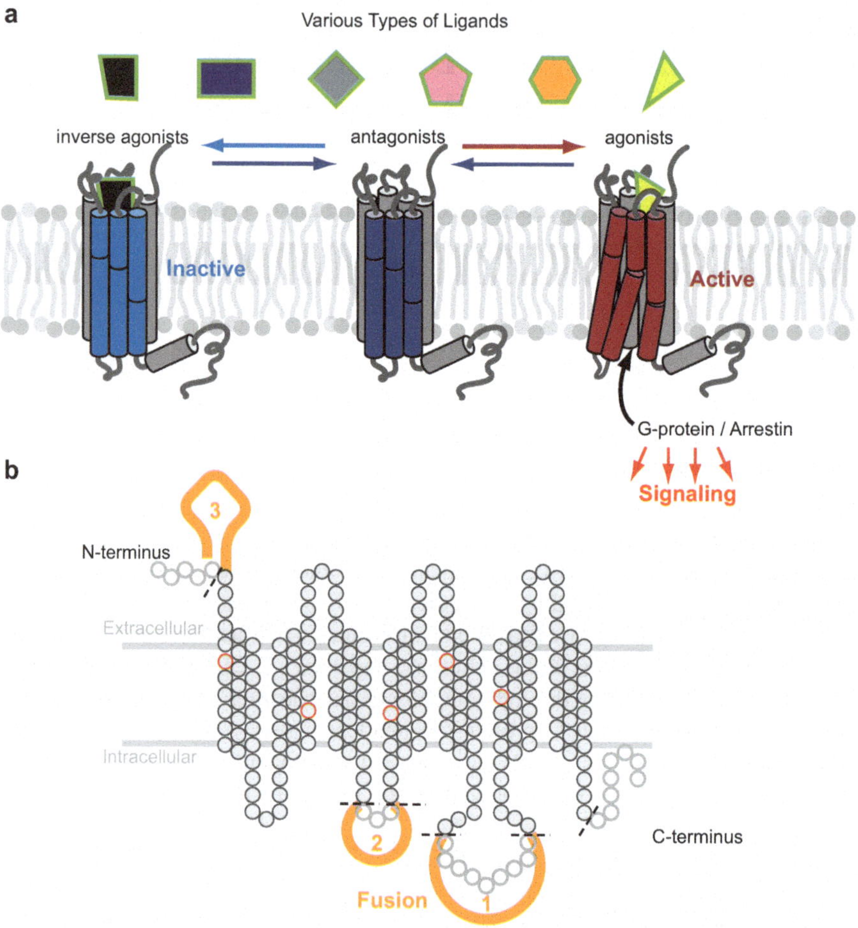

Fig. 10.1 GPCR activation, engineering, and stabilization. (**a**) Conformational plasticity of GPCRs. An unliganded apo receptor can sample multiple conformational states. Inverse agonists stabilize the inactive receptor conformation. Agonist binding triggers series of conformational rearrangements in the 7TM bundle leading to large-scale conformational changes on the intracellular side, most notably an outward movement of helix VI, priming the receptor for G protein binding. (**b**) GPCRs can be conformationally stabilized, for example by site-specific mutations (red circles) and by insertion of fusion partners before the helical bundle, or into ICL3 or ICL2, increasing their yield and stability and making them amenable to crystallization

receptor activity below the basal level are called inverse agonists, and molecules that occupy receptor binding sites without changing its activity level are called (neutral) antagonists. The inherent conformational flexibility of the receptor is an important reason for the difficulty to crystallize them.

High-resolution structure determination is a prerequisite for rational drug design and can provide a structural basis for understanding the molecular determinants of signaling. In recent years, three key developments have facilitated high-resolution structure determination of GPCRs by crystallography: (1) stabilization of the receptor by protein engineering, that is, receptor truncations, point mutations [11], and fusions with soluble protein domains [12] (Fig. 10.1b); (2) crystallization in lipidic cubic phase (LCP), a membrane mimetic, native-like crystallization environment [13]; and (3) advances in micro-crystallography [14]. Furthermore, specific conformational states of the receptors can be stabilized by ligands, allosteric modulators, peptides, antibodies, nanobodies [15, 16], and signaling partners, among others, further increasing their propensity to crystallize [17]. Since the first high-resolution structure of the human β_2-adrenergic receptor bound to a diffusible ligand was published in 2007 [18], overall 45 structures of unique GPCRs have been determined to date (November 2017; Fig. 10.2). Most of these receptors were crystallized in an inactive conformation that is stabilized by an antagonist, or an inverse agonist. Several of the available agonist-bound structures display conformational signatures of an active state; however, only a few of them are captured in a fully active state in complex with a G protein, arrestin, or their mimetics [20–22].

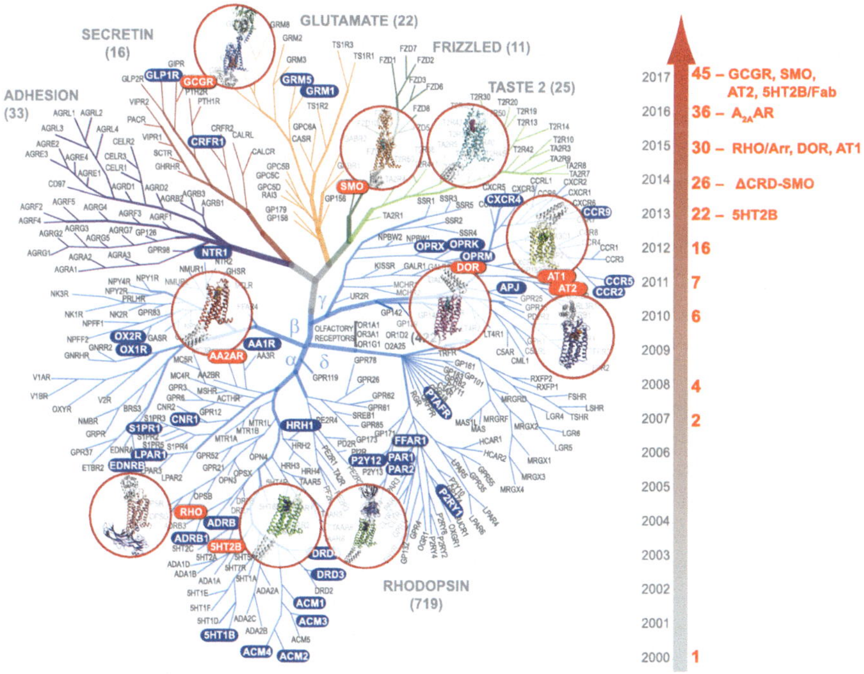

Fig. 10.2 Timeline of GPCR structure determination. A human GPCR sequence homology tree is shown on the left. X-ray crystal structures (highlighted in blue) have been obtained for representatives of all non-olfactory GPCR classes except for Adhesion. There has been a steady increase in the number of GPCR structures over the years (right). GPCR structures obtained by LCP-SFX are highlighted in red and encircled on the tree (left), and their names are shown on the timeline (right). The figure is modified from Ref. [19]

GPCRs share a general topology of a seven transmembrane helical bundle (7TM) with an extracellular N-terminus and an intracellular C-terminus, which often includes a short intracellular amphipathic helix VIII. The transmembrane helices are connected by three intracellular (ICL) and three extracellular loops (ECL), where ECL2 often plays a critical role in ligand binding, while ICL2 and ICL3 engage in G protein and arrestin coupling. Fusion partners are typically inserted in ICL3 or ICL2, or placed at the N-terminus before the 7TM bundle (Fig. 10.1b). Posttranslational modifications of termini and loops are common; for example, the N-terminus and ECLs can be glycosylated, ICLs and C-terminus phosphorylated, and the C-terminus palmitoylated.

While overall sequence identity between different receptors is low, based on their common topology, conserved sequence features, and motifs, as well as presence of extracellular domains, GPCRs are typically grouped into five classes, that is, class A (rhodopsin-like), which constitutes the largest class, class B (secretin-like), class C (glutamate-like), class Frizzled/Taste2, and class Adhesion (GRAFS classification system [23], Fig. 10.2). Typically, class A receptors consist mostly of a 7TM, while class B receptors also contain an extracellular domain that binds peptide ligands [24]. Class C receptors also contain a large extracellular domain and form functional homodimers or heterodimers [25]. Initially, GPCR structure determination was focused on class A receptors and on the 7TM domains of class B, C, and Frizzled/Taste2 GPCRs; recently, structures of full-length non-class A receptors such as the glucagon receptor GCGR [26] and the smoothened receptor [27] have become available. Comparison between class A GPCRs has been facilitated by generic residue numbering schemes, for example the Ballesteros-Weinstein (BW) numbering [28], which predates experimental GPCR structure determination, and for every transmembrane residue denotes helix number X and residue position relative to the most conserved residue for that helix, which is assigned number X.50. More recently, the BW numbering has been updated based on more abundantly available structural information and extended to other GPCR classes [29]. With increasing coverage of structural space, as well as the availability of receptor complexes and full-length structures, rational structure-based drug design will become a routine application over the next few years, and a solid structural foundation will aid the understanding of endogenous signaling processes.

Despite recent technical advances and the impressive rate at which novel GPCR structures have become available, receptor crystallization remains extremely challenging and tedious. Initial GPCR crystal hits obtained via LCP crystallization are typically too small (<10 μm) for data collection at microfocus synchrotron beamlines due to a fast onset of radiation damage. Crystal optimization often takes months to years, and in some cases may fail to improve crystal size and quality [21, 30]. The recently emerged hard X-ray free electron lasers (FELs) have revolutionized structural biology by enabling high-resolution structure determination from micrometer-sized crystals at room temperature with minimal radiation damage and by providing access to ultrafast time-resolved conformational transitions in biological macromolecules [31]. Extremely short duration (femtoseconds) X-ray FEL pulses outrun radiation damage, and, since each extremely bright pulse destroys the sample, the data are typically collected using a serial femtosecond crystallography (SFX) approach, in which crystals are constantly replenished either by streaming them across the beam or by using a fast scanning of crystals deposited on a solid support, also known as a fixed target approach, as discussed in Chap. 5.

GPCR microcrystals grown in LCP are particularly suitable for SFX due to their typically high diffraction quality and small size. The first successful applications of LCP-SFX to GPCR microcrystals were enabled by the development of a special injector capable of streaming LCP [30] and by the introduction of concurrent sample preparation approaches [32, 33]. With structures of ten different receptors determined by LCP-SFX during the last 4 years (Fig. 10.2), the method has proven its strength in acquiring high-resolution structural information from microcrystals of challenging membrane protein targets. In this chapter, we will outline the major steps of GPCR sample preparation for LCP-SFX, summarize successful GPCR structure determination studies including de novo phasing, and conclude with future perspectives of applying X-ray FEL radiation to study structure and dynamics of GPCRs as well as other challenging proteins and their complexes.

10.2 GPCR Sample Preparation and Delivery for LCP-SFX

While in the case of traditional crystallography, a single large, well-ordered crystal is desired for data collection and structure determination, the LCP-SFX sample preparation and optimization are instead focused on obtaining sufficient quantities of micrometer-sized crystals uniformly dispersed in LCP at a high density. Therefore, the LCP-SFX sample preparation is generally conducted in two separate steps (Fig. 10.3). The first step consists of a construct optimization and high-throughput crystal screening and optimization in 96-well plates, while the second step involves scaling up crystallization volume approximately 1000 times by setting up crystallization in gas-tight syringes.

10.2.1 GPCR Construct Optimization and Screening

A major bottleneck in crystallization of GPCRs is related to their low expression level and highly dynamic nature, and therefore, a range of modifications of the wild-type receptors is typically necessary to increase their yield and stability. These modifications include N- and C-terminal truncations, point mutations [11], and fusion partner insertions [12] (Fig. 10.1b and Table 10.1). The resulting modified constructs are expressed in insect or mammalian cells, solubilized in detergent micelles, purified by immobilized metal affinity chromatography (IMAC), and characterized by a number of assays. Several iterations of construct engineering are typically required to obtain a highly stable and monodisperse receptor suitable for crystallization (Fig. 10.3). Another prerequisite for a stable receptor sample

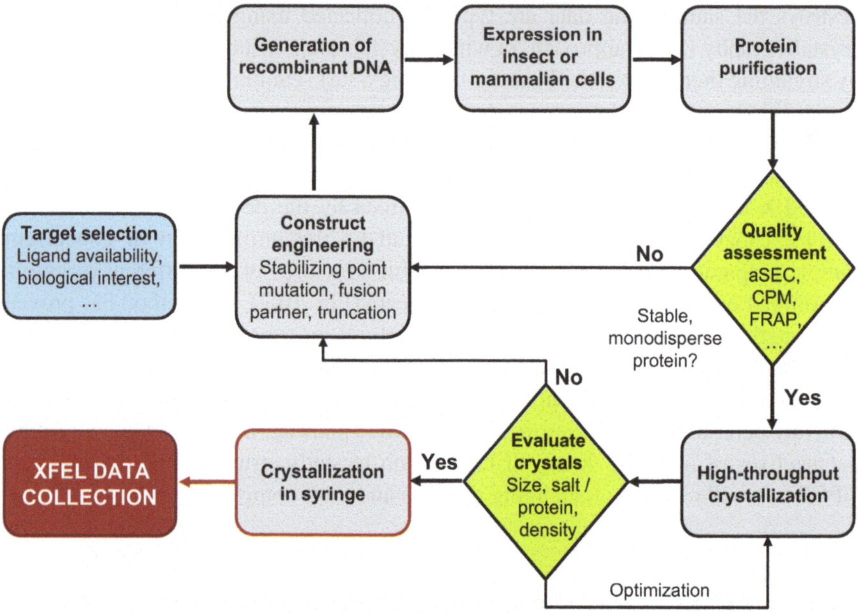

Fig. 10.3 A flow chart for GPCR sample preparation for LCP-SFX data collection. The GPCR structure determination process consists of two general steps (black and red outlines) and involves feedback loops with several stages of evaluation and quality assessment (yellow)

is the identification of a high-affinity ligand that keeps the receptor in either an inactive (antagonist or inverse agonist bound) or an active (active-like) conformation (agonist bound). The influence of ligands on the receptor construct can be evaluated using a fluorescence-based thermal shift assay which utilizes native buried cysteine residues of the receptor for covalent binding of the thiol-specific fluorophore N-[4-(7-diethylamino-4-methyl-3-coumarinyl)phenyl]maleimide (CPM) [39]. As the sample temperature is increased, the receptor unfolds and exposes its previously buried cysteines to binding by the dye. The subsequent increase in fluorescence can be measured as a readout for the overall unfolding of the protein. Typically, ligands that yield higher melting temperatures and therefore more stable receptor-ligand complexes are prioritized for subsequent trials.

Once a stable and monodisperse construct has been identified, the potential for crystallization is examined using the fluorescence recovery after photobleaching (or LCP-FRAP) assay [40]. In this assay, purified protein is labeled with the fluorescent dye 5,5′-disulfato-1′-ethyl-3,3,3′,3′-tetramethylindocarbocyanine (Cy3), reconstituted in LCP, and set up in 96-well glass sandwich plates against a set of crystallization screening solutions. A small, few micrometer-sized spot in the LCP drop is then photobleached with a laser, after which the recovery of fluorescence in this spot is monitored over time, providing information about protein diffusion. If the receptor-ligand complex is found to diffuse well in LCP, it is promoted to

Table 10.1 Construct design and crystallization conditions for GPCR structures determined by LCP-SFX

Receptor/ligand		$5\text{-}HT_{2B}$/ ergotamine	ΔCRD-SMO/ cyclopamine	δ-OR/ DIPP-NH_2	AT_1R/ ZD7155	AT_2R/ Cpd 1	Rhodopsin–arrestin complex	GCGR/ NNC0640-Fab	SMO/ TC114	$A_{2A}AR$/ ZM241385	$5\text{-}HT_{2B}$/ ergotamine-Fab
Construct design	N-term truncation	1–35	CRD (1–189)	1–38	1,7–16	1–34	None	None	1–52	None	1–35
	C-term truncation	406–481	556–787	339–372	320–359	336–363	None	433–477	559–787	317–412	406–481
	Fusion protein	ICL3 BRIL (T249–V313)	ICL3 BRIL (P434–K440)	N-term BRIL	N-term BRIL	N-term BRIL with GSGS linker	N-term T4L	ICL2 T4L (A256–E260)	ICL3 T4L (P434–S443)	ICL3 BRIL (K209–G218)	ICL3 BRIL (T249–V313)
	Mutations	M144 W					N2C,E113Q, M257Y,N282C		E194M		M144 W
	Partner in complex						15 residue linker, Arrestin (L374A, V375A, F376A mutation)	mAb1 Fab			P2C2-Fab

(continued)

Table 10.1 (continued)

Receptor/ligand		5-HT_{2B}/ ergotamine	ΔCRD-SMO/ cyclopamine	δ-OR/ DIPP-NH_2	AT_1R/ ZD7155	AT_2R/ Cpd 1	Rhodopsin–arrestin complex	GCGR/ NNC0640-Fab	SMO/ TC114	$A_{2A}AR$ /ZM241385	5-HT_{2B}/ ergotamine-Fab
Crystallization condition	LCP host lipid[a]	9.9 MAG	9.9 MAG	9.9 MAG	9.9 MAG	9.9 MAG	9.7 MAG	9.9 MAG	9.9 MAG	9.9 MAG	9.7 MAG
	Precipitant	0.1 M Tris pH 8.0, 30%v/v PEG400, 20–80 mM $MgCl_2$	0.1 M Hepes pH 7.0, 30%v/v PEG400, 100 mM NaCl	0.1 M MES pH 6.0, 30–32%v/v PEG400, 50–180 mM Li citrate	0.1 M Na citrate pH 5.0, 28%v/v PEG400. 450 mM NH4H2PO4, 4% v/v DMSO	0.1 M Tris pH 8.0, 25 mM K formate, 25%v/v PEG400, 0.3%v/v MPD	0.15 M NH4PO4 pH 6.4, 32%v/v PEG400	0.1 M HEPES, pH 7.0, 300 mM KH2PO4, 25%v/v PEG500DME, 100 mM Gly-Gly-Gly	0.1 M Na Citrate pH 5.0, 36%v/v PEG400, 150 mM NH4Cl	0.1 M Na citrate pH 5.0, 28%v/v PEG400, 40 mM NaSCN	0.1 M Tris pH 7.7, 60 mM Na/K tartrate, 25%v/v PEG400
Crystal size (μm^3)		5 × 5 × 5	1–5	5 × 2 × 2	10 × 2 × 2	5 × 2 × 2	5–10	~5	5 × 5 × 2	5 × 5 × 2	1–10
Ref		[21]	[30]	[34]	[35]	[36]	[21]	[26]	[27]	[37]	[38]

[a] All lipid host mixtures contained 10%w/w cholesterol

crystallization trials, while if the outcome is negative (no apparent protein diffusion), further optimization of the receptor-ligand combination is required. Additionally, LCP-FRAP experiments have shown to be extremely useful for selecting the most promising precipitant conditions for subsequent crystallization trials and in fact, many initial crystal leads have been found in such setups with high sensitivity and specificity due to the fluorescent labeling.

10.2.2 High-Throughput Crystallization Screening and Optimization

High-throughput crystallization trials are performed in 96-well glass sandwich crystallization plates, where 40 nL LCP boli containing purified protein are overlaid with 800 nL of precipitant solutions (Fig. 10.4a). Typically, a purified receptor at a concentration of 20–50 mg/mL is reconstituted in LCP by mixing with molten host lipid monoolein containing 10% w/w cholesterol at a 2:3 v/v ratio and then dispensed onto glass sandwich plates using either manual setups or an LCP crystallization robot [13]. The resulting drops are incubated at 20 °C and imaged at regular intervals. Conditions exhibiting showers of small crystals (typically in the low micrometer range) are chosen for the subsequent scale-up in syringes (Fig. 10.4b). Since LCP-SFX data collection requires a high density of uniformly sized microcrystals, samples can be further characterized using bright-field and cross-polarized light microscopy, Second-Order Nonlinear Imaging of Chiral Crystals (SONICC) [41], and Transmission Electron Microscopy (TEM) [42] (Fig. 10.4c) (for further reading about detection and characterization of microcrystals, see Chap. 3).

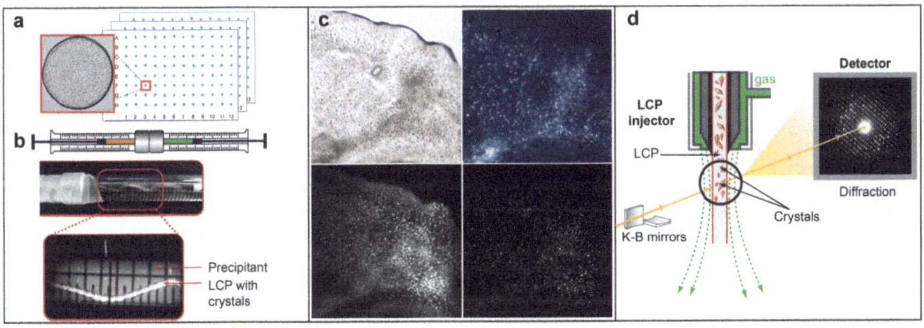

Fig. 10.4 Sample preparation and LCP-SFX data collection. (**a**) High-throughput nanovolume crystallization screening in 96-well glass sandwich plates. (**b**) Scale-up of crystallization in syringes. (**c**) Microcrystal visualization and characterization using different imaging modes (bright-field, cross-polarizers, two-photon UV fluorescence, SONICC). (**d**) A layout for an LCP-SFX experiment. These figures are reproduced from Ref. [33]

10.2.3 Crystallization Scale-Up in Syringes

Structure determination by LCP-SFX requires a dataset typically containing a few tens of thousands indexed diffraction patterns from microcrystals intersecting the X-ray beam in random orientations (Table 10.2). In practice, this translates into a sample volume of 50–100 μL with a crystal density of $\sim 10^5$ μL^{-1} [33], meaning that conditions optimized in 96-well sandwich plates should be scaled up at least 1000 times by volume. Scaling up LCP crystallization setups, however, is not always straightforward because of slow rates of precipitant diffusion through the LCP volume. In the sandwich plates, the LCP bolus is squeezed between two glass slides forming a disk-like shape, ~500 μm in diameter, with precipitant diffusing into LCP around the perimeter of the disk. Therefore, a proper method for scaling up is by mimicking the same geometry as much as possible, which could be accomplished by introducing LCP as an extended filament of the diameter ~500 μm inside a reservoir filled with the precipitant solution. In practice, this is achieved by injecting ~5–8 μL aliquots of protein-laded LCP as a continuous string into 100 μL Hamilton gas-tight syringes prefilled with a 10–15-fold excess of the precipitant solution and incubating them at 20 °C until crystals form (Fig. 10.4b) [33]. However, since the overall geometry of the syringe setup is not completely identical to that of the sandwich plates, small optimizations of the crystallization conditions are often necessary to ensure that a sample with a sufficiently high density of micrometer-sized crystals is produced [33].

Just before starting LCP-SFX data collection, samples from several syringes are combined together by expelling the excess of precipitant solutions from each syringe and consolidating their content into a single syringe. After combining samples, the remaining precipitant solution has to be absorbed to yield a homogeneous LCP sample capable of being run in the injector. This is achieved by adding approximately 10% v/v of 7.9 MAG lipid [30] (a *cis*-monounsaturated 1-monoacylglycerol, referred here by the N. T MAG notation, where N represents the number of carbons between the ester bond and the double bond and T corresponds to the number of carbons between the double bond and the terminal methyl group). The addition of this lipid helps to prevent the formation of a lamellar crystalline phase, which may occur due to evaporative cooling and dehydration upon injection of the sample in a vacuum environment [44]. If the sample is intended to be injected in the chamber at ambient pressure, the same lipid as the one used for crystallization (e.g., monoolein) can be added in this step. At the same time, it is also possible to dilute the sample with freshly prepared LCP, should the crystal density be deemed too high, which should help to avoid recording multiple crystal diffraction patterns in individual detector frames during data collection.

Due to the viscous nature of LCP, a special LCP injector (also known as a High Viscosity Extrusion injector as described in Chap. 5) was designed that is capable of streaming LCP at a wide range of flow rates (0.001–3 μL/min) [30]. LCP sample is typically loaded into a 25 or 40 μL reservoir and extruded through a 15–70 μm diameter capillary using a hydraulic plunger driven by an HPLC (High-Performance

Table 10.2 SFX data collection and processing statistics

Receptor/ ligand	5-HT$_{2B}$/ ergotamine	ΔCRD-SMO/ cyclopamine	δ-OR/ DIPP-NH$_2$	AT$_1$R/ ZD7155	AT$_2$R/Cpd 1		Rhodopsin–arrestin complex	GCGR/ NNC0640-Fab	SMO/ TC114	A$_{2A}$AR/ ZM241385	5-HT$_{2B}$/ ergotamine-Fab
PDB code	4NC3	4O9R	4RWD	4YAY	5UNF	5UNG	5W0P	5XEZ	5 V56	5K2C[a]	5TUD
Resolution (Å)	2.8	3.4/3.2/4.0	2.7	2.9	2.8	2.8	3.6/3.6/3.0	3.0	2.9	1.9	3.0
Space group	C222$_1$	P2$_1$	C2	C2	P2$_1$	P2$_1$22$_1$	P2$_1$2$_1$2$_1$	P2$_1$	P2$_1$	C222$_1$	P2$_1$
Cell parameters											
a, (Å)	61.5	40.5	156.2	72.8	77.4	70.3	109.2	72.6	40.6	40.4	71.2
b, (Å)	122.2	157.3	89.3	41.0	104.3	78.8	109.2	245.3	349.5	180.7	118.8
c, (Å)	168.5	52.4	96.4	167.7	90.1	93.4	452.6	96.2	61.8	142.8	145.6
β, (°)		97.0	92.3	99.4	104.3			90.01	101.1		90.5

(continued)

Table 10.2 (continued)

Receptor/ ligand	5-HT_{2B}/ ergotamine	ΔCRD-SMO/ cyclopamine	δ-OR/ DIPP-NH_2	AT_1R/ ZD7155	AT_2R/Cpd 1		Rhodopsin–arrestin complex	GCGR/ NNC0640-Fab	SMO/ TC114	$A_{2A}AR$/ ZM241385	5-HT_{2B}/ ergotamine-Fab
Crystal size (μm^3)	5 × 5 × 5	1–5	5 × 2 × 2	10 × 2 × 2	5 × 2 × 2		5–10	~5	5 × 5 × 2	5 × 5 × 2	1–10
Data acquisition time (min)	590	490	275	385	375		700	140	300	135	260
No. of collected images	4,217,508	3,510,525	1,967,539	2,764,739	2,701,530		>5,000,000	>1,000,000	2,102,907	948,961	1,877,040
No. of hits/ indexed images	152,651/ 32,819	274,214/ 66,165	125,458/ 36,083	457,275/ 73,130	175,241/ 22,774	175,241/ 15,804	22,462/ 17,730	91,626/ 57,573	320,121/ 65,560	232,283/ 72,735	193,328/ 52,291
No. of total/ unique reflections	18,515,376/ 16,052	20,326,230/ 8082	19,431,440/ 34,699	14,415,424/ 11,190	1,412,692/ 22,934	1,139,069/ 13,330	21,915,320/ 76,360	13,392,502/ 67,598	13,578,966/ 37,101	12,215,692/ 41,882	150,927,26/ 48,643
R/Rfree	0.227/ 0.270	0.232/0.278	0.208/ 0.238	0.228/ 0.274	0.227/ 0.256	0.241/ 0.262	0.234/ 0.272	0.210/ 0.243	0.218/ 0.239	0.173/ 0.208	0.227/ 0.247
Ref	[21]	[30]	[34]	[35]	[36]		[43]	[26]	[27]	[37]	[38]

[a]Data are shown only for the highest resolution (1.9 Å) $A_{2A}AR$/ZM241385_SAD

Liquid Chromatography) pump connected through a tube filled with water. The extruded LCP stream is supported by a sheath of gas, typically helium or nitrogen, to keep it flowing straight. The LCP flow rate can be matched to the X-ray FEL pulse rate to supply fresh crystals for every shot, while simultaneously minimizing the sample waste between pulses. Unlike in the gas dynamic virtual nozzle (GDVN) injector [45], the sheath gas does not focus the LCP stream below the diameter of the capillary nozzle. Therefore, while smaller diameter nozzles can decrease scattering background, they are prone to clogging and require much higher pressures (up to 10,000 psi for a 15 μm nozzle) for successful LCP extrusion. The most optimal nozzles for sub-10 μm crystals have been empirically found to be 30–50 μm in diameter. Larger diameter nozzles can be used to increase hit rates for samples with lower crystal densities, while smaller diameter nozzles can help with reducing sample consumption and background scattering. By keeping the LCP-crystal stream diameter to 30–50 μm, an entire dataset can be collected using <0.3 mg of purified protein [30]. For comparison, in case of a GPCR expression in insect cells, a typical yield is ~1 mg of purified receptor per 1 L of expression media. Additional details on the setup for sample delivery of crystals in LCP are further discussed in Chap. 5.

10.2.4 Sample Selection for LCP-SFX Data Collection

Since beam time at X-ray FEL sources is extremely limited, it is important to pre-screen samples and select those with the best chances for a successful outcome. It is reasonable to expect that conditions yielding the best diffracting crystals at synchrotron sources would also perform comparatively well at an X-ray FEL. Therefore, pre-screening samples at a synchrotron, if feasible, can help with the selection of the most promising conditions for precipitants, salts, pH, and additives. Once such conditions have been identified, it is often sufficient to slightly adjust the concentrations of the main components to produce suitable samples for an X-ray FEL source.

Two of the most important parameters for samples prepared for LCP-SFX data collection are a high crystal density and an optimal crystal size, which are inversely related to each other. In general, GPCR crystals of 5–10 μm size are optimal for LCP-SFX since they produce sufficiently strong diffraction and compatible with relatively high crystal densities, which are required to achieve reasonable crystal hit rates (>5%). Unfortunately, no reliable procedure to concentrate crystals in LCP has been established yet, and, therefore, the crystal density cannot be increased once the crystals have already grown. Another important consideration, especially when collecting data from crystals injected into a vacuum environment, is to avoid precipitant components with low solubility, if possible. Such compounds can readily crystallize upon sample extrusion in vacuum and produce a strong powder diffraction, which could be potentially damaging to sensitive detectors.

In contrast to traditional goniometer-based crystallography, where a crystal is rotated during data collection, SFX data are collected from a large number of randomly oriented still (on the femtosecond time scale) crystals, one shot per crystal, and therefore all recorded reflection intensities are partial. The data are merged using Monte-Carlo approaches, meaning that the accuracy of derived structure factor amplitudes has a strong dependency on the multiplicity of the data, whereas the completeness very quickly reaches 100%. Thus, an SFX dataset is considered "complete" when a sufficient accuracy of the data is achieved, which typically requires an average multiplicity of a few hundred. With rapid advancements in the SFX data processing software [46–52], the required number of diffraction images to reach the desired data quality is constantly decreasing and currently constitutes of 10,000–30,000 indexed patterns. With a 5–10% hit rate, it takes approximately 30–50 μL of crystal-laden LCP to collect enough data at the X-ray FEL pulse repetition rate of 120 Hz, which, at a flow rate of 0.2 μL/min, translates into 2.5–4 h of continuous beamtime (Table 10.2).

10.3 Review of Published Structures

10.3.1 From Validation to Novel GPCR Structure Determination

The LCP-SFX method was first introduced and validated in 2013 through the high-resolution structure determination of the human serotonin 5-HT_{2B} receptor in complex with an agonist ergotamine [53], which was shown to be of comparable quality to the structure previously obtained using synchrotron data [54]. While the synchrotron structure was obtained using relatively large crystals ($80 \times 20 \times 10$ μm^3) at cryogenic temperatures, the data collection with an X-ray FEL was performed using much smaller $5 \times 5 \times 5$ μm^3 crystals and at room temperature. The results confirmed that the LCP-SFX method enables structure determination from sub-10-μm-sized GPCR crystals at room temperature and without apparent radiation damage effects, while providing more accurate insights into receptor structure and dynamics at close to native conditions. The discrepancies between the synchrotron and X-ray FEL data were found predominantly in several side-chain conformations of solvent-exposed amino acids, which supports the view that cryo-cooling of crystals used in synchrotron data collection can trap some side-chains in artificial conformations [53, 55].

After successful validation of LCP-SFX with 5-HT_{2B}/ergotamine, the method was further applied to solve the structure of the human smoothened receptor with the truncated cysteine-rich domain (ΔCRD-SMO) in complex with the teratogen cyclopamine [30]. SMO mediates signal transduction in the hedgehog pathway, which is implicated in normal embryonic cell development and in carcinogenesis. SMO antagonists can suppress the growth of some tumors; however, mutations in SMO have been found to abolish their antitumor effects, a phenomenon known as chemoresistance. Due to poor diffraction with high mosaicity and anisotropy of relatively large cryo-cooled crystals (~120 × 10 × 5 μm^3), the ΔCRD-SMO/cyclopamine structure could not be obtained at synchrotron sources. In contrast to the synchrotron data, the LCP-SFX data collected from sub-10-μm-sized crystals at room temperature were of reasonable quality, allowing for the structure to be solved by molecular replacement after application of an ellipsoidal data truncation at 3.4, 3.2, and 4.0 Å along the three principal crystal axes. The binding pose of the ligand cyclopamine within a narrow elongated binding cavity inside the 7TM domain of SMO was well resolved and provided the structural basis for understanding SMO receptor modulation and chemoresistance [56].

Alkaloid opiates, such as morphine, are effective and widely prescribed against moderate to severe pain. These drugs target μ-opioid receptor (μ-OR), which together with δ-OR and κ-OR play a crucial role in pain management, mood states, consciousness, and other neurophysiological phenomena. However, prolonged administration of opioids often leads to increased tolerance, dependence, and addiction. It was shown that co-administration of morphine with δ-OR antagonists helps to reduce morphine tolerance effects in rodents [57]. The H-Tyr-Tic-Phe-Phe-OH (TIPP) class of endomorphin-derived peptide analogs offers remarkable variety in efficacies with mixed δ-OR and μ-OR profiles. The LCP-SFX method was used to determine the structure of the human δ-OR bound to a bifunctional δ-OR antagonist and μ-OR agonist tetrapeptide H-Dmt-Tic-Phe-Phe-NH_2 (DIPP-NH_2) belonging to the TIPP class [34]. Initially, the X-ray crystal structure of the δ-OR–DIPP-NH_2 complex was determined at 3.3 Å resolution using the synchrotron X-ray diffraction from cryo-cooled crystals, however, the electron density for the DIPP-NH_2 peptide was not of sufficient quality to unambiguously trace it. By using LCP-SFX the resolution was dramatically improved to 2.7 Å, showing excellent density for the peptide ligand (Fig. 10.5a). The structure revealed crucial atomic details of the bifunctional pharmacological profile of DIPP-NH_2. Using a superposition with the previously solved structure of μ-OR, it was observed that DIPP-NH_2 clashes with the side chains of non-conserved residues Trp318$^{7.35}$ and Lys303$^{6.58}$, highlighting the importance of these residues for the bifunctional properties of the peptide. This structure also revealed important details of the peptide recognition by GPCRs, given that the structural information about peptide GPCR complexes is limited, making it valuable for structure-based drug design efforts.

10.3.2 First Novel GPCR Structures Solved Using LCP-SFX

The next important milestone of LCP-SFX was achieved in 2015 with the first structure determination of a novel GPCR, the human angiotensin II receptor type 1 (AT_1R) [35, 36]. AT_1R serves as a primary blood pressure regulator in the cardiovascular system. Although several AT_1R blockers (ARBs) have been developed and approved as antihypertensive drugs, the structural basis for AT_1R ligand-binding and regulation has remained elusive, mostly due to the difficulties of growing

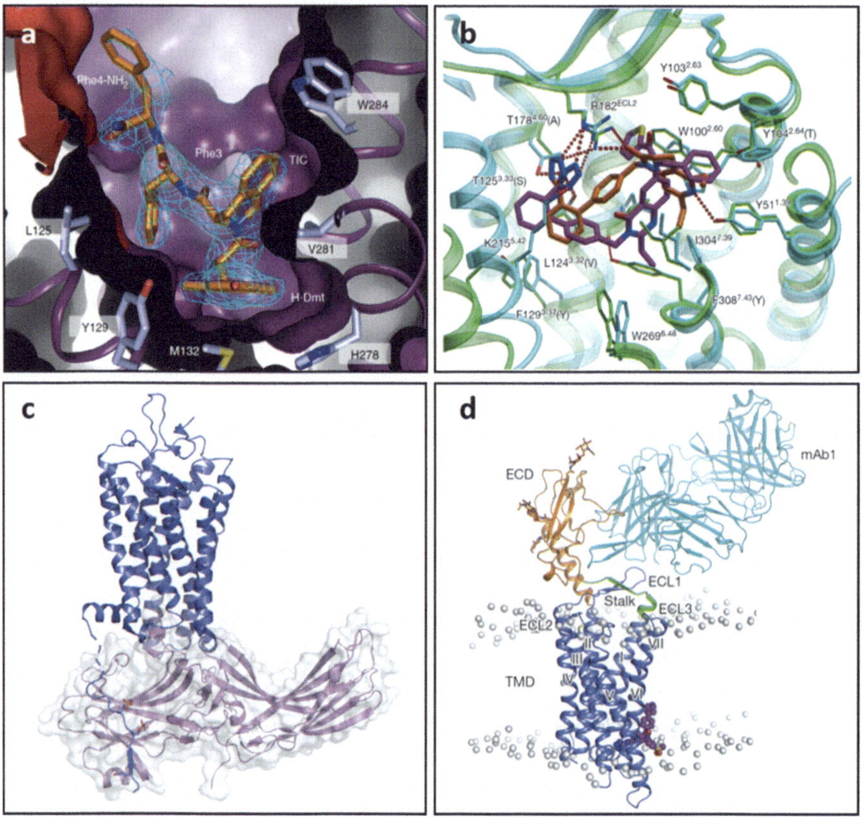

Fig. 10.5 Examples of GPCR structures solved by LCP-SFX. (**a**) The ligand binding pocket of the δ-opioid receptor in complex with DIPP-NH_2 (PDB ID 4RWD). The figure is reused from Ref. [34]. (**b**) A comparison of the ligand binding pose between two angiotensin receptors, AT_1R (orange ligand, PDB ID 4YAY) and AT_2R (purple ligand, PDB ID 5UNG). The figure is reused from Ref. [36]. (**c**) The structure of a complex between rhodopsin (blue) and arrestin (purple) (PDB ID 5W0P). (**d**) The full-length GCGR receptor in complex with a small molecule allosteric modulator (purple) and a monoclonal mAb1 antibody fragment (teal) (PDB ID 5XEZ). The figure is reused from Ref. [26]

high-quality crystals for structure determination using synchrotron radiation. By applying the LCP-SFX method, the first room-temperature crystal structure of the human AT_1R in complex with its selective antagonist ZD7155 was solved at 2.9 Å resolution [35] using crystals with an average size of $10 \times 2 \times 2\ \mu m^3$. This structure revealed the critical interactions between ZD7155 and the receptor and served as a basis for the binding mode determination of other ARBs by means of molecular docking [58].

Successful structure determination of AT_1R was followed by the work on the second angiotensin II receptor type 2 (AT_2R), which is another key component of the renin–angiotensin–aldosterone system. In contrast to the well-defined function of AT_1R, the function of AT_2R is unclear, with a variety of reported effects [59, 60]. The initial crystal hits were optimized to produce a high density of small crystals. Interestingly, analysis of the collected LCP-SFX data revealed that the crystal suspension contained two distinct crystal forms in the same crystallization setup. Consequently, two structures of the human AT_2R bound to an AT_2R-selective ligand (cpd-1) were determined at 2.8 Å resolution in two different space groups (monoclinic $P2_1$ and orthorhombic $P2_12_12_1$) [36]. Both structures captured the receptors in an active-like conformation. Unexpectedly, helix VIII was found in a noncanonical position. In most previously reported GPCR structures helix VIII runs parallel to the membrane on the intracellular side, whereas in the AT_2R structure helix VIII flips over to interact with the intracellular ends of helices III, V, and VI, thus stabilizing the active-like state, but at the same time sterically preventing the recruitment of G proteins or β-arrestins. This finding is in agreement with the absence of signaling responses of AT_2R in standard cellular assays [61, 62]. The AT_2R structure provides insights into the structural basis of the distinct functions of the angiotensin receptors and may guide the design of new selective ligands (Fig. 10.5b).

10.3.3 Rhodopsin–Arrestin Complex Structure

After success with structure determination of novel GPCRs, a phenomenal advantage of LCP-SFX over traditional crystallography was demonstrated with the determination of the first high-resolution structure of a major signaling complex between a GPCR and an arrestin [21]. Most GPCRs upon activation primarily signal via interactions with G proteins, followed by phosphorylation of their C-terminus and ICLs by G protein-coupled receptor kinases (GRKs). Activated and phosphorylated receptors are recognized by arrestins, which, in turn, block interaction with G proteins and induce internalization. Arrestins are also responsible for triggering a variety of G protein-independent signaling cascades, and hence they constitute an essential component of GPCR signaling pathways. Biased GPCR ligands that selectively direct signaling towards specific pathways bear significant therapeutic benefits with fewer side effects compared to unbiased ligands. The first GPCR signaling complex of β_2-adrenergic receptor bound to its cognate Gs protein was solved in 2011 [22]. Structural details of arrestin binding to GPCRs, however, remained undiscovered until 2015, when finally the first structure of the human visual rhodopsin in complex with arrestin was obtained by LCP-SFX [21].

Rhodopsin is a prototypical GPCR that functions as a photon receptor in the visual systems. Crystal structures of rhodopsin have been previously determined in various states: a ground inactive form [63], a partially active form (opsin) [64], and a fully active form in complex with a C-terminal peptide of G_α [65]. The structures of an inactive and preactivated arrestin are also available [66, 67]. Obtaining the structure of the rhodopsin–arrestin complex, however, required overcoming additional challenges. The wild-type complex is characterized by heterogeneous interaction modes between the rhodopsin and arrestin. Therefore, the proteins had to be engineered to stabilize a single state for crystallization [21]. A constitutively active mutant of the human visual rhodopsin (E113Q and M257Y) along with a pre-activated form of the mouse arrestin (3A arrestin with the mutations L374A, V375A, and F376A) were used to increase their mutual affinity. To further improve the stability and shift the equilibrium towards the complex formation, a polypeptide linker was designed to connect the C-terminus of rhodopsin with the N-terminus of arrestin. Initial crystals of the complex ranging in size between 5 and 20 μm were obtained using the LCP crystallization technique. Despite extensive optimization efforts, the diffraction quality could not be improved beyond 6–8 Å at synchrotron microfocus beamlines. Therefore, crystallization conditions were optimized to yield showers of 5–10-μm-sized crystals in syringe crystallization setups. Due to a relatively low hit rate, about 10 h of LCP-SFX data collection were required to solve the structure [21, 68]. Diffraction patterns from 18,874 crystals were indexed and integrated. The data were initially processed in the apparent Laue class 4/mmm with a large unit cell ($a = b = 109.2$ Å, $c = 452.6$ Å). Structure determination was, however, complicated by a pseudo-merohedral twinning, caused by the identical *a* and *b* axes, and a pseudotranslation along the *a* and *b* axes. Finally, the structure was successfully determined by lowering the lattice symmetry in the $P2_12_12_1$ spacegroup with four molecules in the asymmetric unit with perfect twinning using a twin law of $k, h, -l$. The diffraction was anisotropic with resolution limits of 3.8 Å and 3.3 Å along the *a*/*b* and *c* axes, respectively. The obtained structure represented the first crystal structure of a GPCR bound to arrestin and, together with additional biophysical and biochemical data, provided a molecular basis for understanding the mechanism of arrestin-mediated signaling [21].

The key findings of this study include the following observations: (1) rhodopsin, within the complex, adopts a canonical active state conformation, overall highly similar to the conformation of β_2AR in complex with Gs protein [22], except for small deviations of helices I, IV, V, and VII, some of which may be related to the mechanism of arrestin-biased signaling; (2) the phosphorylated C-terminal tail of rhodopsin is paired to the highly cationic N-terminal domain of arrestin, displacing its C-terminus and triggering arrestin activation; (3) activated arrestin undergoes a 20° rotation between its N- and C-domains, consequently opening a cleft between the middle loop and the C-loop to accommodate the ICL2 helix of rhodopsin; (4) additionally, the finger loop of arrestin adopts a short alpha-helical conformation, which fits in the opening created by the outward displacement of helix VI on the intracellular side of rhodopsin and interacts with helices VII and VIII; and (5) finally, a conserved hydrophobic patch at the C-tip of arrestin anchors it in the lipid bilayer helping to stabilize the arrestin-rhodopsin interactions.

More recently, improvements in data processing algorithms made it possible to re-process the LCP-SFX data for rhodopsin–arrestin complex leading to an increased resolution (3.6 Å and 3.0 Å along *a*/*b* and *c* axes respectively) and a better quality electron density maps [43]. The improved maps allowed the C-terminus of the receptor to be traced (Fig. 10.5d), helping to identify a set of phosphorylation codes for arrestin recruitment by GPCRs.

10.3.4 Full-Length Smoothened Receptor Structure

Due to challenges with crystallization of multidomain non-class A receptors, the first GPCR structural studies were focused on their 7TM domains [69, 70]. The progress in receptor stabilization techniques and the development of LCP-SFX facilitated the structure determination of the full-length receptors. While the initial structures of ΔCRD-SMO in complex with several 7TM antagonists and agonists shed light on the ligand binding poses and interactions with the 7TM domain [56, 70], the mutual arrangement of the extracellular domains (ECD), which include a hinge domain (HD) and an extracellular cysteine-rich domain (CRD), remained unknown. Since ECD plays an important role in ligand recognition and receptor activation through allosteric effects, this information was an important missing piece for a full mechanistic understanding of SMO function. Previous biochemical studies have revealed a ligand-binding site that is situated on the surface of the extracellular cysteine-rich domain, targeted by cholesterol-like molecules [71]. It was shown that CRD has an auto-inhibitory effect on SMO [72], whereas oxysterols release CRD suppression and activate the hedgehog pathway. In order to suggest a model of the SMO activation mechanism a structure of the multidomain human SMO in complex with a specially designed super-stabilizing ligand was solved using LCP-SFX and synchrotron data [27]. The structure revealed a hydrophobic pocket that is formed by CRD (residues V107, L108, L112), HD (residue V210), and ECL3 (residues V494, I496) and constitutes an oxysterol binding site. Comparison of these structures with the concomitantly published multidomain SMO structures in complex with vismodegib and cholesterol [73] revealed important structural features, namely, the CRD tilting angle was found to be different in all structures, along with rearrangements of ECL3 supporting this conformation. The structural data combined with hydrogen-deuterium exchange analysis and molecular dynamics simulations suggested a unique mechanism, in which helix VI, ECL3, and HD play a central role in the signal transmission across the receptor.

10.3.5 Full-Length Class B Glucagon Receptor Structure

Class B GPCRs (secretin family) are mostly peptide hormone receptors that are indispensable drug targets for metabolic diseases, like diabetes, cardiovascular disease, neurodegeneration, and some psychiatric disorders [74]. These receptors consist of an N-terminal extracellular domain (ECD) and a 7TM domain, both of which are required for binding to their endogenous peptide ligands and regulation of cell signal transduction. The glucagon receptor GCGR belongs to class B and plays a key role in glucose homeostasis and the pathophysiology of type 2 diabetes. It has long been targeted by structural studies, and the structure of the 7TM domain bound to a small molecule drug was solved by conventional synchrotron crystallography using crystals grown in LCP [69, 75]. Although this structure provided important information about the receptor, the lack of ECD limited our understanding of the receptor function. Crystallization of the full-length receptor required further efforts of construct optimization and, in particular, utilization of a stabilizing Fab (fragment antigen binding) antibody fragment. Eventually, the structure of the full-length human GCGR containing both ECD and 7TM domains in complex with a Fab fragment of an inhibitory antibody mAb1, and a negative allosteric modulator NNC0640 was determined at 3.0 Å resolution using the LCP-SFX method (Fig. 10.5d) [26]. As in most of the previous examples, the data collected from small crystals with an X-ray FEL showed superior quality compared to the synchrotron data, thereby improving the resolution of the dataset from 3.2 Å to 3.0 Å. Despite the challenge of the twinned data, the GCGR/NNC0640–mAb1 complex structure was solved by molecular replacement (MR). No substantial differences were observed between the X-ray FEL and the synchrotron structures with a backbone r.m.s.d. (root-mean-squared deviation) of 0.6 Å. The GCGR/NNC0640–mAb1 structure revealed an unexpected conformation of the stalk region, which links 7TM with ECD. Whereas in the initial structure of the 7TM domain it formed a three-turn α-helical extension of helix I, in the full-length structure the stalk adopts a β-strand conformation that runs across the helical bundle flanked by ECL1 on one side and ECL2 and ECL3 on the other. Given such a dramatic difference in the conformation of the stalk, the relative orientation between ECD and TMD revealed by the full-length structure was observed to be drastically different compared to a predicted model based on the 7TM structure alone. In addition, ECL1 was found to exhibit a β-hairpin conformation interacting with the stalk to form a compact β-sheet structure, potentially playing a critical role in modulating peptide ligand binding and receptor activation.

10.3.6 Structural Basis for GPCR Extracellular Recognition by Antibodies

With their growing success in clinical studies, monoclonal antibodies (mAbs) have become a critically important modality and a powerful alternative to small molecule therapies [76]. Recently developed mAbs have demonstrated a twice higher chance of approval by FDA than conventional small molecule compounds [77]. Despite their significant success compared to other approaches, there is still a considerable rate of failure with 85% of leads falling through the clinical trials, which emphasizes the need of deeper understanding of the underlying biology and interactions with antigens in particular. Due to their potentially high affinity, selectivity, long duration of action and engineered ability to penetrate the blood-brain barrier, mAbs are very suitable for targeting a large variety of GPCRs. Unfortunately, the most abundant class A GPCRs is characterized by relatively small extracellular solvent-exposed surface making the production of high affinity, selective mAbs very challenging. In order to gain insights into the molecular basis of extracellular recognition of GPCRs by mAbs, a complex between the human 5-hydroxytryptamine 2B (5-HT_{2B}) receptor bound to the agonist ergotamine (ERG) and a selective antibody

Fab fragment was crystallized and solved by means of LCP-SFX [38]. While previous structures of 5-HT_{2B}/ERG were captured in a partially activated state with only some of the activation features observed [78], the 5-HT_{2B}/ERG-Fab structure reveals the receptor in a distinct active-like state, with extensive activation-related changes displayed throughout the receptor including conserved activation "microswitches" and large-scale intracellular displacements of helices VI and VII [38]. This work also provided the first insight into structural determinants for extracellular GPCR recognition by mAbs, as all the previous structures of GPCR-antibody complexes contained Fabs/nanobodies bound to the intracellular side of the receptors. The 5-HT_{2B}/ERG-Fab structure, therefore, can be considered an important first step towards a rational development of therapeutic mAbs.

10.4 Experimental Phasing of XFEL Data for GPCRs

All novel GPCR structures obtained by LCP-SFX and discussed in this chapter were solved using the molecular replacement (MR) method that is based on previous knowledge of related structures. The overall conservation of the 7TM fold and the presence of the fusion domains of known structure make it relatively straightforward to generate search models for MR. However, a good search model for MR may not be available for all targets. For instance, in the case of GPCRs, the majority of structural information comprises class A receptors, whereas class B, C, and Frizzled are represented by a few structures only. If it is impossible to create an adequate model for an MR search, experimental phasing methods become necessary. These are typically based on the introduction of anomalous scatterers into crystals that do not change the target structure (i.e., are isomorphous). The first successful experimental phasing of SFX data was demonstrated with lysozyme crystals using single-wavelength anomalous dispersion (SAD) of gadolinium, which exhibits a very strong anomalous signal [79]. Attempts to use experimental phasing of SFX data for another test soluble protein, the luciferin-regenerating enzyme, with a more conventional mercury compound, succeeded by the use of the SIRAS (single isomorphous replacement with anomalous signal) method [80]. A very recent work of Colletier et al., [81] showed that experimental phasing using X-ray FELs could be achieved for crystals with an average size of 500 nm, which corresponds to approximately 50 unit cells per crystal edge. In this work, the experimental phases were derived using multiple isomorphous replacement with anomalous scattering (MIRAS) from combining three heavy-atom derivatives and a native dataset.

These methods rely on the incorporation of heavy atoms into protein crystals, which requires extensive screening of various compounds at different concentrations, while many of them suffer from poor solubility. For example, while derivatization with Ta_6Br_{12} clusters had previously been successful in the case of SMO and mGluR1 at synchrotron sources [70, 82], our attempts of using the same approach for X-ray FEL data collection with crystals of 5-HT_{2B} receptor had limited success. Ta_6Br_{12} precipitated and crystallized immediately upon delivery into the vacuum environment during the LCP-SFX data collection, which resulted in very bright Bragg reflections on the detector. To prevent detector damage the X-ray FEL beam had to be attenuated significantly, resulting in the anomalous signal extending to ~8 Å only. And although the Ta_6Br_{12} cluster was incorporated into crystals and could be located in the anomalous difference electron density map, the phasing attempts were unsuccessful. Moreover, quite often efficient isomorphous incorporation of heavy atom compounds in the crystal lattice is unattainable.

On the other hand, the sulfur SAD (S-SAD) phasing method allows for the determination of protein structures without additional modification of crystals such as heavy-atom derivatization or incorporation of selenomethionine. This method has been proposed by Hendrickson and Teeter in 1981 [83] and is becoming a more and more popular method for de novo experimental phasing due to advances in data collection techniques and data processing algorithms. Challenges, however, remain due to a very small anomalous signal from sulfur atoms requiring a very accurate measurement of the anomalous differences.

The first successful S-SAD phasing of SFX data has been demonstrated for lysozyme crystals [79] followed by another test protein, thaumatin [84]. These proteins are widely used as crystallization standards and in the development of new crystallographic methods, because they are commercially available and inexpensive, their crystal suspensions are stable for years, and their crystals diffract to high resolution. The highest reported resolution for lysozyme crystals is 0.65 Å (PDB ID: 2VB1) and for thaumatin crystals is 0.94 Å (PDB ID: 2VHK). Most GPCR crystals, however, diffract in the range of 2.4–3.4 Å with only several receptors diffracting better than 2.4 Å, for example, the human δ-opioid receptor (PDB ID: 4N6H, 1.8 Å) [85] and the human adenosine A_{2A} receptor (A_{2A}AR) (PDB ID: 4EIY, 1.8 Å) [86], as well as a thermostabilized adenosine A_{2A} receptor (PDB ID: 5NM4, 1.7 Å) [87].

To demonstrate the possibility of S-SAD phasing for GPCRs at X-ray FEL sources, anomalous LCP-SFX data from $\sim 5 \times 2 \times 2\ \mu m^3$ crystals of A_{2A}AR were collected using X-ray energy of 6 keV as a compromise between the strength of anomalous scattering from sulfur atoms (K-edge 2.472 keV) and the detector-size and energy limits on resolution [37]. At this energy the anomalous difference in structure factors is expected to be <1.5%, requiring a very high multiplicity of collected data. To minimize background X-ray scattering the sample was injected into a vacuum chamber, and the X-ray beam was attenuated to 14% to prevent the disruption of the LCP jet by the shockwaves from micro-explosions and to avoid the oversaturation of the CSPAD detector [88].

Within ~17 h of data collection a total of 7,324,430 images were collected, of which 1,797,503 were identified as crystal diffraction patterns using the Cheetah hit finding software [89]. A total of 593,996 of these hits were successfully indexed using the CrystFEL software [90]. The final reflection list, created using Monte Carlo integration and iterative scaling resulted in a dataset at a resolution of 2.5 Å. This resolution was limited by the X-ray energy, detector size, and minimal achievable sample-to-detector distance. To further extend resolution, additional data at an X-ray energy of 9.8 keV (wavelength, 1.27 Å) were collected. This high-resolution data set was assembled from 72,735 indexed patterns and was truncated at 1.9 Å resolution.

Compared to the previously reported S-SAD phasing of SFX data for lysozyme [79] and thaumatin [84], phasing of $A_{2A}AR$ data required approximately four times more indexed patterns. In addition to lower crystal symmetry and lower sulfur content, the diffraction power of $A_{2A}AR$ microcrystals is substantially lower compared to lysozyme crystals of similar size. At the same time, the background scattering from an LCP stream 50 μm in diameter, in which $A_{2A}AR$ microcrystals were delivered, is much greater [91] than the background from a liquid stream 5 μm in diameter used to deliver lysozyme crystals.

These factors, together with a potentially lower isomorphism of $A_{2A}AR$ microcrystals as compared to crystals of soluble test proteins, contribute to the challenge of native sulfur phasing of SFX data for membrane proteins. In this experiment, protein consumption for de novo phasing was very reasonable (~2.7 mg) by merit of the efficient operation of the LCP injector [30, 92]. These results, demonstrating that ~600,000 indexed patterns are sufficient to phase GPCR data starting with 12 ordered sulfur atoms per 447 residues (2.7%) can be placed in perspective with the fact that over 88% of all human proteins have indeed higher than 2.7% of Cys and Met residue content. Thus, this result provides an important reference point reassuring that most human proteins could be phased by S-SAD for de novo structure determination with X-ray FELs, provided sufficient sample quantities are available.

10.5 Conclusions

The nascent technique of LCP-SFX, despite its short history, has already stimulated great progress in the field of GPCR structural biology. With the majority of human GPCR structures still unsolved, it promises to alleviate roadblocks that have traditionally hampered GPCR crystallography, such as the ability to obtain sufficiently large crystals devoid of growth defects. This method has already resulted in a number of important breakthroughs including the structure of rhodopsin bound to arrestin [21], as well as two angiotensin receptors [35, 36], and several full-length non-class A receptors [26, 27, 73]. In the near future, X-ray FELs will likely allow for the study of the highly flexible GPCRs in a dynamic fashion, shedding light on the structural foundation of their activation process. Indeed, X-ray FELs already facilitate crystallography at room temperature, which is closer to physiological conditions compared to cryo-cooled crystals at synchrotrons. Side chain and loop movements, as well as differential water occupancy at room temperature, can be conclusive as to how these proteins function in the native cells [55].

The use of ultrashort X-ray FEL pulses also opened up the field of time-resolved crystallography with very fine temporal resolution and the ability to study irreversible processes. The inherent dynamic nature of GPCRs and the high biological relevance of their conformational response to ligand binding make them very exciting targets for time-resolved studies. For GPCRs, the challenge arises of how to trigger receptor conformational changes precisely, as, except for the case of visual rhodopsins, they are not inherently light sensitive. Current efforts focus on the design of suitable, covalently attached or diffusible photosensitive ligands that can elicit the receptor conformational change upon a flash of laser light, as well as on adapting rapid mixing injectors to microcrystals delivered in LCP, where the high speed and efficiency of diffusion in microcrystals alleviate in part the lower temporal resolution of mixing experiments as compared to light-driven reactions. GPCR activation takes place over several time scales, from fast internal motions and plasticity of the ligand-binding pocket (~nanoseconds), to rearrangements of receptor microswitches (~microseconds), to large-scale helical motions (~milliseconds) [93]. What part of the activation process will be amenable to studies by time-resolved crystallography using X-ray FELs will depend on the efficiency of the mechanism triggering the receptor conformational change and on the ability of crystals to accommodate those conformational changes, but not on the characteristics of the X-ray FEL beam itself, which offers femtosecond temporal resolving power.

Lastly, the LCP-SFX method has potential to facilitate structure-based drug design (SBDD) for GPCRs and other important membrane protein drug targets. SBDD studies rely heavily on the availability of X-ray structures to characterize the ligand binding site and use this information to guide ligand optimization design. Tens to hundreds of ligand-bound structures are normally determined in the course of a drug design program prior to clinical trials. To date, the application of this approach to membrane proteins based on experimentally determined structures, particularly of GPCRs, has been very limited or nonexistent [94]. By enabling high-resolution data collection from micrometer-sized GPCR crystals LCP-SFX can overcome the main bottleneck behind SBDD, namely the need for an extensive crystal optimization for each selected ligand–receptor complex, which often can take months to years. Additionally, micrometer-sized crystals facilitate common procedures of co-crystal generation, such as ligand soaking and exchange, while the overall LCP-SFX protocol further simplifies data collection on a large number of co-crystals by eliminating the need for crystal harvesting and cryo-cooling.

Taking into account all the impressive results reviewed in this chapter we certainly expect that with further progress in X-ray FEL instrumentation, development of novel fast detectors with high dynamic range, designing more efficient sample delivery approaches, and commissioning of new X-ray FEL sources, the impact of LCP-SFX on structural biology of GPCRs will continue to grow.

References

1. Heng, B. C., Aubel, D., & Fussenegger, M. (2013). An overview of the diverse roles of G-protein coupled receptors (GPCRs) in the pathophysiology of various human diseases. *Biotechnology Advances, 31*(8), 1676–1694.
2. Venkatakrishnan, A. J., Deupi, X., Lebon, G., Tate, C. G., Schertler, G. F., & Babu, M. M. (2013). Molecular signatures of G-protein-coupled receptors. *Nature, 494*(7436), 185–194.

3. Audet, M., & Bouvier, M. (2012). Restructuring G-protein- coupled receptor activation. *Cell, 151*(1), 14–23.
4. Galandrin, S., Oligny-Longpre, G., & Bouvier, M. (2007). The evasive nature of drug efficacy: Implications for drug discovery. *Trends in Pharmacological Sciences, 28*(8), 423–430.
5. Rajagopal, S., Rajagopal, K., & Lefkowitz, R. J. (2010). Teaching old receptors new tricks: Biasing seven-transmembrane receptors. *Nature Reviews Drug Discovery, 9*(5), 373–386.
6. Overington, J. P., Al-Lazikani, B., & Hopkins, A. L. (2006). How many drug targets are there? *Nature Reviews Drug Discovery, 5*(12), 993–996.
7. Santos, R., Ursu, O., Gaulton, A., Bento, A. P., Donadi, R. S., Bologa, C. G., et al. (2017). A comprehensive map of molecular drug targets. *Nature Reviews Drug Discovery, 16*(1), 19–34.
8. Hauser, A. S., Attwood, M. M., Rask-Andersen, M., Schioth, H. N., & Gloriam, D. E. (2017). Trends in GPCR drug discovery: New agents, targets and indications. *Nature Reviews Drug Discovery, 16*(12), 829–842.
9. Chung, S., Funakoshi, T., & Civelli, O. (2008). Orphan GPCR research. *British Journal of Pharmacology, 153*(Suppl. 1), S339–S346.
10. Stockert, J. A., & Devi, L. A. (2015). Advancements in therapeutically targeting orphan GPCRs. *Frontiers in Pharmacology, 6*, 100.
11. Heydenreich, F. M., Vuckovic, Z., Matkovic, M., & Veprintsev, D. B. (2015). Stabilization of G protein-coupled receptors by point mutations. *Frontiers in Pharmacology, 6*, 82.
12. Chun, E., Thompson, A. A., Liu, W., Roth, C. B., Griffith, M. T., Katritch, V., et al. (2012). Fusion partner toolchest for the stabilization and crystallization of G protein-coupled receptors. *Structure, 20*(6), 967–976.
13. Caffrey, M., & Cherezov, V. (2009). Crystallizing membrane proteins using lipidic mesophases. *Nature Protocols, 4*(5), 706–731.
14. Smith, J. L., Fischetti, R. F., & Yamamoto, M. (2012). Micro-crystallography comes of age. *Current Opinion in Structural Biology, 22*(5), 602–612.
15. Jahnichen, S., Blanchetot, C., Maussang, D., Gonzalez-Pajuelo, M., Chow, K. Y., Bosch, L., et al. (2010). CXCR4 nanobodies (VHH-based single variable domains) potently inhibit chemotaxis and HIV-1 replication and mobilize stem cells. *Proceedings of the National Academy of Sciences of the United States of America, 107*(47), 20565–20570.
16. Mujic-Delic, A., de Wit, R. H., Verkaar, F., & Smit, M. J. (2014). GPCR-targeting nanobodies: Attractive research tools, diagnostics, and therapeutics. *Trends in Pharmacological Sciences, 35*(5), 247–255.
17. Ghosh, E., Kumari, P., Jaiman, D., & Shukla, A. K. (2015). Methodological advances: The unsung heroes of the GPCR structural revolution. *Nature Reviews Molecular Cell Biology, 16*(2), 69–81.
18. Cherezov, V., Rosenbaum, D. M., Hanson, M. A., Rasmussen, S. G., Thian, F. S., Kobilka, T. S., et al. (2007). High-resolution crystal structure of an engineered human beta2-adrenergic G protein-coupled receptor. *Science, 318*(5854), 1258–1265.
19. Stevens, R. C., Cherezov, V., Katritch, V., Abagyan, R., Kuhn, P., Rosen, H., et al. (2013). The GPCR network: A large-scale collaboration to determine human GPCR structure and function. *Nature Reviews Drug Discovery, 12*(1), 25–34.
20. Carpenter, B., Nehme, R., Warne, T., Leslie, A. G., & Tate, C. G. (2016). Structure of the adenosine A (2A) receptor bound to an engineered G protein. *Nature, 536*(7614), 104–107.
21. Kang, Y., Zhou, X. E., Gao, X., He, Y., Liu, W., Ishchenko, A., et al. (2015). Crystal structure of rhodopsin bound to arrestin by femtosecond X-ray laser. *Nature, 523*(7562), 561–567.
22. Rasmussen, S. G., DeVree, B. T., Zou, Y., Kruse, A. C., Chung, K. Y., Kobilka, T. S., et al. (2011). Crystal structure of the beta2 adrenergic receptor-Gs protein complex. *Nature, 477*(7366), 549–555.
23. Schioth, H. B., & Fredriksson, R. (2005). The GRAFS classification system of G-protein coupled receptors in comparative perspective. *General and Comparative Endocrinology, 142*(1–2), 94–101.

24. Pal, K., Melcher, K., & Xu, H. E. (2012). Structure and mechanism for recognition of peptide hormones by class B G-protein-coupled receptors. *Acta Pharmacologica Sinica, 33*(3), 300–311.
25. Kniazeff, J., Prezeau, L., Rondard, P., Pin, J. P., & Goudet, C. (2011). Dimers and beyond: The functional puzzles of class C GPCRs. *Pharmacology & Therapeutics, 130*(1), 9–25.
26. Zhang, H., Qiao, A., Yang, D., Yang, L., Dai, A., de Graaf, C., et al. (2017). Structure of the full-length glucagon class B G-protein-coupled receptor. *Nature, 546*(7657), 259–264.
27. Zhang, X., Zhao, F., Wu, Y., Yang, J., Han, G. W., Zhao, S., et al. (2017). Crystal structure of a multi-domain human smoothened receptor in complex with a super stabilizing ligand. *Nature Communications, 8*, 15383.
28. Ballesteros, J. A., & Weinstein, H. (1995). Integrated methods for the construction of three-dimensional models and computational probing of structure-function relations in G protein-coupled receptors. In C. S. Stuart (Ed.), *Methods in neurosciences* (pp. 366–428). Cambridge, MA: Academic Press.
29. Isberg, V., de Graaf, C., Bortolato, A., Cherezov, V., Katritch, V., Marshall, F. H., et al. (2015). Generic GPCR residue numbers—Aligning topology maps while minding the gaps. *Trends in Pharmacological Sciences, 36*(1), 22–31.
30. Weierstall, U., James, D., Wang, C., White, T. A., Wang, D., Liu, W., et al. (2014). Lipidic cubic phase injector facilitates membrane protein serial femtosecond crystallography. *Nature Communications, 5*, 3309.
31. Johansson, L. C., Stauch, B., Ishchenko, A., & Cherezov, V. (2017). A bright future for serial femtosecond crystallography with XFELs. *Trends in Biochemical Sciences, 42*(9), 749–762.
32. Ishchenko, A., Cherezov, V., & Liu, W. (2016). Preparation and delivery of protein microcrystals in lipidic cubic phase for serial femtosecond crystallography. *Journal of Visualized Experiments, 115*, e54463.
33. Liu, W., Ishchenko, A., & Cherezov, V. (2014). Preparation of microcrystals in lipidic cubic phase for serial femtosecond crystallography. *Nature Protocols, 9*(9), 2123–2134.
34. Fenalti, G., Zatsepin, N. A., Betti, C., Giguere, P., Han, G. W., Ishchenko, A., et al. (2015). Structural basis for bifunctional peptide recognition at human delta-opioid receptor. *Nature Structural & Molecular Biology, 22*(3), 265–268.
35. Zhang, H., Unal, H., Gati, C., Han, G. W., Liu, W., Zatsepin, N. A., et al. (2015). Structure of the angiotensin receptor revealed by serial femtosecond crystallography. *Cell, 161*(4), 833–844.
36. Zhang, H., Han, G. W., Batyuk, A., Ishchenko, A., White, K. L., Patel, N., et al. (2017). Structural basis for selectivity and diversity in angiotensin II receptors. *Nature, 544*(7650), 327–332.
37. Batyuk, A., Galli, L., Ishchenko, A., Han, G. W., Gati, C., Popov, P. A., et al. (2016). Native phasing of x-ray free-electron laser data for a G protein-coupled receptor. *Science Advances, 2*(9), e1600292.
38. Ishchenko, A., Wacker, D., Kapoor, M., Zhang, A., Han, G. W., Basu, S., et al. (2017). Structural insights into the extracellular recognition of the human serotonin 2B receptor by an antibody. *Proceedings of the National Academy of Sciences of the United States of America, 114*(31), 8223–8228.
39. Alexandrov, A. I., Mileni, M., Chien, E. Y., Hanson, M. A., & Stevens, R. C. (2008). Microscale fluorescent thermal stability assay for membrane proteins. *Structure, 16*(3), 351–359.
40. Fenalti, G., Abola, E. E., Wang, C., Wu, B., & Cherezov, V. (2015). Fluorescence recovery after photobleaching in lipidic cubic phase (LCP-FRAP): A precrystallization assay for membrane proteins. *Methods in Enzymology, 557*, 417–437.
41. Kissick, D. J., Wanapun, D., & Simpson, G. J. (2011). Second-order nonlinear optical imaging of chiral crystals. *Annual Review of Analytical Chemistry, 4*, 419–437.
42. Barnes, C. O., Kovaleva, E. G., Fu, X., Stevenson, H. P., Brewster, A. S., DePonte, D. P., et al. (2016). Assessment of microcrystal quality by transmission electron microscopy for efficient serial femtosecond crystallography. *Archives of Biochemistry and Biophysics, 602*, 61–68.

43. Zhou, X. E., He, Y., de Waal, P. W., Gao, X., Kang, Y., Van Eps, N., et al. (2017). Identification of phosphorylation codes for arrestin recruitment by G protein-coupled receptors. *Cell, 170*(3), 457–469 e13.
44. Boutet, S., & Williams, G. J. (2010). The coherent X-ray imaging (CXI) instrument at the linac coherent light source (LCLS). *New Journal of Physics, 12*(3), 035024.
45. DePonte, D. P., Weierstall, U., Schmidt, K., Warner, J., Starodub, D., Spence, J. C. H., et al. (2008). Gas dynamic virtual nozzle for generation of microscopic droplet streams. *Journal of Physics D-Applied Physics, 41*(19), 195505.
46. Brehm, W., & Diederichs, K. (2014). Breaking the indexing ambiguity in serial crystallography. *Acta Crystallographica. Section D, Biological Crystallography, 70*(1), 101–109.
47. Ginn, H. M., Roedig, P., Kuo, A., Evans, G., Sauter, N. K., Ernst, O. P., et al. (2016). TakeTwo: An indexing algorithm suited to still images with known crystal parameters. *Acta Crystallographica Section D: Structural Biology, 72*(8), 956–965.
48. Mariani, V., Morgan, A., Yoon, C. H., Lane, T. J., White, T. A., O'Grady, C., et al. (2016). OnDA: Online data analysis and feedback for serial X-ray imaging. *Journal of Applied Crystallography, 49*(3), 1073–1080.
49. Sauter, N. K. (2015). XFEL diffraction: Developing processing methods to optimize data quality. *Journal of Synchrotron Radiation, 22*(2), 239–248.
50. Sauter, N. K., Hattne, J., Brewster, A. S., Echols, N., Zwart, P. H., & Adams, P. D. (2014). Improved crystal orientation and physical properties from single-shot XFEL stills. *Acta Crystallographica Section D: Biological Crystallography, 70*(12), 3299–3309.
51. White, T. A. (2014). Post-refinement method for snapshot serial crystallography. *Philosophical Transactions of the Royal Society of London. Series B, Biological Sciences, 369*(1647), 20130330.
52. White, T. A., Mariani, V., Brehm, W., Yefanov, O., Barty, A., Beyerlein, K. R., et al. (2016). Recent developments in CrystFEL. *Journal of Applied Crystallography, 49*(2), 680–689.
53. Liu, W., Wacker, D., Gati, C., Han, G. W., James, D., Wang, D., et al. (2013). Serial femtosecond crystallography of G protein-coupled receptors. *Science, 342*(6165), 1521–1524.
54. Wacker, D., Wang, C., Katritch, V., Han, G. W., Huang, X. P., Vardy, E., et al. (2013). Structural features for functional selectivity at serotonin receptors. *Science, 340*(6132), 615–619.
55. Fraser, J. S., van den Bedem, H., Samelson, A. J., Lang, P. T., Holton, J. M., Echols, N., et al. (2011). Accessing protein conformational ensembles using room-temperature X-ray crystallography. *Proceedings of the National Academy of Sciences of the United States of America, 108*(39), 16247–16252.
56. Wang, C., Wu, H., Evron, T., Vardy, E., Han, G. W., Huang, X. P., et al. (2014). Structural basis for smoothened receptor modulation and chemoresistance to anticancer drugs. *Nature Communications, 5*, 4355.
57. Abdelhamid, E. E., Sultana, M., Portoghese, P. S., & Takemori, A. E. (1991). Selective blockage of delta opioid receptors prevents the development of morphine tolerance and dependence in mice. *The Journal of Pharmacology and Experimental Therapeutics, 258*(1), 299–303.
58. Zhang, H., Unal, H., Desnoyer, R., Han, G. W., Patel, N., Katritch, V., et al. (2015). Structural basis for ligand recognition and functional selectivity at angiotensin receptor. *The Journal of Biological Chemistry, 290*(49), 29127–29139.
59. Guimond, M. O., & Gallo-Payet, N. (2012). How does angiotensin AT(2) receptor activation help neuronal differentiation and improve neuronal pathological situations? *Frontiers in Endocrinology, 3*, 164.
60. Porrello, E. R., Delbridge, L. M., & Thomas, W. G. (2009). The angiotensin II type 2 (AT2) receptor: An enigmatic seven transmembrane receptor. *Frontiers in Bioscience, 14*, 958–972.
61. Nahmias, C., & Strosberg, A. D. (1995). The angiotensin AT2 receptor: Searching for signal-transduction pathways and physiological function. *Trends in Pharmacological Sciences, 16*(7), 223–225.
62. Nouet, S., & Nahmias, C. (2000). Signal transduction from the angiotensin II AT2 receptor. *Trends in Endocrinology and Metabolism, 11*(1), 1–6.

63. Palczewski, K., Kumasaka, T., Hori, T., Behnke, C. A., Motoshima, H., Fox, B. A., et al. (2000). Crystal structure of rhodopsin: A G protein-coupled receptor. *Science, 289*(5480), 739–745.
64. Park, J. H., Scheerer, P., Hofmann, K. P., Choe, H. W., & Ernst, O. P. (2008). Crystal structure of the ligand-free G-protein-coupled receptor opsin. *Nature, 454*(7201), 183–187.
65. Choe, H. W., Kim, Y. J., Park, J. H., Morizumi, T., Pai, E. F., Krauss, N., et al. (2011). Crystal structure of metarhodopsin II. *Nature, 471*(7340), 651–655.
66. Hirsch, J. A., Schubert, C., Gurevich, V. V., & Sigler, P. B. (1999). The 2.8 A crystal structure of visual arrestin: A model for arrestin's regulation. *Cell, 97*(2), 257–269.
67. Kim, Y. J., Hofmann, K. P., Ernst, O. P., Scheerer, P., Choe, H. W., & Sommer, M. E. (2013). Crystal structure of pre-activated arrestin p44. *Nature, 497*(7447), 142–146.
68. Zhou, X. E., Gao, X., Barty, A., Kang, Y., He, Y., Liu, W., et al. (2016). X-ray laser diffraction for structure determination of the rhodopsin-arrestin complex. *Scientific Data, 3*, 160021.
69. Siu, F. Y., He, M., de Graaf, C., Han, G. W., Yang, D., Zhang, Z., et al. (2013). Structure of the human glucagon class B G-protein-coupled receptor. *Nature, 499*(7459), 444–449.
70. Wang, C., Wu, H., Katritch, V., Han, G. W., Huang, X. P., Liu, W., et al. (2013). Structure of the human smoothened receptor bound to an antitumour agent. *Nature, 497*(7449), 338–343.
71. Nachtergaele, S., Mydock, L. K., Krishnan, K., Rammohan, J., Schlesinger, P. H., Covey, D. F., et al. (2012). Oxysterols are allosteric activators of the oncoprotein smoothened. *Nature Chemical Biology, 8*(2), 211–220.
72. Gorojankina, T. (2016). Hedgehog signaling pathway: A novel model and molecular mechanisms of signal transduction. *Cellular and Molecular Life Sciences, 73*(7), 1317–1332.
73. Byrne, E. F. X., Sircar, R., Miller, P. S., Hedger, G., Luchetti, G., Nachtergaele, S., et al. (2016). Structural basis of smoothened regulation by its extracellular domains. *Nature, 535*(7613), 517–522.
74. Bortolato, A., Dore, A. S., Hollenstein, K., Tehan, B. G., Mason, J. S., & Marshall, F. H. (2014). Structure of class B GPCRs: New horizons for drug discovery. *British Journal of Pharmacology, 171*(13), 3132–3145.
75. Jazayeri, A., Dore, A. S., Lamb, D., Krishnamurthy, H., Southall, S. M., Baig, A. H., et al. (2016). Extra-helical binding site of a glucagon receptor antagonist. *Nature, 533*(7602), 274–277.
76. Hutchings, C. J., Koglin, M., Olson, W. C., & Marshall, F. H. (2017). Opportunities for therapeutic antibodies directed at G-protein-coupled receptors. *Nature Reviews Drug Discovery, 16*(9), 787–810.
77. Hay, M., Thomas, D. W., Craighead, J. L., Economides, C., & Rosenthal, J. (2014). Clinical development success rates for investigational drugs. *Nature Biotechnology, 32*(1), 40–51.
78. Wang, C., Jiang, Y., Ma, J., Wu, H., Wacker, D., Katritch, V., et al. (2013). Structural basis for molecular recognition at serotonin receptors. *Science, 340*(6132), 610–614.
79. Barends, T. R., Foucar, L., Botha, S., Doak, R. B., Shoeman, R. L., Nass, K., et al. (2014). De novo protein crystal structure determination from X-ray free-electron laser data. *Nature, 505*(7482), 244–247.
80. Yamashita, K., Pan, D., Okuda, T., Sugahara, M., Kodan, A., Yamaguchi, T., et al. (2015). An isomorphous replacement method for efficient de novo phasing for serial femtosecond crystallography. *Scientific Reports, 5*, 14017.
81. Colletier, J. P., Sawaya, M. R., Gingery, M., Rodriguez, J. A., Cascio, D., Brewster, A. S., et al. (2016). De novo phasing with X-ray laser reveals mosquito larvicide BinAB structure. *Nature, 539*(7627), 43–47.
82. Wu, H., Wang, C., Gregory, K. J., Han, G. W., Cho, H. P., Xia, Y., et al. (2014). Structure of a class C GPCR metabotropic glutamate receptor 1 bound to an allosteric modulator. *Science, 344*(6179), 58–64.
83. Hendrickson, W. A., & Teeter, M. M. (1981). Structure of the hydrophobic protein crambin determined directly from the anomalous scattering of sulphur. *Nature, 290*(5802), 107–113.
84. Nass, K., Meinhart, A., Barends, T. R., Foucar, L., Gorel, A., Aquila, A., et al. (2016). Protein structure determination by single-wavelength anomalous diffraction phasing of X-ray free-electron laser data. *IUCrJ, 3*(3), 180–191.

85. Fenalti, G., Giguere, P. M., Katritch, V., Huang, X. P., Thompson, A. A., Cherezov, V., et al. (2014). Molecular control of delta-opioid receptor signalling. *Nature, 506*(7487), 191–196.
86. Liu, W., Chun, E., Thompson, A. A., Chubukov, P., Xu, F., Katritch, V., et al. (2012). Structural basis for allosteric regulation of GPCRs by sodium ions. *Science, 337*(6091), 232–236.
87. Weinert, T., Olieric, N., Cheng, R., Brunle, S., James, D., Ozerov, D., et al. (2017). Serial millisecond crystallography for routine room-temperature structure determination at synchrotrons. *Nature Communications, 8*(1), 542.
88. Hart, P., Boutet, S., Carini, G., Dragone, A., Duda, B., Freytag, D., et al. (2012). *The cornell-SLAC pixel array detector at LCLS*. Presented at 2012 Nuclear Science Symposium, Medical Imaging Conference Anaheim, CA. SLAC-PUB-15284.
89. Barty, A., Kirian, R. A., Maia, F. R. N. C., Hantke, M., Yoon, C. H., White, T. A., et al. (2014). Cheetah: Software for high-throughput reduction and analysis of serial femtosecond X-ray diffraction data. *Journal of Applied Crystallography, 47*(3), 1118–1131.
90. White, T. A., Kirian, R. A., Martin, A. V., Aquila, A., Nass, K., Barty, A., et al. (2012). CrystFEL: A software suite for snapshot serial crystallography. *Journal of Applied Crystallography, 45*(2), 335–341.
91. Conrad, C. E., Basu, S., James, D., Wang, D., Schaffer, A., Roy-Chowdhury, S., et al. (2015). A novel inert crystal delivery medium for serial femtosecond crystallography. *IUCrJ, 2*(4), 421–430.
92. Weierstall, U. (2014). Liquid sample delivery techniques for serial femtosecond crystallography. *Philosophical Transactions of the Royal Society of London. Series B, Biological Sciences, 369*(1647), 20130337.
93. Latorraca, N. R., Venkatakrishnan, A. J., & Dror, R. O. (2017). GPCR dynamics: Structures in motion. *Chemical Reviews, 117*(1), 139–155.
94. Congreve, M., Dias, J. M., & Marshall, F. H. (2014). Chapter one - structure-based drug design for G protein-coupled receptors. In G. Lawton & D. R. Witty (Eds.), *Progress in medicinal chemistry* (pp. 1–63). New York: Elsevier.

Chapter 11
Time-Resolved Serial Femtosecond Crystallography, Towards Molecular Movies of Biomolecules in Action

Jacques-Philippe Colletier, Giorgio Schirò, and Martin Weik

11.1 Introduction

Biological macromolecules, such as proteins, nucleic acids, and complexes thereof, are characterized by specific structural and dynamic features that are the basis of their respective biological activity, and define their dynamic personalities [29]. Understanding macromolecular activity thus requires studying structural changes over time and on various time-scales, such as equilibrium fluctuations and conformational changes orchestrating enzyme catalysis or enabling signal transduction. The first step in human vision, for instance, is the sub-picosecond time-scale photoisomerization of the retinal pigment in rhodopsin [73], which within microseconds leads to the conformational changes required for activation of transducin, the regulatory protein that initiates the signaling cascade beyond the macromolecular level.

Complementary biophysical techniques each open a window in time on macromolecular dynamics occurring from femtoseconds to minutes (Fig. 11.1). Among those, kinetic X-ray crystallography permits trapping of macromolecular conformational intermediates along a reaction pathway, and their characterization at the atomic level of spatial resolution [8]. The basic concept underlying this ensemble of techniques is that macromolecular activity can be triggered within a crystal, and the structure of intermediate states characterized, provided that (1) the macromolecule is active in the crystalline state; (2) an efficient and synchronous triggering of activity can be attained (at room-temperature) for all (or most) probed molecules;

J.-P. Colletier · G. Schirò · M. Weik (✉)
Institute of Structural Biology, University Grenoble, Alpes, CEA, CNRS, Grenoble, France
e-mail: colletier@ibs.fr; giorgio.schiro@ibs.fr; martin.weik@ibs.fr

S. Boutet et al. (eds.), *X-ray Free Electron Lasers*,
https://doi.org/10.1007/978-3-030-00551-1_11

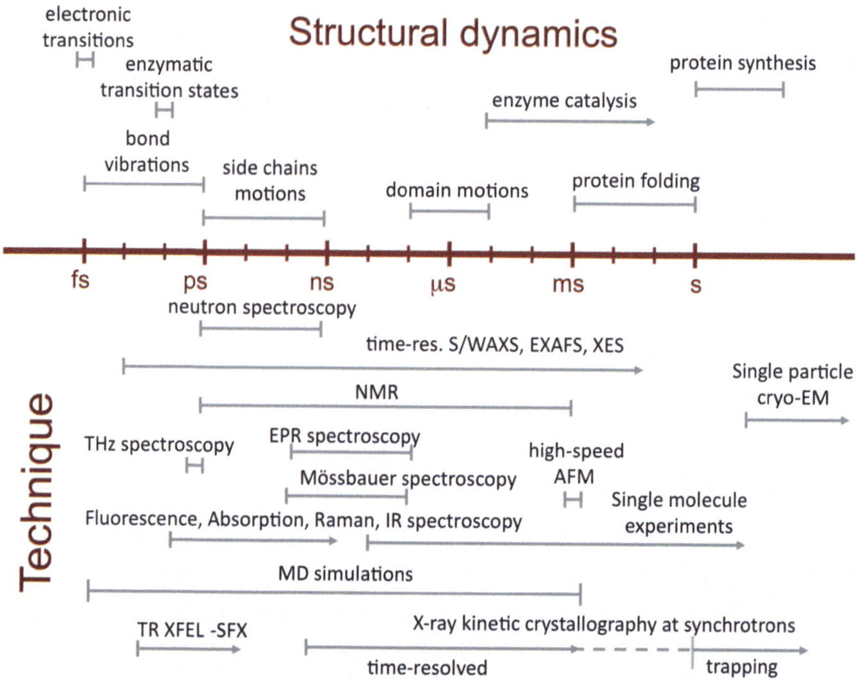

Fig. 11.1 Time-scale of protein motions and biophysical techniques to study them

and (3) the structural information can be recorded on a time scale shorter than the lifetime of the intermediate state of interest. As within a macromolecular crystal neighboring molecules are bathed in solvent (on average around 50% of the crystal volume) and generally make limited direct contacts, they most often preserve their functionally relevant structural dynamics and therefore remain active in the crystalline state [55]. After triggering of the biological reaction within macromolecular crystals, functionally relevant conformational changes are either arrested by flash-cooling the crystal, allowing for characterization of the structure by conventional cryo-crystallography, or followed in real time by time-resolved crystallography. The temporal resolution of the latter is limited to 100 ps if carried out in the form of Laue crystallography at synchrotrons. The advent of X-ray free electron lasers (X-ray FEL) has pushed the resolution to the sub-ps regime, allowing for ultrafast changes to be studied by time-resolved serial femtosecond crystallography. Below, we introduce the principles of time-resolved crystallography and summarize its implementation at synchrotron sources, before offering a review of time-resolved structural studies carried out so far using serial femtosecond crystallography (SFX [11]) at X-ray FELs. Challenges, limitations, and perspectives of time-resolved SFX round out this chapter.

11.2 Principles of Time-Resolved Crystallography

In time-resolved crystallography, a reaction in the crystalline macromolecule is triggered (*pump*) at $t = 0$ and a diffraction pattern (*probe*) collected after a well-defined lapse of time Δt_1 (pump–probe delay). The pump–probe sequence is repeated sufficiently enough for a complete diffraction data set to be produced from which a macromolecular structure can be determined that features conformational changes characteristic of the time delay Δt_1 after reaction initiation. The pump–probe delay is then changed to Δt_2 and the procedure repeated, etc. The structure determined at each Δt_i represents a frame in a molecular movie featuring conformational changes along the reaction pathway (Fig. 11.2). A complete and intelligible time-resolved study of the mode of action of the rhodopsin protein mentioned above, for instance, would require structural data to be collected over ten orders of magnitude, from the fs time scale where the actinic photon is absorbed, up to the μs timescale where transduction is activated.

Efficient and synchronous reaction triggering is key to a successful pump–probe experiment. The pump needs to trigger the reaction faster than the process of interest (synchrony) and in a large fraction of the crystalline molecules (efficiency) so that structural changes can be observed by ensemble-averaged crystallography. The pump method to be used depends on the reaction to be triggered: UV–visible light activates inherently light-sensitive proteins or so-called *caged compounds*

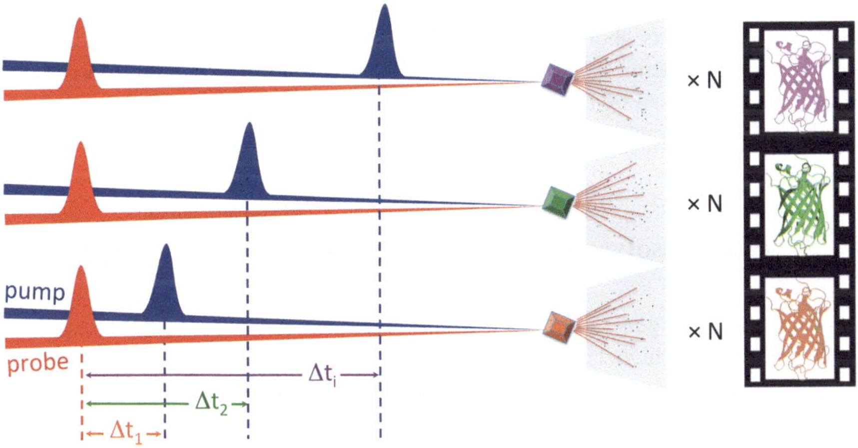

Fig. 11.2 Principle of pump–probe time-resolved crystallography. A crystalline macromolecule is probed by X-ray pulses ("probe," red beam) generated by a synchrotron or an X-ray FEL source. The time domain in the diffraction experiment is defined by the synchronization of X-ray pulses with optical laser pulses ("pump," blue beam). The interaction of optical laser pulses with the sample triggers a reaction and defines the time $t = 0$. The evolution of the protein structure at a generic time Δt_i is monitored by variation of the Δt delay between pump and probe. Every pump–probe sequence is repeated N times to produce a complete dataset from which a structural model can be determined

complexed with light-insensitive proteins; diffusion of substrates initiates enzyme catalysis (see Chap. 12 in this book); a temperature jump shifts conformational equilibria [44]; X-ray irradiation can provide electrons to trigger redox processes [69] or the breakdown of strained intermediate states [12]; electric field pulses allow for studying protein mechanics [28]. By far the most widespread method is optical triggering because of its straightforward technical implementation, the availability of crystals for a large number of light-sensitive proteins with cyclic reactions to be studied, and the accessibility of ultrafast (fs–ps) time scales.

The X-ray probe produces diffraction patterns that allow for structure determination at Δt_i after reaction initiation. The probe needs to be pulsed so that structural changes can be captured, analogous to shutter opening and closing in photography. The probe-pulse length must be shorter than the process of interest and sets, together with the pump-pulse length, the best time resolution that can be attained. At synchrotron sources, the pulse length cannot be shorter than about 100 ps, making it impossible to study processes faster than that limit. Also, time-resolved crystallography at synchrotrons is in practice limited to cyclic reactions because the pump–probe sequence is repeated multiple times on one or several macro-sized crystals. Using time-resolved (TR) SFX at X-ray FELs, however, cyclic as well as noncyclic reactions can be studied since each microcrystal is probed only once by a single X-ray shot. Furthermore, with their femtosecond X-ray pulses, X-ray FELs have extended the time resolution to the sub-ps regime, permitting the structural study of ultrafast processes such as those immediately following photon absorption in light-sensitive processes. Last but not least, the risk of X-ray radiation damage to biological macromolecules, known to compromise structural data collected at synchrotrons [21], is abolished in most SFX studies because diffraction data are collected before chemical and structural damage has had the time to develop [57]. Hence, the advent of X-ray FELs has revived and extended the reach of time-resolved crystallography.

11.3 Time-Resolved Crystallography at Synchrotron Sources

The advent of third-generation synchrotron sources and undulator beamlines, poised with 10^{18} and 10^6 higher peak brilliance than X-ray tubes and second-generation bending magnet sources, respectively, permitted the introduction of time-resolved crystallography by means of the so-called Laue technique [54]. In this approach, crystals of inherently photosensitive proteins or of non-photosensitive proteins complexed with photolabile precursors of their substrates or products—the so-called caged compounds—are activated by exposure to a femtosecond or nanosecond laser (pump) at room temperature. A highly intense polychromatic X-ray beam is then used to probe the structure, at various Δt_i. The advantage of using a white ($\sim$1–5% bandwidth) or pink ($\sim$0.05 to 0.1% bandwidth) polychromatic beam is that enough photons are elastically scattered, even from a single X-ray bunch, to enable collection of a diffraction pattern on which enough information is present to derive a partial dataset. Combining multiple such diffraction images, typically 10–30 exposures to the polychromatic X-ray beam of one or more crystals in different orientations, experimentalists can then produce a full dataset for each Δt_i. The time resolution of Laue crystallography at synchrotrons is inherently limited to 100 ps, corresponding to the pulse length of a single electron bunch generating the X-rays. Furthermore, the methodology requires large (0.5 mm), highly ordered, radiation-resistant protein crystals (to enable multiple exposures to the polychromatic beam without loss of resolution due to X-ray damage) with small unit cells (to minimize Bragg-peak overlaps), and works best with proteins undergoing cyclic photoreactions (see above). In the case of highly radiation sensitive samples, nonreversible photoreactions, or in studies where non-inherently photosensitive proteins are light-functionalized by complexation with caged compounds [68], more crystals are indeed needed. In the case of large proteins with correspondingly large unit cell parameters (e.g., photosystems I and II, or cytochrome c oxidase), data processing is complicated, due to overlaps between diffracted spots even when using a pink beam. Hence the methodology has failed to enable time-resolved crystallography on a large variety of proteins, and has mostly remained limited to a handful of proteins including the small GTPase Ras [68], myglobin [6, 7, 75, 79], the photoactive yellow protein [23, 34, 36, 74], and the photosynthetic reaction center [3, 96]. The largest protein to have been successfully studied by time-resolved Laue crystallography is hemoglobin [39]. The advent of X-ray FELs and monochromatic SFX has allowed for the resolution of time-resolved crystallography to be improved to the fs timescale, and to extend the feasibility of time-resolved studies to larger and radiation sensitive proteins.

11.4 Time-Resolved Serial Femtosecond Crystallography at X-Ray FELs: An Inventory

In time-resolved serial femtosecond crystallography (TR-SFX) following a pump–probe data collection scheme, macromolecular crystals are streamed across the pulsed X-ray FEL beam typically by means of a liquid jet [78, 93], a high-viscosity injector [5, 84, 92] or by acoustic droplet injection coupled with a conveyer-belt drive [20]. When the protein is inherently photosensitive or has been photosensitized by complexation with a caged-compound, crystals can be activated by an optical laser pulse (pump) prior to interaction with the X-ray pulse (probe), generating a diffraction pattern (Fig. 11.3). Repetition of this process on many crystals provides a full data set at a well-defined Δt_i from which the corresponding structural snapshot can be determined. Varying Δt_i then allows for generating frames of a molecular movie. If macromolecular activity is triggered by diffusion of a small molecule into the crystal, two solutions containing the crystals and the small molecule are mixed prior to interaction with an X-ray pulse. This so-called mix-and-inject strategy [71] is extensively discussed in Chap. 12 of this book and has been successfully applied to study ligand-binding in adenine riboswitches [80] and ß-lactamase [46].

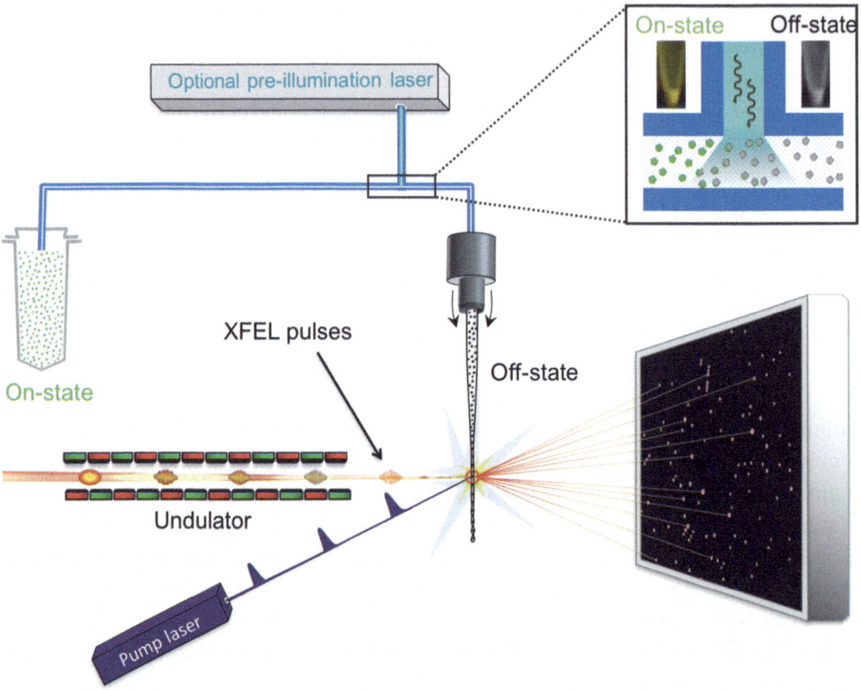

Fig. 11.3 General setup for TR-SFX experiments at X-ray FELs. There are a variety of ways to present crystals to the (probe) X-ray beam in SFX experiments. In the present example, crystals are jetted across the pulsed X-ray beam, and thousands of diffraction patterns are collected for each dataset. In TR-SFX, a pump laser is used to trigger conformational changes in the crystalline protein, and data are collected by the X-ray probe beam at various pump–probe delays. The experiment may require photoconversion of crystals within a pre-illumination device [67], as in the case of the photoswitchable rsEGFP2 that needed to be switched by 488 nm light from the fluorescent *on*-state (resting state) to the nonfluorescent *off*-state that was subsequently activated by a pump laser at 400 nm [15]

Time-resolved crystallography experiments at X-ray FELs remedy several shortcomings that limit those carried out at synchrotron sources. Firstly, the achievable time resolution of maximal 100 ps at a synchrotron is improved to 10 fs, owing to the short length of FEL pulses. Secondly, the high peak brilliance of X-ray FEL pulses generates exploitable diffraction from micron-sized crystals that can be fully activated by the optical pump laser. Indeed, absorption by optically dense protein crystals (Fig. 11.4) limits light penetration to outer layers in the case of large crystals (at least several tens of microns in size) such as those required by time-resolved synchrotron studies. And third, the very nature of (TR)-SFX, wherein the sample is replenished after each shot in a serial manner, enables time-resolved structural studies on nonreversible reactions, provided that enough sample can be produced for the envisaged delivery system.

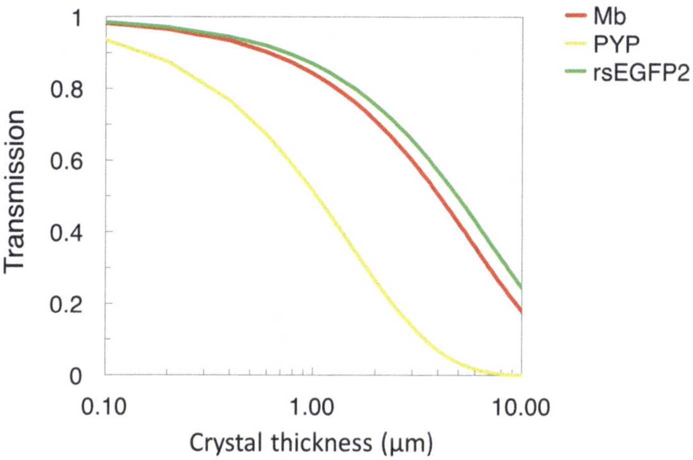

Fig. 11.4 Low pump-light penetration depth in chromophore-containing protein crystals results in only a fraction of the probed molecules being excited and effectively undergoing conformational changes. The figure shows the calculated transmission of pump light across crystals of the three proteins that have so far been studied by TR-SFX on the ps or sub-ps time scale, that is, myoglobin (Mb; [2]), the photoactive yellow protein (PYP; [63]), and the reversibly photoswitchable fluorescent protein rsEGFP2 [15] as a function of crystal thickness. In these studies, concentrations of the crystalline Mb, PYP, and rsEGFP2 were 53, 63, and 27 mM, respectively. The molar extinction coefficients of Mb, PYP, and rsEGFP2 in its *off* state are 13,950, 45,500, and 22,000 M^{-1} cm^{-1} at 532, 450, and 400 nm, respectively. Of note, the molar extinction coefficients used for these calculations were those of the ground states and assumed to be constant across the crystal over the pulse length. A more sophisticated calculation that takes into account the change of penetration depth as time progresses and molecules bleach has been presented for PYP [87]

The first TR-SFX study was carried out on crystals of a photoactive complex of photosystem I (PS I) with ferredoxin [1] with pump–probe delays of 5 μs and 10 μs. Only virtual powder patterns were used in the analysis, since the low amount of indexed patterns per time-delay precluded extraction of structure factors. At 5 μs, a slight increase in diffracted intensities was observed, which was interpreted as a correlated structural change induced by electron transfer in the PSI—ferredoxin complex. At 10 μs, diffraction intensities dropped, rationalized as originating from the disordering of crystals upon dissociation of ferredoxin from the binding pocket, in line with time-resolved optical spectroscopy. Since this proof-of-principle study, several pump–probe TR-SFX studies have been carried out on various proteins (myoglobin, photoactive yellow protein, bacteriorhodopsin, photosystem II, a reversibly photoswitchable green fluorescent protein, cytochrome c oxidase, nitrite oxide reductase), and on timescales ranging from seconds to sub-picoseconds (Fig. 11.6). These results are briefly presented below.

The first high-resolution (1.6 Å) TR-SFX study was carried out on the photoactive yellow protein (PYP) with pump–probe delays of 1 μs and 10 ns using optical pump laser pulses on the ns regime [87]. The difference Fourier maps, computed between data sets of illuminated (light) and non-illuminated (dark) crystals, showed structural changes qualitatively similar to those observed in TR synchrotron Laue studies at comparable time delays, demonstrating validity of the TR-SFX method. Yet peaks in the TR-SFX difference maps were higher than in the corresponding Laue difference maps, owing to the more extensive pump-laser activation of the micron-sized crystals used in the TR-SFX studies, as compared to those used in Laue experiments. As mentioned previously, usability of μm- and sub-μm-sized crystals is one of the advantages of X-ray FEL time-resolved studies. Further steps into the uncharted regime of ps resolution were subsequently taken, as discussed in detail in the next section.

It is of important note that with much of the biological dynamics of interest being in the ns or slower time scales, many TR-SFX studies utilize the unique advantages of X-ray FEL based methods to study relatively slow dynamics compared to the X-ray FEL pulse duration, such as those involved in photosystem II (PS II) activity. The high scientific interest in the water-splitting cycle of PS II has made it so far the macromolecular system most studied by TR-SFX (Fig. 11.6). PSII is a large membrane-bound protein complex (700 kDa) involved in photosynthesis that produces dioxygen by catalyzing light-driven water splitting at a so-called oxygen evolving complex (OEC) containing a Mn_4CaO_5 cluster. The photocycle of PSII is a reversible reaction that involves sequential absorption of four photons generating four consecutive redox states of the cluster that cycles through five S-states (S_0 to S_4; the so-called "Kok cycle") on the μs to ms time scale [40]. Each of the four steps is characterized by a non-unitary quantum yield, so that a mixture of states has accumulated at the end of one photocycle; therefore, the reaction is in practice irreversible and each crystal may be used only once. A high-resolution (1.9 Å) structure of PSII has been solved by synchrotron macromolecular crystallography [90], but determination of the accurate OEC geometry had been hampered by the high X-ray sensitivity of this metalloenzyme [97]. The high-radiation sensitivity, combined with the experimentally nonreversible reaction, makes TR-SFX the method of choice to study the Kok cycle. Using femtosecond FEL pulses, it was indeed possible to determine the radiation-damage free structure of PS II and its OEC at 1.95 Å resolution by rotating and translating a large single crystal at 100 K in a stepwise fashion [82]. Comparison of the synchrotron and X-ray FEL structures revealed a more compact OEC with bond lengths 0.1–0.2 Å shorter in the X-ray FEL structure. Characterizing the radiation-damage free high-resolution structure of PS II [82] is yet only the first step in understanding the atomic details of water splitting, and TR-SFX studies that aimed at resolving time-dependent structural changes following light absorption were then carried out. In the first published TR-SFX study on PS II, simultaneous SFX and X-ray emission spectroscopy (XES) data were collected on crystals in the dark-adapted S_1 state, and in the S_2 state generated by light illumination with a pump–probe delay of 0.4–0.5 s [37]. Electron density maps of the S_2 structure at 5.7 Å resolution did not show differences with respect to the S_1 structure and it was concluded that structural changes, if present, are too small to be seen at that resolution. The simultaneous collection of SFX and XES data not only allowed for assigning the dominant S-state to the structural data collected but also provided evidence that the fs pulses probed the intact electronic structure of metal centers in the OEC. Subsequent studies focused on structural changes that accompany the transition to the S_3 state. While Kupitz and coworkers reported changes in the structure of an S_3 enriched state at 5.5 Å resolution compared to S_1 [45], a similar study by Kern and coworkers came to the conclusion that any structural changes related to the transitions between S_1 and S_3 must be smaller than what can be detected at the resolution achieved (between 4.5 and 5.2 Å) [38, 66]. The resolution of S_3 enriched structures was then improved to 2.25 Å [99] and 2.35 Å [83], yet without reaching a consensus on the nature of associated conformational changes. In one case [99], ~0.1 Å changes in metal distances in the OEC were reported, that is, changes small enough to be at the limit of uncertainty. In the other [83], peaks were observed on the OEC in Fourier difference maps that were interpreted as the appearance of a new oxygen atom and the displacement of a water molecule. Awaiting further improvements in diffraction resolution, the jury is thus still out as to time-resolved changes in the OEC during water splitting.

Another molecular mechanism actively studied by TR-SFX is the structural sequence of photo-intermediates in the light-driven proton pump bacteriorhodopsin (bR). Despite being a well-studied model for membrane proteins, certain structural details of the bR photocycle remain controversially discussed because of a possible convolution with radiation-induced changes in cryo-trapped structural intermediates solved at synchrotron sources [95]. A proof-of-principle that TR-SFX on bR is possible has been provided by Nogly and coworkers [60], using a thick high-viscosity jet to present lipidic-cubic phase (LCP) grown bR crystals to the X-ray beam. Using femtosecond excitation and a pump–probe delay of 1 ms, the structure of the M photointermediate could be solved at 2.3 Å resolution. As multiple consecutive excitations are not possible during the fs pump-pulse duration (see detailed discussion in Sect. 11.5), the intermediate-state occupancy was limited to 13%. The first extensive TR-SFX study on bR presented diffraction data collected following nanosecond excitation and spanning five orders of magnitude in time, from ns to ms [56]. Transient structures solved at 2.1 Å resolution covered photointermediates K to M, and provided a three-dimensional view of the structural changes underlying unidirectional proton pumping. In the earliest time-point studied (16 ns), the chromophore (retinal) had already photoisomerized, so that ultrafast changes right after photon absorption are still to be uncovered.

Two heme proteins other than myoglobin (see below) have been studied by time-resolved crystallography at an X-ray FEL, namely cytochrome c oxidase [77] and NO reductase [89]. In the former, CO release has been followed 20 ns and 100 μs after photolysis with a ns pump-laser, and the characterized structural changes allowed for establishing a mechanism for functionally relevant closing of a water channel [77]. Similar to the characterization of the radiation-damage free high-resolution structure of PS II [82], and at variance with most TR-SFX studies published so far, cytochrome c oxidase crystals were not presented as micron-sized samples in random orientations within a jet or by a conveyer-belt drive, but rather large (up to 500 μm) macrocrystals were rotated and translated through the FEL beam (serial femtosecond rotational crystallography (SF-ROX) [31]). The achievement standing out in the study on NO reductase is the first and successful use of caged compounds in TR-SFX. The enzyme was indeed rendered photosensitive by complexation with a photolabile precursor of NO (caged NO), which was cleaved on the μs time scale by means of ns laser irradiation at 308 nm. X-ray data were collected at a pump–probe delay of 20 ms, allowing for determination of the structure of the ferric NO complex at 2.1 Å resolution. Unlike corresponding synchrotron structures, the TR-SFX structure was devoid of X-ray radiation damage, as corroborated by accompanying QM/MM studies [89].

11.4.1 Ultrafast TR-SFX on the fs—ps Time Scale

Three TR-SFX studies have so far been published beyond the 100 ps resolution-limit of synchrotron-based Laue crystallography. Pump–probe delays of a picosecond and shorter time allowed for entering the time scale of photochemical reactions—a regime inaccessible by X-ray sources other than X-ray FELs—enabling visualization of ultrafast light-induced motions in carbonmonoxy myoglobin [2] and characterization of excited-state photoisomerization intermediates in PYP [63] and in a reversibly photoswitchable variant of the green fluorescent protein (GFP, [15]). In carbonmonoxy (CO) myoglobin (Mb), structural changes following light-activated cleavage of the Fe-CO bond were captured at nominal delays of 0.5 ps, 1 ps, 3 ps, 10 ps, 50 ps and 150 ps [2]. In line with results from other methods, the crystallographic maps showed that within 1 ps after photodissociation, CO has moved to its primary docking site, the heme has domed, and side-chains of neighboring residues have moved (Fig. 11.5a). Significant main-chain conformational changes were observed on the ps time scale with helices moving essentially as rigid bodies, in line with time-resolved solution scattering experiments [47]. The jitter between pump laser and X-ray FEL pulses caused the nominally 500 fs time delay data to contain images with various true time delays, so that they could be sorted into datasets with average time delays of -100 to $+600$ fs, with bin widths of $\sim$100 fs. Refinement against these *ultrafast* datasets revealed that residues in the heme-binding pocket showed an oscillatory behavior with a 500 fs period upon cleavage of the CO. These oscillations were noted to fit very well inside the general *normal mode* model for protein dynamics [53], in which fast coherent motions ultimately couple with larger-scale displacements of entire helices (Fig. 11.5b). In PYP, structural changes of the *p*-coumaric acid chromophore after femtosecond illumination at 450 nm have been followed at nominal pump–probe delays of 0.3 ps, 0.6 ps and 3 ps [63]. A structural transition has been proposed to occur at around 0.59 ps, associated with a *trans*-to-*cis* isomerization of the chromophore that involves some atoms moving by as much as 1.3 Å, on the sub-ps time scale (Fig. 11.5c). At delays shorter than 0.59 ps, the chromophore displays a *trans*, yet distorted, conformation, suggesting that it is still in the electronic excited state. At longer pump–probe delays, a *cis* chromophore is observed, indicating relaxation to the ground state. In the reversibly photoswitchable fluorescent protein rsEGFP2, used for nanoscale imaging of living cells by means of REversible Saturable Optical

Linear Fluorescence Transitions (RESOLFT) microscopy [26], photoisomerization intermediates have been characterized 1 and 3 ps after femtosecond illumination of the crystalline protein in its nonfluorescent *off*-state with the hydroxybenzylidene imidazolinone chromophore in the planar *trans* configuration [15]. One ps after illumination, the chromophore is fully twisted, with its two rings oriented perpendicular to each other (Fig. 11.5d). Quantum-mechanics/molecular-mechanics (QM/MM) and excited-state molecular dynamics simulations and time-resolved absorption spectroscopy in solution indicated that the twisted chromophore conformation after 1 ps corresponds to that of the electronic excited S_1 state, close to the conical intersection where the system relaxes back to the ground state. The entire helix carrying the chromophore shifts along its axis, so as to permit chromophore twisting (Fig. 11.5e). Three ps after illumination, the twisted chromophore conformation is less occupied than at 1 ps, and features indicative of the presence of the *cis* chromophore are present in the difference Fourier map. The twisted chromophore (Fig. 11.5d, e) thus represents a photoisomerization intermediate, half-way between *trans* and *cis* isomers. The maximum displacement of a chromophore atom (phenol OH group) is 4.5 Å in rsEGFP2—that is, much larger than in the case of PYP (1.3 Å)—possibly explaining why photoisomerization takes longer in rsEGFP2 than in the PYP (at 1 ps, rsEGFP2 is still in the excited state, while PYP is already in the ground state again). The excited-state structure of rsEGFP2 (Fig. 11.5d) was the basis for rationally improving this protein used as a molecular label in super-resolution fluorescence microscopy. The twisted chromophore conformation indicated that mutating Val151 (Fig. 11.5d) into an alanine residue should facilitate photoswitching by enlarging the pocket at the tip of the chromophore. Indeed, the V151A mutant turned out to have a twofold increased photoswitching quantum yield [15].

Since TR-SFX on the picosecond time scale requires pumping with a femtosecond pulse, reaction initiation is limited by the primary quantum yield. As a consequence, reaction initiation in the PYP study was low (structurally estimated to be 13% [63]) and the intermediate-states in rsEGFP2 were present at a total occupancy of maximum 7% [15]. An additional caveat of femtosecond excitation is the requirement of using high pump laser power densities (380 GW/cm^2 were used for Mb [2], 570 GW/cm^2 for PYP [63] and 400 GW/cm^2 for rsEGFP2 [15]), which carries the risk of creating artifacts (see more extensive discussion below). Consequently, all three ps TR-SFX studies relied on complementary QM/MM and optical spectroscopy experiments to interpret and validate the intermediate-state structures obtained.

11.5 Challenges and Limitations of TR-SFX

One might face several challenges and limitations when attempting to carry out TR-SFX. The first and most obvious is the necessity to obtain well-diffracting crystals of the protein of interest, or a functional fraction thereof. When crystals

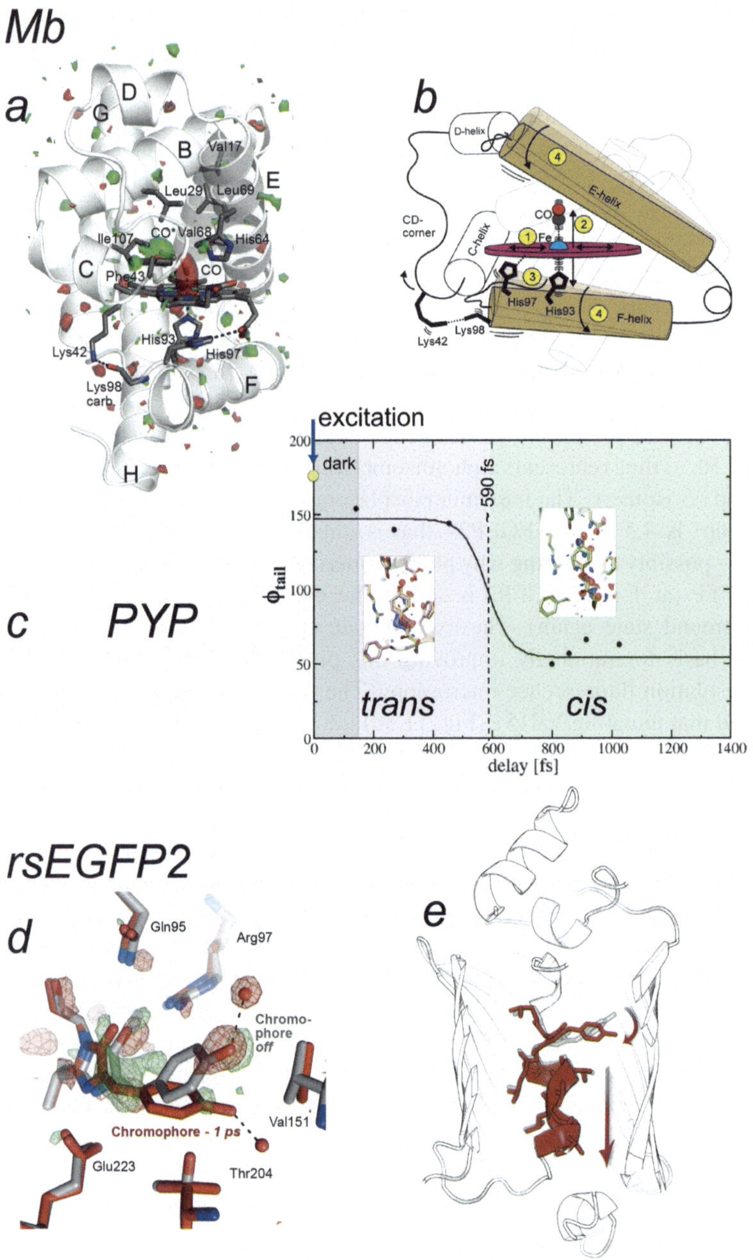

Fig. 11.5 Three TR-SFX studies entered the ultrafast time scale of photochemical reactions with pump–probe delays of a picosecond and shorter. (**a**) Structural changes in myoglobin (Mb) occurring 0.5 ps after photoexcitation [2]. The $F_o^{light} - F_o^{dark}$ difference electron density map is contoured at $+3\sigma$ (green) and -3σ (red). (**b**) Oscillatory side-chain dynamics, excited by

are not available, or when crystal packing hinders or even prevents some of the macromolecular motions required for biological activity, time-resolved solution scattering [10] can be used to follow conformational changes at low resolution after triggering macromolecular activity (see Chap. 15 of this book for further discussions on this matter). Below, we discuss those challenges and limitations that arise when well-diffracting crystals have already been obtained.

11.5.1 Spectroscopic Characterization of Protein Microcrystals

A prerequisite for TR-SFX on photosensitive proteins is to characterize the kinetics of photoproduct formation by spectroscopic methods. The major limiting factor in spectroscopic studies of optically dense protein crystals is the difficulty to grow them in a suitable shape for spectroscopy. Special approaches to produce large, flat crystals for spectroscopic studies were developed in some cases [24, 41] but they cannot be easily generalized to every crystallization condition. Furthermore, it may well be that the crystalline form most suited for spectroscopic studies is not that which diffracts to the highest resolution or is best suited for the TR structural studies, in terms of solvent content, nature of crystal contacts, presence of precipitating agents, unit cell dimensions and space group, radiation sensitivity, etc. An alternative strategy is to characterize the kinetics of photoproduct formation by the protein in solution, and assume the results to be valid for the crystalline state. However, there are several examples showing that not only the time constants but also the kinetic schemes can be altered in protein crystals [18, 58, 89, 98]. A recent work showed the feasibility of single-wavelength spectroscopic characterization of PYP microcrystals after optical femtosecond excitation [33].

11.5.2 Pump Light Penetration in Protein Crystals

The protein concentration in crystals can be as high as 30–60 mM, resulting in a very high optical density in the wavelength range where the chromophore moiety

Fig. 11.5 (continued) photodissociation of the heme, reflect fast coherent motions that ultimately couple with larger-scale displacement of entire helices. (**c**) A structural transition at around 590 fs in the photoactive yellow protein (PYP) has been associated with *trans*-to-*cis* isomerization of the chromophore [63]. (**d, e**) Structural changes in the reversibly photoswitchable fluorescent protein rsEGFP2 occurring 1 ps (red model) after photoexcitation of the nonfluorescent *off*-state (grey model) feature a fully twisted chromophore in the electronically excited state. The $F_o^{light} - F_o^{dark}$ difference electron density map is overlaid in (**d**) and contoured at +3.5σ (green) and −3.5σ (red). (**e**) Chromophore twisting at 1 ps is accompanied by a shift in the entire helix that carries the chromophore. Panels (**a, b**) and (**e**) were extracted from published work ([2, 15], respectively). Panels (**c**) and (**d**) were provided by Marius Schmidt and Nicolas Coquelle, respectively absorbs. As shown in Fig. 11.4 for photosensitive proteins recently studied by TR-SFX, the penetration depth for pump-laser light at the wavelengths used in TR structural studies is limited to only a few microns or less. In case of optical activity, the light penetration can also be dependent on the crystal orientation that is generally randomly distributed in the case of TR-SFX. It is recommended to tune experimental parameters like crystal size, pump-light wavelength, and pump-light polarization in order to ensure sufficient light penetration into the crystal—notably when using an experimental setup where the pump and probe beams are not collinear, that is, when the probability increases of probing a poorly illuminated part of the crystal.

11.5.3 Excitation Efficiency and Pump Power Density Issues

An important parameter to be taken into account when designing a TR-SFX experiment is the expected excitation efficiency by the pump light. The excitation efficiency is defined as the ratio between the number of protein molecules wherein the photoreaction under study has effectively been triggered and the total number of photons impinging on the crystal. This efficiency is the product of the absorption cross-section at the given pump-light wavelength and polarization, and the primary quantum yield of the photoreaction. The primary quantum yield is the probability that an absorbed photon triggers the photoreaction. When the pump-pulse length is shorter than the excited-state lifetime, the extent of reaction initiation is limited by the primary quantum yield. For pump-pulses being longer than the lifetime of the excited state, a protein molecule that after excitation has decayed back to the ground state (instead of progressing further along the photoreaction pathway) can be excited more than once, resulting in an extent of reaction initiation that exceeds the primary quantum yield. For example, in recent TR-SFX experiments on PYP, whose excited state lifetime is about 500 fs [48], the extent of reaction initiation was 40% with nanosecond excitation [87]. With femtosecond excitation, the extent of reaction initiation was only 13%, comparable to the primary quantum yield [63].

To study excited-state structural dynamics on the femtosecond/picosecond time scale, only femtosecond excitation can be used. The usual practice is to excite the sample with a photon density corresponding to an average of at least one absorbed photon per chromophore. Considering typical values of protein concentrations in a crystal, of microcrystal sizes and of extinction coefficients, one absorbed photon per chromophore corresponds to a peak power density of the order of tens of GW/cm^2 for femtosecond excitation. Such a high value is not only far beyond illumination power densities in any biologically relevant condition, but can also produce parasitic processes, like photobleaching, ionization, and radical formation, which will compete with the photophysical process under study and may produce structural artifacts. Moreover, it is common practice to exceed the regime of one absorbed photon per chromophore, in order to compensate for possible pump-laser intensity fluctuations and position drifts affecting the pump–probe spatial overlap. Illustratively, the three recent femtosecond experiments discussed above used power densities of the order of 380 GW/cm^2 [2], 400 GW/cm^2 [15], and 570 GW/cm^2 [63]. The limited access to X-ray FEL beam time has so far precluded the possibility of routinely performing pump power titrations, as required to define the linear regime within which excitation efficiency is maximized. Therefore, it is recommended to perform at least a preliminary thorough spectroscopic characterization of the system upon femtosecond excitation at various pump-power densities, prior to engaging in TR-SFX studies.

11.5.4 Time Resolution and Pump–Probe Synchronization Diagnostics in Ultrafast Experiments

The time resolution of an ultrafast TR-SFX experiment is given by the convolution of the femtosecond pump pulse temporal profile with the femtosecond probe pulse temporal profile, and should in principle also be in the femtosecond range. However, the effective time resolution also depends on the precision of the pump–probe synchronization. Optical laser pulses are synchronized to the X-ray FEL pulses by radiofrequency phase-locking technology. Currently, the precision of temporal synchronization with this technology is limited by the shot-to-shot jitters of optical and X-ray FEL pulses, which is on the order of more than 100 fs [25], and by thermal drifts of the order of a few picoseconds after several hours, which is the typical time scale of a TR-SFX data collection. Although there is an effort to improve the pump–probe synchronization precision and stability [16, 50], the approach currently used is to directly measure, on a shot-to-shot basis, the relative arrival time delay of optical and X-ray pulses close to the experimental interaction region. This real-time diagnostic approach allows for re-sorting data recorded for a given pump–probe delay, by substituting the nominal time delay assigned to each image with the time delay corrected after post-processing. This method was shown to allow for sub-10 fs precision in time delay determination [27] and was successfully applied to process TR-SFX data with time resolution of $\sim$100 fs [2, 63], limited by crystallographic data redundancy rather than by the timing diagnostic itself.

11.5.5 Structure Refinement of Low-Occupancy States

In pump–probe TR-SFX experiments on photosensitive proteins, limited pump-light penetration into protein crystals (Fig. 11.4) and non-unity excitation efficiency together result in only a fraction of the probed molecules being excited and effectively undergoing conformational changes. The probed diffraction signal is thus a mixture of that arising from molecules within which the reaction has been triggered, and from molecules that are still in the resting state. Most often, the $2mF_O\text{-}DF_c$ experimental electron density maps for the datasets without (dark) and with (light) pump excitation hardly differ and straightforward building of a model for the transient structure is not possible. Consequently, insights into conformational changes are obtained by examination of structure-factor amplitude Fourier difference maps, calculated by subtracting observed structure factor amplitudes for the *dark* data set (F_o^{dark}) from those of the *light* data set (F_o^{light}), and then phasing this difference with phases calculated from the *dark* model. Experimental $F_o^{light} - F_o^{dark}$ differences can be weighted prior to map calculation based on the Bayesian likelihood that each $F_o^{light} - F_o^{dark}$ structure factor amplitude difference is observed at a given resolution within measurement errors (*sigmas*)—an approach termed Q-weighting [91]. Q-weighting was introduced as a means to improve the estimates of difference amplitudes and to reduce the noise in difference maps calculated from time-resolved polychromatic Laue data, but it also increases the signal-to-noise ratio of amplitude differences from monochromatic datasets. In $F_o^{light} - F_o^{dark}$ maps, negative and positive peaks indicate positions from which electron density has vanished and appeared upon pumping, respectively. Using this information, and prior knowledge from the resting-state model, the crystallographer may build a model for the transient structure, include it as an alternate conformer in the light structure, and then refine this new model in terms of coordinates and occupancy in the reciprocal space. An initial guess for the validity and occupancy of the transient conformer in the *light* structure can be obtained by computing $F_c^{light} - F_c^{dark}$ maps for various relative occupancies of the resting-state and transient states in the *light* structure, and comparing peak heights (on residues of interest) in these and in the experimental $F_o^{light} - F_o^{dark}$ map. Another approach is to compute extrapolated structure factors for the transient structure at 100% occupancy (F_{ext}^{light}), and then to perform difference refinement against these [88]. By this method, $2mF_{ext}^{light} - DF_c^{dark}$ and $mF_{ext}^{light} - DF_c^{dark}$ electron density maps can be obtained that will guide model-building for the transient structure. The transient structure is then refined by minimizing the residual between the extrapolated and calculated structure factor amplitudes [19, 79, 88]. Again, an initial estimate of the occupancy of the transient state in the *light* dataset is needed to generate extrapolated structure factors, using the following equation: $F_{ext}^{light} = \alpha * Q * (F_o^{light} - F_o^{dark}) + F_o^{dark}$, where Q is the optional Bayesian-based weighting factor and $1/\alpha$ is the occupancy of the time-evolved state in the *light* structure. This estimate may again be obtained by computing $F_c^{light} - F_c^{dark}$ maps and comparing peak heights on residues of interest in these and in the experimental $F_o^{light} - F_o^{dark}$ map [17]. Alternatively, occupancy-scaling can rely on a similar comparison between the experimental $F_o^{light} - F_o^{dark}$ map, and the $mF_{ext}^{light} - DF_c^{dark}$ maps computed using extrapolated F_{ext}^{light} for various occupancies of the transient state in the *light* dataset [15].

11.5.6 Kinetic Analysis of Time-Resolved Diffraction Patterns

A TR-SFX experiment produces a time-dependent set of electron density maps spanning a given time range and possibly containing time-dependent changes in density as the reaction proceeds. The strategy to interpret such a time-resolved dataset depends on the time scale explored, which we divide into dynamic (ultrafast, i.e., fs–ps) and kinetic (several ps and longer) regimes. In the dynamic regime, structural changes reflect the coherent dynamics of atomic motions, allowing for a direct comparison in terms of structure and time scale with results from molecular dynamics simulations. Time-dependent maps each provide a snapshot of the protein in motion and allow for tracking non-exponential structural evolutions such as side chain oscillations in Mb [2]. In the kinetic regime, the time evolution of the crystalline protein ensemble is governed by the energy barriers separating structural intermediates. The time-dependence of density maps then arises from the variation in population of the underlying time-independent intermediate structures, each associated with a reaction intermediate. The objective of a kinetic analysis of time-dependent maps is to extract a set of time-independent maps, each corresponding to a different reaction intermediate that can be used to determine the reaction pathway along with rate coefficients connecting the different reaction intermediates. The two main strategies described in the literature for a kinetic analysis are singular value decomposition (SVD) [72] and cluster analysis [42]. An exhaustive description of the two methods goes beyond the scope of this chapter and we refer to the dedicated literature [42, 64, 72]. Briefly, the SVD method, extensively used in the analysis of any kind of time-resolved data, is a mathematical procedure that represents each data set by two sets of vectors, weighted by corresponding so-called singular values. The first set of singular vectors forms a time-independent orthonormal basis on which all time-dependent data can be decomposed; the second set of singular vectors describes the time-dependence of the first set. Since the basis vectors are weighted by their singular values, the data matrix can be approximated by a subset of vectors which contains signal above the noise level. The obtained reduced representation of the data sets allows for interpreting the time evolution of singular vectors in terms of a kinetic model. The cluster analysis is a general method used in statistics to group objects contained in an ensemble, based on a similarity defined by a given mathematical measure. This method has been proposed to deconvolute time-independent reaction intermediates from an intrinsically noisy and time-dependent electron density datasets produced by time-resolved crystallography, via a so-called "analytical trapping procedure." A complete SVD or cluster analysis has not yet been carried in any of the published TR-SFX studies.

11.5.7 Sample Presentation and Consumption

The ultrashort and highly intense nature of X-ray FEL pulses is what allows for time-resolved studies down to the sub-ps timescale, at the atomic level of resolution. These characteristics are also at the origin of the high sample consumption rate of SFX and TR-SFX experiments. Indeed, because crystals virtually stand still during the fs exposure, all measured reflection intensities are partial, requiring collection from myriads of crystals to obtain a meaningful estimate of structure factor amplitudes. Furthermore, the intense X-ray beam eventually destroys the crystals, requiring the sample to be constantly replenished. Different means to present crystalline samples to the X-ray FEL beam exist, and they are extensively discussed in Chap. 5 of this book. All of them share the prospect of enabling TR-SFX studies, although on different systems and timescales. The first—and to date most successful—approach has been injection of crystalline samples across the X-ray FEL beam by means of gas-focused liquid jets [93]. Sample consumption is the highest when working with liquid jets, with one out of several hundred thousand crystals interacting with an X-ray pulse at a repetition rate of 120 Hz [70], due to the high injection speed (typically 10 m/s) required for the stability and sub-10 μm thickness of the jet (important for minimizing background scattering noise from the jet itself). Liquid jets can be highly stable and enable collection of data at pump–probe delays from sub-ps to a few μs, with pump excitation outside the injector nozzle. Because of the high jet speed, longer pump–probe delays are not easily accessible. Liquid jets remain the first choice in cases where crystalline sample production quantity is not an issue. Illustratively, all three ps timescale pump–probe TR-SFX studies published thus far were conducted using such an injection system, requiring about 1 g of protein crystals for a five-shift experiment at the LCLS [2, 15, 63]. To reduce sample consumption, crystals can be grown or embedded in a viscous carrier matrix prior to injection, enabling a stable jet at a reduced injection speed of typically a few mm/s [5, 84, 92]. The first matrix of the kind was a lipidic cubic phase, in which crystals of the human serotonin receptor where grown before injection across the X-ray FEL beam [49], allowing for structure determination from as little as 0.5 mg of protein. Rapidly after, other inert media were proposed that also form easily extrudable gel-like matrices from crystals that do not grow in a lipidic cubic phase [5, 13, 43, 84–86]. High-viscosity jets have been used in pump–probe TR-SFX experiments on bR [56, 60], where pump–probe delays in the ms range were accessed due to the slow jet speed. Weaknesses of high-viscosity jets for TR-SFX include reduced transparency that limits the excitation efficiency [89], the possibility of fluctuating jet speeds that reduce precision of longer pump–probe delays and the inability to address diffusion-based processes that are prohibited by the high viscosity of the matrix. Furthermore, the slow flow leads to concerns about pre-illumination of the sample from the previous laser pulses, a problem that is worse for irreversible reactions. Recently, the design was reported of a conveyer belt on which crystals may ride following acoustic droplet ejection, allowing for multiple illuminations, and/or exposure to solute or vapors prior to interaction with the X-ray beam [20]. Hence, the conveyer belt system is in theory adapted both for pump–probe and diffusion based TR-SFX experiments. Published work to date used only the pump–probe mode, to provide insights into the photocycle of PS II [99], but it is expected that other experiments will soon take advantage of the versatility and completeness of the setup. Weaknesses are that the size of ejected drops remains high and that the device does not run in vacuum, both generating high background. TR-SFX crystallography data can also be collected from a much smaller number of crystals presented sequentially to the X-ray beam in a cryo-loop, as recently shown for the cytochrome c oxidase [77]. About 40 large crystals (0.5 mm) per time point were used, each of which was exposed at multiple angles and positions while maintained in a humid air flow at 4 °C. Chip-based micro-compartmented systems allowing for sequestration of microcrystals and their presentation to the X-ray beam at a speed matching the repetition rate of most X-ray FEL sources have also been proposed and demonstrated, which could enable TR-SFX on microcrystals with minimal background and high hit rates [61, 62, 65]. Solid-support based-approaches thus hold the promise of allowing for a more efficient use of crystalline material and beamtime, during TR-SFX experiments on photosensitive proteins or proteins complexed with caged compounds. In combination with microfluidic devices, solid-support based-approaches could also provide insight into diffusion-based mechanisms, by means of a variation on the mix-and-inject approach [71].

11.6 Perspective on Future Applications of TR-SFX

Since the first study in 2012 [1], an increasing number of different light-sensitive proteins have been studied by TR-SFX (Fig. 11.6). Even if most of them are model systems that have been extensively studied by various biophysical techniques, TR-SFX experiments, in particular those carried out on the ultrafast time scale, provided new insights. For some model systems (e.g., bR and other rhodopsins), ultrafast TR-SFX is yet to come. For others (e.g., PS II), further insight from TR-SFX awaits increased diffraction resolution. Time-resolved changes along the reaction

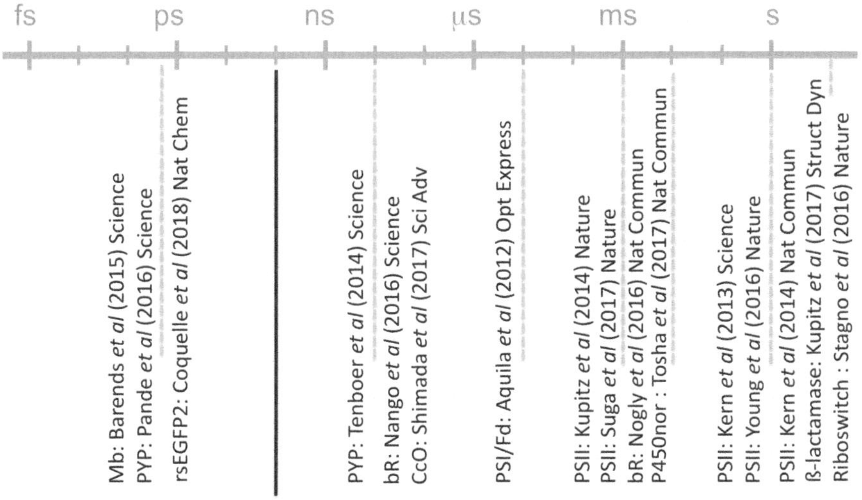

Fig. 11.6 All TR-SFX experiments published by early 2018 and their pump–probe time scales. Except for the riboswitch [80] and ß-lactamase [46] studies, all TR-SFX experiments have been conducted on light-sensitive proteins with optical pump-pulses so far

pathway are still to be uncovered by TR-SFX in a large number of inherently photosensitive proteins (e.g., reaction center, phytochromes, and cryptochromes), as well as in photoenzymes. Also, it is expected that more and more inherently light insensitive proteins will be studied by TR-SFX using caged compounds or synthetic photoswitches [9]. In these cases, the functional relevance of light-induced motions will have to be assessed carefully.

With the proven possibility to be carried out on the ultrafast time scale, TR-SFX provides the unique opportunity to link photochemistry (fs–ps motions) and photobiology (μs-ms motions) for the first time on a structural basis. The prospect of understanding the entire chain of reactions, from photon absorption to signal transduction in the case of rhodopsin, or to proton pumping in the case of bR, is now finally within reach. The complementarity of X-ray FELs and synchrotrons, in particular the highly brilliant fourth generation sources to come up in the next years, is likely to gain shape in the near future. Indeed, serial synchrotron crystallography has gained momentum in the past three years [4, 5, 14, 22, 30, 32, 35, 51, 52, 59, 61, 62, 76, 81, 94] and will be able to cover time-resolved serial crystallography on the μs-ms time scale, should fourth generation sources such as diffraction limited storage rings produce the required photon flux. TR-SFX might then mostly focus on ultrafast studies and on cases where X-ray induced changes need to be avoided.

Thus far, all TR-SFX studies but two [46, 80] have been carried out on light sensitive proteins with an optical laser as the pump method (Fig. 11.6). Other triggering methods mentioned in Sect. 11.2 (T-jump, THz irradiation, X-ray irradiation, . . .) are likely to be incorporated into TR-SFX protocols to study light-insensitive proteins, as will be the mix-and-inject approach discussed in Chap. 12 of this book.

TR-SFX studies published to date "only" provided one or up to a dozen [56] snapshots of a protein structure after reaction initiation, limited by the extremely scarce X-ray FEL beam time and the requirement to collect a large number of indexable diffraction patterns per data set. With three more X-ray FELs having started operation in 2017, on the one hand, and optimization of SFX data collection and data processing schemes, on the other hand, both bottlenecks are expected to be mitigated, so that TR-SFX may soon provide real-time molecular movies of biomolecules in action.

References

1. Aquila, A., Hunter, M. S., Doak, R. B., Kirian, R. A., Fromme, P., White, T. A., et al. (2012). Time-resolved protein nanocrystallography using an X-ray free-electron laser. *Optics Express, 20*(3), 2706–2716. https://doi.org/10.1364/OE.20.002706.
2. Barends, T. R., Foucar, L., Ardevol, A., Nass, K., Aquila, A., Botha, S., et al. (2015). Direct observation of ultrafast collective motions in CO myoglobin upon ligand dissociation. *Science, 350*(6259), 445–450. https://doi.org/10.1126/science.aac5492.

3. Baxter, R. H., Ponomarenko, N., Srajer, V., Pahl, R., Moffat, K., & Norris, J. R. (2004). Time-resolved crystallographic studies of light-induced structural changes in the photosynthetic reaction center. *Proceedings of the National Academy of Sciences of the United States of America, 101*(16), 5982–5987. https://doi.org/10.1073/pnas.0306840101.
4. Beyerlein, K. R., Dierksmeyer, D., Mariani, V., Kuhn, M., Sarrou, I., Ottaviano, A., et al. (2017). Mix-and-diffuse serial synchrotron crystallography. *IUCrJ, 4*(Pt 6), 769–777. https://doi.org/10.1107/S2052252517013124.
5. Botha, S., Nass, K., Barends, T. R., Kabsch, W., Latz, B., Dworkowski, F., et al. (2015). Room-temperature serial crystallography at synchrotron X-ray sources using slowly flowing free-standing high-viscosity microstreams. *Acta crystallographica Section D, Biological crystallography, 71*(Pt 2), 387–397. https://doi.org/10.1107/S1399004714026327.
6. Bourgeois, D., Vallone, B., Arcovito, A., Sciara, G., Schotte, F., Anfinrud, P. A., et al. (2006). Extended subnanosecond structural dynamics of myoglobin revealed by Laue crystallography. *Proceedings of the National Academy of Sciences of the United States of America, 103*(13), 4924–4929. https://doi.org/10.1073/pnas.0508880103.
7. Bourgeois, D., Vallone, B., Schotte, F., Arcovito, A., Miele, A. E., Sciara, G., et al. (2003). Complex landscape of protein structural dynamics unveiled by nanosecond Laue crystallography. *Proceedings of the National Academy of Sciences of the United States of America, 100*(15), 8704–8709. https://doi.org/10.1073/pnas.1430900100.
8. Bourgeois, D., & Weik, M. (2009). Kinetic protein crystallography: A tool to watch proteins in action. *Crystallography Reviews, 15*(2), 87–118.
9. Broichhagen, J., Frank, J. A., & Trauner, D. (2015). A roadmap to success in Photopharmacology. *Accounts of Chemical Research, 48*(7), 1947–1960. https://doi.org/10.1021/acs.accounts.5b00129.
10. Cammarata, M., Levantino, M., Schotte, F., Anfinrud, P. A., Ewald, F., Choi, J., et al. (2008). Tracking the structural dynamics of proteins in solution using time-resolved wide-angle X-ray scattering. *Nature Methods, 5*(10), 881–886.
11. Chapman, H. N., Fromme, P., Barty, A., White, T. A., Kirian, R. A., Aquila, A., et al. (2011). Femtosecond X-ray protein nanocrystallography. *Nature, 470*(7332), 73–77.
12. Colletier, J. P., Bourgeois, D., Sanson, B., Fournier, D., Sussman, J. L., Silman, I., et al. (2008). Shoot-and-trap: Use of specific x-ray damage to study structural protein dynamics by temperature-controlled cryo-crystallography. *Proceedings of the National Academy of Sciences of the United States of America, 105*(33), 11742–11747.
13. Conrad, C. E., Basu, S., James, D., Wang, D., Schaffer, A., Roy-Chowdhury, S., et al. (2015). A novel inert crystal delivery medium for serial femtosecond crystallography. *IUCrJ, 2*(4), 421–430. https://doi.org/10.1107/S2052252515009811.
14. Coquelle, N., Brewster, A. S., Kapp, U., Shilova, A., Weinhausen, B., Burghammer, M., et al. (2015). Raster-scanning serial protein crystallography using micro- and nano-focused synchrotron beams. *Acta Crystallographica Section D, 71*(5), 1184–1196. https://doi.org/10.1107/S1399004715004514.
15. Coquelle, N., Sliwa, M., Woodhouse, J., Schirò, G., Adam, V., Aquila, A., et al. (2018). Chromophore twisting in the excited state of a photoswitchable fluorescent protein captured by time-resolved serial femtosecond crystallography. *Nature Chemistry, 10*, 31–37. https://doi.org/10.1038/nchem.2853.
16. Danailov, M. B., Bencivenga, F., Capotondi, F., Casolari, F., Cinquegrana, P., Demidovich, A., et al. (2014). Towards jitter-free pump-probe measurements at seeded free electron laser facilities. *Optics Express, 22*(11), 12869–12879. https://doi.org/10.1364/oe.22.012869.
17. Duan, C., Adam, V., Byrdin, M., Ridard, J., Kieffer-Jaquinod, S., Morlot, C., et al. (2013). Structural evidence for a two-regime photobleaching mechanism in a reversibly switchable fluorescent protein. *Journal of the American Chemical Society, 135*(42), 15841–15850. https://doi.org/10.1021/ja406860e.
18. Efremov, R., Gordeliy, V. I., Heberle, J., & Büldt, G. (2006). Time-resolved microspectroscopy on a single crystal of bacteriorhodopsin reveals lattice-induced differences in the photocycle kinetics. *Biophysical Journal, 91*(4), 1441–1451. https://doi.org/10.1529/biophysj.106.083345.

19. Fermi, G., Perutz, M. F., Dickinson, L. C., & Chien, J. C. W. (1982). Structure of human deoxy cobalt haemoglobin. *Journal of Molecular Biology, 155*(4), 495–505. https://doi.org/10.1016/0022-2836(82)90483-1.
20. Fuller, F. D., Gul, S., Chatterjee, R., Burgie, E. S., Young, I. D., Lebrette, H., et al. (2017). Drop-on-demand sample delivery for studying biocatalysts in action at X-ray free-electron lasers. *Nature Methods, 14*(4), 443–449. https://doi.org/10.1038/nmeth.4195.
21. Garman, E. F., & Weik, M. (2017). Radiation damage in macromolecular crystallography. *Methods in Molecular Biology, 1607*, 467–489. https://doi.org/10.1007/978-1-4939-7000-1_20.
22. Gati, C., Bourenkov, G., Klinge, M., Rehders, D., Stellato, F., Oberthur, D., et al. (2014). Serial crystallography on in vivo grown microcrystals using synchrotron radiation. *IUCrJ, 1*(Pt 2), 87–94. https://doi.org/10.1107/S2052252513033939.
23. Genick, U. K., Borgstahl, G. E., Ng, K., Ren, Z., Pradervand, C., Burke, P. M., et al. (1997). Structure of a protein photocycle intermediate by millisecond time-resolved crystallography. *Science, 275*(5305), 1471–1475.
24. Gerwert, K., Hess, B., Michel, H., & Buchanan, S. (1988). FTIR studies on crystals of photosynthetic reaction centers. *FEBS Letters, 232*(2), 303–307. https://doi.org/10.1016/0014-5793(88)80758-0.
25. Glownia, J. M., Cryan, J., Andreasson, J., Belkacem, A., Berrah, N., Blaga, C. I., et al. (2010). Time-resolved pump-probe experiments at the LCLS. *Optics Express, 18*(17), 17620–17630. https://doi.org/10.1364/oe.18.017620.
26. Grotjohann, T., Testa, I., Reuss, M., Brakemann, T., Eggeling, C., Hell, S. W., et al. (2012). rsEGFP2 enables fast RESOLFT nanoscopy of living cells. *eLife, 1*, e00248. https://doi.org/10.7554/eLife.00248.
27. Harmand, M., Coffee, R., Bionta, M. R., Chollet, M., French, D., Zhu, D., et al. (2013). Achieving few-femtosecond time-sorting at hard X-ray free-electron lasers. *Nature Photonics, 7*, 215. https://doi.org/10.1038/nphoton.2013.11.
28. Hekstra, D. R., White, K. I., Socolich, M. A., Henning, R. W., Srajer, V., & Ranganathan, R. (2016). Electric-field-stimulated protein mechanics. *Nature, 540*(7633), 400–405. https://doi.org/10.1038/nature20571.
29. Henzler-Wildman, K., & Kern, D. (2007). Dynamic personalities of proteins. *Nature, 450*(7172), 964–972.
30. Heymann, M., Opthalage, A., Wierman, J. L., Akella, S., Szebenyi, D. M., Gruner, S. M., et al. (2014). Room-temperature serial crystallography using a kinetically optimized microfluidic device for protein crystallization and on-chip X-ray diffraction. *IUCrJ, 1*(Pt 5), 349–360. https://doi.org/10.1107/S2052252514016960.
31. Hirata, K., Shinzawa-Itoh, K., Yano, N., Takemura, S., Kato, K., Hatanaka, M., et al. (2014). Determination of damage-free crystal structure of an X-ray-sensitive protein using an XFEL. *Nature Methods, 11*(7), 734–736. https://doi.org/10.1038/nmeth.2962.
32. Huang, C. Y., Olieric, V., Ma, P., Panepucci, E., Diederichs, K., Wang, M., et al. (2015). In meso in situ serial X-ray crystallography of soluble and membrane proteins. *Acta crystallographica Section D, Biological crystallography, 71*(Pt 6), 1238–1256. https://doi.org/10.1107/S1399004715005210.
33. Hutchison, C. D. M., Kaucikas, M., Tenboer, J., Kupitz, C., Moffat, K., Schmidt, M., et al. (2016). Photocycle populations with femtosecond excitation of crystalline photoactive yellow protein. *Chemical Physics Letters, 654*, 63–71. https://doi.org/10.1016/j.cplett.2016.04.087.
34. Ihee, H., Rajagopal, S., Srajer, V., Pahl, R., Anderson, S., Schmidt, M., et al. (2005). Visualizing reaction pathways in photoactive yellow protein from nanoseconds to seconds. *Proceedings of the National Academy of Sciences of the United States of America, 102*(20), 7145–7150 Epub 2005 May 7143.
35. Jaeger, K., Dworkowski, F., Nogly, P., Milne, C., Wang, M., & Standfuss, J. (2016). Serial millisecond crystallography of membrane proteins. *Advances in Experimental Medicine and Biology, 922*, 137–149. https://doi.org/10.1007/978-3-319-35072-1_10.

36. Jung, Y. O., Lee, J. H., Kim, J., Schmidt, M., Moffat, K., Srajer, V., et al. (2013). Volume-conserving trans-cis isomerization pathways in photoactive yellow protein visualized by picosecond X-ray crystallography. *Nature Chemistry, 5*(3), 212–220. https://doi.org/10.1038/nchem.1565.
37. Kern, J., Alonso-Mori, R., Tran, R., Hattne, J., Gildea, R. J., Echols, N., et al. (2013). Simultaneous femtosecond X-ray spectroscopy and diffraction of photosystem II at room temperature. *Science, 340*(6131), 491–495. https://doi.org/10.1126/science.1234273.
38. Kern, J., Tran, R., Alonso-Mori, R., Koroidov, S., Echols, N., Hattne, J., et al. (2014). Taking snapshots of photosynthetic water oxidation using femtosecond X-ray diffraction and spectroscopy. *Nature Communications, 5*, 4371. https://doi.org/10.1038/ncomms5371.
39. Knapp, J. E., Pahl, R., Srajer, V., & Royer Jr., W. E. (2006). Allosteric action in real time: Time-resolved crystallographic studies of a cooperative dimeric hemoglobin. *Proceedings of the National Academy of Sciences of the United States of America, 103*(20), 7649–7654. https://doi.org/10.1073/pnas.0509411103.
40. Kok, B., Forbush, B., & McGloin, M. (1970). Cooperation of charges in photosynthetic O2 evolution-I. A linear four step mechanism. *Photochemistry and Photobiology, 11*(6), 457–475.
41. Kort, R., Ravelli, R. B., Schotte, F., Bourgeois, D., Crielaard, W., Hellingwerf, K. J., et al. (2003). Characterization of photocycle intermediates in crystalline photoactive yellow protein. *Photochemistry and Photobiology, 78*(2), 131–137. https://doi.org/10.1562/0031-8655(2003)0780131copiic2.0.co2.
42. Kostov, K. S., & Moffat, K. (2011). Cluster analysis of time-dependent crystallographic data: Direct identification of time-independent structural intermediates. *Biophysical Journal, 100*(2), 440–449. https://doi.org/10.1016/j.bpj.2010.10.053.
43. Kovacsova, G., Grunbein, M. L., Kloos, M., Barends, T. R. M., Schlesinger, R., Heberle, J., et al. (2017). Viscous hydrophilic injection matrices for serial crystallography. *IUCrJ, 4*(Pt 4), 400–410. https://doi.org/10.1107/S2052252517005140.
44. Kubelka, J. (2009). Time-resolved methods in biophysics. 9. Laser temperature-jump methods for investigating biomolecular dynamics. *Photochemical & Photobiological Sciences, 8*(4), 499–512. https://doi.org/10.1039/b819929a.
45. Kupitz, C., Basu, S., Grotjohann, I., Fromme, R., Zatsepin, N. A., Rendek, K. N., et al. (2014). Serial time-resolved crystallography of photosystem II using a femtosecond X-ray laser. *Nature, 513*(7517), 261–265. https://doi.org/10.1038/nature13453.
46. Kupitz, C., Olmos, J. L., Holl, M., Tremblay, L., Pande, K., Pandey, S., et al. (2017). Structural enzymology using X-ray free electron lasers. *Structural Dynamics, 4*(4), 044003. https://doi.org/10.1063/1.4972069.
47. Levantino, M., Schiro, G., Lemke, H. T., Cottone, G., Glownia, J. M., Zhu, D., et al. (2015). Ultrafast myoglobin structural dynamics observed with an X-ray free-electron laser. *Nature Communications, 6*, 6772. https://doi.org/10.1038/ncomms7772.
48. Lincoln, C. N., Fitzpatrick, A. E., & van Thor, J. J. (2012). Photoisomerisation quantum yield and non-linear cross-sections with femtosecond excitation of the photoactive yellow protein. *Physical Chemistry Chemical Physics, 14*(45), 15752–15764. https://doi.org/10.1039/c2cp41718a.
49. Liu, W., Wacker, D., Gati, C., Han, G. W., James, D., Wang, D., et al. (2013). Serial femtosecond crystallography of G protein-coupled receptors. *Science, 342*(6165), 1521–1524. https://doi.org/10.1126/science.1244142.
50. Löhl, F., Arsov, V., Felber, M., Hacker, K., Jalmuzna, W., Lorbeer, B., et al. (2010). Electron bunch timing with femtosecond precision in a superconducting free-electron laser. *Physical Review Letters, 104*(14), 144801.
51. Martin-Garcia, J. M., Conrad, C. E., Nelson, G., Stander, N., Zatsepin, N. A., Zook, J., et al. (2017). Serial millisecond crystallography of membrane and soluble protein microcrystals using synchrotron radiation. *IUCrJ, 4*(Pt 4), 439–454. https://doi.org/10.1107/S205225251700570X.
52. Meents, A., Wiedorn, M. O., Srajer, V., Henning, R., Sarrou, I., Bergtholdt, J., et al. (2017). Pink-beam serial crystallography. *Nature Communications, 8*(1), 1281. https://doi.org/10.1038/s41467-017-01417-3.

53. Miller, R. J. D. (1994). Energetics and dynamics of deterministic protein motion. *Accounts of Chemical Research, 27*(5), 145–150. https://doi.org/10.1021/ar00041a005.
54. Moffat, K. (1989). Time-resolved macromolecular crystallography. *Annual Review of Biophysics and Biophysical Chemistry, 18*, 309–332. https://doi.org/10.1146/annurev.bb.18.060189.001521.
55. Mozzarelli, A., & Rossi, G. L. (1996). Protein function in the crystal. *Annual Review of Biophysics and Biomolecular Structure, 25*, 343–365.
56. Nango, E., Royant, A., Kubo, M., Nakane, T., Wickstrand, C., Kimura, T., et al. (2016). A three-dimensional movie of structural changes in bacteriorhodopsin. *Science, 354*(6319), 1552–1557. https://doi.org/10.1126/science.aah3497.
57. Neutze, R., Wouts, R., van der Spoel, D., Weckert, E., & Hajdu, J. (2000). Potential for biomolecular imaging with femtosecond X-ray pulses. *Nature, 406*(6797), 752–757.
58. Ng, K., Getzoff, E. D., & Moffat, K. (1995). Optical studies of a bacterial photoreceptor protein, photoactive yellow protein, in single crystals. *Biochemistry, 34*(3), 879–890. https://doi.org/10.1021/bi00003a022.
59. Nogly, P., James, D., Wang, D., White, T. A., Zatsepin, N., Shilova, A., et al. (2015). Lipidic cubic phase serial millisecond crystallography using synchrotron radiation. *IUCrJ, 2*(Pt 2), 168–176. https://doi.org/10.1107/S2052252514026487.
60. Nogly, P., Panneels, V., Nelson, G., Gati, C., Kimura, T., Milne, C., et al. (2016). Lipidic cubic phase injector is a viable crystal delivery system for time-resolved serial crystallography. *Nature Communications, 7*, 12314. https://doi.org/10.1038/ncomms12314.
61. Oghbaey, S., Sarracini, A., Ginn, H. M., Pare-Labrosse, O., Kuo, A., Marx, A., et al. (2016). Fixed target combined with spectral mapping: Approaching 100% hit rates for serial crystallography. *Acta crystallographica Section D, Structural Biology, 72*(Pt 8), 944–955. https://doi.org/10.1107/S2059798316010834.
62. Owen, R. L., Axford, D., Sherrell, D. A., Kuo, A., Ernst, O. P., Schulz, E. C., et al. (2017). Low-dose fixed-target serial synchrotron crystallography. *Acta Crystallographica Section D, 73*(4), 373–378. https://doi.org/10.1107/S2059798317002996.
63. Pande, K., Hutchison, C. D., Groenhof, G., Aquila, A., Robinson, J. S., Tenboer, J., et al. (2016). Femtosecond structural dynamics drives the trans/cis isomerization in photoactive yellow protein. *Science, 352*(6286), 725–729. https://doi.org/10.1126/science.aad5081.
64. Rajagopal, S., Kostov, K. S., & Moffat, K. (2004). Analytical trapping: Extraction of time-independent structures from time-dependent crystallographic data. *Journal of Structural Biology, 147*(3), 211–222. https://doi.org/10.1016/j.jsb.2004.04.007.
65. Roedig, P., Ginn, H. M., Pakendorf, T., Sutton, G., Harlos, K., Walter, T. S., et al. (2017). High-speed fixed-target serial virus crystallography. *Nature Methods, 14*(8), 805–810. https://doi.org/10.1038/nmeth.4335.
66. Sauter, N. K., Echols, N., Adams, P. D., Zwart, P. H., Kern, J., Brewster, A. S., et al. (2016). No observable conformational changes in PSII. *Nature, 533*(7603), E1–E2.
67. Schiro, G., Woodhouse, J., Weik, M., Schlichting, I., & Shoeman, R. L. (2017). Simple and efficient system for photoconverting light-sensitive proteins in serial crystallography experiments. *Journal of Applied Crystallography, 50*(3), 932–939. https://doi.org/10.1107/S1600576717006264.
68. Schlichting, I., Almo, S. C., Rapp, G., Wilson, K., Petratos, K., Lentfer, A., et al. (1990). Time-resolved X-ray crystallographic study of the conformational change in Ha-Ras p21 protein on GTP hydrolysis. *Nature, 345*(6273), 309–315.
69. Schlichting, I., Berendzen, J., Chu, K., Stock, A. M., Maves, S. A., Benson, D. E., et al. (2000). The catalytic pathway of cytochrome p450cam at atomic resolution. *Science, 287*(5458), 1615–1622.
70. Schlichting, I., & Miao, J. (2012). Emerging opportunities in structural biology with X-ray free-electron lasers. *Current Opinion in Structural Biology, 22*(5), 613–626. https://doi.org/10.1016/j.sbi.2012.07.015.

71. Schmidt, M. (2013). Mix and inject: Reaction initiation by diffusion for time-resolved macromolecular crystallography. *Advances in Condensed Matter Physics, 2013*, 10. https://doi.org/10.1155/2013/167276.
72. Schmidt, M., Rajagopal, S., Ren, Z., & Moffat, K. (2003). Application of singular value decomposition to the analysis of time-resolved macromolecular X-ray data. *Biophysical Journal, 84*(3), 2112–2129. https://doi.org/10.1016/S0006-3495(03)75018-8.
73. Schoenlein, R. W., Peteanu, L. A., Mathies, R. A., & Shank, C. V. (1991). The first step in vision: Femtosecond isomerization of rhodopsin. *Science, 254*(5030), 412–415.
74. Schotte, F., Cho, H. S., Kaila, V. R., Kamikubo, H., Dashdorj, N., Henry, E. R., et al. (2012). Watching a signaling protein function in real time via 100-ps time-resolved Laue crystallography. *Proceedings of the National Academy of Sciences of the United States of America, 109*(47), 19256–19261. https://doi.org/10.1073/pnas.1210938109.
75. Schotte, F., Lim, M., Jackson, T. A., Smirnov, A. V., Soman, J., Olson, J. S., et al. (2003). Watching a protein as it functions with 150-ps time-resolved x-ray crystallography. *Science, 300*(5627), 1944–1947. https://doi.org/10.1126/science.1078797.
76. Schubert, R., Kapis, S., Gicquel, Y., Bourenkov, G., Schneider, T. R., Heymann, M., et al. (2016). A multicrystal diffraction data-collection approach for studying structural dynamics with millisecond temporal resolution. *IUCrJ, 3*(6), 393–401. https://doi.org/10.1107/S2052252516016304.
77. Shimada, A., Kubo, M., Baba, S., Yamashita, K., Hirata, K., Ueno, G., et al. (2017). A nanosecond time-resolved XFEL analysis of structural changes associated with CO release from cytochrome c oxidase. *Science Advances, 3*(7), e1603042. https://doi.org/10.1126/sciadv.1603042.
78. Sierra, R. G., Laksmono, H., Kern, J., Tran, R., Hattne, J., Alonso-Mori, R., et al. (2012). Nanoflow electrospinning serial femtosecond crystallography. *Acta crystallographica Section D, Biological crystallography, 68*(Pt 11), 1584–1587. https://doi.org/10.1107/S0907444912038152.
79. Šrajer, V., T-y, T., Ursby, T., Pradervand, C., Ren, Z., Adachi, S.-i., et al. (1996). Photolysis of the carbon monoxide complex of myoglobin: Nanosecond time-resolved crystallography. *Science, 274*(5293), 1726.
80. Stagno, J. R., Liu, Y., Bhandari, Y. R., Conrad, C. E., Panja, S., Swain, M., et al. (2017). Structures of riboswitch RNA reaction states by mix-and-inject XFEL serial crystallography. *Nature, 541*(7636), 242–246. https://doi.org/10.1038/nature20599.
81. Stellato, F., Oberthur, D., Liang, M., Bean, R., Gati, C., Yefanov, O., et al. (2014). Room-temperature macromolecular serial crystallography using synchrotron radiation. *IUCrJ, 1*(Pt 4), 204–212. https://doi.org/10.1107/S2052252514010070.
82. Suga, M., Akita, F., Hirata, K., Ueno, G., Murakami, H., Nakajima, Y., et al. (2015). Native structure of photosystem II at 1.95 Å resolution viewed by femtosecond X-ray pulses. *Nature, 517*(7532), 99–103. https://doi.org/10.1038/nature13991.
83. Suga, M., Akita, F., Sugahara, M., Kubo, M., Nakajima, Y., Nakane, T., et al. (2017). Light-induced structural changes and the site of O=O bond formation in PSII caught by XFEL. *Nature, 543*(7643), 131–135. https://doi.org/10.1038/nature21400.
84. Sugahara, M., Mizohata, E., Nango, E., Suzuki, M., Tanaka, T., Masuda, T., et al. (2015). Grease matrix as a versatile carrier of proteins for serial crystallography. *Nature Methods, 12*(1), 61–63. https://doi.org/10.1038/nmeth.3172.
85. Sugahara, M., Nakane, T., Masuda, T., Suzuki, M., Inoue, S., Song, C., et al. (2017). Hydroxyethyl cellulose matrix applied to serial crystallography. *Scientific Reports, 7*(1), 703. https://doi.org/10.1038/s41598-017-00761-0.
86. Sugahara, M., Song, C., Suzuki, M., Masuda, T., Inoue, S., Nakane, T., et al. (2016). Oil-free hyaluronic acid matrix for serial femtosecond crystallography. *Scientific Reports, 6*, 24484. https://doi.org/10.1038/srep24484.
87. Tenboer, J., Basu, S., Zatsepin, N., Pande, K., Milathianaki, D., Frank, M., et al. (2014). Time-resolved serial crystallography captures high-resolution intermediates of photoactive yellow protein. *Science, 346*(6214), 1242–1246. https://doi.org/10.1126/science.1259357.

88. Terwilliger, T. C., & Berendzen, J. (1995). Difference refinement: Obtaining differences between two related structures. *Acta Crystallographica. Section D, Biological Crystallography, 51*(Pt 5), 609–618.
89. Tosha, T., Nomura, T., Nishida, T., Saeki, N., Okubayashi, K., Yamagiwa, R., et al. (2017). Capturing an initial intermediate during the P450nor enzymatic reaction using time-resolved XFEL crystallography and caged-substrate. *Nature Communications, 8*(1), 1585. https://doi.org/10.1038/s41467-017-01702-1.
90. Umena, Y., Kawakami, K., Shen, J.-R., & Kamiya, N. (2011). Crystal structure of oxygen-evolving photosystem II at a resolution of 1.9 Å. *Nature, 473*, 55. https://doi.org/10.1038/nature09913.
91. Ursby, T., & Bourgeois, D. (1997). Improved estimation of structure-factor difference amplitudes from poorly accurate data. *Acta Crystallogr. Sect. A, 53*(5), 564–575. https://doi.org/10.1107/S0108767397004522.
92. Weierstall, U., James, D., Wang, C., White, T. A., Wang, D., Liu, W., et al. (2014). Lipidic cubic phase injector facilitates membrane protein serial femtosecond crystallography. *Nature Communications, 5*, 3309. https://doi.org/10.1038/ncomms4309.
93. Weierstall, U., Spence, J. C., & Doak, R. B. (2012). Injector for scattering measurements on fully solvated biospecies. *The Review of Scientific Instruments, 83*(3), 035108. https://doi.org/10.1063/1.3693040.
94. Weinert, T., Olieric, N., Cheng, R., Brunle, S., James, D., Ozerov, D., et al. (2017). Serial millisecond crystallography for routine room-temperature structure determination at synchrotrons. *Nature Communications, 8*(1), 542. https://doi.org/10.1038/s41467-017-00630-4.
95. Wickstrand, C., Dods, R., Royant, A., & Neutze, R. (2015). Bacteriorhodopsin: Would the real structural intermediates please stand up? *Biochimica et Biophysica Acta, 1850*(3), 536–553. https://doi.org/10.1016/j.bbagen.2014.05.021.
96. Wohri, A. B., Katona, G., Johansson, L. C., Fritz, E., Malmerberg, E., Andersson, M., et al. (2010). Light-induced structural changes in a photosynthetic reaction center caught by Laue diffraction. *Science, 328*(5978), 630–633.
97. Yano, J., Kern, J., Irrgang, K. D., Latimer, M. J., Bergmann, U., Glatzel, P., et al. (2005). X-ray damage to the Mn4Ca complex in single crystals of photosystem II: A case study for metalloprotein crystallography. *Proceedings of the National Academy of Sciences of the United States of America, 102*(34), 12047–12052.
98. Yeremenko, S., van Stokkum, I. H. M., Moffat, K., & Hellingwerf, K. J. (2006). Influence of the crystalline state on photoinduced dynamics of photoactive yellow protein studied by ultraviolet-visible transient absorption spectroscopy. *Biophysical Journal, 90*(11), 4224–4235. https://doi.org/10.1529/biophysj.105.074765.
99. Young, I. D., Ibrahim, M., Chatterjee, R., Gul, S., Fuller, F. D., Koroidov, S., et al. (2016). Structure of photosystem II and substrate binding at room temperature. *Nature, 540*(7633), 453–457.

Chapter 12
Towards Molecular Movies of Enzymes

Christopher Kupitz and Marius Schmidt

12.1 Introduction

Biological macromolecules can be thought of as molecular machines. The components of these machines are in constant motion, even in their "rest" states. This constant motion causes structural heterogeneity visible in high resolution measurements [1], revealed as regions of disorder or as multiple conformations. The nature and relevance of this heterogeneity and its underlying fluctuations is of large scientific interest, and it is widely investigated as they have functional importance [1–5]. In order to understand macromolecular function, "static" structures are simply not sufficient. As a reaction proceeds, the macromolecular structure has to change in order to promote function. The determination of both structure and dynamics is possible through the use of time-resolved crystallography [6–9]. This requires that a reaction is triggered successfully inside crystals. For this to be feasible, the duration of reaction initiation must be significantly shorter than the fastest process of interest in the biomolecule. If time-resolved methods are not applicable, reactions can be examined on longer time scales. Molecules can accumulate into a steady state regime [10] or they can be trapped in intermediate states by lowering the temperature [11]. For example, trap-freeze is a technique in which a reaction in a crystal is started at room temperature, and the crystal is subsequently flash frozen after a targeted intermediate state has been occupied. The introduction of amino acid mutations into the protein may also be an effective method to trap intermediates [12]. However, the time-scale of the population formation and its decay is lost.

C. Kupitz (✉) · M. Schmidt
Physics Department, University of Wisconsin-Milwaukee, Milwaukee, WI, USA
e-mail: kupitz@uwm.edu; smarius@uwm.edu

S. Boutet et al. (eds.), *X-ray Free Electron Lasers*,
https://doi.org/10.1007/978-3-030-00551-1_12

The goal of time-resolved crystallography is to create a “movie” of reacting protein molecules, which proceed along a reaction pathway. Using the term “movie” can be slightly misleading, so it is important to understand what type of information is actually being obtained. Rather than showing the trajectory of the molecules in the ensemble, it shows a population increase or decrease in a succession of intermediate states along the reaction pathway. An example of a chemical kinetic mechanism of a reaction catalyzed by an enzyme is shown below:

$$\mathrm{E}+\mathrm{S} \underset{k_{-1}}{\overset{k_1}{\rightleftarrows}} \mathrm{ES} \overset{k_2}{\rightarrow} \mathrm{EP} \overset{k_3}{\rightarrow} \mathrm{E}+\mathrm{P}$$

To start this reaction, the free enzyme (E), is mixed with substrate (S). The substrate binds to the enzyme forming the enzyme-substrate complex (ES). Here, k_1 is a second order rate coefficient for the binding, while k_{-1} accounts for the dissociation of ES. The reaction then proceeds along the catalytic pathway leading to an enzyme-product (EP) complex. Finally, free enzyme (E) is recovered, and the product (P) is released. ES and EP are reaction intermediates while k_2 and k_3 are first order rate coefficients.

A reaction can also be successfully triggered by light provided that a photosensitive group (a chromophore) is bound to the protein. After activation, the molecules within the crystal travel along the reaction pathway and may occupy intermediate states. Intermediate states correspond to local energy minima in configurational space [13–15] (See Fig. 12.1a). Intermediate states are characterized by distinct structures. In Fig. 12.1a, two intermediates denoted as I_1 and I_2 are shown. Molecules only briefly (transiently) populate these states before moving on to the next state. Structures of the intermediate states can only be determined when they are occupied by reacting molecules. To determine their structures, the concentrations of molecules in these states must build up to detectable levels. Even if the chemical kinetic mechanism allows for a sufficient buildup in a particular intermediate state, other intermediate states are occupied at the same time, hence they might mix in. There might, however, exist time points where only a single intermediate is occupied. Moreover, there might be periods of time when concentrations of molecules in intermediate states do not change (see Fig. 12.1). Then a single time point may be sufficient to determine the structure of an intermediate. In general, however, multiple intermediates contribute to any point in time, and methods for deconvoluting the X-ray data into pure constituents are needed in order to determine the individual structures [17, 18]. In contrast to intermediate states, transition states are located at the top of the barriers of activation; an example is shown in Fig. 12.1b. A protein molecule may pass from one minimum to another minimum by acquiring sufficient energy from the thermal bath to surmount this barrier. Transition states are only visited very shortly. Molecules do not persist there long enough that they can accumulate sufficiently (build up sufficient occupancy) for time-resolved experiments to detect them.

Synchrotron-based time-resolved crystallography has been highly successful at elucidating cyclic (reversible) reactions, which are activated by short light pulses

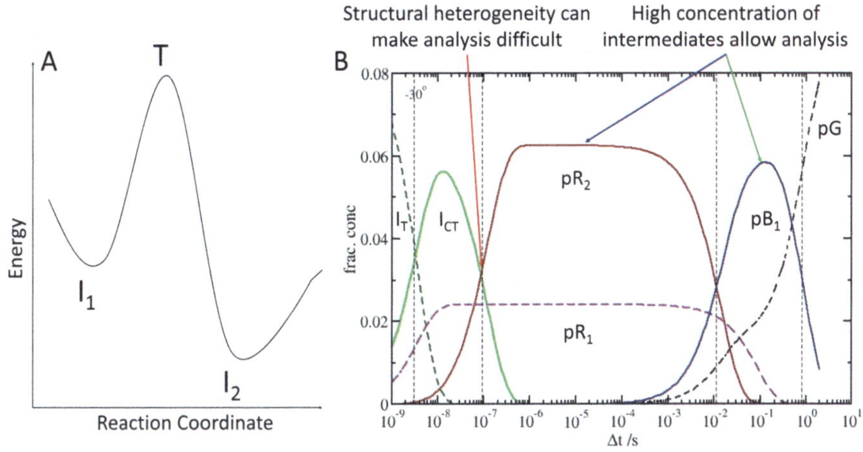

Fig. 12.1 Chemical kinetics. (**a**) Diagram of intermediate states with the energy minima corresponding to intermediate states (I_1 and I_2) and the transitional state (T) at the peak of the activation barrier. Reaction coordinate can refer to many variables such as distance, or torsional angle. (**b**) Time-dependent concentrations as determined for Photoactive Yellow Protein at −30 °C [16]. The red arrow indicates a time point where several intermediate states are occupied at the same time (I_{CT}, pR_1, and pR_2). The blue arrow points out a time regime where the concentration of molecules in two intermediate states does not change. Since the time dependence of pR_1 and pR_2 are similar it is difficult to separate them using the time information. The green arrow shows a point in time where only one intermediate is occupied (pB_1). The structure of this intermediate can be solved at this time point

[16, 19–27], but irreversible (non-cyclic) reactions are a challenge to study at synchrotrons. A major difficulty lies in having to reset the reaction before every X-ray pulse in order to ensure an accumulation of signal from a single X-ray pulse. While light activated reactions can also be non-cyclic, many of the interesting systems that contain irreversible processes do not involve light sensitive molecules. This implies that general methods to start reactions other than by light activation are developed. With the appearance of the first free electron lasers (FELs) for hard X-rays, time-resolved crystallography is on the verge of a revolution [28–32], which opens the door to optically and non-optically triggered non-cyclic (single path) reactions. Several prominent features of X-ray FELs promote a new paradigm for time-resolved crystallography: (1) The "diffraction-before-destruction" principle [33] states that the X-ray pulses are so short that a diffraction pattern is collected before radiation damage destroys the sample [28, 29, 34]. Early SFX experiments have demonstrated that the data sets are essentially free of radiation damage [28]. (2) The high intensity allows the use of micron-sized crystals that are exposed to the X-ray pulse only once at a given time delay, a requirement to investigate irreversible reactions. Crystals are injected in random orientations as a serial data collection strategy is employed [29, 30, 32, 35, 36]. (3) Micron-sized crystals make it easier to excite the sample consistently with light or trigger the reaction chemically by diffusion based methods. (4) The ultrafast pulses allow the crystallographic data to be collected at ambient (i.e., biologically relevant) temperatures [37], and (5) the fs X-ray pulses available at the X-ray FEL, as opposed to the 100 ps pulse duration at synchrotrons, improves the time-resolution that can be achieved, if needed.

12.2 Reaction Initiation

The ability to trigger or initiate a reaction is essential for any time-resolved experiment. The reaction initiation must occur essentially simultaneously across as many molecules as possible in a crystal. In addition, the trigger must be significantly shorter than the lifetime of the intermediates of interest. A number of methods can be employed to trigger reactions (Table 12.1), and two methods are widely used: light activation and chemical activation via adding substrates, ligands, and other small molecules. Reaction initiation techniques can cause large structural changes, which might compromise crystal integrity. For example, crystals of the Photosystem I–ferredoxin complex quickly fall apart upon light excitation [35, 42]. Therefore, these experiments are particularly difficult to perform at synchrotrons, when multiple pump–probe sequences on the same crystal are required to produce a single diffraction pattern. At the X-ray FEL, however, experiments are performed in a different way. (1) Microcrystals are used, which might support larger structural changes. (2) The experiment is not impaired if the crystal fails or cracks after the single shot measurement, because crystal cracking takes longer than most time-points of interest. As a consequence, the dynamics can be measured up to the time point where the crystal is falling apart (see below).

12.2.1 *Activation by Light*

Reaction initiation by short laser pulses provides undoubtedly the highest temporal resolution for time-resolved crystallography. The photo-dissociation and rebinding of CO to heme proteins [26, 43], and the isomerization of the photoactive yellow protein (PYP) chromophore from trans to cis [19] are both examples. Both

Table 12.1 Methods of reaction initiation in protein crystals

Method	Time resolution	Experimental complexity
Light driven proteins [31]	Ultrafast, ~100–280 fs achievable	Low
Caged substrates, macrocrystallography [38]	>100 ns	High
Diffusive mixing, serial crystallography [39, 40]	~100 μs	Low
Electric fields, temperature jump, etc. [41]	Variable but typically moderate	Variable

myoglobin and PYP were essential to develop new techniques in time-resolved crystallography [16, 18, 20, 22, 23, 25, 26, 31, 32, 44–50]. Protein macrocrystals contain between 10^{13} and 10^{14} individual protein molecules, depending on the size and the packing in the crystal. The optical density in the absorption region of the chromophore is unusually high due to the tight packing of molecules in a crystal. Consequently, the penetration depth into a crystal can be very small, on the order of a few micrometers [16, 32]. Therefore, wavelength and illumination geometry have to be carefully considered [32]. Microcrystals are clearly advantageous as the optical path through a microcrystal is, by definition, small. This allows the excitation wavelength to be closer to the absorption maximum, which results in a more effective reaction initiation [51, 52]. However, the laser peak power needs to be carefully selected. Too high a laser power could lead to damaging effects such as two photon absorption and radical formation, especially when ultrashort laser pulses are used. In contrast, if the laser power is too low, not enough molecules are excited and the occupation of molecules in intermediate states might be too small to be detected. A broad range of laser pulse energy densities have been used, typically ranging from 0.5–5 mJ/mm^2. This should be tailored to the individual protein and experiment for best results, by examining the reaction spectroscopically ahead of time [30, 32, 35, 43].

12.2.2 Activation by Caged Substrates

The majority of proteins are not naturally photosensitive. In these cases, there are other options that need to be explored to trigger the protein dynamics. Photoreactive caged substrates, also known as photoactivatable bio-agents can be used. These agents are composed of small molecules and compounds, for example a cofactor, substrate, or even a protein residue, which has been chemically inactivated by a photosensitive protecting group [11, 53, 54]. Oxygen, phosphate, adenosine triphosphate, and nucleotides are just a few examples that are available in caged form [55]. In time-resolved crystallography, caged substrates are soaked into crystals and then activated using a light pulse with some notable examples. Caged guanosine triphosphate was successfully used to study Ras P21 [56], the reaction catalyzed by isocitrate dehydrogenase was investigated using caged nicotinamide adenine dinucleotide phosphate [57], and acetylcholinesterase was examined with caged choline [11]. However, wide spread use of caged compounds in time-resolved crystallography is challenging due to several factors: (1) the design and fabrication of caged substrates requires significant expertise, (2) each substrate is specifically tailored to the protein and the reaction to be investigated, (3) the photo-removable protecting groups need to meet high specifications such as high quantum yield of the photoreaction, and high absorbance above the 300 nm wavelength to limit protein photo-damage [55], (4) once the protecting group has been removed, the byproducts should not harm the protein nor inhibit the reaction with the substrate, (5) activation must occur on a faster time-scale than the reaction being probed (Table 12.1), and

(6) cleavage of the protecting group induced by X-ray exposure must be kept to a minimum when longer or multiple X-ray pulses are required. All of these largely impede the routine application to time-resolved crystallographic investigations.

12.2.3 Other Activation Mechanisms

By recombinantly attaching a light activated domain to an enzyme its activity can be controlled by light [58, 59]. The time delay between light activation and signal transmission to the effector domain is in the range of microseconds, which allows the observation of enzymatic processes that occur on the millisecond time-scale [60]. Recently, a reaction has been triggered by an electric field [41], which would alter charges, dipoles, and polarizability. This will affect any movement of charge within a protein, and therefore is a general method to study structural dynamics. However, it is not clear how the vast number of very specific enzymatic reactions can be triggered by this method.

12.2.4 Activation by Diffusion

A versatile option to initiate a reaction for non-photosensitive proteins is through diffusion. The crystals are simply mixed with substrate such as small molecules, redox reagents, or other cofactors. Although this is conceptually a simple method, in practice diffusive studies with crystals are difficult because diffusion is an inherently slow process. Depending on crystal size, protein packing density in the crystal, size of the substrate, and the type of chemicals that are used for crystallization, the substrate can take seconds to even minutes to diffuse throughout the crystal [61–64].

For previous diffusion-based work at synchrotrons, flow cells were used to collect data [65–67]. A flow cell works by immobilizing a crystal inside of a capillary while flowing a solution with the substrate over the crystal [67]. This approach has been successfully used to determine the structure of rate-limited species that accumulate enough in the steady state [10]. Common challenges when working with flow cells are crystal mounting, keeping the crystal centered, and keeping it immobilized in the X-ray beam even as the substrate solution is flowing [6, 68]. On the other hand, serial femtosecond crystallography seems to be almost tailor made for a diffusion-based method, as microcrystals allow for rapid diffusion. Microcrystals mixed with substrate can be utilized at high-intensity X-ray FEL sources, or with micro-focused pulsed X-ray beams from synchrotrons, using a technique called "mix-and-inject serial crystallography" (MISC) [40]. MISC works through rapid diffusion of a substrate into a slurry of microcrystals or nanocrystals just prior to the X-ray interaction region [69]. The method to accomplish this is discussed below. With sizes approximately 10 μm or less, microcrystals allow millisecond or even sub-millisecond diffusion times [69] with estimated diffusion times for several

Table 12.2 Diffusion times into crystals as determined by calculation, simulation and experiment

Crystal size (μm^3)	0.5 × 0.5 × 0.5	1 × 2 × 3	3 × 4 × 5	10 × 20 × 30	400 × 400 × 1600
Diffusion time	17 μs^a	150 μs^a	1 ms^a	15 ms^a	16 s^a24 s^b<1 min^c

[a]Times from [69]
[b]Time from [61]
[c]Time from [62]

crystal sizes shown in Table 12.2. Fortunately, very fast enzyme catalysis is still in the range of tens of microseconds [70], while most enzymatic reactions occur in the millisecond time regime.

12.3 Sample Delivery

In synchrotron-based crystallography a single large crystal can be used to collect multiple data sets [22, 24–26] or even a complete time-series [16, 19, 21]. In most room temperature synchrotron experiments, crystals are individually mounted and sealed in a capillary with a small amount of mother liquor to prevent them from drying out. This is very different from serial crystallography, in which tens of thousands of microcrystals are introduced into the X-ray interaction region one at a time in random orientation. Serial crystallography can be performed at X-ray FELs and at synchrotrons. At synchrotrons, somewhat large crystals (>10 μm) are needed, which ameliorate radiation damage and scatter with higher intensities. The velocity at which the crystals move through the X-ray interaction volume needs to be adjusted so that there is negligible crystal motion during the longer (100 ps) synchrotron pulse. In the case of X-ray FELs, crystals sizes can be smaller, allowing for more rapid diffusion and all crystal displacements are negligible during the femtosecond X-ray pulses. In recent years, numerous microcrystalline injection methods have been developed. Depending on factors like time scale and sample consumption, sample delivery systems such as the gas dynamics virtual nozzle [71], viscous jet [72, 73], electrospun jet [74, 75], fixed target system [76–79], and the droplet on demand system [80] are all viable options. These systems are compatible with time-resolved studies triggered by light, however none of these were designed for introducing small molecules for mixing. Therefore for mix-and-inject experiments special injectors had to be developed such as the one shown in Fig. 12.2. Mixing times must be as quick as possible, and diffusion times fast to allow for sufficient time-resolution. The time delay is given by the difference in distance between the mixing region and the X-ray interaction volume divided by the jet velocity. New generations of injector technology allows for the rapid mixing of substrate with enzyme in crystal form [39, 40]. Current capabilities support mixing times on the microsecond to millisecond time scales, significantly faster than the majority of enzymatic reactions. If longer time scales are of interest, a simple T-junction mixer can also be used, as shown in Fig. 12.3.

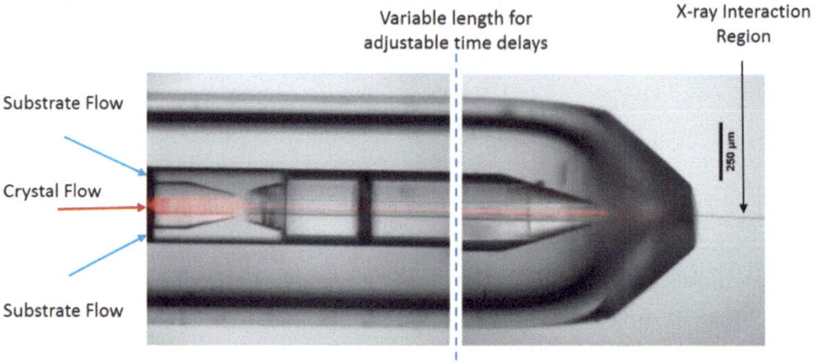

Fig. 12.2 Mixing injector, picture modified from Calvey et al., Structural Dynamics, 3, 054301, 2016. Crystal suspension, simulated by central red flow, is focused by the substrate flow, blue arrows. The mixture is then gas focused to produce a jet to be probed by the X-ray FEL pulse. The time delay is variable depending on the distance between the mixing region and the X-ray interaction region. This allows for the collection of multiple time points [39]

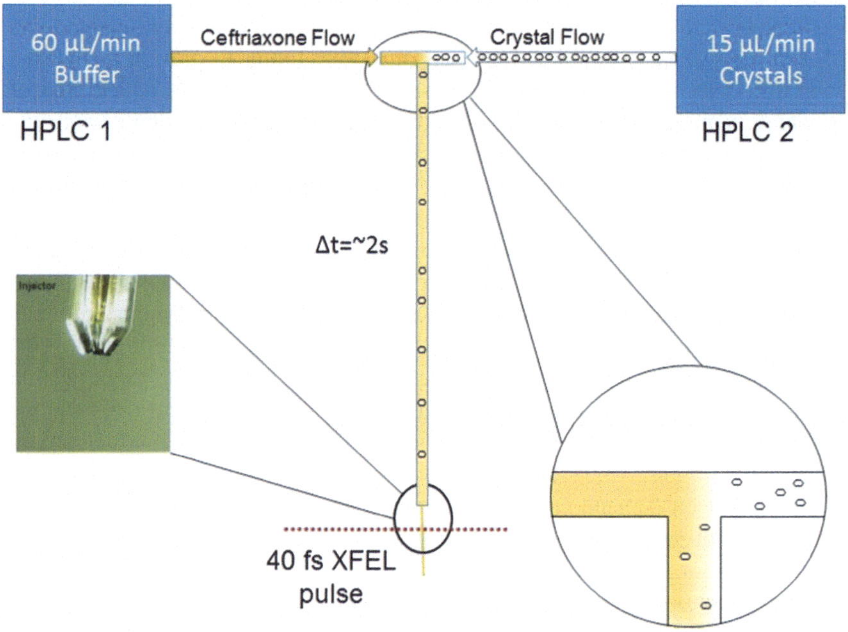

Fig. 12.3 Schematic example of the T-junction mixer used at LCLS for mix-and-inject serial crystallography. Substrate is pumped by HPLC 1, crystals are pumped by HPLC 2. The distance from the mixer to the X-ray interaction zone along with the velocity of crystal flow determines the time-delay. In this setup, the crystals are injected into the X-ray interaction region via a standard gas dynamic virtual nozzle

12.4 Data Analysis of Serial and Time-Resolved Crystallography

Data analysis of X-ray FEL data has come a long way since the first days of X-ray FEL based crystallography. All the standard methods for hit finding and Bragg intensity integration developed for SFX apply to time-resolved SFX as well. It is a simple matter of properly tagging which shots are "delayed" after the initiation of a reaction in the crystals, and what the delay times are for each X-ray exposure. Separate data sets of structure factor amplitudes for each time delay (including the reference with $t = 0$) are then available. Details on SFX data analysis can be found in Chap. 7 of this book. An overview of the entire data analysis process is shown in Fig. 12.4.

12.4.1 *Difference Maps*

A time-series of difference electron density maps lies at the core of time-resolved crystallography. If the unit cell does not change during the reaction, isomorphous difference maps can be calculated from reference amplitudes collected without reaction initiation ($|F_r|$) and the time-dependent amplitudes ($|F_t|$) collected some time interval (t) after reaction initiation. Phases (ϕ^{calc}) are obtained from an accurate reference model. Difference amplitudes are calculated by subtracting the F_t^{obs} from the F_r^{obs}: $\Delta F_t = |F_t^{obs}| - |F_r^{obs}|$. ΔF_t can be weighted [32, 49, 81] in order to account for errors in the measured amplitudes that might impair the proper interpretation of the maps. The weighting factor ensures that those difference structure factor amplitudes, which are erroneously large or contain large experimental error values, are down-weighted appropriately. An example of a difference map is shown in Fig. 12.4.

If unit cell parameters largely change [30], or even the space group changes [82] during a reaction, the reference and the time-dependent crystallographic data are

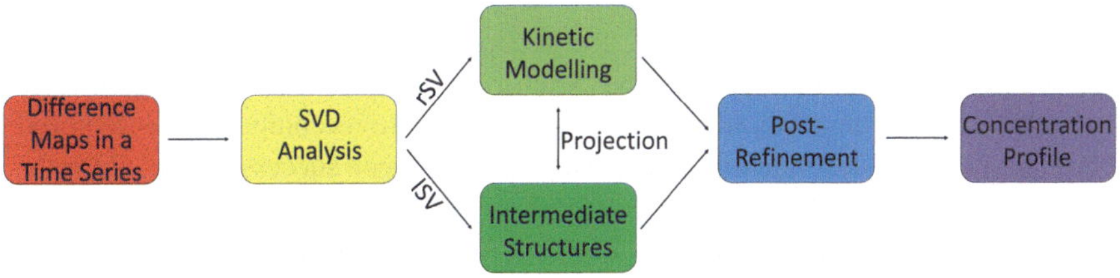

Fig. 12.4 SVD-based analysis of experimental difference maps allows the determination of the structures of the intermediates by applying a kinetic model. The kinetic mechanism is further post-refined (see text). As a result the concentrations of the intermediates at all time points (the concentration profile) can be determined

no longer isomorphous. Difference maps cannot be calculated by subtracting the observed reference data. In this case omit maps, preferentially generated using a simulated annealing (SA) refinement strategy, can circumvent this problem. Before the refinement, parts of the model where changes are expected are deliberately removed. Then this structural model is heated in the computer to several thousand Kelvin, gradually cooled and simultaneously refined against the time-resolved structure factor amplitudes [83, 84]. This strategy eliminates model phase bias which would otherwise impair the reliability of the difference map [84]. After the refinement, structure factors F^{calc} with amplitude $|F^{calc}|$ and phase ϕ^{omit} can be determined. Maps calculated with $\Delta F_t^{omit} = |F_t^{obs}|$-$|F^{calc}|$ and model phases ϕ^{omit} show unbiased difference electron density, which accounts for the local structural changes during a reaction. This SA omit map can then be structurally explained either by reinterpreting the missing part that was omitted in the refinement, or by adding ligands that coordinate during the reaction, or both. Either way, omit maps (when there is non-isomorphism) or isomorphous difference maps are equally suitable to extract the kinetics.

The methods discussed above in conjunction with high-resolution time-resolved serial femtosecond crystallographic experiment at the LCLS were demonstrated to be highly successful. Difference maps were generated that contain features with high signal to noise ratio and show chemically sensible structural changes of PYP [32]. An example of the difference maps generated for PYP can be seen in Fig. 12.4. Here the difference density clearly shows changes between the reference model (PG, yellow) and the pR_2 model (red) and pR_1 model (magenta), with the negative features of the electron density map primarily on top of the atoms of the reference model. These structures are part of the photocycle as shown in Fig. 12.4a.

12.4.2 Kinetic Mechanism and Intermediates and Detangling Multiple States

Chemical kinetics is governed by a kinetic mechanism with one or more intermediate states connected by rate coefficients (see above). Intermediate states are occupied (or visited) by macromolecules whose concentrations then vary as a function of time. This means multiple states may be occupied at the same time (see Fig. 12.5 for an example). Thus, a time-resolved experiment will probe this mixture of states. As mentioned earlier, a way to separate this mixture into its components is required. The singular value decomposition (SVD) method has been successfully used for this purpose [16, 18, 32, 85]. The SVD method separates out the spatial and temporal dependencies in the time-series of difference maps [18]. Two sets of vectors are obtained, the left singular vector (lSV) and the right singular vector (rSV). The lSV contains the spatial components, which are difference maps, while the rSVs contain the corresponding time evolutions. The significant rSVs are then interpreted with a kinetic model. This requires a kinetic mechanism with a corresponding set of coupled differential equations that contain rate coefficients [18]. A compatible

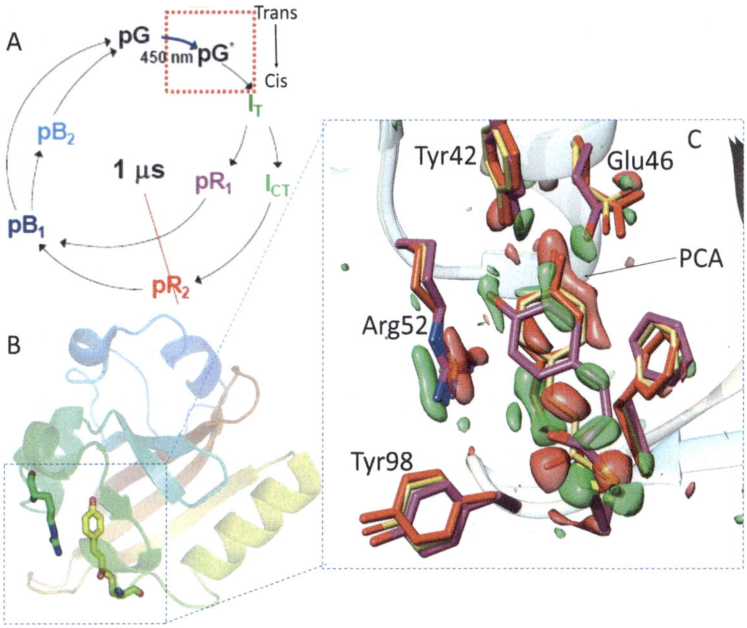

Fig. 12.5 Photoactive yellow protein (PYP). (**a**) Overview of the PYP photocycle. The reaction can be triggered by 450 nm light. I_T is an early intermediate already in the cis configuration. pR_1 is an intermediate also identified by spectroscopy. Its absorption maximum is red-shifted. I_{CT} is slightly changed from I_T. pR_2 is the second red-shifted intermediate. The red box denotes a part of the photo cycle that is faster than the time-resolution of synchrotrons. The red line denotes the 1 μs time point that is observed in (**c**). (**b**) An overview of PYP structure. (**c**) An isomorphous difference map of the chromophore pocket of PYP calculated from a 1 μs time delay data set collected at the LCLS. Green denotes positive difference density, red denotes negative difference density, both are contoured to 3σ and −3σ respectively. The structures of pR_1, pR_2, and pG (reference) are shown in magenta, red and yellow respectively. The chromophore has undergone photo isomerization from trans to cis. Several nearby residues have changed their conformation

mechanism must match the observed relaxation times in the rSV. Globally fitting sums of exponentials to the significant rSVs provides this number of relaxation times. The number of significant right singular vectors, significant singular values, and the exponentials should all be compatible with the number of distinguishable intermediates. After fitting the right singular vectors with concentration profiles of a compatible chemical kinetic mechanism, time-independent difference maps can then be calculated for each intermediate state using a projection algorithm [18, 86].

12.4.3 Structures of the Intermediate

Once the mixture-free, time-independent difference electron densities of the intermediates have been extracted, the structures of the intermediates can be determined.

A convenient way is to use extrapolated, conventional electron density maps for this purpose. To calculate these maps, difference structure factors ΔF_j^{ind} for all intermediates, $j = 1 \ldots J$ are determined by Fourier transforming the time-independent difference maps. A multiple, N, of these difference structure factors need to be added to the calculated structure factors F_r^{calc} derived from a reference model.

Extrapolated maps for all the intermediates, ρ_j^{ext}, are calculated from the F_j^{ext}. The multiplier N must be optimized so that the electron density in a particular extrapolated map vanishes where negative electron density is present in the corresponding difference map. Extrapolated maps can be interpreted just like conventional electron density maps. Since structural changes are usually small, most of the electron density in the extrapolated map can be explained already by the reference model. This will also allow for a direct structural refinement. In areas where there is a strong deviation between the reference model and the electron density, a program like "Coot" [87] is used to adjust the model by hand to better match the electron density. The resulting model can be refined in real space against the electron density before performing the final refinement in reciprocal space using the $|F_j^{ext}|$ with standard programs such as "Refmac" [88] or "Phenix" [83].

12.4.4 Post-Analysis

The purpose of this analysis is to refine a candidate kinetic mechanism by making use of the experimentally determined intermediate structures. Calculated time-dependent difference maps ($\Delta\rho(t)^{calc}$) must agree with the observed time-dependent difference maps ($\Delta\rho(t)^{obs}$). In order to obtain the $\Delta\rho(t)^{calc}$, several steps are required. (1) Structure factors for each of the J intermediates are computed from their respective structural models. From these, time-independent difference structure factors, ΔF_j^{calc}, can be determined by subtracting the structure factors of the reference model. (2) The ΔF_j^{calc} are used to determine time-independent difference electron density maps ($\Delta\rho_j^{calc}$). (3) Time-dependent concentrations, $c(t,k)_j$, for each intermediate are calculated by integrating the coupled differential equations of the candidate kinetic mechanism [18]. Note that $c(t,k)$ is determined by the magnitude of the rate coefficients of the mechanism. Finally, (4) $\Delta\rho(t)^{calc}$ is calculated by adding together individual time-independent difference electron density maps $\Delta\rho_j^{calc}$, which are weighted by their time-dependent concentrations $c(t)_j$ at all time-points.

The time-series of $\Delta\rho(t)^{calc}$ can then be compared to the observed $\Delta\rho(t)^{obs}$. The $\Delta\rho(t)^{calc}$ are dependent on both the rate coefficients and a global scaling factor (sc). The sc accounts for the difference in scale between the observed and calculated difference electron densities. The rate coefficients and the scale factor are iteratively changed to optimize the agreement between the difference density maps at all time-points. The final result of this analysis is a series of refined rate coefficients, which can be used to determine a refined concentration profile (Fig. 12.1b) for all intermediates. The scale factor is a measure for the extent of reaction initiation. It accounts for the number of molecules that enter the reaction in the crystal. In some cases the extent of the reaction initiation is small, only around 5–10%, however activation yields as high as 40–50% have been reported [21, 25, 32].

12.5 Results of Mix-and-Inject Serial Crystallography

Results using MISC have recently been reported [40, 82], in which a T-junction (Fig. 12.3) was used to mix substrate and microcrystals. In one study, an overall reaction time of 2 s [40] was selected, and in the other reaction times of 10 s as well as 10 min were reported [82]. These results are summarized below.

Successful experiments on a ß-lactamase (BlaC) mixed with the third generation lactam antibiotic Ceftriaxone were published by Kupitz et al. [40]. BlaC from *M. tuberculosis* is responsible for broad resistance of the bacteria to lactam antibiotics, such as penicillin and cephalosporin derivatives [89, 90]. BlaC microcrystals were grown via ammonium phosphate in thin plates, and then crushed using glass beads to small shards. These shards, which contain ~16 mmol/L of BlaC, were mixed with 200 mmol/L ceftriaxone. The crystal-substrate suspension was probed by an X-ray FEL pulse approximately 2 s after mixing (Fig. 12.3). The results from this experiment were maps generated at approximately 2.4 Å resolution. The electron density was found to have changed in the catalytic cleft of BlaC in two out of four subunits of the asymmetric unit (Fig. 12.6).

Since the turnover time of Ceftriaxone is 1.2 s [91], most likely a steady state was observed. In the steady state, three states can contribute to the electron density. The enzyme-substrate complex (ES) in which the full ceftriaxone molecule is intact prior to any attack by Ser70 is the first contributor. The second contributor is the covalently bound serine-adduct in which ceftriaxone is bound to Ser70. The lactam ring is cleaved opened, and a portion of the ceftriaxone molecule has been cleaved off [92]. It might also be that the third contributor is the product form in which the ceftriaxone molecule is hydrolyzed from Ser70 after the lactam ring has been opened. The full length model of ceftriaxone fits the electron density best, although the two other models both might provide slight perturbations, which improve the fit to the electron density. Moreover it is both interesting and important to note that the electron density reported here only shows ceftriaxone in two (subunits B and D) of the four subunits present in the asymmetric unit. Not all subunits may react equally, depending on accessibility of the binding pocket and its chemical environment. In addition, crystal lattice constraints might influence the dynamics of the enzyme molecules. Therefore, reaction rates may be different in the crystal compared to those in solution.

MISC has also been used to determine structures during the binding of adenine to a RNA-riboswitch [82] at 10 s and approximately 10 min. For the 10 s time point, the injection method was similar to the double HPLC T-junction setup shown

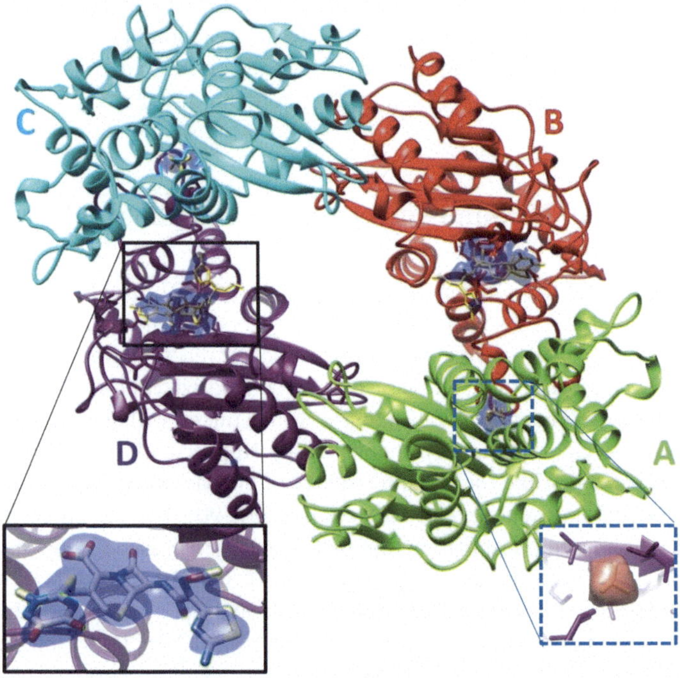

Fig. 12.6 BlaC tetramer after 2 s mixing with ceftriaxone, subunits (**a–d**) labelled. (**b**) and (**d**) subunits show ceftriaxone electron density (blue density inside solid black box), (**a**) and (**c**) subunits show only phosphate density (red density inside blue dashed box). The catalytic clefts of subunits (**a**) and (**c**) are blocked from interaction by the neighboring subunits

in Fig. 12.3. For the 10 min time point substrate was directly added to crystals and allowed to react before injection. During this time, unit cell parameters changed. For example, the length of cell parameter b increased from 46.7 to 154.9 Å. Even the crystallographic space group changed from $P2_1$ to $P2_12_12$. Remarkably, the crystals accommodated these large changes, and still diffracted to 3.0 Å.

As discussed above, mixtures can only be separated when a time-series with shortly space time-points is used. The ability to collect X-ray data at various time-delays is a major goal at X-ray FEL facilities. Enzymatic intermediates along a catalytic pathway can then be extracted. The mix-and-inject method, when applied to medically or pharmaceutically important targets, will become an integral tool in the field of structural based drug design. The above examples demonstrate that it is necessary to collect full time-series on catalytically active targets routinely. With continued effort in injector technology [39, 93] to decrease mixing times, the time resolution can be improved, and faster reactions can be visualized on the sub-millisecond time scale.

12.6 Moving Forward

Time-resolved crystallography is currently entering a new era. Previously, time-resolved crystallography was mostly limited to light-activated reversible reactions, such as those observed in PYP [17, 19, 27, 81] or in various heme proteins [22, 25, 26, 43]. Most biological, pharmaceutical and medical important reactions have not been explored.

The advent of the X-ray FEL allows the use of microscopic crystals. This provides new opportunities to investigate enzymes. Several studies have already been reported that have successfully used diffusion-based techniques at an X-ray FEL to explore structural changes on the second time scale. Continuing work on injector technology promises mix-and-inject delivery with time resolution as fast as microseconds [39]. Through the use of slightly larger micron sized crystals, mix-and-inject experiments can also be performed at synchrotrons. A 5 μm crystal still allows for an achievable time resolution of better than 10 ms, which is sufficient for most enzymatic reactions, while still tolerating the radiation dose deposited by the synchrotron pulses [94]. Beamtimes at XFELs are limited. Synchrotron light sources may provide additional beamtime for these investigations, so that more biologically and biomedically important biomolecules can be examined this way.

With all of these new techniques and sources on the horizon, the future of structural biology and specifically structural enzymology is bright. Moving forward, the goals of crystallographers will change from the determination of static images and stills of protein molecules to the collection of comprehensive time-courses. From one well-planned, essentially routine experiment not only the three-dimensional structure of the biological macromolecule can be solved, but also its function, kinetics, and dynamics may be extracted.

From one well-planned, essentially routine experiment not only the three-dimensional structure of the biological macromolecule can be solved, but also it's function, kinetics, and dynamics may be extracted, as most recently demonstrated by Olmos et al, 2018

References

1. DePristo, M. A., de Bakker, P. I., & Blundell, T. L. (2004). Heterogeneity and inaccuracy in protein structures solved by X-ray crystallography. *Structure, 12*, 831–838.
2. Botha, S., Nass, K., Barends, T. R., Kabsch, W., Latz, B., Dworkowski, F., et al. (2015). Room-temperature serial crystallography at synchrotron X-ray sources using slowly flowing free-standing high-viscosity microstreams. *Acta Crystallographica. Section D, Biological Crystallography, 71*, 387–397.
3. Frauenfelder, H., Sligar, S. G., & Wolynes, P. G. (1991). The energy landscapes and motions of proteins. *Science, 254*, 1598–1603.
4. McCammon, J. A., & Harvey, S. C. (1987). *Dynamics of proteins and nucleic acids*. Cambridge, UK: Cambridge University Press.

5. Steinbach, P. J., Ansari, A., Berendzen, J., Braunstein, D., Chu, K., Cowen, B. R., et al. (1991). Ligand binding to heme proteins: Connection between dynamics and function. *Biochemistry, 30*, 3988–4001.
6. Moffat, K. (1989). Time-resolved macromolecular crystallography. *Annual Review of Biophysics and Biophysical Chemistry, 18*, 309–332.
7. Moffat, K. (2001). Time-resolved biochemical crystallography: A mechanistic perspective. *Chemical Reviews, 101*, 1569–1581.
8. Schmidt, M. (2008). Structure based enzyme kinetics by time-resolved X-ray crystallography. In *Ultrashort laser pulses in medicine and biology*. Berlin, Germany: Springer.
9. Schmidt, M. (2015). Time-resolved crystallography at X-ray free Electron lasers and synchrotron light sources. *Synchrotron Radiation News, 28*, 25–30.
10. Stoddard, B. L. (2001). Trapping reaction intermediates in macromolecular crystals for structural analyses. *Methods, 24*, 125–138.
11. Bourgeois, D., & Weik, M. (2009). Kinetic protein crystallography: A tool to watch proteins in action. *Crystallography Reviews, 15*, 87–118.
12. Flint, A. J., Tiganis, T., Barford, D., & Tonks, N. K. (1997). Development of "substrate-trapping" mutants to identify physiological substrates of protein tyrosine phosphatases. *Proceedings of the National Academy of Sciences of the United States of America, 94*, 1680–1685.
13. Buch, I., Giorgino, T., & De Fabritiis, G. (2011). Complete reconstruction of an enzyme-inhibitor binding process by molecular dynamics simulations. *Proceedings of the National Academy of Sciences of the United States of America, 108*, 10184–10189.
14. Schmidt, M., & Saldin, D. K. (2014). Enzyme transient state kinetics in crystal and solution from the perspective of a time-resolved crystallographer. *Structural Dynamics, 1*, 1–14.
15. Steinfeld, J. I., Francisco, J. S., & Hase, W. L. (1985). *Chemical kinetics and dynamics*. Upper Saddle River, NJ: Prentience Hall.
16. Schmidt, M., Srajer, V., Henning, R., Ihee, H., Purwar, N., Tenboer, J., et al. (2013). Protein energy landscapes determined by five-dimensional crystallography. *Acta Crystallographica Section D, 69*, 2534–2542.
17. Rajagopal, S., Anderson, S., Srajer, V., Schmidt, M., Pahl, R., & Moffat, K. (2005). A structural pathway for signaling in the E46Q mutant of photoactive yellow protein. *Structure, 13*, 55–63.
18. Schmidt, M., Rajagopal, S., Ren, Z., & Moffat, K. (2003). Application of singular value decomposition to the analysis of time-resolved macromolecular X-ray data. *Biophysical Journal, 84*, 2112–2129.
19. Ihee, H., Rajagopal, S., Srajer, V., Pahl, R., Anderson, S., Schmidt, M., et al. (2005). Visualizing reaction pathways in photoactive yellow protein from nanoseconds to seconds. *Proceedings of the National Academy of Sciences of the United States of America, 102*, 7145–7150.
20. Jung, Y. O., Lee, J. H., Kim, J., Schmidt, M., Moffat, K., Srajer, V., et al. (2013). Volume-conserving trans-cis isomerization pathways in photoactive yellow protein visualized by picosecond X-ray crystallography. *Nature Chemistry, 5*, 212–220.
21. Knapp, J. E., Pahl, R., Srajer, V., & Royer, W. E., Jr. (2006). Allosteric action in real time: Time-resolved crystallographic studies of a cooperative dimeric hemoglobin. *Proceedings of the National Academy of Sciences of the United States of America, 103*, 7649–7654.
22. Schmidt, M., Nienhaus, K., Pahl, R., Krasselt, A., Anderson, S., Parak, F., et al. (2005). Ligand migration pathway and protein dynamics in myoglobin: A time-resolved crystallographic study on L29W MbCO. *Proceedings of the National Academy of Sciences of the United States of America, 102*, 11704–11709.
23. Schotte, F., Cho, H. S., Kaila, V. R., Kamikubo, H., Dashdorj, N., Henry, E. R., et al. (2012). Watching a signaling protein function in real time via 100-ps time-resolved Laue crystallography. *Proceedings of the National Academy of Sciences of the United States of America, 109*, 19256–19261.
24. Schotte, F., Lim, M., Jackson, T. A., Smirnov, A. V., Soman, J., Olson, J. S., et al. (2003). Watching a protein as it functions with 150-ps time-resolved X-ray crystallography. *Science, 300*, 1944–1947.

25. Srajer, V., Ren, Z., Teng, T. Y., Schmidt, M., Ursby, T., Bourgeois, D., et al. (2001). Protein conformational relaxation and ligand migration in myoglobin: A nanosecond to millisecond molecular movie from time-resolved Laue X-ray diffraction. *Biochemistry, 40*, 13802–13815.
26. Srajer, V., Teng, T. Y., Ursby, T., Pradervand, C., Ren, Z., Adachi, S., et al. (1996). Photolysis of the carbon monoxide complex of myoglobin: Nanosecond time-resolved crystallography. *Science, 274*, 1726–1729.
27. Tripathi, S., Srajer, V., Purwar, N., Henning, R., & Schmidt, M. (2012). pH dependence of the photoactive yellow protein photocycle investigated by time-resolved crystallography. *Biophysical Journal, 102*, 325–332.
28. Boutet, S., Lomb, L., Williams, G. J., Barends, T. R., Aquila, A., Doak, R. B., et al. (2012). High-resolution protein structure determination by serial femtosecond crystallography. *Science, 337*(6092), 362–364.
29. Chapman, H. N., Fromme, P., Barty, A., White, T. A., Kirian, R. A., Aquila, A., et al. (2011). Femtosecond X-ray protein nanocrystallography. *Nature, 470*, 73–77.
30. Kupitz, C., Basu, S., Grotjohann, I., Fromme, R., Zatsepin, N. A., Rendek, K. N., et al. (2014). Serial time-resolved crystallography of photosystem II using a femtosecond X-ray laser. *Nature, 513*, 5.
31. Pande, K., Hutchison, C. D. M., Groenhof, G., Aquila, A., Robinson, J. S., Tenboer, J., et al. (2016). Femtosecond structural dynamics drives the trans/Cis isomerization in photoactive yellow protein. *Science, 352*, 725–729.
32. Tenboer, J., Basu, S., Zatsepin, N., Pande, K., Milathianaki, D., Frank, M., et al. (2014). Time-resolved serial crystallography captures high-resolution intermediates of photoactive yellow protein. *Science, 346*, 1242–1246.
33. Neutze, R., Wouts, R., van der Spoel, D., Weckert, E., & Hajdu, J. (2000). Potential for biomolecular imaging with femtosecond X-ray pulses. *Nature, 406*, 752–757.
34. Chapman, H. N., Barty, A., Bogan, M. J., Boutet, S., Frank, M., Hau-Riege, S. P., et al. (2006). Femtosecond diffractive imaging with a soft-X-ray free-electron laser. *Nature Physics, 2*, 839–843.
35. Aquila, A., Hunter, M. S., Doak, R. B., Kirian, R. A., Fromme, P., White, T. A., et al. (2012). Time-resolved protein nanocrystallography using an X-ray free-electron laser. *Optics Express, 20*, 2706–2716.
36. Barty, A., Caleman, C., Aquila, A., Timneanu, N., Lomb, L., White, T. A., et al. (2012). Self-terminating diffraction gates femtosecond X-ray nanocrystallography measurements. *Nature Photonics, 6*, 35–40.
37. Parak, F. G., Achterhold, K., Croci, S., & Schmidt, M. (2007). A physical picture of protein dynamics and conformational changes. *Journal of Biological Physics, 33*, 371–387.
38. Bourgeois, D., & Weik, M. (2005). *New perspectives in kinetic protein crystallography using caged compounds*. Hoboken, NJ: Wiley.
39. Calvey, G. D., Katz, A. M., Schaffer, C. B., & Pollack, L. (2016). Mixing injector enables time-resolved crystallography with high hit rate at X-ray free electron lasers. *Structural Dynamics, 3*, 054301.
40. Kupitz, C., Olmos, J., Holl, M., Tremblay, L. W., Pande, K., Pandey, S., et al. (2017). Structural enzymology using X-ray free Electron lasers. *Structural Dynamics, 4*, 044003.
41. Hekstra, D. R., White, K. I., Socolich, M. A., Henning, R. W., Srajer, V., & Ranganathan, R. (2016). Electric-field-stimulated protein mechanics. *Nature, 540*, 400–405.
42. Spence, J. C., Weierstall, U., & Chapman, H. N. (2012). X-ray lasers for structural and dynamic biology. Reports on progress in physics. *Physical Society, 75*, 102601.
43. Key, J. M., Srajer, V., Pahl, R., & Moffat, K. (2004). Time-resolved crystallographic studies of the heme-based sensor protein FixL. *Biophysical Journal, 86*, 246a–246a.
44. Barends, T. R., Foucar, L., Ardevol, A., Nass, K., Aquila, A., Botha, S., et al. (2015). Direct observation of ultrafast collective motions in CO myoglobin upon ligand dissociation. *Science, 350*(6259), 445–450.

45. Barty, A., Kirian, R. A., Maia, F. R. N. C., Hantke, M., Yoon, C. H., White, T. A., et al. (2014). Cheetah: Software for high-throughput reduction and analysis of serial femtosecond X-ray diffraction data. *Journal of Applied Crystallography, 47*, 1118–1131.
46. Bourgeois, D., Vallone, B., Arcovito, A., Sciara, G., Schotte, F., Anfinrud, P. A., et al. (2006). Extended subnanosecond structural dynamics of myoglobin revealed by Laue crystallography. *Proceedings of the National Academy of Sciences of the United States of America, 103*, 4924–4929.
47. Nienhaus, K., Ostermann, A., Nienhaus, G. U., Parak, F. G., & Schmidt, M. (2005). Ligand migration and protein fluctuations in myoglobin mutant L29W. *Biochemistry, 44*, 5095–5105.
48. Schmidt, M., Graber, T., Henning, R., & Srajer, V. (2010). Five-dimensional crystallography. *Acta crystallographica. Section A, Foundations of crystallography, 66*, 198–206.
49. Schmidt, M., Ihee, H., Pahl, R., & Srajer, V. (2005). Protein-ligand interaction probed by time-resolved crystallography. *Methods in Molecular Biology, 305*, 115–154.
50. Van Brederode, M. E., Hoff, W. D., Van Stokkum, I. H., Groot, M. L., & Hellingwerf, K. J. (1996). Protein folding thermodynamics applied to the photocycle of the photoactive yellow protein. *Biophysical Journal, 71*, 365–380.
51. Hutchison, C. D. M., Tenboer, J., Kupitz, C., Moffat, K., Schmidt, M., & van Thor, J. J. (2016). Photocycle populations with femtosecond excitation of crystalline photoactive yellow protein. *Chemical Physics Letters, 654*, 63–71.
52. Lincoln, C. N., Fitzpatrick, A. E., & van Thor, J. J. (2012). Photoisomerisation quantum yield and non-linear cross-sections with femtosecond excitation of the photoactive yellow protein. *Physical Chemistry Chemical Physics: PCCP, 14*, 15752–15764.
53. Bourgeois, D., & Royant, A. (2005). Advances in kinetic protein crystallography. *Current Opinion in Structural Biology, 15*, 538–547.
54. Pelliccioli, A. P., & Wirz, J. (2002). Photoremovable protecting groups: Reaction mechanisms and applications. *Photochemical & Photobiological Sciences, 1*, 441–458.
55. Goelder, M., & Givens, R. (2005). *Dynamic studies in biology: Phototriggers, photoswitches and caged biomolecules*. Hoboken, NJ: Wiley.
56. Schlichting, I., Almo, S. C., Rapp, G., Wilson, K., Petratos, K., Lentfer, A., et al. (1990). Time-resolved X-ray crystallographic study of the conformational change in ha-Ras P21 protein on Gtp hydrolysis. *Nature, 345*, 309–315.
57. Stoddard, B. L., Cohen, B. E., Brubaker, M., Mesecar, A. D., & Koshland, D. E., Jr. (1998). Millisecond Laue structures of an enzyme-product complex using photocaged substrate analogs. *Nature Structural Biology, 5*, 891–897.
58. Moglich, A., & Hegemann, P. (2013). Biotechnology: Programming genomes with light. *Nature, 500*, 406–408.
59. Moglich, A., & Moffat, K. (2010). Engineered photoreceptors as novel optogenetic tools. *Photochemical & Photobiological Sciences, 9*, 1286–1300.
60. Crosson, S., & Moffat, K. (2002). Photoexcited structure of a plant photoreceptor domain reveals a light-driven molecular switch. *The Plant Cell, 14*, 1067–1075.
61. Geremia, S., Campagnolo, M., Demitri, N., & Johnson, L. N. (2006). Simulation of diffusion time of small molecules in protein crystals. *Structure, 14*, 393–400.
62. Hajdu, J., Acharya, K. R., Stuart, D. I., Mclaughlin, P. J., Barford, D., Oikonomakos, N. G., et al. (1987). Catalysis in the crystal - synchrotron radiation studies with glycogen phosphorylase-B. *The EMBO Journal, 6*, 539–546.
63. Sjogren, T., Svensson-Ek, M., Hajdu, J., & Brzezinski, P. (2000). Proton-coupled structural changes upon binding of carbon monoxide to cytochrome cd1: A combined flash photolysis and X-ray crystallography study. *Biochemistry, 39*, 10967–10974.
64. Sluyterman, L. A., & de Graaf, M. J. (1969). The activity of papain in the crystalline state. *Biochimica et Biophysica Acta, 171*, 277–287.
65. Kim, T. H., Mehrabi, P., Ren, Z., Sljoka, A., Ing, C., Bezginov, A., et al. (2017). The role of dimer asymmetry and protomer dynamics in enzyme catalysis. *Science, 355*, 28104837.
66. Kurisu, G., Sugimoto, A., Kai, Y., & Harada, S. (1997). A flow cell suitable for time-resolved X-ray crystallography by the Laue method. *Journal of Applied Crystallography, 30*, 555–556.

67. Petsko, G. A. (1985). Diffraction methods for biological macromolecules. Flow cell construction and use. *Methods in Enzymology, 114*, 141–146.
68. Chupas, P. J., Chapman, K. W., Kurtz, C., Hanson, J. C., Lee, P. L., & Grey, C. P. (2008). A versatile sample-environment cell for non-ambient X-ray scattering experiments. *Journal of Applied Crystallography, 41*, 822–824.
69. Schmidt, M. (2013). Mix and inject, reaction initiation by diffusion for time-resolved macromolecular crystallography. *Advances on Condensed Matter Physics, 2013*, 1–10.
70. Purwar, N., McGarry, J. M., Kostera, J., Pacheco, A. A., & Schmidt, M. (2011). Interaction of nitric oxide with catalase: Structural and kinetic analysis. *Biochemistry, 50*, 4491–4503.
71. Weierstall, U., Spence, J. C., & Doak, R. B. (2012). Injector for scattering measurements on fully solvated biospecies. *The Review of Scientific Instruments, 83*, 035108.
72. Nogly, P., James, D., Wang, D., White, T. A., Zatsepin, N., Shilova, A., et al. (2015). Lipidic cubic phase serial millisecond crystallography using synchrotron radiation. *IUCrJ, 2*, 168–176.
73. Weierstall, U., James, D., Wang, C., White, T. A., Wang, D., Liu, W., et al. (2014). Lipidic cubic phase injector facilitates membrane protein serial femtosecond crystallography. *Nature Communications, 5*, 3309.
74. Sierra, R. G., Gati, C., Laksmono, H., Dao, E. H., Gul, S., Fuller, F., et al. (2016). Concentric-flow electrokinetic injector enables serial crystallography of ribosome and photosystem II. *Nature Methods, 13*, 59–62.
75. Sierra, R. G., Laksmono, H., Kern, J., Tran, R., Hattne, J., Alonso-Mori, R., et al. (2012). Nanoflow electrospinning serial femtosecond crystallography. *Acta Crystallographica. Section D, Biological Crystallography, 68*, 1584–1587.
76. Hunter, M. S., Segelke, B., Messerschmidt, M., Williams, G. J., Zatsepin, N. A., Barty, A., et al. (2014). Fixed-target protein serial microcrystallography with an X-ray free electron laser. *Scientific Reports, 4*, 6026.
77. Mueller, C., Marx, A., Epp, S. W., Zhong, Y., Kuo, A., Balo, A. R., et al. (2015). Fixed target matrix for femtosecond time-resolved and in situ serial micro-crystallography. *Structural Dynamics, 2*(5), 054302.
78. Perry, S. L., Guha, S., Pawate, A. S., Henning, R., Kosheleva, I., Srajer, V., et al. (2014). In situ serial Laue diffraction on a microfluidic crystallization device. *Journal of Applied Crystallography, 47*, 1975–1982.
79. Roedig, P., Vartiainen, I., Duman, R., Panneerselvam, S., Stube, N., Lorbeer, O., et al. (2015). A micro-patterned silicon chip as sample holder for macromolecular crystallography experiments with minimal background scattering. *Scientific Reports, 5*, 10451.
80. Roessler, C. G., Agarwal, R., Allaire, M., Alonso-Mori, R., Andi, B., Bachega, J. F., et al. (2016). Acoustic injectors for drop-on-demand serial femtosecond crystallography. *Structure, 24*, 631–640.
81. Ren, Z., Perman, B., Srajer, V., Teng, T. Y., Pradervand, C., Bourgeois, D., et al. (2001). A molecular movie at 1.8 A resolution displays the photocycle of photoactive yellow protein, a eubacterial blue-light receptor, from nanoseconds to seconds. *Biochemistry, 40*, 13788–13801.
82. Stagno, J. R., Liu, Y., Bhandari, Y. R., Conrad, C. E., Panja, S., Swain, M., et al. (2017). Structures of riboswitch RNA reaction states by mix-and-inject XFEL serial crystallography. *Nature, 541*(7636), 242–246.
83. Adams, P. D., Afonine, P. V., Bunkoczi, G., Chen, V. B., Davis, I. W., Echols, N., et al. (2010). PHENIX: A comprehensive python-based system for macromolecular structure solution. *Acta Crystallographica. Section D, Biological Crystallography, 66*, 213–221.
84. Brunger, A. T., & Rice, L. M. (1997). Crystallographic refinement by simulated annealing: Methods and applications. *Methods in Enzymology, 277*, 243–269.
85. Romo, T. D., Clarage, J. B., Sorensen, D. C., & Phillips, G. N., Jr. (1995). Automatic identification of discrete substates in proteins: Singular value decomposition analysis of time-averaged crystallographic refinements. *Proteins, 22*, 311–321.
86. Henry, E. R., & Hofrichter, J. (1992). Singular value decomposition - application to analysis of experimental-data. *Methods in Enzymology, 210*, 129–192.

87. Emsley, P., Lohkamp, B., Scott, W. G., & Cowtan, K. (2010). Features and development of coot. *Acta Crystallographica. Section D, Biological Crystallography, 66*, 486–501.
88. Murshudov, G. N., Skubak, P., Lebedev, A. A., Pannu, N. S., Steiner, R. A., Nicholls, R. A., et al. (2011). REFMAC5 for the refinement of macromolecular crystal structures. *Acta Crystallographica. Section D, Biological Crystallography, 67*, 355–367.
89. Hugonnet, J. E., Tremblay, L. W., Boshoff, H. I., Barry, C. E., 3rd, & Blanchard, J. S. (2009). Meropenem-clavulanate is effective against extensively drug-resistant *Mycobacterium tuberculosis*. *Science, 323*, 1215–1218.
90. Tremblay, L. W., Hugonnet, J. E., & Blanchard, J. S. (2008). Structure of the covalent adduct formed between Mycobacterium tuberculosis beta-lactamase and clavulanate. *Biochemistry, 47*, 5312–5316.
91. Hugonnet, J. E., & Blanchard, J. S. (2007). Irreversible inhibition of the *Mycobacterium tuberculosis* beta-lactamase by clavulanate. *Biochemistry, 46*, 11998–12004.
92. Boyd, D. B., & Lunn, W. H. (1979). Electronic structures of cephalosporins and penicillins. 9. Departure of a leaving group in cephalosporins. *Journal of Medicinal Chemistry, 22*, 778–784.
93. Deponte, D. P., McKeown, J. T., Weierstall, U., Doak, R. B., & Spence, J. C. (2011). Towards ETEM serial crystallography: Electron diffraction from liquid jets. *Ultramicroscopy, 111*, 824–827.
94. Holton, J. M., & Frankel, K. A. (2010). The minimum crystal size needed for a complete diffraction data set. *Acta Crystallographica Section D, 66*, 393–408.

Chapter 13
X-Ray Spectroscopy with XFELs

Roberto Alonso-Mori and Junko Yano

13.1 Introduction

Many important redox-active metalloenzymes employ 3d transition metals (TM) in their active sites to catalyze multi-electron reactions at ambient temperature and pressure in aqueous solution. While these catalysts cannot simply be transferred into industrial processes, they provide important information on how to spatially and temporally control electron and proton flow and product/substrate transport during chemical transformations. X-ray spectroscopy at synchrotron radiation (SR) facilities has contributed significantly to the structural and functional studies of those metalloenzymes in the past decades. Owing to its element sensitivity and less-restrictive sample environment, it has been an essential method to study geometric and electronic structures of metal catalytic centers in the complicated biological environment. However, to minimize or eliminate radiation-induced changes during data collection has been challenging, and the measurements are typically carried out at cryogenic temperature (liquid nitrogen or liquid helium) to prevent the destruction of samples from the diffusion of radicals [1–3]. At the same time, there is increasing desire to understand biological phenomena as the enzymes function, i.e., under physiological conditions.

X-ray free electron lasers (XFELs) have shown the potential for being a transformative tool for studying biological systems under physiological conditions. The X-ray pulses generated by an XFEL are extremely intense with short pulse

R. Alonso-Mori (✉)
Linac Coherent Light Source, SLAC National Accelerator Laboratory, Menlo Park, CA, USA
e-mail: robertoa@slac.stanford.edu

J. Yano
Molecular Biophysics and Integrated Bioimaging Division, Lawrence Berkeley National Laboratory, Berkeley, CA, USA
e-mail: jyano@lbl.gov

S. Boutet et al. (eds.), *X-ray Free Electron Lasers*,
https://doi.org/10.1007/978-3-030-00551-1_13

widths, containing as many photons ($>10^{12}$) in one pulse (typical pulse length <100 fs) as most SR beamlines generate in 1 s. Moreover, the focused X-ray beam is generally very small in size, typically 1–5 μm and in some instruments even down in the nm range, enabling the study of very small samples and providing very high fluence. These unique properties of the XFEL pulses enable, within a set of conditions, shot-by-shot radiation damage-free data collection (see Sect. 13.2 "Radiation Damage" below); the ultrafast pulse (fs) is faster than the damage caused by the diffusion of hydroxo radicals or solvated electrons (ps). This results in collection of spectroscopic data, diffraction patterns, or other measurements at room temperature without relying on cryo-techniques that are the norm for X-ray studies at SR sources. This capability leads to the possibility of following structural and electronic changes in biological systems in real time, as the reaction proceeds. XFEL pulses can be used to access fs to early ps time regime phenomena, which is not possible at the current SR sources, except for laser-electron slicing facilities that provide orders of magnitude fewer photons. Owing to advancements in the timing systems (time tools based on crystals and gratings at the LCLS and SACLA, respectively [4]), the effect of the XFEL timing jitter on temporal resolution can be reduced down to tens of fs. On the other hand, the velocity mismatch between the X-ray "probe" and optical "pump" beams is still the limiting factor in many cases, and the resolution of most time-resolved studies is currently ~50–100 fs.

To take full advantage of the benefits of the XFELs for the mechanistic studies of biological samples, there are additional challenges that include preparing and delivering the samples efficiently, triggering chemical reactions in situ, and the data analysis in shot-by-shot mode. The sample needs to be replenished at the repetition rate of the X-ray laser as it is destroyed after each pulse, making solutions in liquid jets being the best candidates for spectroscopy experiments. New analysis routines to handle the shot-to-shot characteristics of the signal and the instabilities of the probed sample and the XFEL beam have to be established. In addition, one needs to consider other unique characters of the XFEL X-rays such as the SASE (Spontaneous Amplified Stimulated Emission beam) [5] bandwidth, shot-by-shot fluctuations of the beam (in intensity, spectral energy, timing, and space), and possible nonlinear effects due to the extreme fluence deposited on the sample (see Sect. 13.2 below).

In this chapter, we discuss various X-ray spectroscopy methods that are useful for studying biological systems. While a large part of the current biological XFEL research is centered on crystallography, the spectroscopic techniques are required to obtain electronic structural information at specific metal sites. The element sensitivity of X-ray spectroscopy uniquely provides geometric and electronic information of the metal site in both crystalline and solution samples. Such information will help bridging between protein dynamics and chemical changes in metalloenzymes, *e.g.*, how the protein environment modulates the site for the catalytic reaction to occur through sequential events during the reaction. Furthermore, simultaneous detection of X-ray crystallography and spectroscopy is possible under some experimental conditions in XFELs, and the method contributes to a seamless understanding of structural dynamics of proteins and chemical dynamics at the metal centers.

We discuss here X-ray absorption spectroscopy (XAS), nonresonant X-ray emission spectroscopy (XES), and resonant inelastic X-ray scattering (RIXS), and how to adapt them to exploit the XFELs' particularities described above. Thereafter, we will summarize the application of these spectroscopic techniques to the study of biologically relevant systems reported in the past few years since XFEL sources became available, in particular at LCLS (Linac Coherent Light Source) [6] and SACLA (SPring-8 Angstrom Compact Free Electron Laser) [7].

13.2 Radiation Damage

The term "radiation damage" can be used to describe different phenomena, each having a unique impact and requiring a different response when performing spectroscopy experiments. Radiation damage can be due to: (1) the diffusion of radicals or solvated electrons created by the interaction of X-rays and water, (2) Coulomb explosion when the fluence is high enough to break the bonds holding the molecules together, and (3) nonlinear effects when the fluence is high enough to enhance these low cross section electronic phenomena.

Traditionally, the main cause of "radiation damage" at SR-based experiments in biological systems is the diffusion of solvated electrons and radicals created by the interaction of X-rays and water. The interaction of these species with the sample causes the reduction of highly oxidized metals, the breakage of disulfide bonds and the decarboxylation of carboxylate residues among others. Therefore, at SR facilities data collection for crystallography and spectroscopy experiments is usually carried out at cryogenic temperatures to minimize or eliminate radiation-induced changes. Lowering the temperature reduces the diffusion of radicals, thus decreasing damage to the proteins and lipid environment. The diffusion of radicals/solvated electrons occurs in the sub-ps to ps range, so another way to avoid their effects is to outrun this process by collecting data in a faster timescale. The ultrashort fs X-rays at XFELs thus allow collecting data from the intact form of a system, free from this secondary radiation damage due to radical diffusion, even at room temperature, permitting the study of biological systems in their functional form in native-like conditions.

At XFELs though, one needs to consider the consequences of the intense X-ray pulses depositing its energy into a small sample volume. The photoionization of the irradiated sample, which also causes the formation of the free radicals, can cause a residual positive charge to accumulate at the sample. When the X-ray pulse fluence is high enough such that the electric charge due to the photoionization can break the molecule bonds, a Coulomb explosion occurs [8–10]. In this chapter, we will refer to this as "electronic damage," to differentiate it from the diffusion-based radiation damage described earlier. The first step of the phenomenon involves stripping off multiple electrons. Its repercussions for structural studies are prominent and are discussed in more detail in Chap. 6 in this book. The intense X-ray pulses can increase the probability of nonlinear effects (which have generally a very low cross section), where the relation between the incoming X-ray fluence and the outgoing signal intensity is not linear anymore. Nonlinear effects include many phenomena (e.g., two-photon absorption, double core-hole excitation, sequential ionization, X-ray-induced transparency, or stimulated emission) and their study may lead to novel applications. However, for standard spectroscopy experiments, one consequence is that the spectra of interest may be distorted as compared with the spectra observed in the linear-response regime. This effect has been observed at XFELs, with tightly focused hard X-ray beams, and even more pronounced in the soft X-ray regime in which the absorption cross section is larger. To avoid this, one needs to reduce the focus or attenuate the beam. This is a dilemma, in particular, for some of the photon-hungry experiments like RIXS on dilute systems. In general, hard X-ray spectroscopy measurements carried out with "standard" XFEL fluences (few μm focus and $\sim 10^{11}$ ph/pulse) still preserve the spectral shape and yield valid chemical information. This has been demonstrated by early XES studies at LCLS, conducted with Mn inorganic compounds in solution, and the comparison with spectra collected at SR at cryogenic temperature [11]. For experiments or techniques requiring tighter focus, or more prone to this type of damage, having a higher repetition rate rather than increasing the fluence is a preferable approach.

13.3 Methods

13.3.1 Sample Delivery

A single XFEL pulse can damage or destroy the probed volume, especially when using a focused beam, making it necessary to replenish the sample between shots. Therefore, developing methods for consistently delivering a homogeneous sample to the X-ray beam in a synchronous way is critical for XFEL applications. A few different approaches have been used based on the needs of the application, e.g., sample type, volume available, sample environment, or reaction-triggering method. Most spectroscopic applications at XFELs rely on using liquid jets, where a solution (or suspended small particle/crystal) is continuously transported from a reservoir through a capillary and ejects at the interaction region. At low repetition rates, only a small fraction of the sample volume is probed by the pulsed XFEL beam. In some cases, this allows recirculating the sample through a closed-loop system if the sample is limited. For spectroscopy experiments, the stability of the jet is fundamental, in particular for resonant techniques where normalization of the signal for each shot is critical. Moreover, these photon-hungry techniques require thick liquid jets, ideally on the order of the absorption length. When sample availability is not an issue, round Rayleigh or sheet jets are preferred, providing excellent stability for diameters above 10 μm, with sample consumption in the ml/min range. Round jets with hundreds of nm diameter and sheet jets with thickness above 100 μm have also been used at LCLS. More details can be found here [12].

When the sample is limited, other injection systems can be used, including droplet systems. The Rayleigh flow can be broken up by compression (e.g., by a piezoelectric device) to create a stable train of drops [13], thus eliminating the sample volume loss in between X-ray pulses. Another drop-based system is the acoustic droplet injector (ADE) [14, 15] where liquid drops are ejected from a well reservoir by means of focused sound waves synchronized with the XFEL repetition rate. For time-resolved experiments that require longer time delays of ms to few s time range, the ADE can be combined with a drop on tape (DOT) system, where a conveyor belt system transports the droplets to the X-ray beam through the reaction region (e.g., light or gas activated) [16]. The flow rate of these systems is in the 10 μl/min range.

Finally, for applications where the sample is in the solid state, fixed targets continuously scanned through the XFEL beam have been used at LCLS [17–19], allowing up to 120 Hz data collection. For more details on sample delivery systems tailored for XFEL applications, see Chap. 5 in this book.

13.3.2 X-Ray Absorption Spectroscopy

X-ray absorption spectroscopy (XAS) as we know it today has been utilized since the mid-1970s, concomitantly with the advent of the first SR sources. During all these years, it has greatly contributed to the advancement of many fields, from material science to biology, due to its element specificity and sensitivity to changes in the electronic structure and local coordination of the selected element. XAS studies photoinduced transitions of an electron from a core state to low-energy, empty molecular orbitals above the Fermi energy, probing the density of unoccupied electronic states [20] (see Fig. 13.1). The first tens of eV of an XAS spectrum, the X-ray absorption near edge structure (XANES) region, arises from transitions close to the Fermi level that may include bound states (resonant excitations). XANES provides chemical information, and the position of the absorption edge indicates the valence states, while its shape yields information about the coordination environment of the absorber. As shown in Fig. 13.2a, the region including the following few hundreds of eV is called extended X-ray absorption fine structure (EXAFS), and it is dominated by scattering processes. Here, the highly energetic photons liberate photoelectrons that propagate from the absorber atom and are backscattered by neighboring atoms producing interference and the characteristic EXAFS oscillations. From the EXAFS signal, it is possible to determine the atomic number, distance, and coordination number of the atoms surrounding the absorber.

Transmission detection is the only direct approach to measure the absorption coefficient, where the XAS spectrum is obtained by measuring the beam intensity going through the sample while scanning the incident energy across the edge. However, this approach is not straightforward for soft X-rays, where the attenuation length of few micrometers implies that the sample has to be extremely thin. The absorption coefficient can also be approximated indirectly with total fluorescence

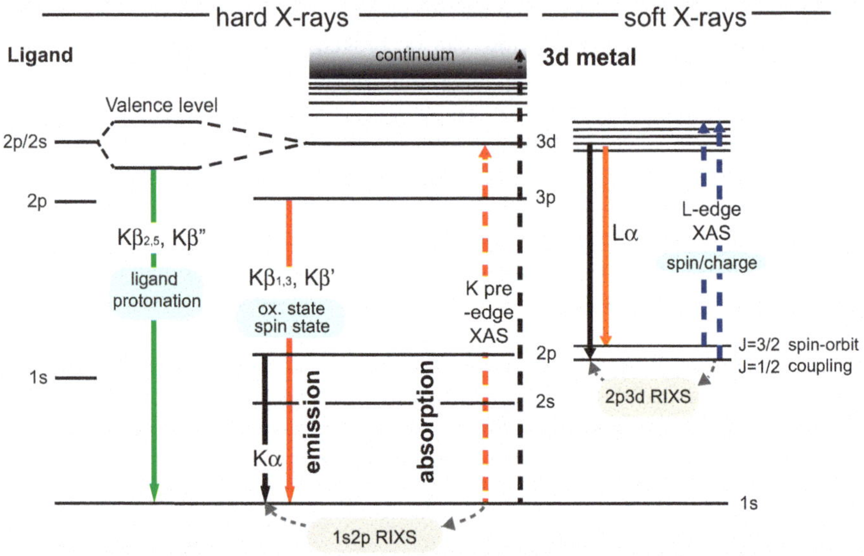

Fig. 13.1 Energy diagram levels for 3d transition metals showing the characteristic transitions relevant for hard X-ray and soft X-ray spectroscopy techniques

yield (TFY) detection, where the spectrum is obtained by recording the radiation produced in the decay process after the occurrence of the absorption event. TFY is generally applied in the soft X-ray range as well as in the hard X-ray range for diluted samples that do not absorb enough photons to obtain a XAS spectrum in transmission mode. While transmission and TFY are routinely applied at SR sources, various factors have to be taken into account to transfer these spectroscopic approaches to XFEL experiments.

XAS requires a highly monochromatized incident beam (typically $< \Delta E = 10^{-4}$) and the incident X-ray energy needs to be scanned in a wide energy range (XANES: 50 eV, EXAFS: 500 eV). As the SASE radiation from XFELs has a spectral width on the order of $\Delta E/E = 10^{-3}$, the use of an incident beam monochromator, to scan specific energies across the absorption edge, is necessary. The use of SASE presents some drawbacks. Due to the broad SASE spectral bandwidth and also its structure (composed of multiple peaks fluctuating in position and intensity from pulse to pulse), the average transmission through the monochromator is only few percent with 100% intensity variations pulse to pulse. As an alternative to the use of SASE in combination with a beamline monochromator, the use of a more monochromatic self-seeded beam (demonstrated for soft and hard X-rays at LCLS with a bandwidth down to $\Delta E = 10^{-4}$) [21] has been explored. This could in principle eliminate the need of a beamline monochromator and provide higher monochromatic flux. However, in some cases the use of a monochromator is still required to filter out the significant SASE background present in the spectrum [22]. Improvements in accelerator technology could improve the performance of self-

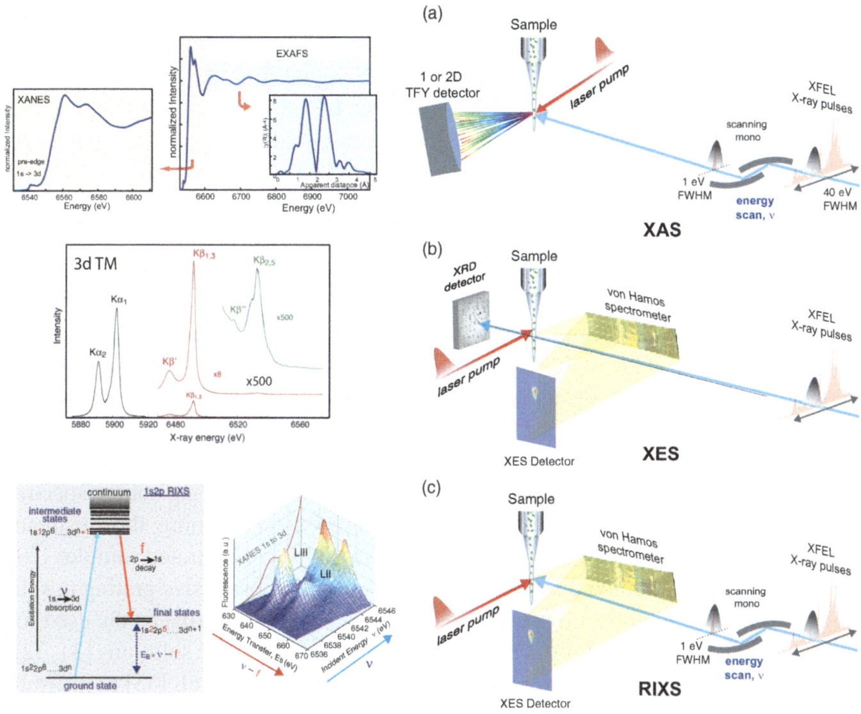

Fig. 13.2 Schematic of hard X-ray XAS, XES, and RIXS spectroscopic experimental setups for a typical XFEL pump-probe liquid jet experiment, and the spectroscopic signals are collected at variable time delays after laser pump. **(a)** In an XAS experiment, the TFY signal is recorded with a diode or a 2D detector while scanning the incident energy by means of a monochromator. XANES (scanning the first ~50 eV) to obtain chemical information or EXAFS (scanning the first ~500 eV) spectra to obtain structural information can be collected with this setup. **(b)** The XES signal is collected by means of a dispersive spectrometer with no scanning elements for either the incident or the emitted beams. This technique is compatible with the simultaneous collection of XRD/XDS data in forward scattering geometry. **(c)** RIXS setup requiring both the scanning of the incident energy and the high-resolution spectrometer to record the emitted spectra. XES-type spectra are recorded while scanning the incident beam through the absorption edge to reconstruct a RIXS map. **(Left)** Energy level diagram showing the electronic levels and decay channels involved in a 1s2p measurement and the corresponding RIXS plane

seeding in the future. Additionally, the thermal load in the upstream optics can cause spatial and spectral drifts and induce unreliable measurements. All these experimental aspects inherent to XFELs make it clear that accurate beam diagnostics and normalization protocols, both for the incident X-ray properties (intensity and spectrum) and for sample variations (probed volume and concentration), are critical to obtain a reliable spectrum and reduce systematic errors.

Despite the above challenges, several pump-probe XAS experiments have been realized at XFELs in the past few years. In the hard X-ray regime, Lemke et al. [23] demonstrated that the inherent instabilities of the SASE beam can be overcome and high-quality time-resolved XANES data can be measured on a spin-crossover system. The setup is depicted in Fig. 13.2a, in which they used a photodiode placed at 90 degrees from the sample to capture the TFY signal from the aqueous solution of $[Fe(bpy)_3]^{2+}$. More recent measurements at LCLS have shown that the use of a fast 2D detector with high-energy resolution could improve data quality by collecting the spectra in partial fluorescence yield (PFY) mode, i.e., approximating the absorption coefficient by measuring a single emission line, usually the stronger $K\alpha$, discriminated against other signals. A liquid jet in He environment was used to replace the solution sample after each shot and avoid electronic damage by the intense XFEL pulses. A similar setup was used in subsequent XANES experiments to study the early dynamics of photoexcited reactions of molecular complexes and proteins in solution at the LCLS [24–28]. Though most of the XFEL XANES studies were based on TFY methods, a potential alternative relies on the use of transmissive spectrometers. By subtracting the spectral signal between two spectrometers located upstream and downstream of the sample, the absorption spectrum can be obtained on a single shot basis, allowing self-normalization of the fluctuating source spectral content and more efficient data collection compared to scanning the upstream monochromator. This method has been demonstrated for concentrated samples [29] at LCLS. A similar method developed at SACLA uses a transmissive grating to split the XFEL beam and collect dispersive XAS data [30]. This approach was used for time-resolved XAS studies of 0.1 and 1 M ammonium ferroxalate solutions [31, 32]. Yet another dispersive approach was used at LCLS by Gaudin et al. [33] to measure the Mo L3 edge from a solid sample. No EXAFS study has been yet realized at an XFEL, mainly due to the requirement of the long-range energy scan (about 500 eV), but also to the high sensitivity required to capture changes in the EXAFS region, on the order of few percent changes. This makes it very challenging to measure EXAFS spectra on very diluted biological systems. The foreseen average flux increase, accomplished via higher repetition rates, provided by LCLS-II and the European XFEL will make EXAFS studies possible.

At SR sources, metal K-edge spectroscopy has been widely used for biological spectroscopy of 3d TM. On the other hand, metal L-edge spectroscopy, which falls into the soft X-ray energy range (approximately <2000 eV), has been rarely used for biological systems. This is largely because of severe radiation damage that occurs even at cryogenic temperatures, as the absorption cross section is about 100 times larger than that of K-edges. The larger absorption cross section also implies that the X-ray penetration depth is much shallower (for example, the attenuation length is 0.8 μm at 650 eV) than that of hard X-rays (500 micrometer at 6500 eV), making the soft X-ray spectroscopy methods a semi-surface sensitive tool. Furthermore, the requirement of ultrahigh vacuum, which dehydrates samples, makes the experiments with biological samples more difficult. Despite such challenges, there are several advantages in using metal L-edge spectroscopy. It has significantly better resolution, as the natural line widths at the L2 and L3 edges are approximately one-fourth of those at the K-edge due to the longer core-hole lifetime. This makes L-edge spectra more informative about the electronic structure [20, 34]. The 2p to 3d transitions are allowed under dipole selection rules (the atomic orbital angular momentum

$\Delta l = +/-1$). On the other hand, the pre-edge peak (1s–3d transitions) at K-edge spectroscopy that is widely used for extracting the electronic structure is weak as it is quadrupole-allowed ($\Delta l = 0, +/-2$), and often overwhelmed by the dipole-allowed main edge intensity. Therefore, L-edge spectroscopy has a greater sensitivity to the occupancy and interactions of the metal 3d-derived orbitals and can provide a better indication of the bonding, oxidation/spin states, and symmetry of the complex involved.

Biological soft X-ray spectroscopy at XFELs is just at the stage of its birth, but the collection of X-ray damage-free data at room temperature would be of particular advantage for biological systems if a suitable detection scheme is developed. Such a detection scheme needs to probe the spectroscopic signal in the dilute sample arising from the metal L-edges at 400–1000 eV, while separating it from the strong $K\alpha$ background signal of the light elements in the sample (C, N, O with absorption K-edges at 280–540 eV). This can be realized with an energy discriminating scheme as the element-specific PFY detection. Grating spectrometers have been commonly used for PFY detection techniques at SR sources. However, these instruments have not been successfully used so far for dilute samples due to their small subtended solid angle. Another approach is to use a superconducting tunnel junction (STJ) detector [35], which presently provides ~20 eV resolution; efforts are undergoing to improve the resolution and solid angle of these detectors. A more robust way of detecting fluorescence signals has been recently developed, using a high-reflection zone-plate spectrometer to spatially separate the metal $L\alpha$ signal from the background [36]. The solid angle of the zone-plate spectrometer is several times higher than that of gratings or STJs. This method [36, 37] (see Fig. 13.3a) has been demonstrated to overcome the experimental difficulties of metal L-edge spectroscopy, and opened the way to the use of soft X-ray spectroscopy as a unique and powerful tool to study metalloenzymes in physiological conditions and catalytic reactions in dilute solution environments. More details can be found in Sect. 13.4.

In addition to the efficient signal collection described above, soft XAS on dilute samples at XFELs often faces the challenge of a limited number of incoming photons. In standard soft XAS experiments, the incoming X-rays are monochromatized with gratings and the absorption energy range is scanned continuously or step-wise, significantly attenuating the beam in the process (~20% reflectivity). As a consequence of the use of the monochromator and its related losses, the monochromatic peak flux on the sample is reduced down to a few percent of the original peak flux (depending on the selected resolving power by the bandwidth narrowing exit slit of the monochromator). This reduced monochromatic photon flux is a limitation for XAS data collection for dilute samples like metalloenzymes (at the mM level) and specially for more photon-hungry techniques like RIXS that require a high average flux. One way to address this limitation at XFELs is using the self-seeding mode of operation described above, where the monochromatization is performed before the laser amplification. As a consequence, one can obtain a higher spectral brightness as compared to SASE where the monochromatization is done after lasing. One can therefore expect an improvement for soft XAS by

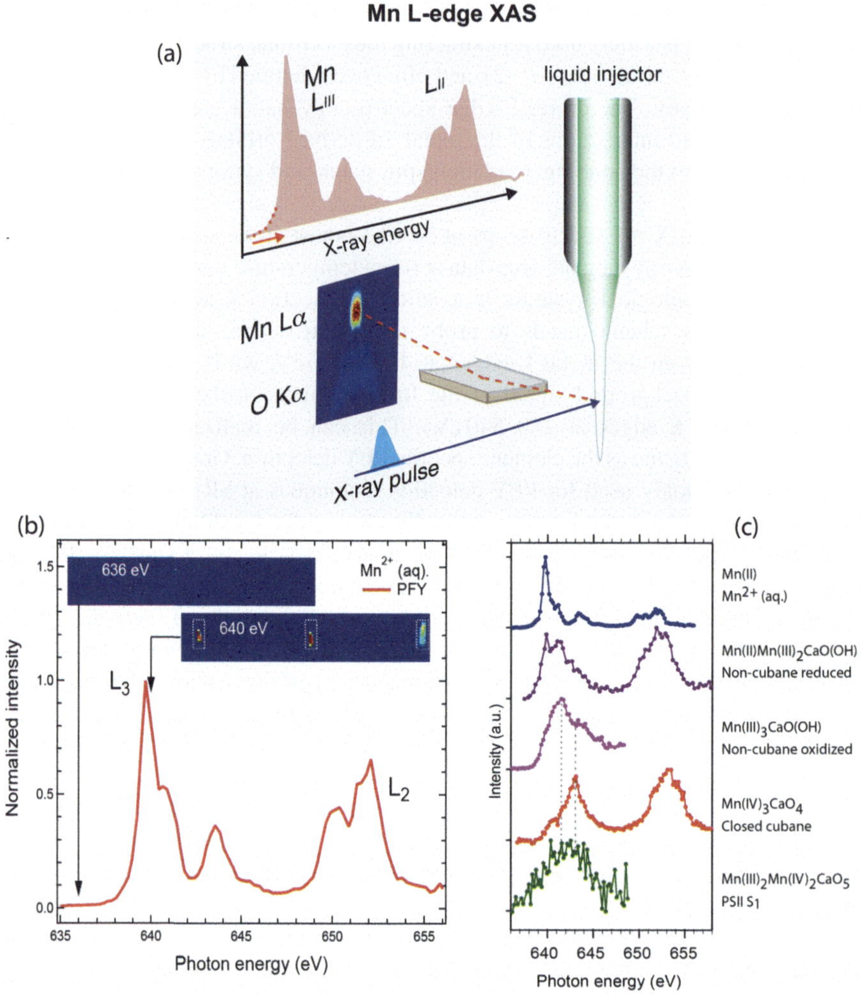

Fig. 13.3 **(a)** Soft XAS setup based on zone plates allowing the geometric discrimination of the metal signal from the Oxygen background from a solution sample. Reproduced with permission from [36]. **(b)** Mn L-edge partial fluorescence yield (PFY) XAS spectrum from $MnCl_2$ (500 mM) aqueous solution (above two CCD images in the pre-edge and L3-edge regions). Reproduced with permission from [36]. **(c)** Mn L-edge PFY XAS spectra of solutions of inorganic models (500 mM Mn_{aq}^{2+} solution, three inorganic Mn_3CaOx model complexes with Mn concentrations of 6–15 mM) and from Mn_4CaO_5 cluster in PSII (S1 dark resting state, Mn concentration of 0.8 mM). Reproduced with permission from [37]

a narrower bandwidth which leads to a higher peak and average flux in the self-seeding mode. The use of the self-seeding scheme for soft XAS data collection has been shown by Kroll et al. [22]. The study shows that the spectrum of the seeded pulse is narrow and clean enough to allow direct use for XAS data collection without the additional beamline monochromator, and that it could lead to a factor of 3–5 increase in average flux. In the soft X-ray energy regime, one also needs to be aware of a higher possibility of sequential absorption of multiple X-ray photons by the same atom due to the large absorption cross section, which reduces the peak intensity [38]. For this reason, a higher repetition rate of the XFEL pulses is often more desirable than higher fluence per pulse for efficient data collection.

13.3.3 Nonresonant XES

Subsequent to the X-ray absorption process, the atom, in an excited state, spontaneously decays by emitting XES signal (also called fluorescence). Nonresonant XES, where the excitation energy is well above the absorption edge, gives access to different information by probing different decay channels. K emission lines of 3d TM (Kα, K$\beta_{1,3}$, $K\beta'$, $K\beta_{2,5}$, and $K\beta''$) in the hard X-ray regime, shown in Fig. 13.1, provide information on the oxidation state, effective spin, and the metal bonding orbitals [39]. In the soft X-ray range, L emission lines are a direct probe of the 3d electronic structure, providing chemical sensitivity to oxidation states, symmetry, and covalency of the system under study (see Fig. 13.1). XAS and XES are complementary techniques, providing information about unoccupied and occupied states, respectively. XES is not as ubiquitous as XAS at SR sources. This is in part because it is more complex experimentally, requiring additional instrumentation. High-resolution emission spectrometers, generally based on perfect Si/Ge crystal analyzers with reflections matching emission lines in a close-to-backscattering geometry, are required. Nonresonant XES techniques are more favorable at XFELs since the broad bandwidth of the pulse and its spectral and intensity fluctuations have no effect on the XES spectrum, and therefore no monochromator is necessary, avoiding the associated problems described in the previous section. Moreover, the use of a dispersive detection method enables the collection of spectra in single shots, which is beneficial when dealing with the pulse-to-pulse fluctuations of the SASE beam, and eliminates normalization issues. In the hard X-ray regime, dispersive setups can be realized based on the von Hamos geometry [40]. In this approach, one or more cylindrically bent dispersive crystal analyzers diffract the spectral energy range (by means of Bragg's law) onto the detector. The axis along the sample and detector positions and the axis along the crystal analyzer(s) surface are parallel to each other and separated by a distance equal to the crystal(s) bending radius. A multi-crystal von Hamos spectrometer deployed at LCLS is described here [41] and adaptations of this design are being deployed in other XFEL sources. Dispersive instrumentation requires a 2D detector to capture the extended signal reflected by the crystal analyzers, which is then integrated over the dispersive direction to obtain the XES spectrum. Various studies have been performed at XFELs using this type of instrumentation (see Fig. 13.2b). Alonso-Mori et al. [11] demonstrated that it is feasible to use hard X-ray spectroscopy at LCLS to provide reliable information on the intact electronic structure of redox-active compounds without being affected by electronic damage caused by the intense XFEL pulses. Subsequent studies at LCLS and SACLA include measurements of photoinduced charge transfer and spin dynamics in coordination complexes [42–47] and of the redox state of photoactive proteins [16, 48–50]. In some of these experiments, the von Hamos dispersive approach was complemented by simultaneous collection of a single spectral energy from the same or a different emission line with a point-to-point Rowland setup (using spherically bent crystal analyzers) [51], which is convenient to follow the kinetics of the reaction. An important advantage of dispersive XES is that it can be easily combined with forward scattering data collection since both setups are located in orthogonal geometries. Both signals can therefore be collected simultaneously from the same pulse, and many of the previously cited examples have exploited these features to monitor the electronic structure of proteins, such as Photosystem II through its reaction cycle, while collecting X-ray diffraction data [48, 49] and to follow spin state changes while measuring X-ray diffuse scattering [44, 45, 47]. To date, no studies using the weaker $K\beta_{2,5}$ have been performed at XFELs. $K\beta_{2,5}$ is ~50 to 80 weaker than $K\beta_{1,3}$ and new methods/improved setups need to be developed to collect this valence to core transition on diluted samples. LCLS-II, with its increased average incident flux, will make these studies on biological systems possible. For a more detailed discussion on hard X-ray XES instrumentation and experiments at the LCLS, see here [52].

The tender X-ray regime (typically between 2 and 5 keV) has historically not been widely accessible at X-ray sources. However, this is an interesting range since it enables probing the electronic structure of elements that are ubiquitous in biological systems, including 4d TM L-edges (e.g., Ru, Mo, and Ag), and the K-edges of lower Z elements (e.g., P, S, and Cl). Not many SR facilities have spectrometers covering this energy range and therefore only few spectroscopic studies have been realized up to now [53, 54], in particular for resonant techniques [55]. This is because the transmission of hard X-ray beamlines decays very rapidly when decreasing the energy to this range, and because of the difficulties of developing emission spectrometers with convenient crystal reflections for these energies. SwissFEL will be able to access this energy range, and efforts are also undergoing to make instruments at LCLS-II compatible with the tender X-ray regime. Von Hamos-based spectrometers with newly developed crystal analyzers are being explored for LCLS and will be able to probe the tender energy range in a close-to-backscattering geometry, improving the efficiency of previous designs by about tenfold. These developments will enable fs and ps time-resolved experiments to study the electronic structure of elements that are not accessible elsewhere. This setup can be applicable not only to XES but also to other spectroscopic techniques like XAS and RIXS.

For the soft X-ray regime, XES has not been applied to biologically relevant systems at XFELs. This is due to the difficulties of collecting emission signals efficiently, in addition to the current limitation of the incoming X-ray photons at XFELs (namely, the average flux is not high enough to collect data within a reasonable time for dilute samples). The details can be referred to in Sect. 13.3.2, in which we described the PFY-detected XAS methods. In the nonbiological molecular systems, however, the method has been applied to study ultrafast phenomena of a transiently populated state in CO desorption from Ru(0001) with pump-probe technique [56, 57]. The detection of photoelectrons has also been applied to study various chemical processes in the soft X-ray regime [57–59].

In general, XES data collection is not very efficient as even advanced multi-analyzer spectrometers can only cover a fraction of the signal emitted isotropically by the sample through spontaneous processes. However, the high fluence provided by XFELs can be used to stimulate the X-ray emission process. This is achieved when the incoming X-ray pulses produce sufficient ions to invert the population to an excited state along the path of the X-ray beam. The subsequent decay along the beam direction results in stimulated emission processes leading to a strong amplification gain. The advantage of this method is twofold: (1) the strong directionality of the stimulated process makes it very easy to collect the emission signal, for example, by a flat crystal downstream of the sample in the hard X-ray regime, removing the need of complicated X-ray emission spectrometers, and (2) the amplification increases the signal strength by many orders of magnitude. Taken together, these effects can dramatically decrease data collection time and increase sensitivity. This approach has been already demonstrated both in the hard [60] and soft [61] X-ray regimes.

13.3.4 RIXS

Beyond more conventional XAS and XES approaches, resonant inelastic X-ray scattering (RIXS) or resonant X-ray emission spectroscopy (RXES) offers extended electronic structural information through photon-in photon-out 2-dimensional scans (Fig. 13.2c). In RIXS, the emitted photon energy is scanned as a function of the incident photon energy across the absorption edges[62]. The approach is beneficial in several ways as: (1) it provides the ability to probe occupied (XES) and unoccupied orbitals (XAS) together, and (2) RIXS spectra are better energy-resolved than conventional XAS, where the core-hole lifetime broadens the features of the spectrum. This effect is more significant in the K-edge (1s core hole) than in the L-edge (2p core hole) [39, 62]. So far, RIXS has not been applied as frequently as XAS and XES at XFELs due to the experimental difficulties related with scanning both the X-ray energy of the incoming beam and the emission energy with high-energy resolution. This requires a stable sample delivery for an accurate normalization of signals over the RIXS plane and very good normalization of the scanned incident beam. Below, the possibility of measuring 3d metal RIXS in the hard (1s2p RIXS) and soft X-ray (2p3d RIXS) energy range at XFELs is discussed.

In 1s2p RIXS, hard X-rays, scanned across the K pre-edge, are used to excite a 1s electron into an unoccupied valence orbital (1s to 3d transition). The emission due to the decay of a 3p or 2p electron into the 1s shell is measured as a function of the excitation energy by means of an emission spectrometer (see Fig. 13.1). The energy difference between excitation and emission corresponds to the energy difference between 2p or 3p and 3d orbitals and, therefore, one can get L-edge-like XAS spectra with the advantages of using hard X-rays (high-excitation energy and therefore high X-ray penetration depth and larger attenuation length) to probe these transitions indirectly. In addition, the RIXS 2D plot makes the background separation of the pre-edge structure from the main K-edge feature easier. At SR facilities, 1s2p RIXS has been used to study several biological systems and proven to be a powerful tool for extracting charge and spin state information of metal centers [63, 64]. In order to enable RIXS measurements at XFELs, however, one needs to overcome several challenges, many of which are common with XAS data collection described above, such as accurate incident beam diagnostics and normalization protocols, both for the incident X-ray intensity and for the sample variations (probing volume and concentration). With the on-going development of sample delivery methods, improved efficiency of the spectrometers, better beamline diagnostics, and the availability of high-repetition rate XFELs, hard X-ray RIXS measurements of dilute biological samples will be more easily accessible in the near future.

Another way of collecting hard X-ray RIXS spectra is by using a transmissive spectrometer to record the XFEL SASE beam spectra (input spectra) shot-by-shot in combination with a dispersive XES spectrometer to collect the emission spectra (output spectra). The XFEL pink beam is composed of multiple peaks exhibiting profound fluctuations from shot to shot in both amplitude and energy. With proper data treatment, these random shot-to-shot spectral fluctuations can be exploited to extract the incidence energy dependence of a signal. The input and output spectra are measured for each shot, and the RIXS map can be reconstructed from input–output correlations. This so-called stochastic spectroscopy approach has been used in other non-X-ray spectroscopic methods [65]. Moreover, a XANES spectrum can be extracted as the integral of the RIXS spectrum over the detected (output) energy axis. This technique eliminates the need for an upstream monochromator, and the corresponding flux loss (up to 99%), which is particularly problematic for dilute samples and photon-hungry experiments like RIXS, in which the photon count rates are very low.

In the soft X-ray regime, pursuing 2p3d RIXS (see electronic transitions in Fig. 13.1) shares common issues with its hard X-ray counterpart (described above), including accurate incident beam diagnostics and normalization protocols. While hard X-ray RIXS is feasible with the current technology, soft X-ray RIXS of dilute biological systems is not presently within reach. The RIXS data collection requires a high-energy resolution (resolving power better than 1000 (i.e., better than 0.6 eV)) on the emitted photons, which can be achieved by a grating spectrometer or by the projected resolution of the so-called TES detector (transition-edge sensor) [66]. The grating spectrometer, however, has yet to become efficient for collecting signals from dilute samples, as it typically covers only 10^{-5} sr of the solid angle and its quantum efficiency is around 10%. While the method has been applied to more concentrated molecular systems [67–69], further development of the spectrometer and optimizing detection efficiency [70, 71] is necessary for the data collection of dilute systems. In addition, a higher repetition rate of the XFEL pulses such as planned for LCLS-II and European XFEL is required to realize such photon-hungry experiments.

13.4 Applications

For the past few decades, SR-based X-ray spectroscopy has provided valuable insights into the structure and mechanism of metalloenzymes. Many of the SR measurements are carried out at cryogenic temperatures to minimize X-ray radiation damage to the sample. Only with the advent of XFELs has it become possible to collect damage-free X-ray data at room temperature. XFELs have also opened up the possibility of capturing reaction intermediates and chemical dynamics through the catalytic cycle by incorporating in situ reaction-triggering approaches. Yet, many of the methods are still under development and the potential power of X-ray spectroscopy at XFELs has not been fully utilized. We expect to see more studies as the technology advances and more XFEL beamlines become available. Below, we summarize some examples, reported up to now, with a focus on metalloenzymes.

13.4.1 Metalloenzymes

In nature, the water oxidation reaction is accomplished effectively by the oxygen-evolving complex (OEC) in Photosystem II (PSII), a multi-subunit membrane protein in green plants, algae, and cyanobacteria (Fig. 13.4a). This protein uses sunlight to oxidize water, generating most of the oxygen in the atmosphere, while producing reduced compounds that are a major source of our biological and fossil fuel energy:

$$2H_2O \text{ ——> } O_2 + 4e^- + 4H^+$$

The OEC in PSII that catalyzes this reaction contains a heteronuclear Mn_4Ca cluster that couples the four-electron oxidation of water with the one-electron photochemistry occurring at the PSII reaction center by acting as the locus of charge accumulation. The OEC cycles through a series of five intermediate S-states (S_0 to S_4), representing the number of oxidizing equivalents stored on the OEC, and releases O_2 before the formation of the S_0-state (Fig. 13.4b). The recent advancement in the PSII crystal structure determination is significant [72] but the geometric information of the Mn_4CaO_5 cluster has to be complemented by electronic structural information to address several key questions of the mechanism of photosynthetic water oxidation. One of these questions is whether Mn-centered or ligand-centered oxidation occurs, which triggers the O–O bond formation at the last step of catalysis. Previous SR-based studies suggested that the electrons are highly

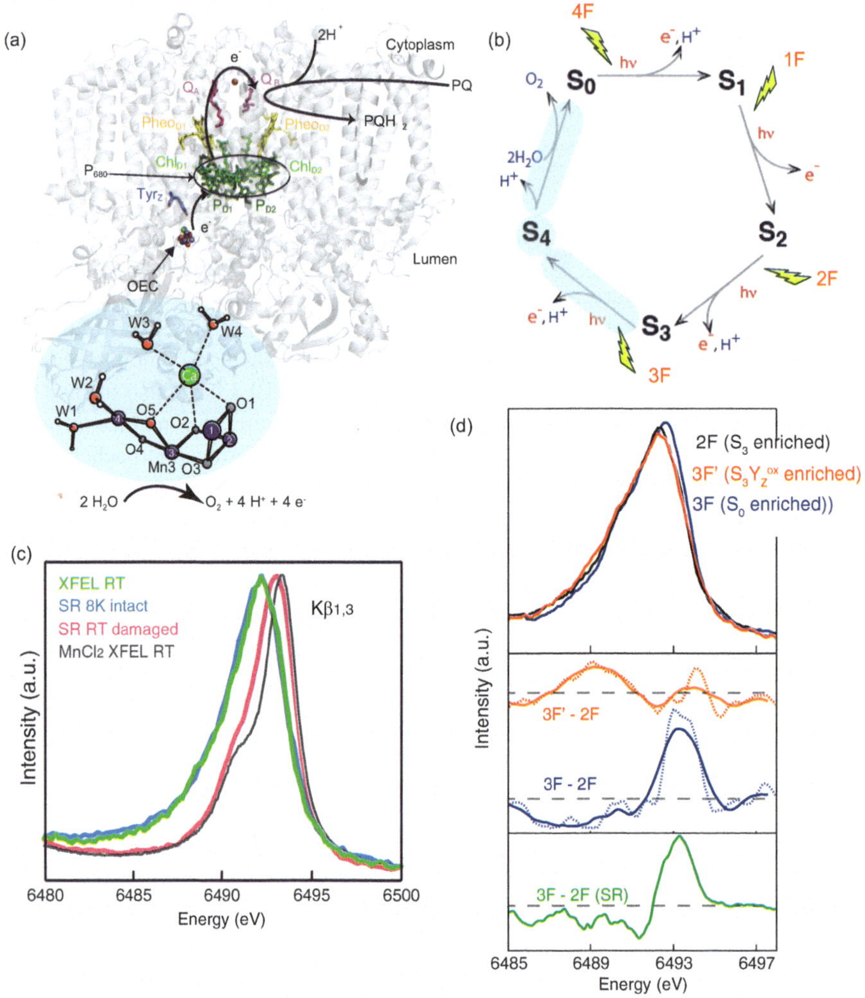

Fig. 13.4 XFEL studies on Photosystem II. **(a)** Structure of PSII and the catalytic site of water oxidation, showing the flow of electrons through the enzyme complex. **(b)** Kok-cycle of photosynthetic oxygen evolution, with states S_0–S_3 being stable intermediates and S_4 a transient state. **(c)** Mn XES of PSII measured at LCLS at room temperature and at SR source at cryogenic temperature, both spectra overlap showing that the LCLS spectra are from the intact sample. Reproduced with permission from [48] **(d)** Mn XES spectra from PSII at room temperature for different flash states, indicating turnover of the samples into the S_3 and the S_0 state and possible indication of a slight oxidation 250 μs into the S_3–S_0 transition. Reproduced with permission from [49]

delocalized in the Mn_4CaO_5 cluster, and that the ligands are intimately involved in charge delocalization, by modulating the covalency of the Mn–O bonds. Therefore, understanding the electronic structure beyond the framework of formal oxidation states under physiological conditions is critical.

Mn $K\beta$ XES from PSII has been collected at multiple hard X-ray instruments of LCLS from solution and crystalline samples [48, 49] (Fig. 13.4c). In these studies, the valence state of the Mn in the OEC was followed through the reaction cycle including time points in the critical S_3 to S_0 step (Fig. 13.4d). As the XES measurement can be performed simultaneously with downstream XRD collection, this method provides a diagnostic capability for X-ray crystallography, since it can also be used as a control to estimate the degree of S-state advancement by the number of laser flashes. The results demonstrate the intactness of the protein during data collection, and the feasibility of measuring X-ray crystallography and spectroscopy to characterize metalloenzymes using the high fluence provided by the XFELs. In parallel to the XES data collection with hard X-rays, there is also an effort to collect Mn L-edge XAS of PSII. In this case, the main challenge is to discriminate the Mn $L\alpha$, β (2p→3d) fluorescence at ~640 eV from 4 Mn atoms from the overwhelming O $K\alpha$-edge fluorescence at ~525 eV from ~25,000 O-atoms. Mitzner et al. [36] have introduced a spectrometer based on high transmission reflection zone plates (RZPs) for XAS with PFY detection providing a bandwidth of 20 eV (FWHM), which is enough to discriminate the unwanted signal (see Fig. 13.3a). RZPs have the potential for high-photon detection efficiency with a large solid angle. This setup has been used at LCLS, where a solution sample was delivered with a liquid jet into a vacuum chamber, preventing dehydration of samples and providing a constant sample volume delivery necessary for the XAS data normalization (see Fig. 13.3b). Kubin et al. have shown that it is possible to collect data using a metal concentration of mM-range samples with this approach [37] (see Fig. 13.3c), demonstrating that PFY-XAS with RZPs is a robust approach for collecting dilute metal signals at XFELs.

Cytochrome C, a heme-containing enzyme from the cytochrome family, serves as an electron transfer protein in biological processes as well as a peroxidase enzyme in cellular apoptosis. A single chemical bond, Fe-S(Met), regulates the function of the protein, from transferring electrons to produce energy for cellular respiration when the bond is intact, to triggering the breakdown of the mitochondria when it is broken. The strength of this bond was studied at the XPP [73] endstation of LCLS by photoexciting ferrous cytochrome C with a 520-nm laser, which results in thermal ultrafast photodissociation and recombination of the S(Met) [50]. The protein was delivered to the beam in solution form by a liquid jet and the Fe K-edge XAS and Fe $K\beta_{1,3}$ XES signals were collected using a diode and a von Hamos spectrometer with a 2D detector, respectively (see Fig. 13.5a). The study followed the bond rupture and formation of a five-coordinated high-spin active site with Met recombination at 6.3 ps. This study quantified the protein contribution to keeping the Fe–S(Met) bond under physiological conditions to be 4 kcal/mol, which is derived from an adjacent hydrogen bond network. This contribution is weak enough to allow bond rupture in the presence of cardiolipin, a lipid in the mitochondria's membrane, opening a catalytic ligand-binding site that triggers the programmed cellular death.

In another recent XFEL-based experiment, vitamin B_{12}, or cobalamins, were studied. The vitamin B_{12} family is vital to ensure the proper function of many processes including the production of energy, red blood cells, and the operation

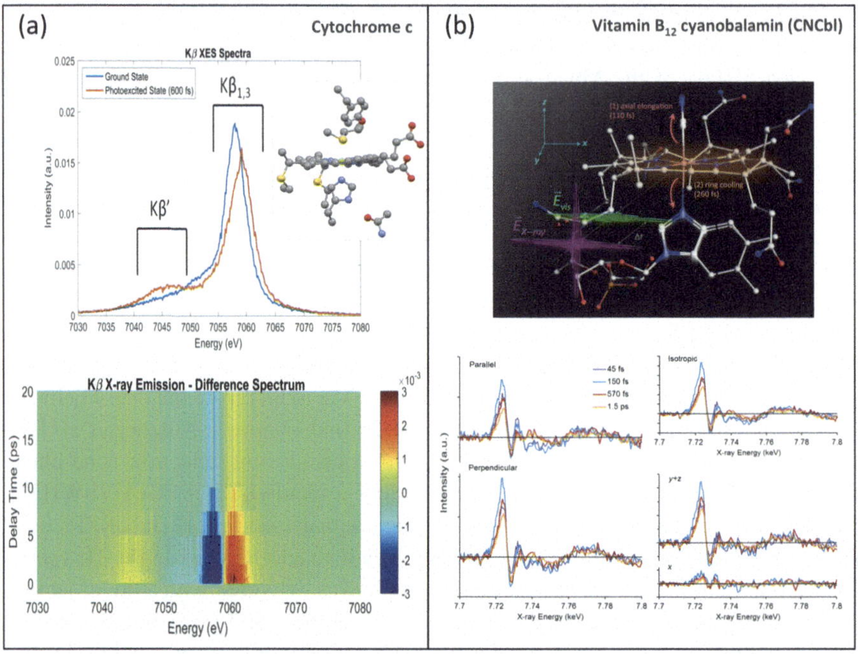

Fig. 13.5 **(a)** Fe $K\beta$ XES spectra cytochrome c, ground (blue) and excited state (red), collected at the LCLS following 520 nm excitation **(top)** and difference spectra showing time dependence up to 20 ps **(bottom)**. Reproduced with permission from [50]. **(b)** Diagram of structural and electronic changes in vitamin B_{12}, cyanocobalamin (CNCbl) following photoexcitation by 550 nm laser wavelength **(top)**. Co K-edge XANES spectra collected at the LCLS. Parallel and perpendicularly polarized time-dependent difference spectra in four representative time regions **(bottom, left)**. Isotropic, x and $y + z$ directions difference spectra derived from the polarization data **(bottom, right)**. Reproduced with permission from [28]

of the nervous system through methyl transfer and radical rearrangement. They also function as gene regulators triggered via photolysis of the Co–C bond. The dynamics of this photochemical reaction mechanism were studied by time-resolved polarized Co XANES and UV/vis spectroscopy by Miller et al. [28] (see Fig. 13.5b). They measured the vitamin B_{12} cofactor cyanocobalamin (CNCbl) in solution photoexcited by a 550-nm fs laser. They were able to separate the directional-dependent components of the reaction dynamics by varying the laser/X-ray polarization angle (Fig. 13.5b). The Co–CN and Co–NIM bonds were shown to elongate in the axial direction of the corrin ring, on a 110-fs timescale. Subsequently, the corrin ring relaxation of the equatorial ligand happens at 260 fs followed by ground state internal conversion on a 6.2-ps timescale. This reaction mechanism describes reactivity, stability, and deactivation of electronically excited cobalamins, which, if understood in detail, will facilitate the development of new targeted medications controlled through light.

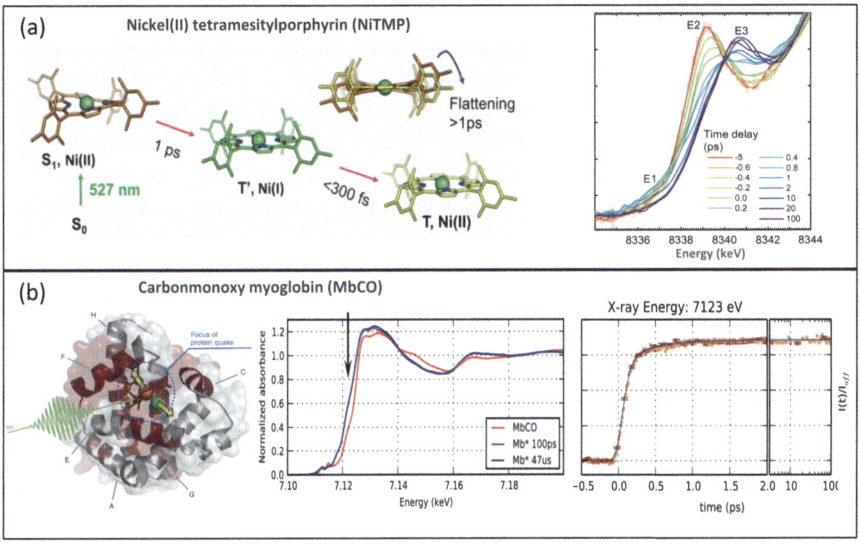

Fig. 13.6 **(a)** Diagram of structural and electronic changes in Nickel(II) tetramesitylporphyrin (NiTMP) following photoexcitation by 527 nm laser wavelength **(left)**. Ni K-edge XANES spectra of NiTMP collected at the LCLS showing the time evolution between 5 and 100 ps after excitation **(right)**. Reproduced with permission from [26] **(b)** Carbonmonoxy myoglobin (MbCO) structure **(left)**. Time-resolved Fe XANES measurements showing the overall spectrum **(center)** and the time trace for absorption changes at 7123 eV **(right)**. Reproduced with permission from [25]

13.4.2 Inorganic Systems Relevant to Biological Systems

Another example of biological relevance was the study of the excited state dynamics of Ni(II) tetramesitylporphyrin (NiTMP) by Shelby et al. [26]. Ni metalloporphyrins undergo a cascade relaxation through multiple electronic states following photoexcitation (see Fig. 13.6a). Some of the states involving metal centers could be regulated catalytically to control the excited state relaxation. It is uncertain if a transient Ni(I) charge transfer state is present in the few ps time domain. Transient Ni K-edge XANES spectra were collected after fs photoexcitation (527 nm) of an NiTMP solution (Fig. 13.6a). Additionally to the spectra of the initial S0 state and the final T(d,d) state (formed within 20 ps), the study shows an intermediate state forming in the sub-ps timescale, which was ascribed by DFT calculations to an Ni(I)(pi,d) electronic state (T'). These results enable the determination of the excited state structural dynamics of NiTMP before thermal relaxation and to resolve intermediates of potential photocatalytic relevance.

The dissociation of the Fe–CO bond in carbonmonoxy myoglobin (MbCO), an Fe- and O-binding protein responsible for the transport of oxygen in muscle tissue, was studied by Levantino and coworkers [25] (see Fig. 13.6b). They collected the

XAS signal in TFY mode after 538 nm laser excitation from a 5.6-mM solution of MbCO transported to the beam through a liquid jet. They monitored changes in the Fe K-edge during the ultrafast (sub ps) photolysis process. By measuring the Fe XANES spectrum at three different energy points using 30 fs short X-ray pulses, they were able to fit the dynamics of this process by two components with characteristic times of 70 fs and 400 fs (Fig. 13.6b). The former was interpreted as the first step of the protein quake (a propagation of the strain released at the active site during the movement of the protein away from the active site), consisting of an Fe out-of-plane motion coupled to the motion of the helix F. The 400 fs process was explained as a residual elongation of the Fe–N bond length in the heme, connected to a contraction of the Fe–His bond. This work serves as a model for studying structural and electronic dynamic processes occurring after photoexcitation of proteins in general.

13.5 Conclusions

In this chapter, we have summarized the current status of X-ray spectroscopic methods and their application to biological systems at XFELs. These new X-ray sources have made X-ray data collection of biological samples under physiological conditions possible, both in the soft and hard X-ray energy regimes, as well as enabling the time-resolved study of fs/ps processes. As mentioned throughout the chapter, further method developments to take full advantage of the XFELs novel characteristics are needed and underway. We expect that high-repetition rate XFELs currently planned will make a big impact in X-ray spectroscopy-based studies, in particular for photon-hungry experiments. We also expect nonlinear X-ray spectroscopy and/or multidimensional X-ray spectroscopy to be realized in the near future, likely expanding the possibility of collecting X-ray spectroscopy data on biological samples. X-ray spectroscopy methods, in combination with structural studies, have the potential to lead to the understanding on how nature controls complicated multi-electron reactions in metalloenzymes by using metal clusters embedded in protein environments.

Acknowledgements R. A.-M. and J.Y. acknowledge all the collaborators of the research presented in this chapter. Use of the Linac Coherent Light Source (LCLS), SLAC National Accelerator Laboratory, is supported by the US Department of Energy, Office of Science, Office of Basic Energy Sciences under Contract No. DE-AC02-76SF00515. J.Y. thanks the Director, Office of Science, Office of Basic Energy Sciences, Division of Chemical Sciences, Geosciences, and Biosciences of the Department of Energy under contract DE-AC02-05CH11231, and the NIH Grants GM110501 and GM055302, which contribute to supporting some of the research presented in this chapter.

References

1. Yano, J., Kern, J., Irrgang, K. D., Latimer, M. J., Bergmann, U., Glatzel, P., et al. (2005). *Proceedings of the National Academy of Sciences of the United States of America, 102*(34), 12047–12052.
2. Garman, E., & Weik, M. (2017). *Journal of Synchrotron Radiation, 24*(1), 1–6.
3. Holton, J. (2009). *Journal of Synchrotron Radiation, 16*(2), 133–142.
4. Lemke, H., Weaver, M., Chollet, M., Robinson, J., Glownia, J. M., Bionta, M. R., et al. (2013). *Proceedings of SPIE, 8778*, 87780S.
5. Margaritondo, G., & Ribic, P. R. (2011). *Journal of Synchrotron Radiation, 18*, 101–108.
6. Emma, P., Akre, R., Arthur, J., Bionta, R., Bostedt, C., Bozek, J., et al. (2010). *Nature Photonics, 4*, 641–647.
7. Ishikawa, T., Aoyagi, H., Asaka, T., Asano, Y., Azumi, N., Bizen, T., et al. (2012). *Nature Photonics, 6*, 540–544.
8. Chapman, H. N., Fromme, P., Barty, A., White, T., Kirian, R., Aquila, A., et al. (2011). *Nature, 470*(7332), 73–77.
9. Nass, K., Foucar, L., Barends, T. R. M., Hartmann, E., Botha, S., Shoeman, R. L., et al. (2015). *Journal of Synchrotron Radiation, 22*(2), 225–238.
10. Boutet, S., Lomb, L., Williams, G. J., Barends, T. R. M., Aquila, A., Doak, R. B., et al. (2012). *Science, 337*, 362–364.
11. Alonso-Mori, R., Kern, J., Gildea, R. J., Sokaras, D., Weng, T.-C., Lassalle-Kaiser, B., et al. (2012). *Proceedings of the National Academy of Sciences of the United States of America, 109*(47), 19103–19107.
12. Weierstall, U. (2014). *Philosophical Transactions of the Royal Society B, 369*, 20130337.
13. Weierstall, U., Doak, R. B., Spence, J. C. H., Starodub, D., Shapiro, D., Kennedy, P., et al. (2008). *Experiments in Fluids, 44*, 675.
14. Roessler, C. G., Kuczewski, A., Stearns, R., Ellson, R., Olechno, J., Orville, A. M., et al. (2013). *Journal of Synchrotron Radiation, 20*, 805–808.
15. Roessler, C. G., Agarwal, R., Allaire, M., Alonso-Mori, R., Andi, B., Bachega, J. F. R., et al. (2016). *Structure, 24*(4), 631–640.
16. Fuller, F. D., Gul, S., Chatterjee, R., Burgie, E. S., Young, I. D., Lebrette, H., et al. (2017). *Nature Methods, 14*(4), 443–449.
17. Milathianaki, D., Boutet, S., Williams, G. J., Higginbotham, A., Ratner, D., Gleason, A. E., et al. (2013). *Science, 342*(6155), 220–223.
18. Wittenberg, J. S., Miller, T. A., Szilagyi, E., Lutker, K., Quirin, F., Lu, W., et al. (2014). *Nano Letters, 14*(4), 1995–1999.
19. Roedig, P., Ginn, H. M., Pakendorf, T., Sutton, G., Harlos, K., Walter, T. S., et al. (2017). *Nature Methods, 14*, 805–810.
20. Koningsberger, D., & Prins, R. (1988). *X-Ray absorption: Principles, applications, techniques of EXAFS, SEXAFS and XANES*. New York: Wiley.
21. Amann, J., Berg, W., Blank, V., Decker, F. J., Ding, Y., Emma, P., et al. (2012). *Nature Photonics, 6*, 693.
22. Kroll, T., Kern, J., Kubin, M., Ratner, D., Gul, S., Fuller, F. D., et al. (2016). *Optics Express, 24*(20), 22469–22480.
23. Lemke, H. T., Bressler, C., Chen, L. X., Fritz, D. M., Gaffney, K. J., Galler, A., et al. (2013). *The Journal of Physical Chemistry A, 117*, 735–740.
24. Cammarata, M., Bertoni, R., Lorenc, M., Cailleau, H., Di Matteo, S., Mauriac, C., et al. (2014). *Physical Review Letters, 113*, 227402.
25. Levantino, M., Lemke, H. T., Schirò, G., Glownia, M., Cupane, A., Cammarata, M., et al. (2015). *Structural Dynamics, 2*(4), 041713.
26. Shelby, M. L., Lestrange, P. J., Jackson, N. E., Haldrup, K., Mara, M. W., Stickrath, A. B., et al. (2016). *Journal of the American Chemical Society, 138*, 8752–8764.

27. Marino, A., Cammarata, M., Matar, S. F., Ltard, J.-F., Chastanet, G., Chollet, M., et al. (2016). *Structural Dynamics, 3*(2), 023605.
28. Miller, N. A., Deb, A., Alonso-Mori, R., Garabato, B. D., Glownia, J. M., Kiefer, L., et al. (2017). *Journal of the American Chemical Society, 139*(5), 1894–1899.
29. Zhu, D., Cammarata, M., Feldkamp, J., Fritz, D., Hastings, J., Lee, S., et al. (2013). *Journal of Physics Conference Series, 425*, 052033.
30. Obara, Y., Katayama, T., Ogi, Y., Suzuki, T., Kurahashi, N., Karashima, S., et al. (2013). *Optics Express, 22*(1), 1105–1113.
31. Katayama, T., Inubushi, Y., Obara, Y., Sato, T., Togashi, T., Tono, K., et al. (2013). *Applied Physics Letters, 103*, 131105.
32. Ogi, Y., Obara, Y., Katayama, T., Suzuki, Y.-I., Liu, S. Y., Bartlett, N. C.-M., et al. (2015). *Structural Dynamics, 2*, 034901.
33. Gaudin, J., Fourment, C., Cho, B. I., Engelhorn, K., Galtier, E., Harmand, M., et al. (2014). *Scientific Reports, 4*, 4724.
34. Stöhr, J. (1992). *NEXAFS spectroscopy*. Berlin: Springer.
35. Friedrich, S., Funk, T., Drury, O., Labov, S., & Cramer, S. (2002). *Review of Scientific Instruments, 73*(3), 1629.
36. Mitzner, R., Rehanek, J., Kern, J., Gul, S., Hattne, J., Taguchi, T., et al. (2013). *Journal of Physical Chemistry Letters, 4*, 3641.
37. Kubin, M., Kern, J., Gul, S., Kroll, T., Chatterjee, R., Löchel, H., et al. (2017). *Structural Dynamics, 4*, 054307.
38. Schreck, S., Beye, M., Sellberg, J. A., McQueen, T., Laksmono, H., Kennedy, B., et al. (2014). *Physical Review Letters, 113*, 153002.
39. Glatzel, P., & Bergmann, U. (2005). *Coordination Chemistry Reviews, 249*(7), 65.
40. Hamos, L. V. (1932). *Naturwiss, 20*, 705–706.
41. Alonso-Mori, R., Kern, J., Sokaras, D., Weng, T.-C., Nordlund, D., Tran, R., et al. (2012). *Review of Scientific Instruments, 83*, 073114.
42. Zhang, W., Alonso-Mori, R., Bergmann, U., Bressler, C., Chollet, M., Galler, A., et al. (2014). *Nature, 509*, 345–348.
43. Zhang, W., Kjr, K. S., Alonso-Mori, R., Bergmann, U., Chollet, M., Fredin, L. A., et al. (2017). *Chemical Science, 8*, 515–523.
44. Canton, S. E., Kjr, K. S., Vank, G., van Driel, T. B., ichi Adachi, S., Bordage, A., et al. (2015). *Nature Communications, 6*, 6359.
45. Haldrup, K., Gawelda, W., Abela, R., Alonso-Mori, R., Bergmann, U., Bordage, A., et al. (2016). *The Journal of Physical Chemistry B, 120*, 1158–1168.
46. Alonso-Mori, R., Asa, K., Bergmann, U., Brewster, A. S., Chatterjee, R., Cooper, J. K., et al. (2016). *Faraday Discussions, 194*, 621–638.
47. Kjr, K. S., Zhang, W., Alonso-Mori, R., Bergmann, U., Chollet, M., Hadt, R. G., et al. (2017). *Structural Dynamics, 4*(4), 044030.
48. Kern, J., Alonso-Mori, R., Tran, R., Hattne, J., Gildea, R. J., Echols, N., et al. (2013). *Science, 340*(6131), 491–495.
49. Kern, J., Tran, R., Alonso-Mori, R., Koroidov, S., Echols, N., Hattne, J., et al. (2014). *Nature Communications, 5*, 4371.
50. Mara, M. W., Hadt, R. G., Reinhard, M. E., Kroll, T., Lim, H., Hartsock, R., et al. (2017). *Science, 356*(6344), 1276–1280.
51. Stojanoff, V., Hamalainen, K., Siddons, D. P., Hastings, J. B., Berman, I. E., Cramer, S., et al. (1992). *Review of Scientific Instruments, 63*(1), 1125–1127.
52. Alonso-Mori, R., Sokaras, D., Zhu, D., Kroll, T., Chollet, M., Feng, Y., et al. (2015). *Journal of Synchrotron Radiation, 22*(3), 612–620.
53. Mori, R. A., Paris, E., Giuli, G., Eeckhout, S. G., Kavcic, M., Zitnik, M., et al. (2009). *Analytical Chemistry, 81*(15), 6516–6525.
54. Mori, R. A., Paris, E., Giuli, G., Eeckhout, S. G., Kavc, M., Zitnik, M., et al. (2010). *Inorganic Chemistry, 49*(14), 6468–6473.

55. Thomas, R., Kas, J., Glatzel, P., Samarai, M. A., de Groot, F. M., Mori, R. A., et al. (2015). *Journal of Physical Chemistry C, 119*(5), 2419–2426.
56. Beye, M., Anniyev, T., Coffee, R., DellAngela, M., Föhlisch, A., Gladh, J., et al. (2013). *Physical Review Letters, 110*, 186101.
57. Nilsson, A., LaRue, J., Öberg, H., Ogasawara, H., Dell'Angela, M., Beye, M., et al. (2017). *Chemical Physics Letters, 675*, 145–173.
58. Dell'Angela, M., Anniyev, T., Beye, M., Coffee, R., Föhlisch, A. Gladh, J., et al. (2013). *Science, 339*(6125), 1302–1305.
59. Shavorskiy, A., Cordones, A., Vura-Weis, J., Siefermann, K., Slaughter, D., Sturm, F., et al. (2013). *AIP Conference Proceedings, 1525*, 475.
60. Kroll, T., Weninger, C., Alonso-Mori, R., Sokaras, D., Zhu, D., Mercadier, L., et al. (2017). *Physical review letters, 120*(13), 133203.
61. Rohringer, N., Ryan, D., London, R. A., Purvis, M., Albert, F., Dunn, J., et al. (2012). English. *Nature, 481*(7382), 488–491.
62. Kotani, A., & Shin, S. (2001). *Reviews of Modern Physics, 73*, 203.
63. Kroll, T., Hadt, R. G., Wilson, S. A., Lundberg, M., Yan, J. J., Weng, T.-C., et al. (2014). *Journal of the American Chemical Society, 136*(52), 18087–18099.
64. Glatzel, P., Schroeder, H., Pushkar, Y., BoronIII, T., Mukherjee, S. G., Christo, V. L., et al. (2013). *Inorganic Chemistry, 52*(10), 5642–5644.
65. Franz, M. O., & Schölkopf, B. (2006). *Neural Computation, 18*(12), 3097–3118.
66. Titus, C. J., Baker, M. L., Lee, S. J., mei Cho, H., Doriese, W. B., Fowler, J. W., et al. (2017). arXiv 1706.09878.
67. Wernet, P., Kunnus, K., Josefsson, I., Rajkovic, I., Quevedo, W., Beye, M., et al. (2015). *Nature, 520*, 78–81.
68. Kunnus, K., Josefsson, I., Rajkovic, I., Schreck, S., Quevedo, W., Beye, M., et al. (2016). *Structural Dynamics, 3*(4), 043204.
69. Eckert, S., Norell, J., Miedema, P.S., Beye, M., Fondell, M., Quevedo, W., et al. (2017). *Angewandte Chemie, International Edition, 56*, 6088.
70. Zhong, Y., Rehanek, J., Löchel, H., Braig, C., Buck, J., Firsov, A., et al. (2017). *Optics Express, 25*(10), 10984–10996.
71. Qiao, R., Li, Q., Zhuo, Z., Sallis, S., Fuchs, O., Blum, M., et al. (2017). *Review of Scientific Instruments, 88*(3), 033106.
72. Umena Y., Kawakami, K., Shen, J.-R., & Kamiya, N. (2011). *Nature, 473*, 55–60.
73. Chollet, M., Alonso-Mori, R., Cammarata, M., Damiani, D., Defever, J., Delor, J. T., et al. (2015). *Journal of Synchrotron Radiation, 22*(3), 503–507.

Chapter 14
Single Molecule Imaging Using X-ray Free Electron Lasers

Andrew Aquila and Anton Barty

14.1 Introduction

Newly developed X-ray free electron lasers operating today including FLASH [1], FERMI [2], LCLS [3], SACLA [4], and the European XFEL [5] are capable of producing single X-ray pulses of shorter than 40 fs in duration containing more than 10^{12} transversely coherent photons at X-ray photon energies ranging from ~280 eV to ~13 keV and within a relatively narrow spectral bandwidth. Focusing these pulses using almost perfect optics into micron sized focal spots, or smaller, enables focused irradiances in excess of 10^{18} W cm^{-2}. These pulses are intense enough to destroy anything in their path, with an effective shutter speed measured in the tens of femtoseconds. X-ray FELs promise to greatly impact many scientific disciplines by opening up the study of materials at the length scale of interatomic distances and at the corresponding timescales of atomic motion, providing new opportunities for imaging time-resolved phenomena such as biochemical reactions in progress on femtosecond timescales and angstrom length scales.

Short wavelength probes such as X-rays, high energy electrons or neutrons are needed to obtain atomic resolution images of objects. Biological structure determination has been one of the outstanding success stories of X-ray FELs, enabling a range of measurements that were not previously possible including the study of radiation sensitive membrane proteins at room temperature [6–8], light induced

A. Aquila
Linac Coherent Light Source, SLAC National Accelerator Laboratory, Menlo Park, CA, USA
e-mail: aquila@slac.stanford.edu

A. Barty (✉)
Centre for Free-Electron Laser Science, DESY, Hamburg, Germany
e-mail: anton.barty@desy.de

S. Boutet et al. (eds.), *X-ray Free Electron Lasers*,
https://doi.org/10.1007/978-3-030-00551-1_14

time-resolved structural studies down to sub-picosecond temporal resolution [9–14], irreversible solution-triggered enzymatic reactions [15, 16], solution scattering [12, 17, 18] and spectroscopic studies combined with crystallography [19, 20]. To date, the vast majority of molecular structures solved at X-ray FELs have involved some variant of crystallography, ensemble scattering, or inferences from spectroscopy. The success of femtosecond crystallography, as described in most of this book, has been in part due to the relative ease of measuring the more intense Bragg peaks formed by crystals, and in part due to the established knowledge base from decades of protein crystallography at other X-ray sources.

While crystallography has been extremely productive of molecular structures, there is a strong desire to break free from the need for crystallization in biological structure determination. Solution scattering methods, covered in Chap. 15, do not require crystallization but present limitations as it is an ensemble approach that results in measurements averaged across particle orientations. Ideally, it would be possible to image individual molecules. For quite some time, theoretical studies of X-ray damage have indicated that sufficiently short duration pulses could collect meaningful diffraction from the sample before it was destroyed, as discussed in Chap. 6 [21–23]. Thus was born the idea that X-ray pulses of femtosecond duration could provide a mechanism for circumventing the dose limitations found in conventional techniques such as electron microscopy and synchrotron-based X-ray imaging. This is the principle of diffraction-before-destruction, first demonstrated using soft X-rays [24] and later at higher resolution [25] through to atomic resolution [26]. Understanding the range of validity of the assumptions of no damage during an X-ray FEL pulse is still the subject of active research and is the subject of Chap. 6 of this book.

The prospect of imaging biological samples such as individual macromolecules to near atomic resolution without the need for crystalline periodicity was one of the science goals motivating the development of high peak brightness X-ray FELs [27]. Single molecule imaging, as it became known, would open the door to determining the structure of proteins and protein complexes that refuse to crystallize, or to studying molecular dynamics at physiological temperatures free from the restraints of a crystal lattice. Short pulses enable the capture of transient states without the need for artificially halting reactions in intermediate states, room temperature measurements give access to reactions under physiological conditions, and femtosecond exposure times enable the use of a dose far in excess of conventional radiation damage limits. The field of single particle X-ray imaging is a vibrant one which, although not yet a mature or routine technique for structural biology, nevertheless has a strong following. In this chapter we review the principles of single particle imaging, detail successes to date, and highlight developments just around the corner which will bring this promising technique closer to reality.

14.2 Early Results

The principle of single particle imaging is elegantly simple: a coherent X-ray beam illuminates the entire sample, which is typically non-periodic in structure, producing far field diffraction on an area detector located some distance downstream. Diffraction from the isolated object gives rise to a continuous diffraction pattern, which can be phased to recover an image of the object. Rather than using an X-ray lens to produce an image on the detector, the far field intensity is measured directly with a resolution equivalent to the numerical aperture subtended by the detector. Because only the intensity of the wave field is measured and not the phase, computational lenses must be used to reconstruct an image of the object. This is necessary to achieve high resolution because manufacturing lenses for X-ray measurements is challenging, with the best resolution achieved to date using X-ray lenses limited to 15 nm [28]. For three-dimensional imaging, the three dimensional diffraction volume is first assembled from measurements at many orientations of the sample then phased in 3D to directly reconstruct the object volume.

The principle of recovering an image by directly phasing the continuous diffraction pattern, which has become known as coherent diffractive imaging (CDI), was first demonstrated using X-rays at synchrotron sources for planar objects [29, 30] and soon thereafter extended to three dimensional structure determination [31, 32] (Fig. 14.1(b)). These early experiments demonstrated the feasibility of obtaining structures without prior information using only diffraction from an isolated, non-crystalline object. Extending these results to the regime in which one can capture images before the sample explodes, diffraction before destruction, was first demonstrated in 2006 using the then newly operational FLASH free electron laser in Hamburg, Germany for inorganic fixed targets (Fig. 14.1(a)) [24] and biological samples [33, 34] using femtosecond duration soft X-ray pulses in the 7–32 nm wavelength range (50–200 eV photon energy range).

Access to higher resolution using X-ray photons with energies from $\sim$280 eV to $\sim$13 keV became possible with the opening of the Linac Coherent Light Source (LCLS) in California in 2009 [3] and soon after the Spring8 Angstrom Compact LAser (SACLA) in Japan in 2012 [4]. Building off the success of previous experiments at lower energies, single particle imaging experiments at these facilities quickly succeeded in producing images of the Mimivirus, a giant virus of $\sim$450 nm size, which was imaged first in two dimensions (Fig. 14.1(c)) [35], and later in three dimensions [36] (Fig. 14.1(d)) at LCLS. Experiments rapidly progressed to ever smaller particles including cell organelles of $\sim$100 nm size [37] and viruses of 70 nm dimensions [38]. At the other end of the scale, other experiments used the penetrating power of X-rays to achieve high resolution on larger objects such as micron-sized cells [39, 40]. Examples of some of these results are shown in Fig. 14.1. These results used X-rays in the 1–1.7 keV photon energy range where the higher scattering cross sections and the larger number of photons per pulse

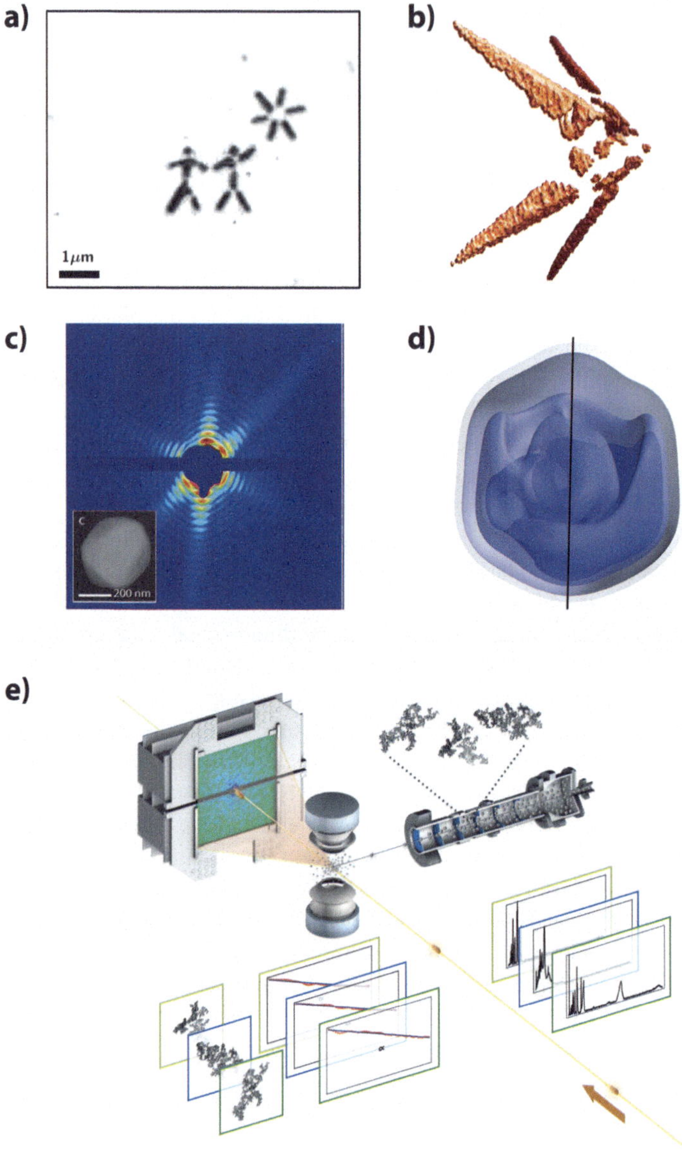

Fig. 14.1 A selection of published CDI results (**a**) First diffraction before destruction. Reproduced with permission from [24] and (**b**) 3D coherent diffractive imaging. Reproduced with permission from [32]. (**c**) Single mimivirus diffraction pattern collected at LCLS. Reproduced with permission from [35] and (**d**) 3D reconstruction at low resolution. Reproduced with permission from [36]. (**e**) Correlative diffraction imaging and spectroscopy of airborne soot and nanoparticles in flight through the XFEL beam, enabling structural studies of the natural soot morphology not possible when the particles are not in flight from [41]

make it easier for diffraction signal to rise above noise and can, in principle, reach resolutions below 1 nm based on the experiment geometry and the accessible wavelength range [25].

Outside of the field of biology, single shot X-ray FEL imaging has enabled the study (Fig. 14.1(e)) of airborne aerosols in flight to nanometer resolution [41]. The structure of airborne aerosols is highly important in fields ranging from climate research to health, and previous studies had relied on collecting particles on substrates, which alters their morphology. Measurements at LCLS provided the first demonstration of the morphology of soot in flight and are a good example of the type of measurements for which single particle diffractive imaging in flight is ideally suited.

14.3 Towards Molecular Single Particle Imaging

Moving beyond early successes with micron to 100 nm-sized particles at nanometer resolution towards atomic resolution imaging of single biological molecules is conceptually easy but in practice requires overcoming significant technical challenges. It is no accident that results to date have been on larger particles: nanometer resolution studies can be performed using soft X-rays where scattering signal is higher due to a higher scattering cross section, and large particles produce a stronger scattering signal relative to instrument and extraneous noise sources. This makes particle hits both easier to detect and easier to interpret into a reconstructed image. Moving to smaller particles, and ultimately single molecules, demands that several technical requirements are achieved simultaneously:

(1) Delivery of X-ray pulses of femtosecond to tens of femtosecond duration containing of order 10^{12} photons in the 3–7 keV range and in a focal spot approximately 5–10 times larger than the particle of interest to ensure uniform and consistent illumination across the sample;
(2) Negligible measured stray photons from extraneous sources such as fluorescence or scattering from beamline optics, sample supports, vacuum windows, or carrier gas, currently a limiting issue;
(3) A detector capable of reliably detecting both single and multiple photon events, read out at either the machine repetition rate or for each sample hit; and
(4) A sample delivery system capable of delivering reproducible individual particles into the X-ray with low background at a useful hit rate.

These requirements amount to bringing experimental reality as close as possible to the theoretical assumptions on which single molecule imaging predictions are based: low background sample delivery, a tightly focused X-ray FEL beam with negligible stray scattering, and a near perfect detector. Many of the practical realities and technical aspects have been addressed during the course of the LCLS Single Particle Imaging initiative [42] and are discussed in a recent article [43]. Results to date are promising and demonstrate steady progress. Scattering to 5 Å resolution has been seen from viruses of less than 100 nm size using hard X-rays [44], however there was insufficient data for a 3D reconstruction. Nevertheless, current accomplishments remain a long way from the ultimate goal of routine *non-crystalline molecular imaging* at resolutions approaching 2–3 Å. The challenge of steadily moving to higher resolution has been compounded by a gap in available energy range. To date, the brightest and most intense X-ray beams, required for single particle imaging to high resolution, are available at LCLS where the AMO instrument accesses X-rays in the 280–2000 eV range while the next higher energy available is in the CXI instrument in the 5–12 keV hard X-ray energy range necessitating the use of different technologies for each instrument.

14.3.1 Achieving the Required X-ray Focus

Relatively few X-ray photons are scattered from sub-100 nm single particles even with the best-focused X-ray beams. Obtaining the maximum peak focused intensity requires not only a powerful free electron laser, but also efficient optics capable of transporting the produced photons into a well-controlled focal spot distribution. To date, diffraction patterns from biological samples, measured using soft X-ray instruments at LCLS [45] in the 280 eV to 2 keV range and the CXI nanofocus instrument [46] using harder X-rays in the 5–12 keV range, have been significantly weaker than predicted from computational models of the same particles, suggesting a lower focal spot power density than expected, or greater than expected loss of scattering power due to rapid ionization [47], or both. Finite optic size and contamination on mirrors from window-less operation account for some of these losses, as does a larger than intended focal spot size: the brightest part of the 100 nm focus at CXI may be closer to 170 × 120 nm [48] depending on alignment, system drifts, and possibly high order aberrations in the focal system. Inevitable limits on mirror fabrication result in a less than perfect focal spot distribution [49]. Further upstream, the grazing incidence optics required to transport the LCLS X-ray beam have been barely long enough, at 0.5 m, to avoid clipping the beam and introducing artifacts into the intensity. These mirrors were upgraded to meter long mirrors at LCLS in 2017 but the benefits of this upgrade for single particle imaging have not yet been determined experimentally. The European XFEL's SPB/SFX optics are expected to be 1 m in length and require sub-nanometer height errors across the entire surface of the optics [50]. Achieving a near perfect focal spot with minimum photon losses is in practice highly challenging.

14.3.2 Background Scattering and Background Characterization

A further complication is that anything intersecting the beam results in scattering and/or fluorescence, a reality familiar to anyone working on beamline design.

Apertures designed to prevent beamline scatter reaching the detector themselves cause scattering and must be placed with care, be manufactured of proper materials so they do not fluoresce and have smooth edges to limit undesirable scattering to one direction. With 10^{12} photons per pulse, only a tiny fraction of those photons need be scattered onto the detector or cause fluorescence in the chamber to overwhelm the photons scattered from the sample, which may be as weak as a few thousand photons in total or less. Sufficient diagnostics must be included in beamline designs to routinely find and maintain optimum focus over extended periods of time. The task of reducing unwanted photon scatter to levels that enable detection of signal from the sample requires considerable care in beamline design and operation.

Theoretical studies suggest that diffraction pattern orientation determination should be possible even with very dilute photon counts of as little as 100 scattered photons per image [51]. A variety of image reconstruction algorithms have been proposed for analyzing such sparse diffraction data, including the Expand-Maximize-Compress (EMC) algorithm [51, 52], the manifold embedding algorithms [53], simple best-fit models [54], and multi-particle cross-correlation analysis [55–57]. Simulations included in these papers have gone so far as to include Poisson photon counting statistics, and recent experiments, using a low-power copper anode X-ray tube instead of an X-ray FEL source, have demonstrated the successful recovery of random "particle" orientations when photon statistics do indeed closely approximate pure Poisson photon counting noise [58].

As already mentioned, everything near the beam contributes to measured signal including beamline scatter, fluorescence, and scattering from the sample delivery medium. For samples consisting of a few hundred thousand atoms, there may easily be as many non-sample atoms intersecting the entire beam path as atoms in the sample itself, both of which act as a source of scattered photons. Simulations including the effects of realistic experimental noise indicate that the situation in a realistic diffractive imaging experiment will add complexity to the analysis conducted on the diffraction patterns [21, 59]. Image reconstruction may still be possible in the presence of additional noise sources but requires knowledge of the statistical distribution of the background signal to be incorporated into the orientation–alignment–reconstruction algorithm. Reconstruction will become particularly challenging, if not impossible, when the magnitude of statistical fluctuations in the background signal is greater than the expected particle scattering signals. Low background is desirable also for hit detection, since the most common method of detecting particle hits is an elevated level of photons on the detector; although hit detection using external diagnostics, such as a time-of-flight spectrometer measuring ion fragments from the exploding sample, appears possible [60].

Instrument background contributions are complex and are difficult to reliably simulate because the background is frequently dominated by unexpected and unaccounted for imperfections not included in models. Instrument noise is intimately linked to both detector performance and parasitic X-ray scattering through the entire instrument/X-ray optical system. In practice, instrument background can only be determined through an experiment, and then used to build a model. The highest resolution single biological particle scattering measured above background levels to date was from Rice Dwarf Virus [44]. In this experiment etched Germanium (which does not fluoresce at the 7 keV X-ray energy used) slits were used upstream of the sample to reduce low angle scatter, a large steel plate with a 5 mm diameter hole was also used upstream of the sample to block instrument scatter that could look like a high angle signal and finally a 3 mm Tantalum aperture was placed downstream of the sample (called a post-sample aperture). This post-sample aperture additionally limited the scatter that was not originating from the interaction region. A downside of a post-sample aperture is that it can limit the highest scattering angle achievable, but this can easily be outweighed by dramatically reducing the instrument scatter that can reach the detector. This indicates that careful preparation of the beamline, with a view to reducing background, can reduce measured photon background to levels consistent with feasibility for single particle X-ray imaging.

Further improvements may be achieved through the use of newer generation detectors. These detectors operate at the high repetition rates of current and future X-ray FELs [61, 62], are capable of improved single photon discrimination, and have dynamic ranges into the thousands to tens of thousands of photons; such as the Adaptive Gain Integrating Pixel Detector (AGIPD) [63], the PERCIVAL detector [64], the ePix [65, 66] family of detectors, or the Jungfrau detector [67]. Ultimately, the detector requirement for atomic-scale single particle imaging is the reliable identification of single photons and discrimination from noise at the highest resolution shell and this drives a need for low noise, high efficiency detectors with the highest possible frame rates.

14.3.3 Delivering the Sample

As already mentioned, everything in the X-ray beam scatters photons. This includes substrates and dust in the case of fixed targets, or residual gas in the case of aerosol injection. The lowest background sample delivery method to date has been aerosol injection [35, 36, 68]. Aerosol injection is a method where a sample, in a liquid buffer, is aerosolized into tiny drops with each drop containing ideally a single sample particle. These droplets are formed with either an electrospray source or a gas-dynamic virtual nozzle commonly used in X-ray FEL crystallography. The drops then flow into a skimmer/relaxation chamber where they evaporate due to a lower gas pressures in the chamber but also via the use of a volatile buffer. From there the particles are dragged along flow lines through a focusing apparatus called an aerodynamic lens stack [68] that focuses the particles down to a few tens of microns. This stream of sample overlaps the X-ray FEL where particles intersect the beam in random orientations. The predominant source of background from aerosol injection comes from scattering from the carrier gas. A modest pressure of 10^{-5} Torr of Helium has been demonstrated to show a detectable background level with incident intensities of $\sim 10^{12}$ photons per pulse.

Fixed target sample delivery can be an alternative, which simplifies the challenges for reliably hitting the sample with the beam but at the cost of extra scattering from the target substrate that supports the sample of interest. Fixed target methods have been successful for 2D crystalline samples of relatively few unit cells [69, 70] but, to date, the background has been too high to be viable for single molecule imaging. The possibility of using graphene substrates, such as those used in cryo-electron microscopy [71], to reduce background is an active research topic possibly suitable for fibers, larger particles, or inorganic samples [72].

A critical factor limiting single particle imaging is the rate at which useful data frames are collected, commonly referred to as the "hit rate." Each of the reconstruction methods mentioned above assumes the availability of tens of thousands, if not millions, of diffraction snapshots from identical objects in order to achieve high resolution. For the aerosol injection of the sample, particle densities are often low to the point where the majority of X-ray FEL pulses do not hit any sample. This is due to a mismatch in the focal size of the X-ray FEL, which is between 1 and 0.1 μm, and the focal size of the sample jet, which is on the size of 10 μm or more. The experimentally observed hit rate in the nanofocus instrument at the CXI instrument of LCLS is currently 1% at best, and more realistically closer to 0.1% [44]. A 1% hit rate generates at most 4000 hits per hour, or just over 50,000 hits per 12-h shift, at the 120 Hz LCLS operating rate. Measuring the needed one million diffraction patterns would take on the order of 20 shifts of continuous data collection at the highest peak hit rates observed to date, which is prohibitively high for the limited availability of beamtime.

Higher repetition rate X-ray FELs under construction such as the European XFEL and LCLS-II will increase the data acquisition rate by 10–1000 times. Detection of particle hits may be possible using external diagnostics such as elevated ion signals [60], potentially enabling the external vetoing of blank frames prior to detector readout [62]. This would enable high hit rate collection without the need to record all data, which would be a major advance since data transfer and storage represent a massive challenge at MHz repetition rates for multi-megapixel detectors. Additionally, sample delivery methods are being investigated to increase hit rates. These include optical trapping, the use of multiple electrosprays to increase particle densities at the interaction region, charging the sample and then holding them in the interaction region with electrostatic fields, and finally synchronizing the sample injector to the X-ray FEL to maximize chances of hitting the sample particles and minimizing sample consumption. Fixed target methods are also being investigated but the suitability of these methods at MHz repetition rates is questionable.

14.3.4 The LCLS Single Particle Imaging Initiative

In 2014 the LCLS user community launched a collaborative and long term set of experiments aimed at overcoming the technical challenges involved in realizing single particle imaging. Known as the "Single Particle Imaging initiative" (SPI initiative) [42], this project has made significant progress towards bringing experimental reality closer to the requirements needed for single molecule imaging. One

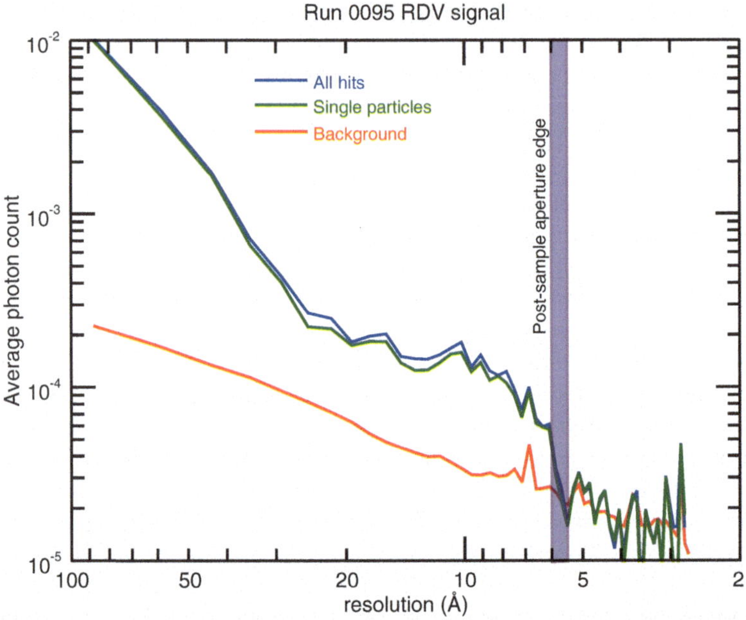

Fig. 14.2 Elevated photon counts from the sample are visible up to an angle commensurate with 5.9 Å resolution, this being the resolution limit set by the angular acceptance of the post-sample aperture. Figure from the Single Particle Imaging Initiative [42], reproduced with permission from [44]

of the aims was to carefully study sources of instrument background and detector limitations to establish detection limits, as well as developing the instrumentation required to increase the event/hit rate to useful levels, and reports on technical progress are starting to appear in the literature [38, 44]. Starting from large viruses that produce ample signal, one achievement has been the measuring of single virus scattering from individual Rice Dwarf Virus particles above instrument background levels to 5.9 Å resolution in the CXI nanofocus instrument at 7 keV photon energy, as shown in Fig. 14.2. The photon energy was chosen to provide the highest signal to noise ratio on the Cornell-SLAC Pixel Array Detectors (CSPAD) [73], which improves at higher photon energies, while also staying below the 7.1 keV iron absorption edge. Operating above the iron absorption edge increases background due to fluorescence from the steel vacuum chamber, with the task of disentangling photons scattered from the sample from other sources of experimental background an important element of data analysis [43, 44]. Although too few patterns were measured to enable three-dimensional reconstruction, this represents the first experimental verification that signal from single molecules can be identified at resolution better than 1 nm.

Additionally, the SPI initiative has produced a soft X-ray data set for algorithm testing. This data set was collected using the 1.5 μm^2 focus of the LAMP endstation of the Atomic, Molecular & Optical science beamline at LCLS [74] with a photon energy of 1.6 keV, just below the Silicon K edge [38] (for similar reasons described above). While the resolution at the lower photon energies is modest, 11.6 nm to the edges of the detector and 8.3 nm in the corner, the data set contains between 8000 and 37,000 single diffraction patterns of the 70 nm Coliphage PR772 virus, making it a good dataset for algorithm and analysis technique development [75, 76]. A large number of these datasets are openly available on the Coherent X-ray Imaging Data Bank (http://cxidb.org) [77].

14.3.5 Algorithm Development

The development of robust algorithms is critical to the success of single molecule imaging. The far field diffraction pattern from a randomly oriented single molecule is the Fourier transform of the modulus square of a projection of the electron density consistent with the orientation of the molecule [78]. A more mathematically detailed discussion of coherent diffraction theory and how a diffraction pattern relates to the electron density is found in Chap. 9 of this book. Significant progress has been made in sorting, orientation and phasing algorithms that invert the recorded diffraction patterns into a 3D diffraction volume, which in turn is transformed into a 3D electron density [79].

The first pass through the data is usually a hit finding step, which identifies frames with scattering elevated above background levels, indicating that some particle has been hit. A first pass is often a simple metric summing all the photons in a frame and calculating a chi-squared value with respect to a running background [59]. External diagnostics such as an ion time-of-flight spectrometer may also be used [60]. Hit finding algorithms have to be fast enough to run in near real time to optimize sample injection [43]. Additionally, these algorithms also play a role in data reduction, enabling blank frames to be promptly discarded.

Hit finding is followed by other sorting algorithms implemented to further down select diffraction patterns belonging to a single class; they sort out single hits from multiple particle hits, and other background contamination. As can be seen in Fig. 14.3, a fraction of the diffraction patterns measured in a real experiment is of multiple particles and other contaminations. Sorting algorithms for single molecule imaging categorize the collected patterns into different classes for future processing. Additionally, sorting algorithms try to reject contamination, and select only single hit particles from clusters of similar hits based on techniques like diffusion map analysis.

During the first single particle imaging experiments at LCLS in 2009, a human sorting the data into classes was necessary. This manual method was slow, subjective, and prone to error. Clearly, such a task could be highly improved using statistical computing or machine learning methods such as Principle Component Analysis [80], Manifold Embedding Methods [81], or diffusion maps [82]. The statistical algorithms, such as diffusion maps, are based on the techniques of feature

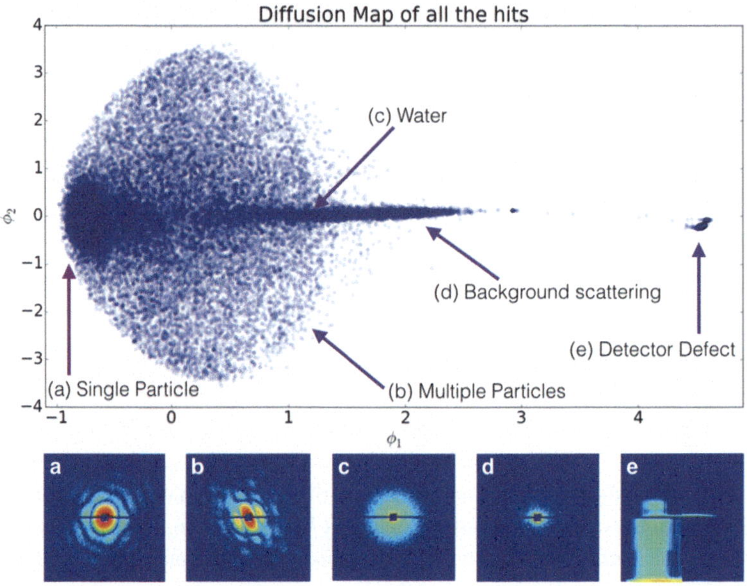

Fig. 14.3 An example of a diffusion map analysis to sort single particle data frames and representative diffraction patterns [82] (**a**) from frames that contain multiple particles (**b**) from various background sources of water/contaminant residue, higher than average background scatter, and defective detector readout (**c**–**e** respectively). Each blue dot in the image represents single frames in a reduced dimension space were clustering is used to sort the data. The Figure is from the Single Particle Imaging Initiative [42], reproduced with permission from [38]

extraction and dimensional reduction. The algorithms first project each image as a point in an N-dimensional Euclidean vector space where each pixel (total number N) in the image represents one orthogonal direction of that space. For example, if an image consisted of only two pixels, with 3 photons in pixel one and with 5 in pixel two, then a point would be placed at [3, 5] in this space. Each image then proceeds to add more and more points to fill this vector space. After all the data has been added to the vector space, it is normalized (for example the mean value is subtracted so that the space is centered around the origin) and then either the linear principle components are found, clusters are produced using k nearest neighbors, or the dominant eigenvectors and eigenvalues of the space are used to reduce the dimensionality of the space for classification purposes. These recently developed classification methods have different degrees of human intervention requirements, from unsupervised [82, 83] to semi-supervised in which a human operator assists in assigning groups and classifying particle type. Nonetheless, these algorithms are obviously faster and less subjective then their human counterpart.

After sorting, the individual single molecule patterns need to be oriented into a 3D diffraction volume. One of the initial ideas was to use the common-line approach [84] where it was recognized that all diffraction patterns are 2D curved projections in reciprocal space that cross the origin of the 3D Ewald sphere. This implies that any two diffraction patterns share a common-line or common arc, depending on their extent in reciprocal space. The primary limiting factor for the common-line approach is that as the molecule becomes smaller the diffraction pattern becomes sparser with fewer photons per diffraction pattern. This makes identifying a common line impossible with too few detected photons within each line. To overcome these limitations, other methods were developed and are now commonly used, such as the Expand-Maximize-Compress (EMC) algorithm [51, 85]. The EMC algorithm is an expectation maximization algorithm that iteratively guesses the orientation of the diffraction patterns based on its log-likelihood best fit into the diffraction volume, with the initial diffraction volume consisting of randomly oriented patterns. Such a method uses the whole diffraction pattern instead of a common line to determine the orientation and therefore is functional at significantly lower signal levels. With each successive iteration of the algorithm, the guesses for best fit in orientation improve and a self-consistent diffraction pattern emerges. The EMC algorithm has been shown to work on sparse data [58] in real space. Additionally, depending on symmetry, the diffusion map of an ensemble of single particles can be approximated as Wigner-D functions which form an SO(3) rotation group [53]. Intensity correlation [86] algorithms have also been used to successfully orient particles. Pictorial examples of the EMC and manifold embedding algorithms are shown in Fig. 14.4.

Once the 3D diffraction volume is oriented, phase retrieval algorithms are used to iteratively determine a set of self-consistent phases for each pixel. Iterative phase retrieval algorithms [89–90] are well known in the field of coherent X-ray imaging, and have been extensively developed and demonstrated for 3D structures [32, 91, 92]. Of particular importance for single molecule imaging is the need to make the phase retrieval algorithms robust against noise [93]. These algorithms start by randomly assigning phases to each diffraction intensity, project the diffraction volume into real space via a Fourier transform, and remove any intensity outside the constrained volume. The algorithms then project back into diffraction space and keep the new phases but replace the amplitudes with the measured amplitude from the intensity. The algorithm then repeats the process until the phases in diffraction space are stable and the real space object is reconstructed. More detailed discussion of phase retrieval algorithms is found in Chap. 9.

Lastly it is worth noting the existence of newly developed algorithms that parse the problem in a different way. An extension of the multitiered iterative phasing (M-TIP) algorithm designed for fluctuation scattering experiments [55] simultaneously determines the orientations, 3D intensities, and phases in a single iterative process [94]. Through a set of projection operators on the real space object, measured diffraction data, and orientation estimates, the modified M-TIP algorithm iteratively

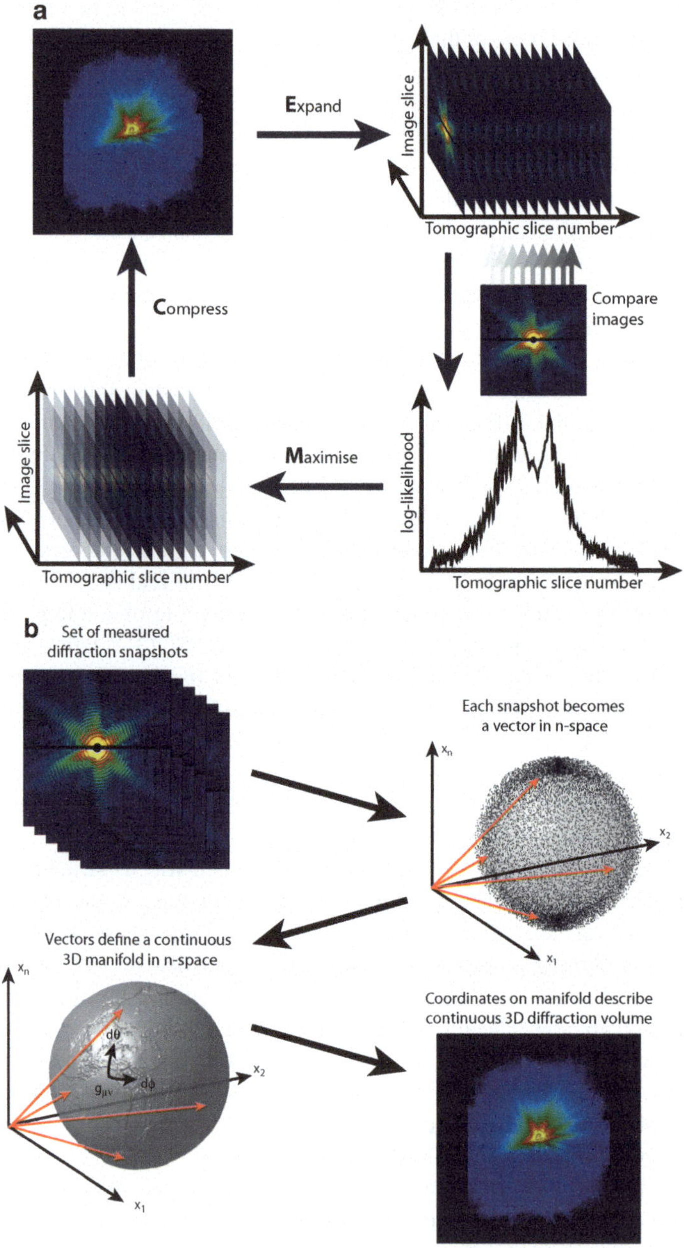

Fig. 14.4 Two approaches to single particle orientation determination: (**a**) The expand-maximization-compression orientation-classification algorithm employs a Bayesian approach to orientation determination, based on the probability that each diffraction pattern fits at a given angle into an estimated 3D volume (**b**) Manifold embedding on the other hand represents images as individual vectors in an abstract n-dimensional pixel space and seeks to find the smoothest connection between adjoining vectors. Reproduced with permission from [87]

refines a set of models for the density, intensity function, orientations, and states to be consistent with specified constraints and the measured data. The use of a spherical harmonic basis derived from fluctuation scattering approaches enables the building of an accurate 3D intensity model from sparse images without the need to cover every voxel in reciprocal space.

In most cases, the practical success of an algorithmic approach depends critically on its robustness to noise and unavoidable experimental artifacts. Testing with experimental data is crucial. Care needs to be paid to preparing the raw experimental data for analysis in the form of photon finding and hit detection/sorting but as of yet there are no obvious standards for this first step. While many algorithms work on theoretical data, a common choice of algorithms for structure determination is a combination of EMC for orientation determination combined with difference map based iterative phasing.

14.4 Enhancing the Molecular Signal

The weak scattering from a single molecule could be enhanced if a large number of identical (to the resolution of interest) and identically oriented particles could be arranged in the beam in a single exposure. Under ideal conditions exposing n oriented molecules in the beam should result in n times more signal.

Solution scattering adopts this approach, delivering molecules in solution in random orientations, but results in angularly averaged data. In order to measure the 3D molecular transform, identical particles must be aligned with respect to each other so that the summed molecular scattering is not averaged over orientations. The use of lasers to align gas phase molecules for single molecule imaging has been proposed [95–97] but has been challenging to execute in practice, particularly for large molecules such as proteins, and to date has not lead to sufficiently perfect alignment to allow atomic level resolution.

Exposing a small number of molecules at once may enable the molecular transform to be retrieved using angular correlation methods [57] as has been demonstrated on simple test systems [86, 98]. However it has been pointed out that the signal-to-noise does not necessarily increase proportionally to the number of particles in each exposure in vacuum [56]. Arranging molecules into a two dimensional crystal produces Bragg spots in two planes and extended transforms perpendicular to the two dimensional lattice [69, 70], which may be phased using novel methods [99].

Protein crystals consist of millions or even trillions of individual molecules aligned into a regular lattice with a high degree of angular and spatial regularity. This results in strong X-ray scattering and for well-ordered crystals can lead to atomic resolution in protein crystallography measurements. Suites of protein crystallography programs have arisen to turn these measurements into molecular structures, resulting in the vast majority of protein structures deposited to date in the Protein Data Bank (www.PDB.org) [100]. However, a regular and perfect three-dimensional crystalline lattice gives rise to intensity at only the Bragg peak locations—those points in reciprocal space where unit cell diffraction interferes constructively. While amplifying the molecular signal to useful levels, the Bragg peak locations undersample the molecular transform required for direct molecular imaging [101]. This gives rise to the well-known phase problem of crystallography, whereby other knowledge or experimental phasing must be used in order to determine the reflection phases and thus the structure. It was realized early on that direct phasing from the molecular transform itself requires measurements in between Bragg peak locations [101]. As discussed in Chap. 8 of this book, methods have been proposed to reveal the molecular transform and make of use of finite size crystals [102, 103] and have recently been demonstrated in a two-dimensional model system [104], but are yet to be demonstrated on a three dimensional protein crystal.

A recent breakthrough involved noting that protein crystals are often not perfect. In particular, under certain conditions otherwise rigid molecules in the unit cell can be randomly translated from locations on a perfect crystal lattice, and this random molecular displacement can make the continuous scattering from the molecule within the unit cell directly measurable. At low resolution, translational disorder of the molecules gives rise to Bragg peaks because the scattering is still coherent enough to constructively interfere. However, at high resolution the molecular scattering adds incoherently as shown in Fig. 14.5. The measured signal corresponds to the Fourier transform of the rigid unit symmetrized according to the number of rigid unit orientations, making it possible to get million-fold, or greater, amplification of the molecular transform signal at high resolution. The translation of asymmetric units in a protein crystal gives rise to loss of Bragg intensity at high resolution and reveals the more desirable molecular transform itself. 2D diffraction patterns can be merged directly in 3D using image orientations obtained from indexing the Bragg peaks. It has been shown that combining molecular transform and Bragg data enables extension of resolution [105] and may enable ab initio model-free phasing under certain circumstances. The 3D molecular transform is made measurable by exploiting crystals as a means to align millions of molecules in space at which point the phasing techniques of single particle imaging described earlier in this chapter can then be applied directly for structure determination. A detailed and somewhat more mathematical description is found in Chap. 9.

14.5 Holography

An additional coherent diffraction technique worth discussing with respect to single molecule imaging is that of Fourier transform holography [23, 106–108]. The method holds the promise of quickly recovering the amplitudes and phases of the object of interests without the use of an iterative phase retrieval algorithm. The method relies on a holographic reference point source, typically a pinhole, located close to the sample. The reconstruction is achieved by calculating the autocorrelation of the object from the Fourier transform of the measured intensities.

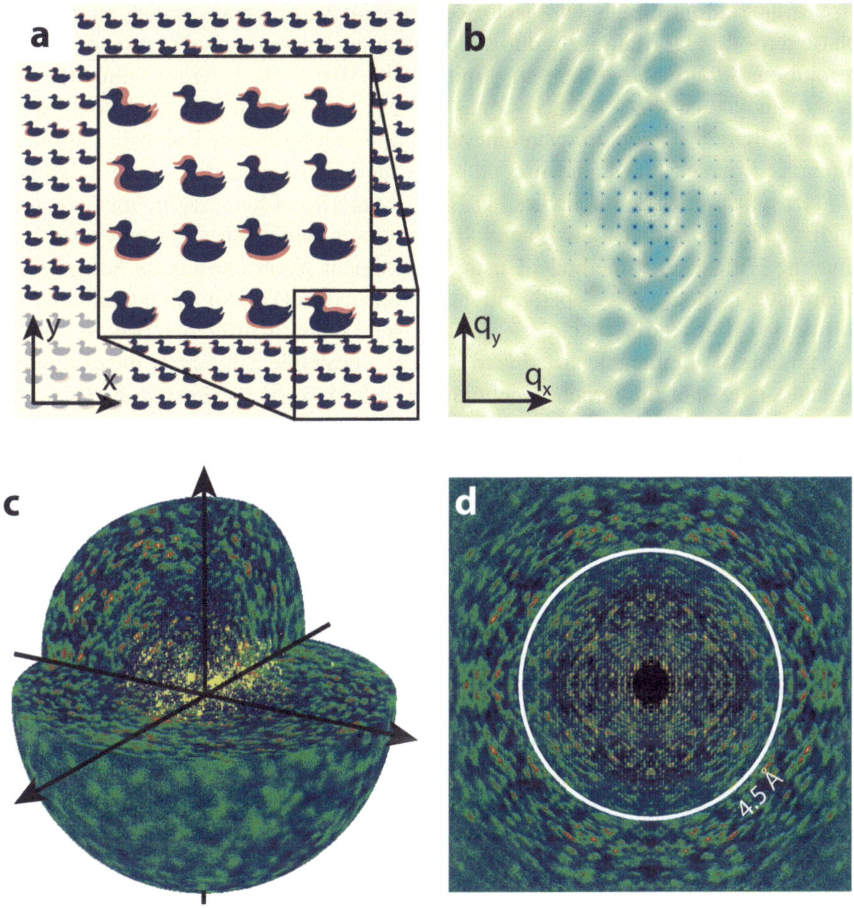

Fig. 14.5 Diffraction from unit cells with random displacements from a perfect crystalline lattice (**a**) reveals the molecular transform in addition to Bragg peaks at lower resolution (**b**). Continuous molecular transform diffraction can be seen in 3D merges of diffraction data from Photosystem II (**c**). Bragg peak resolution extends to the white line at 4.5 Å, while continuous diffraction extends to higher resolution (**d**). Reproduced with permission from [105]

The hologram, thus created by the interference of the sample and the reference, yields a pair of images of the object in the autocorrelation, as long as the reference pinhole is farther away from the object than the maximum size of the object in the given direction of the pinhole. The method has successfully been employed at X-ray FELs to study magnetic samples [109].

The technique has limitations in terms of obtainable resolution, and also provides limited depth information about the sample. Typically, only a 2D reconstruction is possible unless the reference is in a different plane then that of the sample in the propagation direction of the beam. The resolution is often limited to that of the size of the reference pinhole/object used to create the hologram, unless an iterative phase retrieval algorithm is used to supplement the reconstruction to provide higher resolution. Additionally, there are limits to the size of the reference pinhole; if it is too small then the reference signal is weak resulting in a lack of contrast of the hologram. Attempts to overcome this limitation have relied on using multiple reference sources [110], the use of binary uniformly redundant array reference [111], or customizable references [112]. The majority of holographic methods rely on fixed targets, which could potentially limit their applicability to single molecule imaging. However, a method of using a spherical Xeon clusters as holographic references is being developed and has produced successful reconstructions of viruses [113]. As the Xeon cluster is opaque it produces a diffraction pattern similar to a pinhole of comparable size due to Babinet's principle. Ultimately the resolution of any single shot holographic technique will be limited by shot noise. This will occur when there are insufficient photons, at the desired resolution, in the diffraction pattern to enable the contrast necessary to reconstruct the hologram.

14.6 Electron Microscopy

No discussion of single molecule imaging would be complete without noting the phenomenal progress made in cryo-electron microscopy over the past decade. Developments in electron detector technology combined with maximum likelihood image processing algorithms have steadily improved resolution to the point where, depending on the sample, near-crystallographic resolution of better than 5 Å is regularly achieved [115–116]. Better than 3 Å resolution remains a challenge and may not be possible for all samples, but steady progress in the use of phase plates for in-focus phase-contrast imaging of unstained samples [117, 118] combined with further improvements in detector and image processing technology, continue to expand the range of samples to which cryo-EM can be applied [114]. The development of electron microscopes with automated sample handling, automated data acquisition and largely automatic data processing has made the instrumentation accessible to the broader biological community, with the overall impact on biological science being significant enough to warrant a Nobel Prize in 2017. Nevertheless, high resolution cannot be achieved for all samples and results are highly dependent on the sample, thus there remains a need for alternative and complementary single particle imaging methodologies with different capabilities.

Single particle imaging using femtosecond X-rays pulses from X-ray FELs differs from current cryo-EM techniques in several ways. First, exposure times of a few femtoseconds duration are shorter than even thermal vibrations. Thus, sample motion during exposure is not an issue. Secondly, femtosecond pulses can outrun conventional radiation damage processes. Instead of traditional radiation damage effects [119] one is instead concerned with multiple photoelectron absorption and Auger decay during the femtosecond duration pulse [47], nonuniform ionization and ultimately Coulomb explosion during the pulse [10, 22, 120, 121] as discussed in Chap. 6. The different damage mechanisms enable the use of X-ray doses far in excess of conventional radiation dose limits, which in turn facilitates studies at room temperature and the study of radiation sensitive samples. Femtosecond pulses therefore make it possible to measure much more signal from weakly scattering and radiation damage prone objects, such as individual molecules or viruses, than possible using the same dose spread out over longer time periods. Space-charge limitations prevent the use of such short exposures with the relatively low-energy electron beams used in conventional imaging electron microscopes [122, 123]. Thirdly, the ability to perform experiments at physiological temperatures with X-ray FELs enables direct study of molecular machines, the femtosecond pulses providing both time resolution and a shutter-speed with which to freeze molecular reactions in their tracks. Dynamics close to natural conditions and of radiation sensitive samples is where single particle imaging at X-ray FELs is expected to have the largest impact.

14.6.1 Outlook

Experiments to date indicate that molecular imaging at sub-nm resolution using X-ray free electron lasers should be feasible in the near future given sufficient effort invested into matching experimental conditions to theoretical requirements. A reduction of instrument background levels has been achieved, and detector technology is improving with better signal to noise ratios, increased dynamic ranges, and higher frame rates. Perhaps the biggest bottleneck remaining in single particle imaging is sample delivery—technology for the delivery of isolated particles into the X-ray FEL focal spot in sufficient density to obtain a useful hit rate. These aspects are currently under development, and continue to be pursued at both LCLS and the European XFEL [124]. Even with the rapid progress in single particle Cryo-EM, X-ray single molecule imaging will find application in the study of molecular machines and protein conformational states at physiologically relevant temperatures. In the coming years it is not unreasonable to expect that further developments in injector technology and low-background fixed targets, combined with higher repetition rate X-ray FEL sources such as the European XFEL and LCLS-II coming on line, will establish the viability of single particle X-ray diffractive imaging as a valuable additional tool for structural biology.

References

1. Ayvazyan, V., Baboi, N., Bähr, J., Balandin, V., Beutner, B., Brandt, A., et al. (2005). First operation of a free-electron laser generating GW power radiation at 32 nm wavelength. *European Physical Journal D: Atomic, Molecular, Optical and Plasma Physics, 37*(2), 297–303.

2. Allaria, E., Badano, L., Bassanese, S., Capotondi, F., Castronovo, D., Cinquegrana, P., et al. (2015). The FERMI free-electron lasers. *Journal of Synchrotron Radiation, 22*, 485–491.
3. Emma, P., Akre, R., Arthur, J., Bionta, R., Bostedt, C., Bozek, J., et al. (2010). First lasing and operation of an ångstrom-wavelength free-electron laser. *Nature Photonics, 4*(9), 641–647.
4. Ishikawa, T., Aoyagi, H., Asaka, T., Asano, Y., Azumi, N., Bizen, T., et al. (2012). A compact X-ray free-electron laser emitting in the sub-ångström region. *Nature Photonics, 6*(8), 540–544.
5. Altarelli, M. (2011). The European X-ray free-electron laser facility in Hamburg. *Nuclear Instruments and Methods in Physics Research B, 269*, 2845.
6. Fenalti, G., Zatsepin, N. A., Betti, C., Giguere, P., Han, G. W., Ishchenko, A., et al. (2015). Structural basis for bifunctional peptide recognition at human δ-opioid receptor. *Nature Structural & Molecular Biology, 22*(3), 265–268.
7. Kang, Y., Zhou, X. E., Gao, X., He, Y., Liu, W., Ishchenko, A., et al. (2015). Crystal structure of rhodopsin bound to arrestin by femtosecond X-ray laser. *Nature, 523*(7562), 561–567.
8. Liu, W., Wacker, D., Gati, C., Han, G. W., James, D., Wang, D., et al. (2013). Serial femtosecond crystallography of G protein-coupled receptors. *Science, 342*(6165), 1521–1524.
9. Aquila, A., Hunter, M. S., Doak, R. B., Kirian, R. A., Fromme, P., White, T. A., et al. (2012). Time-resolved protein nanocrystallography using an X-ray free-electron laser. *Optics Express, 20*(3), 2706–2716.
10. Barends, T. R. M., Foucar, L., Ardevol, A., Nass, K., Aquila, A., Botha, S., et al. (2015). Direct observation of ultrafast collective motions in CO myoglobin upon ligand dissociation. *Science, 350*(6259), 445–450.
11. Kupitz, C., Basu, S., Grotjohann, I., Fromme, R., Zatsepin, N. A., Rendek, K. N., et al. (2014). Serial time-resolved crystallography of photosystem II using a femtosecond X-ray laser. *Nature, 513*(7517), 261–265.
12. Neutze, R. (2014). Opportunities and challenges for time-resolved studies of protein structural dynamics at X-ray free-electron lasers. *Philosophical Transactions of the Royal Society, B: Biological Sciences, 369*(1647), 20130318.
13. Pande, K., Hutchison, C. D. M., Groenhof, G., Aquila, A., Robinson, J. S., Tenboer, J., et al. (2016). Femtosecond structural dynamics drives the trans/cis isomerization in photoactive yellow protein. *Science, 352*(6286), 725–729.
14. Spence, J. (2008). X-ray imaging - ultrafast diffract-and-destroy movies. *Nature Photonics, 2*(7), 390–391.
15. Kupitz, C., Olmos, J. L., Holl, M., Tremblay, L., Pande, K., Pandey, S., et al. (2017). Structural enzymology using X-ray free electron lasers. *Structural Dynamics, 4*(4), 044003.
16. Stagno, J. R., Liu, Y., Bhandari, Y. R., Conrad, C. E., Panja, S., Swain, M., et al. (2017). Structures of riboswitch RNA reaction states by mix-and-inject XFEL serial crystallography. *Nature, 541*(7636), 242–246.
17. Arnlund, D., Johansson, L. C., Wickstrand, C., Barty, A., Williams, G. J., Malmerberg, E., et al. (2014). Visualizing a protein quake with time-resolved X-ray scattering at a free-electron laser. *Nature Methods, 11*(9), 923–926.
18. Cammarata, M., Levantino, M., Schotte, F., Anfinrud, P. A., Ewald, F., Choi, J., et al. (2008). Tracking the structural dynamics of proteins in solution using time-resolved wide-angle X-ray scattering. *Nature Methods, 5*(10), 881–886.
19. Alonso-Mori, R., Kern, J., Gildea, R. J., Sokaras, D., Weng, T. C., Lassalle-Kaiser, B., et al. (2012). Energy-dispersive X-ray emission spectroscopy using an X-ray free-electron laser in a shot-by-shot mode. *Proceedings of the National Academy of Sciences, 109*(47), 19103–19107.
20. Kern, J., Alonso-Mori, R., Tran, R., Hattne, J., Gildea, R. J., Echols, N., et al. (2013). Simultaneous femtosecond X-ray spectroscopy and diffraction of photosystem II at room temperature. *Science, 340*(6131), 491–495.

21. Ayyer, K., Geloni, G., Kocharyan, V., Saldin, E., Serkez, S., Yefanov, O., et al. (2015). Perspectives for imaging single protein molecules with the present design of the European XFEL. *Structural Dynamics, 2*(4), 041702–041711.
22. Neutze, R., Wouts, R., van der Spoel, D., Weckert, E., & Hajdu, J. (2000). Potential for biomolecular imaging with femtosecond X-ray pulses. *Nature, 406*(6797), 752–757.
23. SOLEM, J. C. (1986). Imaging biological specimens with high-intensity soft X-rays. *Journal of the Optical Society of America B: Optical Physics, 3*(11), 1551–1565.
24. Chapman, H. N., Barty, A., Bogan, M. J., Boutet, S., Frank, M., Hau-Riege, S. P., et al. (2006). Femtosecond diffractive imaging with a soft-X-ray free-electron laser. *Nature Physics, 2*(12), 839–843.
25. Chapman, H. N., Fromme, P., Barty, A., White, T. A., Kirian, R. A., Aquila, A., et al. (2011). Femtosecond X-ray protein nanocrystallography. *Nature, 470*(7332), 73–77.
26. Boutet, S., Lomb, L., Williams, G. J., Barends, T. R. M., Aquila, A., Doak, R. B., et al. (2012). High-resolution protein structure determination by serial femtosecond crystallography. *Science, 337*(6092), 362–364.
27. Galayda, J. N. (2003). LCLS the first experiments. SLAC-R-611. http://www-public.slac.stanford.edu/scidoc/index.asp
28. Chao, W., Harteneck, B. D., Liddle, J. A., Anderson, E. H., & Attwood, D. T. (2005). Soft X-ray microscopy at a spatial resolution better than 15 nm. *Nature, 435*(7046), 1210–1213.
29. Marchesini, S., He, H., Chapman, H. N., Hau-Riege, S. P., Noy, A., Howells, M. R., et al. (2003). X-ray image reconstruction from a diffraction pattern alone. *Physical Review B, 68*(14), 140101.
30. Miao, J., Charalambous, P., Kirz, J., & Sayre, D. (1999). Extending the methodology of X-ray crystallography to allow imaging of micrometre-sized non-crystalline specimens. *Nature, 400*(6742), 342–344.
31. Barty, A., Marchesini, S., Chapman, H. N., Cui, C., Howells, M. R., Shapiro, D. A., et al. (2008). Three-dimensional coherent X-ray diffraction imaging of a ceramic nanofoam: Determination of structural deformation mechanisms. *Physical Review Letters, 101*(5), 055501.
32. Chapman, H. N., Barty, A., Marchesini, S., Noy, A., Hau-Riege, S. P., Cui, C., et al. (2006). High-resolution ab initio three-dimensional x-ray diffraction microscopy. *Journal of the Optical Society of America. A, 23*(5), 1179–1200.
33. Boutet, S., Bogan, M. J., Barty, A., Frank, M., Benner, W. H., Marchesini, S., et al. (2008). Ultrafast soft X-ray scattering and reference-enhanced diffractive imaging of weakly scattering nanoparticles. *Journal of Electron Spectroscopy and Related Phenomena, 166-167*, 65–73.
34. Seibert, M. M., Boutet, S., Svenda, M., Ekeberg, T., Maia, F. R. N. C., Bogan, M. J., et al. (2010). Femtosecond diffractive imaging of biological cells. *Journal of Physics B, 43*(19), 194015.
35. Seibert, M. M., Ekeberg, T., Maia, F. R. N. C., Svenda, M., Andreasson, J., Jonsson, O., et al. (2011). Single mimivirus particles intercepted and imaged with an X-ray laser. *Nature, 470*(7332), 78–81.
36. Ekeberg, T., Svenda, M., Abergel, C., Maia, F. R. N. C., Seltzer, V., Claverie, J.-M., et al. (2015). Three-dimensional reconstruction of the giant mimivirus particle with an x-ray free-electron laser. *Physical Review Letters, 114*(9), 098102.
37. Hantke, M. F., Hasse, D., Maia, F. R. N. C., Ekeberg, T., John, K., Svenda, M., et al. (2014). High-throughput imaging of heterogeneous cell organelles with an X-ray laser. *Nature Photonics, 8*(12), 943–949.
38. Reddy, H. K. N., Yoon, C. H., Aquila, A., Awel, S., Ayyer, K., Barty, A., et al. (2017). Coherent soft X-ray diffraction imaging of coliphage PR772 at the Linac Coherent Light Source. *Scientific Data, 4*, 170079.
39. Kimura, T., Joti, Y., Shibuya, A., Song, C., Kim, S., Tono, K., et al. (2013). Imaging live cell in micro-liquid enclosure by X-ray laser diffraction. *Nature Communications, 5*, 1–7.

40. van der Schot, G., Svenda, M., Maia, F. R. N. C., Hantke, M., DePonte, D. P., Seibert, M. M., et al. (2015). Imaging single cells in a beam of live cyanobacteria with an X-ray laser. *Nature Communications, 6*, 5704.
41. Loh, N. D., Hampton, C. Y., Martin, A. V., Starodub, D., Sierra, R. G., Barty, A., et al. (2012). Fractal morphology, imaging and mass spectrometry of single aerosol particles in flight. *Nature, 486*(7404), 513–517.
42. Aquila, A., Barty, A., Bostedt, C., Boutet, S., Carini, G., de Ponte, D., et al. (2015). The Linac Coherent Light Source single particle imaging road map. *Structural Dynamics, 2*(4), 041701–041713.
43. Daurer, B. J., Okamoto, K., Bielecki, J., Maia, F. R. N. C., Mühlig, K., Seibert, M. M., et al. (2017). Experimental strategies for imaging bioparticles with femtosecond hard X-ray pulses. *IUCrJ, 4*(3), 251–262.
44. Munke, A., Andreasson, J., Aquila, A., Awel, S., Ayyer, K., Barty, A., et al. (2016). Coherent diffraction of single Rice Dwarf virus particles using hard X-rays at the Linac Coherent Light Source. *Scientific Data, 3*, 160064.
45. Strüder, L., Epp, S., Rolles, D., Hartmann, R., Holl, P., Lutz, G., et al. (2010). Large-format, high-speed, X-ray pnCCDs combined with electron and ion imaging spectrometers in a multipurpose chamber for experiments at 4th generation light sources. *Nuclear Instruments and Methods in Physics Research Section A, 614*, 483–496.
46. Liang, M., Williams, G. J., Messerschmidt, M., Seibert, M. M., Montanez, P. A., Hayes, M., et al. (2015). The coherent X-ray imaging instrument at the Linac Coherent Light Source. *Journal of Synchrotron Radiation, 22*, 514–519.
47. Young, L., Kanter, E. P., Krässig, B., Li, Y., March, A. M., Pratt, S. T., et al. (2010). Femtosecond electronic response of atoms to ultra-intense X-rays. *Nature, 466*(7302), 56–61.
48. Nagler, B., Aquila, A., Boutet, S., Galtier, E. C., Hashim, A., Hunter, M. S., et al. (2017). Focal spot and wavefront Sensing of an X-ray free electron laser using Ronchi shearing interferometry. *Scientific Reports, 7*(1), 13698.
49. Barty, A., Soufli, R., McCarville, T., Baker, S. L., Pivovaroff, M. J., Stefan, P., et al. (2009). Predicting the coherent X-ray wavefront focal properties at the Linac Coherent Light Source (LCLS) X-ray free electron laser. *Optics Express, 17*(18), 15508–15519.
50. Bean, R. J., Aquila, A., Samoylova, L., & Mancuso, A. P. (2016). Design of the mirror optical systems for coherent diffractive imaging at the SPB/SFX instrument of the European XFEL. *Journal of Optics, 18*(7), 074011.
51. Loh, N.-T. D., & Elser, V. (2009). Reconstruction algorithm for single-particle diffraction imaging experiments. *Physical Review E, 80*(2), 026705.
52. Ayyer, K., Philipp, H. T., Tate, M. W., Wierman, J. L., Elser, V., & Gruner, S. M. (2015). Determination of crystallographic intensities from sparse data. *IUCrJ, 2*(1), 29–34.
53. Giannakis, D., Schwander, P., & Ourmazd, A. (2012). The symmetries of image formation by scattering. I. Theoretical framework. *Optics Express, 20*(12), 12799–12826.
54. Tegze, M., & Bortel, G. (2012). Atomic structure of a single large biomolecule from diffraction patterns of random orientations. *Journal of Structural Biology, 179*, 41–45.
55. Donatelli, J. J., Zwart, P. H., & Sethian, J. A. (2015). Iterative phasing for fluctuation X-ray scattering. *Proceedings of the National Academy of Sciences of the United States of America, 112*(33), 10286–10291.
56. Kirian, R. A., Schmidt, K. E., Wang, X., Doak, R. B., & Spence, J. C. H. (2011). Signal, noise, and resolution in correlated fluctuations from snapshot small-angle x-ray scattering. *Physical Review E, 84*(1), 011921.
57. Saldin, D. K., Poon, H. C., Shneerson, V. L., Howells, M., Chapman, H. N., Kirian, R. A., et al. (2010). Beyond small-angle x-ray scattering: Exploiting angular correlations. *Physical Review B, 81*(17), 174105.
58. Philipp, H. T., Ayyer, K., Tate, M. W., Elser, V., & Gruner, S. M. (2012). Solving structure with sparse, randomly-oriented X-ray data. *Optics Express, 20*(12), 13129–13137.

59. Loh, N. D. (2012). Effects of extraneous noise in cryptotomography. *Proceedings of SPIE, 8500*, 85000K.
60. Andreasson, J., Martin, A. V., Liang, M., Timneanu, N., Aquila, A., Wang, F., et al. (2014). Automated identification and classification of single particle serial femtosecond X-ray diffraction data. *Optics Express, 22*(3), 2497–2510.
61. Denes, P. (2014). Two-dimensional imaging detectors for structural biology with X-ray lasers. *Philosophical Transactions of the Royal Society, B: Biological Sciences, 369*(1647), 20130334.
62. Hatsui, T., & Graafsma, H. (2015). X-ray imaging detectors for synchrotron and XFEL sources. *IUCrJ, M2*, 371–383.
63. Becker, J., Greiffenberg, D., Trunk, U., Shi, X., Dinapoli, R., Mozzanica, A., et al. (2012). The single photon sensitivity of the adaptive gain integrating pixel detector. *Nuclear Instruments and Methods in Physics Research A, 694*, 82–90.
64. Wunderer, C. B., Marras, A., Bayer, M., Correa, J., Lange, S., Shevyakov, I., et al. (2014). Percival: An international collaboration to develop a MAPS-based soft X-ray imager. *Synchrotron Radiation News, 27*(4), 30–34.
65. Blaj. G., Caragiulo, P., Carini, G., Dragone, A., Haller, G. (2015). Design and performance of the ePix camera systems.
66. Carini, G. A., Alonso-Mori, R., Blaj, G. (2016). ePix100 camera: Use and applications at LCLS. Available from https://www.researchgate.net/profile/Angelo_Dragone/publication/305685847_ePix100_camera_Use_and_applications_at_LCLS/links/57c6e09908aec24de042a16a.pdf
67. Mozzanica, A., Bergamaschi, A., Cartier, S., Dinapoli, R., Greiffenberg, D., Johnson, I., et al. (2014). Prototype characterization of the JUNGFRAU pixel detector for SwissFEL. *Journal of Instrumentation, 9*(5), C05010.
68. Bogan, M. J., Boutet, S., Chapman, H. N., Marchesini, S., Barty, A., Benner, W. H., et al. (2010). Aerosol imaging with a soft X-ray free electron laser. *Aerosol Science and Technology, 44*(3), i–vi.
69. Frank, M., Frank, M., Carlson, D. B., Carlson, D. B., Hunter, M. S., Hunter, M. S., et al. (2014). Femtosecond X-ray diffraction from two-dimensional protein crystals. *IUCrJ, 1*(Pt 2), 95–100.
70. Pedrini, B., Tsai, C.-J., Capitani, G., Padeste, C., Hunter, M. S., Zatsepin, N. A., et al. (2014). 7 Å resolution in protein two-dimensional-crystal X-ray diffraction at Linac Coherent Light Source. *Philosophical Transactions of the Royal Society, B: Biological Sciences, 369*(1647), 20130500.
71. Yuk, J. M., Park, J., Ercius, P., Kim, K., Hellebusch, D. J., Crommie, M. F., et al. (2012). High-resolution EM of colloidal nanocrystal growth using graphene liquid cells. *Science, 336*(6077), 61–64.
72. Seuring, C., Ayyer, K., Filippaki, E., Barthelmess, M., Longchamp, J.-N., Ringler, P., et al. (2018). Femtosecond X-ray coherent diffraction of aligned amyloid fibrils on low background graphene. *Nature Communications, 9*, 1–10. https://doi.org/10.1038/s41467-018-04116-9.
73. Hart, P., Boutet, S., Carini, G., Dubrovin, M., Duda, B., Fritz, D., et al. (2012). The CSPAD megapixel x-ray camera at LCLS. *Proceedings of SPIE, 8504*, 85040C.
74. Ferguson, K. R., Bucher, M., Bozek, J. D., Carron, S., Castagna, J. C., Coffee, R., et al. (2015). The atomic, molecular and optical science instrument at the Linac Coherent Light Source. *Journal of Synchrotron Radiation, 22*, 492–497.
75. Hosseinizadeh, A., Mashayekhi, G., Copperman, J., Schwander, P., Dashti, A., Sepehr, R., et al. (2017). Conformational landscape of a virus by single-particle X-ray scattering. *Nature Methods, 5*(9), 4061–4881.
76. Kurta, R. P., Donatelli, J. J., Yoon, C. H., Berntsen, P., Bielecki, J., Daurer, B. J., et al. (2017). Correlations in scattered X-ray laser pulses reveal nanoscale structural features of viruses. *Physical Review Letters, 119*(15), 158102.
77. Maia, F. R. N. C. (2012). The coherent X-ray imaging data bank. *Nature Methods, 9*(9), 854–855.

78. Miao, J., & Sayre, D. (2000). On possible extensions of X-ray crystallography through diffraction-pattern oversampling. *Acta Crystallographica. Section A, 56*, 596–605.
79. Loh, N. D. (2014). A minimal view of single-particle imaging with X-ray lasers. *Philosophical Transactions of the Royal Society, B: Biological Sciences, 369*(1647), 20130328–20130328.
80. Bobkov, S. A., Teslyuk, A. B., Kurta, R. P., Gorobtsov, O. Y., Yefanov, O. M., Ilyin, V. A., et al. (2015). Sorting algorithms for single-particle imaging experiments at X-ray free-electron lasers. *Journal of Synchrotron Radiation, 22*(6), 1345–1352.
81. Hosseinizadeh, A., Schwander, P., Dashti, A., Fung, R., D'Souza, R. M., & Ourmazd, A. (2014). High-resolution structure of viruses from random diffraction snapshots. *Philosophical Transactions of the Royal Society, B: Biological Sciences, 369*(1647), 20130326–20130326.
82. Yoon, C. H., Schwander, P., Abergel, C., Andersson, I., Andreasson, J., Aquila, A., et al. (2011). Unsupervised classification of single-particle X-ray diffraction snapshots by spectral clustering. *Optics Express, 19*(17), 16542–16549.
83. Park, H. J., Loh, N. D., Sierra, R. G., Hampton, C. Y., Starodub, D., Martin, A. V., et al. (2013). Toward unsupervised single-shot diffractive imaging of heterogeneous particles using X-ray free-electron lasers. *Optics Express, 21*(23), 28729–28742.
84. Shneerson, V. L., Ourmazd, A., & Saldin, D. K. (2008). Crystallography without crystals. I. The common-line method for assembling a three-dimensional diffraction volume from single-particle scattering. *Acta Crystallographica, Section A: Foundations of Crystallography, 64*(2), 303–315.
85. Ayyer, K., Lan, T. Y., Elser, V., & Loh, N. D. (2016). Dragonfly: An implementation of the expand-maximize-compress algorithm for single-particle imaging. *Journal of Applied Crystallography, 49*, 1320–1335.
86. Starodub, D., Aquila, A., Bajt, S., Barthelmess, M., Barty, A., Bostedt, C., et al. (2012). Single-particle structure determination by correlations of snapshot X-ray diffraction patterns. *Nature Communications, 3*, 1276.
87. Barty, A., Küpper, J., & Chapman, H. N. (2013). Molecular imaging using X-ray free-electron lasers. *Annual Review of Physical Chemistry, 64*, 415–435.
88. Chapman, H. N., & Nugent, K. A. (2010). Coherent lensless X-ray imaging. *Nature Photonics, 4*(12), 833–839.
89. Marchesini, S. (2007). Phase retrieval and saddle-point optimization. *Journal of the Optical Society of America. A, 24*(10), 3289–3296.
90. Shechtman, Y., Eldar, Y. C., Cohen, O., & Segev, M. (2013). Efficient coherent diffractive imaging for sparsely varying objects. *Optics Express, 21*(5), 6327–6338.
91. Jiang, H., Song, C., Chen, C.-C., Xu, R., Raines, K. S., Fahimian, B. P., et al. (2010). Quantitative 3D imaging of whole, unstained cells by using X-ray diffraction microscopy. *Proceedings of the National Academy of Sciences, 107*(25), 11234–11239.
92. Robinson, I. (2008). Coherent diffraction - giant molecules or tiny crystals? *Nature Materials, 7*(4), 275–276.
93. Martin, A. V., Loh, N. D., Hampton, C. Y., Sierra, R. G., Wang, F., Aquila, A., et al. (2012). Femtosecond dark-field imaging with an X-ray free electron laser. *Optics Express, 20*(12), 13501–13512.
94. Donatelli, J. J., Sethian, J. A., & Zwart, P. H. (2017). Reconstruction from limited single-particle diffraction data via simultaneous determination of state, orientation, intensity, and phase. *Proceedings of the National Academy of Sciences, 114*(28), 7222–7227.
95. Kierspel, T., Wiese, J., Mullins, T., Robinson, J., Aquila, A., Barty, A., et al. (2015). Strongly aligned gas-phase molecules at free-electron lasers. *Journal of Physics B: Atomic, Molecular and Optical Physics, 48*(20), 1–7.
96. Starodub, D., Doak, R. B., Schmidt, K., Weierstall, U., Wu, J. S., Spence, J. C. H., et al. (2005). Damped and thermal motion of laser-aligned hydrated macromolecule beams for diffraction. *The Journal of Chemical Physics, 123*(24), 244304.

97. Stern, S., Holmegaard, L., Filsinger, F., Rouzee, A., Rudenko, A., Johnsson, P., et al. (2014). Toward atomic resolution diffractive imaging of isolated molecules with X-ray free-electron lasers. *The Royal Society of Chemistry, 171*, 393–418.
98. Pedrini, B., Menzel, A., Guizar-Sicairos, M., Guzenko, V. A., Gorelick, S., David, C., et al. (2013). Two-dimensional structure from random multiparticle X-ray scattering images using cross-correlations. *Nature Communications, 4*, 1647.
99. Gipson, B., Masiel, D., Browning, N., Spence, J., Mitsuoka, K., & Stahlberg, H. (2011). Automatic recovery of missing amplitudes and phases in tilt-limited electron crystallography of two-dimensional crystals. *Physical Review E, 84*(1), 011916.
100. Berman, H. M., Westbrook, J., Feng, Z., Gilliland, G., Bhat, T. N., Weissig, H., et al. (2000). The protein data bank. *Nucleic Acids Research, 28*(1), 235–242.
101. Sayre, D. (1952). Some implications of a theorem due to Shannon. *Acta Crystallographica, 5*(6), 843–843.
102. Kirian, R. A., Bean, R. J., Beyerlein, K. R., Yefanov, O. M., White, T. A., Barty, A., et al. (2014). Phasing coherently illuminated nanocrystals bounded by partial unit cells. *Philosophical Transactions of the Royal Society, B: Biological Sciences, 369*(1647), 20130331–20130331.
103. Spence, J. C. H., Kirian, R. A., Wang, X., Weierstall, U., Schmidt, K. E., White, T., et al. (2011). Phasing of coherent femtosecond X-ray diffraction from size-varying nanocrystals. *Optics Express, 19*(4), 2866–2873.
104. Kirian, R. A., Bean, R. J., Beyerlein, K. R., Barthelmess, M., Yoon, C. H., Wang, F., et al. (2015). Direct phasing of finite crystals illuminated with a free-electron laser. *Physical Review X, 5*(1), 011015.
105. Ayyer, K., Yefanov, O. M., Oberthür, D., Roy-Chowdhury, S., Galli, L., Mariani, V., et al. (2016). Macromolecular diffractive imaging using imperfect crystals. *Nature, 530*(7589), 202–206.
106. Eisebitt, S., Lüning, J., Schlotter, W. F., Lorgen, M., Hellwig, O., Eberhardt, W., et al. (2004). Lensless imaging of magnetic nanostructures by X-ray spectro-holography. *Nature, 432*(7019), 885–888.
107. McNulty, I., Kirz, J., Jacobsen, C., Anderson, E. H., Howells, M. R., & KERN, D. P. (1992). High-resolution imaging by Fourier-transform X-ray holography. *Science, 256*(5059), 1009–1012.
108. Solem, J. C., & Baldwin, G. C. (1982). Micro-holography of living organisms. *Science, 218*(4569), 229–235.
109. Wu, B., Wang, T., Graves, C. E., Zhu, D., Schlotter, W. F., Turner, J. J., et al. (2016). Elimination of X-ray diffraction through stimulated X-ray transmission. *Physical Review Letters, 117*(2), 027401.
110. Schlotter, W. F., Rick, R., Chen, K., Scherz, A., Stöhr, J., Lüning, J., et al. (2006). Multiple reference Fourier transform holography with soft x rays. *Applied Physics Letters, 89*(16), 163112.
111. Marchesini, S., Boutet, S., Sakdinawat, A. E., Bogan, M. J., Bajt, S., Barty, A., et al. (2008). Massively parallel X-ray holography. *Nature Photonics, 2*(9), 560–563.
112. Martin, A. V. (2014). The correlation of single-particle diffraction patterns as a continuous function of particle orientation. *Philosophical Transactions of the Royal Society, B: Biological Sciences, 369*(1647), 20130329.
113. Gorkhover, T., Ulmer, A., & Ferguson, K. (2017). Femtosecond X-ray Fourier holography imaging of free-flying nanoparticles. *Nature Photonics, 12*, 150–153.
114. Bai, X.-C., McMullan, G., & Scheres, S. H. W. (2015). How cryo-EM is revolutionizing structural biology. *Trends in Biochemical Sciences, 40*(1), 49–57.
115. Cheng, Y. (2015). Single-particle cryo-EM at crystallographic resolution. *Cell, 161*(3), 450–457.
116. Henderson, R. (2015). Overview and future of single particle electron cryomicroscopy. *Archives of Biochemistry and Biophysics, 581*, 19–24.

117. Danev, R., & Baumeister, W. (2016). Cryo-EM single particle analysis with the Volta phase plate. *eLife, 5*, e13046.
118. Glaeser, R. M. (2013). Invited Review Article: Methods for imaging weak-phase objects in electron microscopy. *The Review of Scientific Instruments, 84*(11), 111101–111117.
119. Garman, E. F. (2010). Radiation damage in macromolecular crystallography: What is it and why should we care? *Acta Crystallographica. Section D, Biological Crystallography, 66*(4), 339–351.
120. Caleman, C., Ortiz, C., Marklund, E., Bultmark, F., Gabrysch, M., Parak, F. G., et al. (2009). Radiation damage in biological material: Electronic properties and electron impact ionization in urea. *Europhysics Letters, 85*(1), 18005.
121. Nagler, B., Zastrau, U., Fäustlin, R. R., Vinko, S. M., Whitcher, T., Nelson, A. J., et al. (2009). Turning solid aluminium transparent by intense soft X-ray photoionization. *Nature Physics, 5*(8), 1–4.
122. Armstrong, M. R., Boyden, K., Browning, N. D., Campbell, G. H., Colvin, J. D., DeHope, W. J., et al. (2007). Practical considerations for high spatial and temporal resolution dynamic transmission electron microscopy. *Ultramicroscopy, 107*(4-5), 356–367.
123. LaGrange, T., Armstrong, M. R., Boyden, K., Brown, C. G., Campbell, G. H., Colvin, J. D., et al. (2006). Single-shot dynamic transmission electron microscopy. *Applied Physics Letters, 89*(4), 044105.
124. Altarelli, M., & Mancuso, A. P. (2014). Structural biology at the European X-ray free-electron laser facility. *Philosophical Transactions of the Royal Society, B: Biological Sciences, 369*(1647), 20130311–20130311.

Chapter 15
The Use of Angular Correlations to Extract Three-Dimensional Structural Information from X-Ray Solution Scattering

Sebastian Doniach

15.1 Introduction

In this chapter we focus on scattering from non-crystalline solutions of molecules or nanoparticles in which the scattering objects are rotationally disordered.

This field of solution scattering dates back to the classic work of Guinier [10] and Kratky [15] in the 1930s and 1940s. At that time it was shown that the radius of gyration of the macromolecule could be extracted from the small angle scattering profile as a model-free parameter (see Doniach 2001 [8] for details).

Following the earlier work by Svergun and Stuhrmann [22], a significant development in the field was made by P. Chacon et al. in 1998 [5] who were working in electron microscopy at the time. They utilized a basic principle of feature extraction from complex data whereby the application of prior knowledge, in the form of constraints, to the fitting of simple models to data (in this case scattering data) can help the ability to relate the scattering data to molecular models of the molecules being examined.

As shown by Chacon et al., models of three-dimensional molecular shapes at nanometer-scale resolution can be extracted from one-dimensional $I(q)$ Small and Wide Angle X-ray Scattering (SAXS/WAXS) data by imposing suitable constraints of continuity and positivity of sample density.

This approach was further developed by Svergun [21] and by Walther et al. [23]. As discussed below, the features extracted from scattering data are represented by the pair distribution functions of atoms in the molecules being examined. These

S. Doniach (✉)
Departments of Applied Physics, Physics, and Photon Science, Stanford University, Stanford, CA, USA

SLAC National Accelerator Center, Menlo Park, CA, USA
e-mail: sxdwc@slac.stanford.edu.

S. Boutet et al. (eds.), *X-ray Free Electron Lasers*,
https://doi.org/10.1007/978-3-030-00551-1_15

one-dimensional pair distribution functions coming from the 2-point correlation functions represented by the data can then provide low resolution three-dimensional shape models when suitable constraints are applied.

The advent of X-ray FELs has now made possible the extraction of 4-point correlation functions, from scattering data on solutions of rotationally disordered molecules, yielding data in a three-dimensional representation. In this chapter we focus on the experimental problems to be dealt with in the extraction of three-dimensional models from the 4-point correlation functions obtained from X-ray FEL scattering data. This three-dimensional representation is obtained by computing angular correlations of the deviations away from the angular means $I(q)$.

15.2 Extraction of 3d Molecular Shapes from 1d SAXS/WAXS Solution Scattering Data

S/WAXS measurements from a mono-disperse solution of macromolecules result in one-dimensional data sets which may be related to the three-dimensional structure of the scattering molecules via

$$I(q) \propto \int d\phi \; d\cos\theta \; |F(q\cos\theta, \phi)|^2 \tag{15.1}$$

where F is the scattering form factor of the molecule

$$F(q\cos\theta, \phi) = \sum_i f_i(q) exp(i\mathbf{q}\cdot\mathbf{r}_i) \tag{15.2}$$

with atomic positions $\mathbf{r}_i$ and atomic scattering factors $f_i(q)$. Averaging over orientations gives the Debye formula [8]

$$\begin{aligned} I(q) &= \left\langle \int d\phi \; d\cos\theta \; |F|^2 \right\rangle_{\text{orientations}} \\ &= \sum_i |f_i|^2 + \sum_{i<j} f_i f_j \frac{\sin qr_{ij}}{qr_{ij}} \end{aligned} \tag{15.3}$$

Equation (15.3) can be written in terms of the pair distribution function within the molecule,

$$g(r) = \sum_{ij} f_i f_j \delta(r - r_{ij}) \tag{15.4}$$

as

$$I(q) = \sum_i |f_i|^2 + \int dr g(r) \frac{\sin(qr)}{qr} \tag{15.5}$$

Although one-dimensional, this data, supplemented by prior knowledge about molecules occupying a finite volume (support), contains significant information about the three-dimensional structure of the scattering molecule.

Application of positivity and smoothness constraints beyond the Guinier small angle region can extract the 2-point correlation function shown in Eq. (15.5), giving the atomic pair distribution in the molecule [22] (see Doniach [8] for details).

15.3 Angular Correlations of Solution X-Ray Scattering Lead to Three-Dimensional Data

Kam's 1977 paper [12] showed how the measurement of azimuthal angular correlations in the scattering data yields three-dimensional information about the structure of the scattering molecules. At that time, the use of Dynamic Light Scattering (DLS) to obtain size information for mixtures of molecules had recently been developed. DLS involves the correlations of fluctuations in the scattered light intensity as a function of time. When Kam extended the application of these ideas to X-ray scattering, he referred to the resulting measurement of correlations as "fluctuation" scattering although the time-dependent correlation aspect was replaced by correlations of the angular dependence of the scattering.

The correlator for the scattering as defined by Kam [12] is

$$C(q_1, q_2, \psi) \equiv \int d\phi \, \delta I(q_1, \phi) \times \delta I(q_2, \phi + \psi) \tag{15.6}$$

Here, $\delta I = I(\mathbf{q}, \phi) - I(q)$ represents the deviation of $I(\mathbf{q}, \phi)$ from the azimuthal mean, $I(q)$, and ψ is the angle between vectors $\mathbf{q}_1$ and $\mathbf{q}_2$. As may be seen by comparison with Eq. (15.1), this involves the product of two form factors, hence is a measurement of a 4-point correlation function. This then needs to be averaged over orientations of the scattering molecules.

Since this is a 4-point correlation function, the rotational averaging involves 2 pairs of atoms $\mathbf{r}_{ij}$, and $\mathbf{r}_{kl}$. In order for the scattering to be correlated, $\mathbf{r}_{ij,kl}$ must form a tetrahedron of atoms in the *same* individual molecule in the irradiated sample. The correlation requirement therefore leads to two kinds of rotational averaging: the correlator, Eq. (15.4) averaged over the set of random rotations $\mathcal{R}_\omega$ applied to the *same* molecule containing both $\mathbf{r}_{ij}$ and $\mathbf{r}_{kl}$; and a second set of rotations, averaging over *independent* rotationally random ensembles for each of the molecules containing $\mathbf{r}_{ij}$ and $\mathbf{r}_{kl}$.

The rotationally averaged correlator is thus decomposed into two terms:

$$C^{\mathcal{C}}(q_1, q_2, \psi) = \sum_{\omega_i} \int d\phi \, \delta I^{\mathcal{R}_i}(q_1, \phi) \times \delta I^{\mathcal{R}_i}(q_2, \phi + \psi) \tag{15.7}$$

and

$$C^{\mathscr{U}} = \sum_{\omega_{i \neq j}} \int d\phi \, \delta I^{\mathscr{R}_i}(q_1, \phi) \times \delta I^{\mathscr{R}_j}(q_2, \phi + \psi) \tag{15.8}$$

Here, $\delta I^{\mathscr{R}_i}$ denotes the value of $\delta I(q, \phi)$ after rotation by rotation matrix $\mathscr{R}_{\omega_i}$.

15.4 Extraction of Structural Features by Averaging Many X-Ray FEL Shots

Since the correlator $C^{\mathscr{U}}$ involves independent sets of random rotations for each of the molecules contributing to $\mathbf{r}_{ij}$ and to $\mathbf{r}_{kl}$, the product of the deviations from the mean will contain as many positive as negative terms. Hence, the mean of $C^{\mathscr{U}}$ is expected to $\Rightarrow 0$ when summed over many different ensembles of randomly oriented molecules.

When collecting data using an X-ray FEL, samples of the solution are typically illuminated by X-ray beam at rates varying from 30–120 Hz to date, with potential for faster rates at the recently operational European XFEL and the future LCLS-II. Thus, removal of the $C^{\mathscr{U}}$ term, which involves averaging over a sufficient number of shots to lead to adequate convergence, can take place over a suitable length of beam time due to the high repetition rate.

The signal/noise for X-ray FEL measurements of randomly oriented ensembles of molecules has been studied in detail by Kirian et al. [14]. Here we use their results to give a qualitative summary.

In the high X-ray fluence limit in which many photons per shot scatter off a given molecule, we can treat the statistics of photons falling on a given pixel in the Gaussian limit. The probability distribution for the product of a pair of Gaussian ransodm variables has been studied rigorously [3] and is non-Gaussian. Here we simplify the discussion by approximating this distribution by the product of two Gaussians. In this approximation the magnitude of C for a given pair of pixels at $\mathbf{q}_1$ and $\mathbf{q}_2$ then scales as the sum of a set of products of two independent Gaussian variables. The mean, μ and standard deviation, σ of a product, Y, of two Gaussian variables, X, with same means and σ's, scales as μ^2 with standard error scaling as σ^2.

An atomic scale estimate of the autocorrelator will be measured at large angles $q \sim 2\pi/a$, where a is an average interatomic distance. The standard deviation of the autocorrelator

$$C(q, q, \psi) \approx \sum_{\phi} \delta I(q, \phi) \times \delta I(q, \phi + \psi) \tag{15.9}$$

of a macromolecule will be summed over the number $m_\phi(q)$ of pixels which define the scattering ring q. Since the variance of a sum of independent gaussian variables scales as the sum of the individual variances, this leads to scaling of the signal/noise ratio on the estimate of C as $\pm 1/\sqrt{m(q)n(q)}$, where $n(q)$ is the number of photons falling on pixel $m(q)$.

Because of the Gaussian nature of the distribution at large flux, this is then further narrowed by $\sqrt{M}$ when averaged over M shots. This finally leads to an S/N error bar for $C^{\mathscr{C}}(q, q, \psi)$ of $\pm 1/\sqrt{Mm(q)n(q)}$.

15.5 Dealing with Artifacts

In order to observe correlated scattering from a solution of randomly ordered macromolecules, their Brownian motion needs to be eliminated. This may be achieved either by freezing in a gas stream from liquid nitrogen, in which case the sample is in a capillary, or by generating images from the 20–50 fs bursts of X-rays scattered from a jet of liquid sample injected into the X-ray FEL beam.

In a recent study, a suspension of gold nanoparticles in LCP (lipid cubic phase) a continuous jet of the sample was injected into the X-ray beam, which was running at 30 Hz [19].

An immediate observation was that the images on the detector are contaminated by parasitic X-rays, some of which arise from scattering from the detector nozzle. This parasitic scattering presents a basic problem for the measurement of angular correlations in the scattered beam since the image contamination is anisotropic around the azimuth normal to the incoming beam. This asymmetry causes large artifactual distortions of the angular correlations computed from the images, inhibiting the measurement of the correlated scattering from the molecules in the sample. There can also be artifactual anisotropies arising from the detector geometry and gain variations between the detector subunits.

A solution to this problem, originally proposed by Kam and collaborators [13], is to take the difference between pairs of images for which the artifactual anisotropy is the closest. In the case of the gold nanoparticle measurements [19], this was done by fitting a polynomial to the scattering intensity on the principal $q = [111]$-ring and using its coefficients to compare different samples. In each run of 5000 shots, pairs of images for which the polynomials were the closest were selected. For more complex samples such as biomolecules, pairing algorithms which sample over more of the detector need to be developed.

On calculation of the autocorrelations, the subtracted images were found to have the artifactual contributions considerably reduced. See Fig. 15.1 below. (For more details, see the Supplementary Information published in IUCrJ [19]).

When plotted on a log plot, the anisotropy effects on the raw correlator (Fig. 15.1 upper left panel) are found to be $\sim$2 orders of magnitude larger than the signal (Fig. 15.1 upper right panel). We fit 6th degree polynomials (dashed yellow) to the

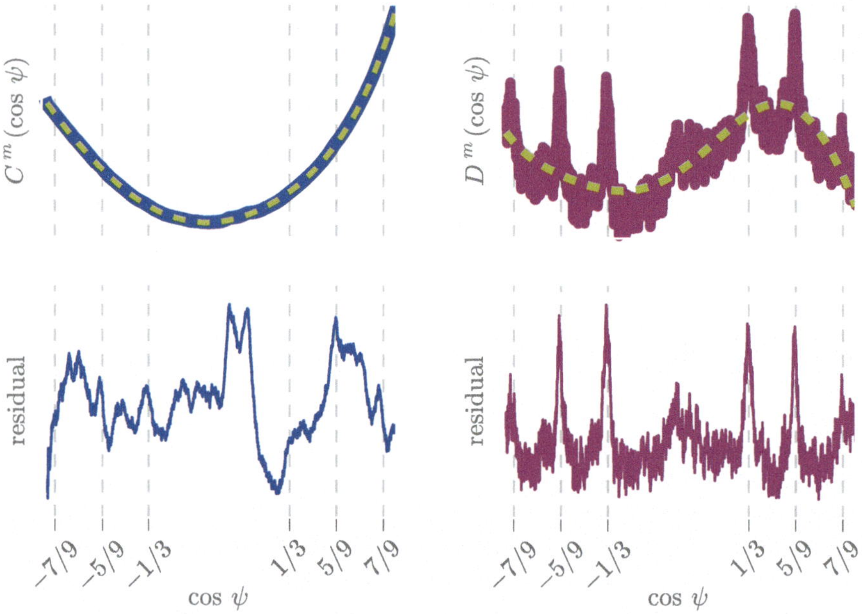

Fig. 15.1 Comparing the raw correlation of selected moderate intensities $C^m(\cos\psi)$ to the difference correlation of the moderate intensities $D^m(\cos\psi)$
Reproduced with permission of the International Union of Crystallography

paired data and subtracted them (lower panels) to emphasize that, even after pair differencing, the correlations still contain residual anisotropies which are certainly artifactual.

These data represent averages over tens of thousands of exposures. Expected CXS signals for gold nanoparticles are marked on the axis and shown with grid lines. Apparent in the figure, the difference correlation is a critical step in the analysis. Without it we would not be able to distinguish the gold nanoparticle CXS signal from the artifactual CXS signal. Low frequency variation in the difference correlation (Fig. 15.1 right panel) persists, and is probably due to extreme detector artifacts.

As an additional Filter against artifacts, we employ a Friedel symmetry constraint. Friedel's law states that $I(q) = I(-q)$ (in the absence of anomalous scattering). Hence, if one measures a physical correlation at an angle $\psi = \arccos(\mathbf{q}_1 \cdot \mathbf{q}_2/q_1q_2)$, one should measure the same correlation at an angle $\pi - \psi = \arccos(-\mathbf{q}_1 \cdot \mathbf{q}/q_1q_2)$. This implies that a pure CXS function should be mirror-symmetric about $\pi/2$ ($\cos\psi = 0$). Any signal violating this symmetry is likely to contain an artifactual component. We define the Friedel difference correlation $D_F(\cos\psi) = 1/2\{D(\cos\psi) + D(-\cos\psi)\}$ in order to enhance the true CXS information while minimizing false correlation peaks that defy Friedel symmetry.

15.6 Measurement of Molecular Details on the Atomic Scale Using Correlated Scattering

An electron density map allowing extraction of molecular details requires a range of scattering vectors up to $q \simeq 2\pi/a$, where a is on the order of interatomic distances, generally $\approx$3.5 Å. In recent publications ([18] and [19]), wide angle scattering measurements were made of suspensions of silver and gold nanoparticles.

As is well established, the internal structure of these nanoparticles forms a face centered cubic lattice [11], which may be distorted by twinning defects. Calculation of the powder diffraction pattern for X-ray scattering $I(q)$ for a model nanoparticle in the form of an fcc lattice truncated to a spherical shape gives a principal scattering peak corresponding to the Bragg vector $q(111) = 2.674\,\text{Å}^{-1}$ of the fcc lattice (lattice spacing 2.35 Å).

Raw images from both capillary scattering from Ag nanoparticles [18], and jet scattering from Au nanoparticles exhibit bright rings at $q = 2.67\,\text{Å}^{-1}$ superposed on artifactual parasitic anisotropic scattering [19].

In the recent work of Mendez et al. [19], X-ray FEL images were first categorized according to the total average exposure. Out of a total of roughly 3.8×10^5 usable exposures in runs of 5000 shots each, pairing was done on exposures that occurred during the same experimental run.

Each exposure i was paired with an exposure j according to their azimuthal anisotropies, quantified by the fitted polynomials y_i^*, y_j^*. The squared Euclidean distance $d_{ij} = \sum_\phi ||y_i^*(\phi) - y_j^*(\phi)||$ was used as a metric of comparison between two exposures (for further details, see [19]-supplementary information). Before computing angular correlations, images were sorted based on the [111] Bragg ring intensities. These were separated into two components: images with the brightest Bragg spots (Fig. 15.2a) and images with moderate intensities (Fig. 15.2b). This separation indicated the existence of two populations of nanoparticles in the samples.

After removal of the artifactual scattering by pair selection as described in Sect. 15.5 additional peaks were found in the autocorrelations around the q[111] ring for the moderate component but not for the brightest component.

After pairing, the difference of the angular correlations of the moderate intensities between pairs of images was denoted by $D^m(\cos\psi) = \delta C_i^m(\cos\psi) - \delta C_j^m(\cos\psi)$ for pairs i, j of images (see Eqs. (15.7) and (15.8)).

It was found that $D^m(\cos\psi)$ showed peaks at $\cos\psi = \pm 1/3$, $\pm 5/9$, and $\pm 7/9$ indicating the presence of twinning (Fig. 15.3c). On the other hand, the CXS of the images with bright Bragg spots, $D^b(\cos\psi)$, only showed peaks at $\cos\psi = \pm 1/3$ (Fig. 15.3d), implying that the domains which scattered the brightest Bragg spots were most likely not twinned. The absence of pronounced peaks at $\cos\psi = \pm 5/9$ and $\pm 7/9$ further indicates that this signal possibly arises from a population of non-twinned scattering domains.

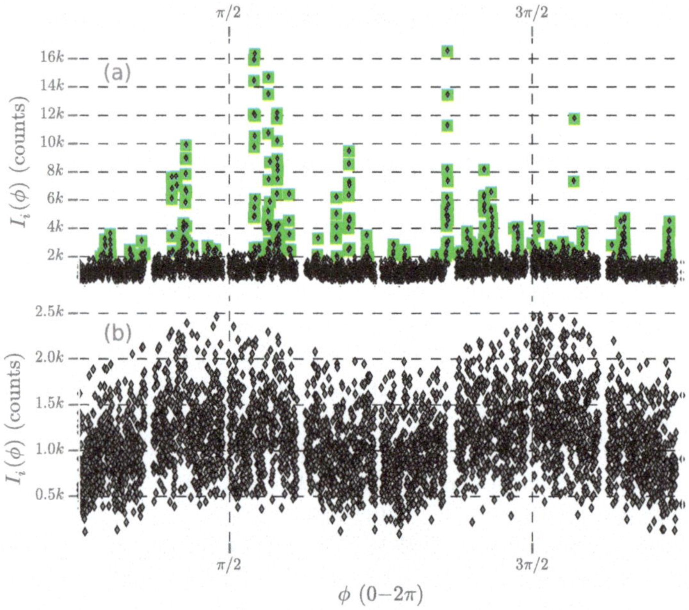

Fig. 15.2 Separation of bright Bragg spots in the angular intensity profile. (**a**) The 111 Bragg ring intensity of a single snapshot exposure i. Highlighted in green are the brightest intensities. (**b**) The same as (**a**), but the bright Bragg spots are removed, leaving behind the moderate intensity, which forms a relatively noisy signal.

Because the peak width is inversely proportional to the domain size, it was inferred that the relatively sharp width of the CXS peaks coming from the bright Bragg spots at $\cos\psi = 1/3$ indicates that the corresponding nanoparticle domains are larger than the twinned domains which produced the CXS signal shown in Fig. 15.3b.

These conclusions are emphasized by simulation of CXS from single- and multi-domain nanoparticle models shown in Fig. 15.4.

In Fig. 15.4a, the single domain cuboctahedron shows only a single peak in the correlated signal, while the correlator for the multi domain nearest-neighbor tetrahedron (NNT) model in Fig. 15.4b shows three twinning peaks on each side of $\psi = 90°$. The angular gap seen in the NNT model (Fig. 15.4b) of the dcahedron results because the tetrahedra are each close-packed fcc domains [24]. The twinning gives rise to the additional CXS peaks. The ability to distinguish these two classes of model based on angular correlations of the data demonstrates that these observations are able to distinguish model differences on an atomic length scale.

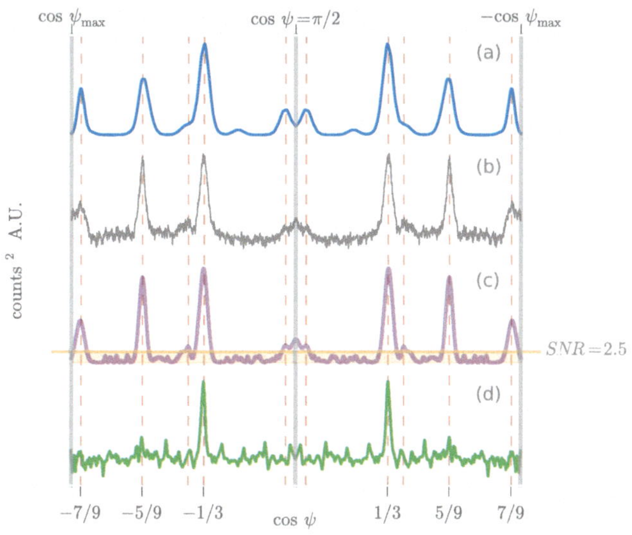

Fig. 15.3 (**a**) Simulated CXS for the gold decahedron in Fig. 15.4b The horizontal line marks an SNR of 2.5; (**d**) The mirror-symmetric difference correlation of the bright Bragg intensities, $D^b(\psi)$. The absence of pronounced peaks at $\cos(\psi) = \pm 5/9$ and $\pm 7/9$ suggests that this signal arises from a population of non-twinned scattering domains.
Reproduced with permission of the International Union of Crystallography

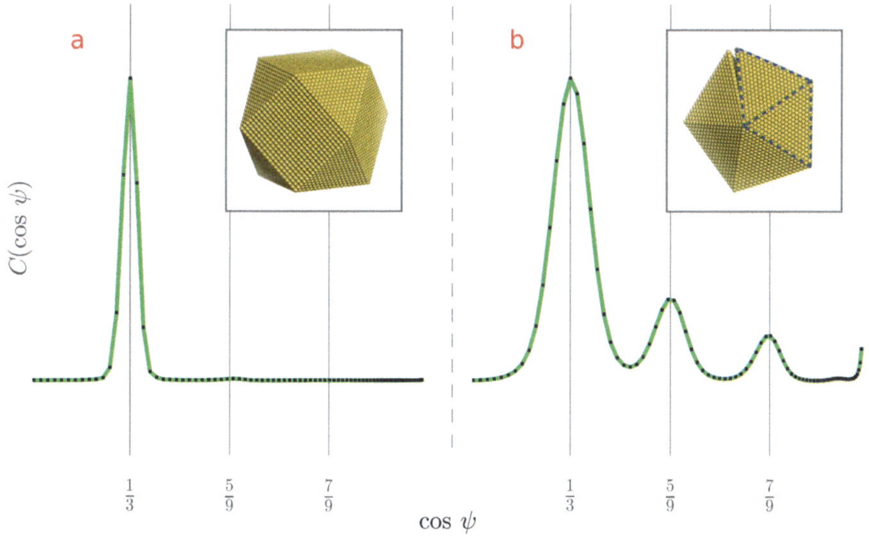

Fig. 15.4 (**a**) (left panel): Simulated CXS autocorrelation signals for a non-twinned cuboctahedron gold nanoparticle atomic model. (**b**) (right panel): Simulated CXS autocorrelation signals for a nearest-neighbor tetrahedron (NNT) model.
Reproduced with permission of the International Union of Crystallography

15.7 Time-Resolved Solution Scattering: One-Dimensional Data

At the macromolecular level, living systems are characterized by structure—function relationships in which deviations from thermodynamic equilibrium result in biochemical reactions which modify cellular information at the chemical or physical level. Examples include polymerases using DNA templates to synthesize messenger RNA, molecular motors converting chemical to mechanical energy, and G-protein coupled receptors transmitting chemical information across cellular membranes.

A unique feature of X-ray FEL radiation is that the X-rays are delivered in bursts lasting a few 10 s of femtoseconds. From the perspective of solution scattering this implies that pump-probe types of experiment can be carried out in which a modifying macromolecular reaction is started at an initial time, then the samples are exposed to the X-ray FEL beam at a series of later times t_i leading to the possibility of generating "molecular" movies.

In recent years X-ray FEL technology has been used in pump-probe experiments to characterize the kinetics of macromolecular conformation change for a couple of well-studied systems—photolysis of the CO-myoglobin complex [16], and photo-excitation of the *Blastochloris viridis* photosynthetic reaction center, RC_{vir} [2].

In each case, X-ray FEL radiation was used to measure one-dimensional solution X-ray scattering SAXS/WAXS profiles at a time-series of observations following a stimulus by an optical laser pulse.

For the CO-myoglobin experiments, Cammarata [16] and his team measured changes in the radius of gyration, R_g and the molecular volume V_p over the first 10 picoseconds (ps) following the optical excitation pulse. After an initial rapid rise in R_g and volume, V_p, these quantities are observed to relax to close to their initial value. The global conformational change reflected in the WAXS difference signal suggested that the internal secondary structure motions may initially occur through a quasi-ballistic mechanism as opposed to the more usual overdamped Brownian motion associated with thermally excited conformational changes.

The dynamics of energy dissipation in proteins have traditionally been referred to as "protein quakes" [1]. These new time-resolved X-ray FEL experiments provide valuable insight into the ultrafast dynamics of proteins in solution, but they also raise new questions: a recent paper by Brinkman and Hub [4], based on systematic molecular dynamics simulations, suggests a very different interpretation from that suggested by Cammarata et al. [16].

In the Cammarata work [16], underdamped oscillations in the ΔR_{Guin} and ΔV_{Guin} after CO photodissociation in Mb were interpreted as underdamped oscillations of the protein. However, Brinkman and Hub [4] find that ΔR_{Guin} and ΔV_{Guin} are dominated by modulations of the solvent density. These results lead to the conclusion that the small angle S/WAXS data from Mb report on protein dynamics only up to $\sim$500 fs, after which the signals are dominated by the propagation of the pressure wave into the solvent.

These findings highlight the importance of detailed simulations which accurately include solvent effects when interpreting solution scattering data.

Although a similar study has not yet appeared for the photosynthetic reaction center study [2], it seems likely that similar conclusions would also be reached for that case.

The overall conclusion is that one-dimensional X-ray FEL solution scattering experiments, coupled with detailed molecular dynamics simulations can provide atomic insights into the ultrafast dynamics of proteins. This ability provides a new perspective with which to examine the role of molecular dynamics in the structure-function relationships needed for understanding living systems.

15.8 Time-Resolved Solution Scattering: Three-Dimensional Data

In contrast to time-dependent S/WAXS measurements, the ability to obtain three-dimensional data by measuring 4-point angular correlations of X-ray FEL solution scattering data has the potential to reveal changes in molecular structure at resolutions close to atomic and over a large range of time scales. Development of this methodology will provide an independent test of simulations based on molecular dynamics calculations, while at the same time providing detailed insight into the structural biology of macromolecular interactions.

The papers of Donatelli et al. [6, 7] provide the ability to generate density maps from correlated solution scattering.

Here the application of suitable constraints on the scattering data is effected by application of spherical harmonic projections of density maps to give an iterative solution of the phase problem. To do this, Donatelli et al. combine the Hybrid Input-Output (HIO) methods of Fienup [9] with the method of shrink-wrap of Marchesini [17].

To combine phase retrieval with shrink-wrap, Marchesini et al. use the current real space estimate of the autocorrelation function density distribution of the scattering molecules, then perform a thresholding of the intensity. The result is a blurred estimate of the boundary of the object at a given intensity contour. This is then repeated every 50 steps or so of the HIO phase retrieval optimization.

The authors find that application of the shrink-wrap method acts to smooth out noise and provides an improved support constraint that gives rise to a successively better estimate of the object in real space as the iterative refinement proceeds to convergence.

Donatelli et al. have used simulated data to demonstrate that this approach has the power to extract structural details from three-dimensional X-ray FEL scattering from non-crystalline molecular solutions at close to atomic resolution. In this way, the time resolution of X-ray FEL scattering can eventually lead to atomic scale models of biomolecular reactions on physiological time and length scales.

15.9 Summary

The measurements discussed in this chapter demonstrate that angular correlation of X-ray scattering data may be used to reveal the internal structure of macromolecules in non-crystalline solutions on atomic length scales. The advent of X-ray FEL sources thus opens up a new frontier in the science of X-ray scattering from non-crystalline samples.

In particular, X-ray FEL radiation has the potential to reveal structural molecular dynamic biology under physiological conditions and at time scales and spatial resolution not accessible to other methodologies.

Acknowledgements Thanks to Derek Mendez, Shenglan Qiao, Kevin Raines, Gundolf Schenk and Herschel Watkins who contributed to the work reported here. Partial support came from NIH grant 1R01GM09746301A1.

References

1. Ansari, A., Berendzen, J., Bowne, S.F., Frauenfelder, H., Iben, I. E. T., Sauke, T. B., et al. (1985). Protein states and protein quakes. *Proceedings of the National Academy of Sciences of the United States of America, 82*(15), 5000–5004.
2. Arnlund, D., Johansson, L. C., Wickstrand, C., Barty, A., Williams, G. J., Malmerberg, E., et al. (2014). Visualizing a protein quake with time-resolved X-ray scattering at a free-electron laser. *Nature Methods, 11*(9), 923–926.
3. Aroian, L. A., Taneja, V. S., & Cornwell, L. W. (1978). Mathematical forms of the distribution of the product of two normal variables: Mathematical forms of the distribution. *Communications in Statistics - Theory and Methods, 7*, 165–172.
4. Brinkmann, L. U. L., & Hub, J. S. (2016). Ultrafast anisotropic protein quake propagation after CO photodissociation in myoglobin. *Proceedings of the National Academy of Sciences of the United States of America, 113*(38), 10565–10570.
5. Chacon, P., Moran, F., Diaz, J. F., Pantos, E., & Andreu, J. M. (1998). Low-resolution structures of proteins in solution retrieved from X-ray scattering with a genetic algorithm. *Biophysical Journal, 74*(6), 2760–2775.
6. Donatelli, J. J., Sethian, J. A., & Zwart, P. H. (2017). Reconstruction from limited single-particle diffraction data via simultaneous determination of state, orientation, intensity, and phase. *Proceedings of the National Academy of Sciences of the United States of America, 114*(28), 7222–7227.
7. Donatelli, J. J., Zwart, P. H., & Sethian, J. A. (2015). Iterative phasing for fluctuation X-ray scattering. *Proceedings of the National Academy of Sciences of the United States of America*, 112(33), 10286–10291.
8. Doniach, S. (2001). Changes in biomolecular conformation seen by small-angle x-ray scattering. *Chemical Reviews, 101*(6), 1763–1778.
9. Fienup, J. R. (1978). Reconstruction of an object from modulus of its Fourier-transform. *Optics Letters, 3*(1), 27–29.
10. Guinier, A. (1939). la diffraction des rayons x aux tr\`es petits angles: application *à* l'*é*tude de phenomènes ultramicroscopiques. *Annals of Physics, 12*, 161–237.
11. Howie, A., & Marks, L. D. (1984). Elastic Strains and the energy-balance for multiply twinned particles. *Philosophical Magazine A-Physics of Condensed Matter Structure Defects and Mechanical Properties, 49*(1), 95–109.

12. Kam, Z. (1977). Determination of macromolecular structure in solution by spatial correlation of scattering fluctuations. *Macromolecules, 10*(5), 927–934.
13. Kam, Z., Koch, M. H. J., & Bordas, J. (1981). Fluctuation X-ray-scattering from biological particles in frozen solution by using synchrotron radiation. *Proceedings of the National Academy of Sciences of the United States of America-Biological Sciences, 78*(6), 3559–3562.
14. Kirian, R. A., Schmidt, K. E., Wang, X., Doak, R. B., & Spence, J. C. H. (2011). Signal, noise, and resolution in correlated fluctuations from snapshot small-angle x-ray scattering. *Physical Review E, 84*(1), 011921.
15. Kratky, O., & Sekora, A. (1943). Regulation of the shape and size of separated particles from the smallest lower angle diffuse inflected x-rays. *Naturwissenschaften, 31*, 46–47.
16. Levantino, M., Schiro, G., Lemke, H. T., Cottone, G., Glownia, J. M., Zhu, D., et al. (2015). Ultrafast myoglobin structural dynamics observed with an X-ray free-electron laser. *Nature Communications, 6*, Article number: 6772.
17. Marchesini, S., He, H., Chapman, H. N., Hau-Riege, S. P., Noy, A., Howells, M. R., et al. (2003). X-ray image reconstruction from a diffraction pattern alone. *Physical Review B, 68*(14), 140101–140104
18. Mendez, D., Lane, T. J., Sung, J., Sellberg, J., Levard, C., Watkins, H., et al. (2014). Observation of correlated X-ray scattering at atomic resolution. *Philosophical Transactions of the Royal Society B-Biological Sciences, 369*(1647). https://doi.org/10.1098/rstb.2013.0315
19. Mendez, D., Watkins, H., Qiao, S., Raines, K. S., Lane, T. J., & Schenk, G. (2016). Angular correlations of photons from solution diffraction at a free-electron laser encode molecular structure. *IUCRJ, 3*(6), 420–429.
20. Qiao, S., Hilger, D., Fonseca, R., Mendez, D., & Doniach S. (submitted). Structural details of a protein in a non-crystalline solution revealed by angular correlations of scattered xFEL photons. Optics Express.
21. Svergun, D. I. (1999). Restoring low resolution structure of biological macromolecules from solution scattering using simulated annealing. *Biophysical Journal, 76*(6), 2879–2886.
22. Svergun, D. I., & Stuhrmann, H. B. (1991). New developments in direct shape determination from small-angle scattering . 1. Theory and model-calculations. *Acta Crystallographica A, 47*, 736–744.
23. Walther, D., Cohen, F. E., & Doniach, S. (2000). Reconstruction of low-resolution three-dimensional density maps from one-dimensional small-angle X-ray solution scattering data for biomolecules. *Journal of Applied Crystallography, 33*(2), 350–363.
24. Yang, C. Y. (1979). Crystallography of decahedral and Icosahedral Particles .1. Geometry of twinning. *Journal of Crystal Growth, 47*(2), 274–282.

Chapter 16
Future Directions of High Repetition Rate X-Ray Free Electron Lasers

Mike Dunne and Robert W. Schoenlein

16.1 Introduction

A new scientific frontier opened in 2009 when the world's first X-ray free electron laser (FEL), the Linac Coherent Light Source (LCLS) facility, began operations at SLAC National Accelerator Laboratory. The scientific start of LCLS has arguably been one of the most vigorous and successful of any new research facility, with a dramatic effect on a broad cross section of scientific fields, ranging from atomic and molecular science, ultrafast chemistry and catalysis, fluid dynamics, clean energy systems, structural biology, high energy-density science, photon science, and advanced materials [1].

This success was closely followed by the rapid development and planning of X-ray FEL facilities around the world, including SACLA (Japan), the European XFEL (Germany), PAL-FEL (Republic of Korea), Swiss-FEL (Switzerland), SCLF (China), and is complemented by soft X-ray facilities such as FLASH (Germany), Fermi@Elettra (Italy), and SXFEL (China). See Table 16.1.

These facilities provide ultrashort pulses (from sub-femtosecond (fs) to >100 fs), with unprecedented peak brightness (typically 10^{32} ph/s/mm^2/mrad^2/0.1% BW), in SASE or seeded-mode operation, over an energy range from ~250 to ~25,000 eV. They can provide dual-pulses with relatively arbitrary separation in time, with the option of dual color and variable linear/circular polarization.[1]

[1]For example, see details of the operating modes of LCLS at: https://lcls.slac.stanford.edu/machine-faq and the summary table at: https://portal.slac.stanford.edu/sites/lcls_public/machinefaqpix/MultiColorModes-8-22-16.pdf.

M. Dunne (✉) · R. W. Schoenlein
Linac Coherent Light Source, SLAC National Accelerator Laboratory, Menlo Park, CA, USA
e-mail: mdunne@slac.stanford.edu; rwschoen@slac.stanford.edu

S. Boutet et al. (eds.), *X-ray Free Electron Lasers*,
https://doi.org/10.1007/978-3-030-00551-1_16

Table 16.1 Hard X-ray XFEL facilities and capabilities

Hard X-ray FEL facility	Number of undulator sources	X-ray energy range	Repetition rate	Accelerator	Status for users
LCLS-I (U.S.)	1	0.25–12.8 keV	120 Hz	Warm Cu 15 GeV	Operational (2009)
LCLS-II (U.S.)	2	0.25–5 keV	1 MHz	CW-SCRF 4 GeV	2021
LCLS-II-HE (U.S.)	2	0.25 to 12.8–20 keV	1 MHz	CW-SCRF 8 GeV	Design
SACLA (Japan)	2	5–20 keV	60 Hz	Warm Cu 8 GeV	Operational (2011)
PAL-FEL (Republic of Korea)	2	0.3–20 keV	60 Hz	Warm Cu 10 GeV	Operational (2017/18)
Swiss-FEL (Switzerland)	2	0.2–12 keV	100 Hz	Warm Cu 5.8 GeV	Operational (2017/18)
European XFEL (Germany)	3 (initially) 5 (future)	0.2–25 keV	27 kHz (effective) in 10 Hz bursts of 4.5 MHz	Pulsed-SCRF 17.5 GeV	Operational (2017)
SCLF (China)	3	0.25–12 keV	1 MHz	CW-SCRF 8 GeV	Design

The number of undulator sources is quoted in terms of initial operation. The proposed upgrade path of the LCLS facility is broken out in the first three rows for clarity, as the sources will be independently delivered to users once complete

Glossary: *CW-SCRF* continuous wave superconducting radio frequency accelerator, *Pulsed-SCRF* pulsed superconducting RF

The initial suite of X-ray FEL sources operates at a relatively modest repetition rate of 60 to 120 Hz, based on normal conducting copper accelerator technology. With the commissioning of the European XFEL in 2017, the repetition rate took a step jump to 27,000 Hz (in bunches of 4.5 MHz pulses delivered at 10 Hz), covering a broad spectral range from 0.25 to 25 keV.

Similarly, by 2021 the LCLS-II Project will be able to generate coherent X-rays from 0.2 to 5 keV at up to one million pulses per second using a new superconducting accelerator in the first kilometer of the SLAC Linac Tunnel. Beyond this, the LCLS-II-HE ("High Energy") project, which is currently in the design and evaluation stage, will double the energy of the superconducting Linac to 8 GeV, and thus extend the MHz repetition rate X-ray beam performance from a current cutoff of energy of $\sim$5 keV to at least 13 keV and approaching 20 keV, dependent on projected improvements in electron beam emittance. A similar approach is being adopted for the SCLF facility in Shanghai.

These sources have the ability to tune the X-ray energy via variable-gap undulators, and LCLS-II will also have the ability to direct X-rays from two independently tunable undulators onto a single sample. Complementary to this, the SACLA facility can combine synchrotron radiation with its X-ray FEL beam, and all of the facilities offer multiple permutations of pump/probe capabilities, ranging from XUV to THz sources; high intensity short pulse lasers; high magnetic fields and extreme pressure environments. As such, they open up entirely new measurement possibilities and enable many innovative experiments that are simply not possible today.

The development of user facilities such as these is being driven in large part by the urgent need for a deeper understanding of the phenomena underpinning the future of the energy sector, including advanced materials and chemical processes (that are environmentally benign), and the biological sciences discussed in this book. In particular, the development of high X-ray energy, high repetition rate sources will qualitatively change the way that X-ray imaging, scattering and spectroscopy can be used to study how natural and artificial systems function. They will enable new ways to capture rare chemical events, characterize fluctuating heterogeneous complexes, and reveal quantum phenomena in matter, using nonlinear, multidimensional and coherent X-ray techniques that are possible only with X-ray lasers. As an example, the extremely high photon flux and high repetition rate of the emerging generation of X-ray FEL sources will enable the study of structural dynamics of biomolecular reactions by selectively observing the reacting partners at critical points in their approach, opening up new classes of investigation.

Advanced hard X-ray sources are being developed with the following characteristics:

1. *High-repetition-rate, hard X-rays*: Coupled atomic and electronic dynamics can be revealed in unprecedented detail. Advanced X-ray techniques will be able to measure simultaneously the electronic structure and subtle nuclear displacements of systems at the atomic scale, and in operating environments that require the penetrating capabilities of hard X-rays coupled to the ability of high repetition rates to provide high sensitivity for dilute and rare species.

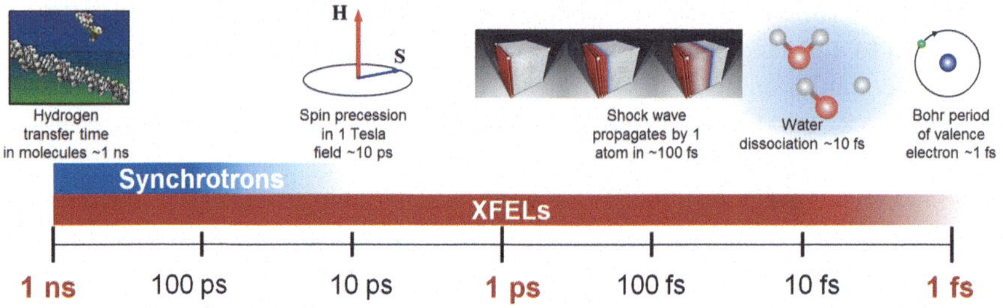

Fig. 16.1 A unique feature of X-ray free-electron lasers is the ability to probe the fastest timescales, providing over three orders of magnitude higher temporal resolution than a synchrotron source

2. *Temporal resolution*: Coherent X-rays delivered on femtosecond timescales open up experimental opportunities that were previously unattainable due to low signal-to-noise at current repetition rates, or that are simply not possible on non-laser sources. The typical temporal resolution limit for synchrotron sources is ∼100 ps (100,000 fs), whereas the performance of LCLS has progressed from initial pulse durations of 300 fs down to 5 fs (see Fig. 16.1), coupled to the capability for generating double pulses with independent control of energy, bandwidth, and timing. Ongoing development programs have recently demonstrated <0.2 fs pulses, as described below.
3. *Temporal coherence*: Control over the X-ray FEL spectral bandwidth will be a major advance for high-resolution inelastic X-ray scattering and spectroscopy in the hard X-ray range. The present scientific impact is substantially limited by the available spectral flux (photons/s/meV) from temporally incoherent synchrotron sources. The European XFEL and planned sources such as SCLF and LCLS-II-HE will provide a tens to hundreds fold increase in average spectral flux compared to synchrotron sources, opening new areas of science and exploiting high-energy resolution and dynamics near the Fourier transform limit.
4. *Spatial coherence*: Delivery of high average coherent power in the hard X-ray range, with programmable pulses at high repetition rate, will enable studies of spontaneous ground-state fluctuations and heterogeneity at the atomic scale from microseconds (or longer) down to fundamental femtosecond timescales using powerful time-domain approaches such as X-ray photon correlation spectroscopy (XPCS). This will provide a qualitative advance for understanding non-equilibrium dynamics and fluctuations via time-domain inelastic X-ray scattering and X-ray Fourier-transform spectroscopy approaches using Bragg crystal interferometers.
5. *Structural dynamics and complete time sequences*: LCLS achieved early success in the determination of high-resolution structures of biological systems and nanoscale matter before the onset of radiation damage, as discussed earlier in this book. X-ray scattering with ultrashort pulses represents a step-change in the field of protein crystallography. An important scientific challenge is to understand

function as determined by structural dynamics—at the atomic scale (requiring ~1 Å resolution) and in physiologically relevant (operating) environments (e.g., aqueous, room temperature). The potential of dynamic pump–probe structural studies has been demonstrated in model systems, but the much higher repetition rates of emerging X-ray FEL sources are needed in order to extract complete time sequences from biologically relevant complexes. Here, small differential scattering signals that originate from dilute concentrations of active sites and low photolysis levels are essential in order to provide interpretable results.

6. *Heterogeneous sample ensembles and rare events*: The high repetition rate and variable time structure of emerging facilities provide a transformational capability to collect 10^8–10^{10} scattering patterns (or spectra) per day with sample replacement between pulses. By exploiting revolutionary advances in data science (e.g., Bayesian analysis, pattern recognition, manifold maps, or machine learning algorithms), it will be possible to characterize heterogeneous ensembles of particles or identify and extract new information about rare transient events from comprehensive data sets.

16.2 Development of New Experimental Techniques

The technical capabilities of the European XFEL (along with SCLF and LCLS-II-HE when completed) will enable new experimental approaches in the hard X-ray range that are qualitatively different from anything available from current sources, or from any other X-ray source in the foreseeable future. This section presents a brief introduction to some of these new scientific methods and tools, as well as brief notes on their anticipated impact. The selected techniques stretch beyond those currently utilized for biological system studies, since it is very often the case that awareness of opportunities and progress in neighboring domains can drive innovation and new insights in other fields.

16.2.1 Coupled Atomic and Electronic Structure of Real Systems on Fundamental Time and Length Scales

High-repetition-rate ultrafast hard X-rays will enable techniques that simultaneously measure electronic structure and subtle nuclear displacements at the atomic scale, on fundamental timescales (femtoseconds and longer), and in operating environments that require the penetrating capabilities of hard X-rays. The high scientific potential of this approach has been demonstrated in recent experiments at LCLS, in which hard X-ray scattering and X-ray emission from a model transition-metal charge-transfer complex in solution have been measured simultaneously. These experiments revealed the coupling of charge dynamics and atomic structure on the sub-Å scale and on the 100 fs timescale for the first time [2].

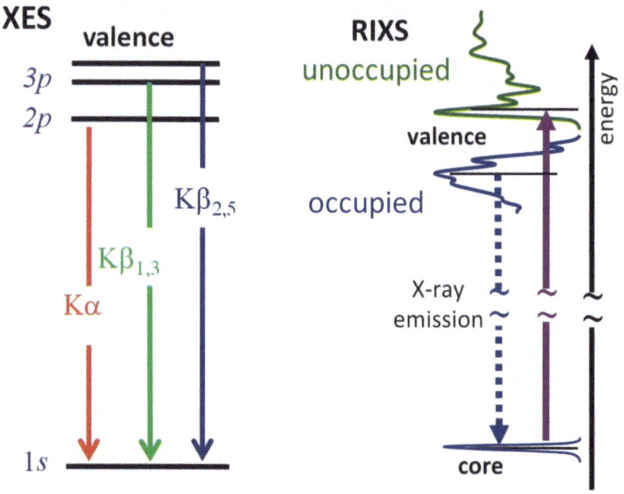

Fig. 16.2 Left: X-ray emission spectroscopy (XES). Right: Resonant inelastic X-ray scattering (RIXS)

Applying these techniques to much more complex assemblies (e.g., light-harvesting and photosynthetic complexes) will require a much higher repetition rate than LCLS (or other normal conducting accelerator facilities) can currently provide, in order to extract quantitative information that can be used to inform design and synthesis. For example, studies to date at 120 Hz have been limited to relatively simple molecules in high concentrations with high photolysis yields, whereas functioning assemblies typically have low concentrations of active sites and must be investigated at low photolysis levels to properly understand how they operate.

The newly emerging superconducting accelerator sources will enable element-specific probing of the fine details of the chemical structure, such as valence excited-state charge dynamics, by exploiting sensitive valence-to-core ($K_{\beta 2,5}$) X-ray emission spectroscopy (XES) and resonant inelastic X-ray scattering (RIXS) in the 5–12 keV range, as illustrated in Fig. 16.2. Time-resolved X-ray FEL studies to date have been limited by the average spectral flux (photons/s/meV) to less informative X-ray emission lines (K_{α}, $K_{\beta 1,3}$), and to RIXS probing at single time snapshots of large chemical changes in simple molecules at high concentration [3, 4].

The new high repetition rate sources will provide a similarly dramatic advance for differential extended X-ray absorption fine structure (EXAFS) studies of changes in local structure at the 0.01 Å scale. The hard X-ray range encompasses the core resonances of the Earth-abundant 3*d* transition metals (e.g., Fe and Cu K-edges at 7.1 and 9.0 keV) that are the basis for many biological enzymes and photocatalysts, and the 4*d* and 5*d* metals (e.g., Pt L_3-edge at 11.6 keV) commonly used in artificial catalysts and photosynthetic assemblies.

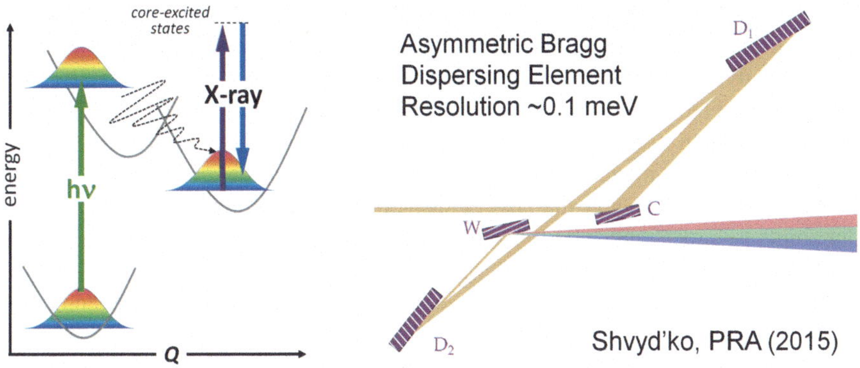

Fig. 16.3 Element-specific excited-state RIXS (left) and example of high-resolution spectrograph design (right) [4]

16.2.2 Temporal Coherence: Dynamics, and High-Resolution X-ray Spectroscopy at the Fourier Limit

Achieving an orders of magnitude increase in average spectral flux (photons/s/meV) via self-seeding will be a major advance for high-resolution inelastic X-ray scattering and spectroscopy in the hard X-ray range (both resonant, RIXS, and non-resonant, IXS), where the available spectral flux from temporally incoherent sources substantially limits the scientific impact. New instrument designs for ultrahigh-resolution (sub-meV) IXS [5, 6] (Fig. 16.2 right), will enable the use of quartz- and sapphire-based analyzers for hard X-ray RIXS at the few meV scale [7], and will make possible entirely new approaches for time-resolved resonant inelastic scattering.

For example, element-specific mapping of the excited-state potential energy surfaces of reactive molecular complexes (Fig. 16.3 left) will provide entirely new scientific insight for chemistry, materials science, and biology. Time-resolved RIXS is the X-ray analog of powerful femtosecond visible Raman spectroscopy [8], with tunable ultrafast X-rays providing element specificity [9, 10] along with dynamics.

The new source capabilities, combined with advanced spectrometer designs [7, 11, 12], will enable this approach to be applied to complex reactive systems for the first time. Of particular interest is the opportunity to distinguish local bonding environments of complex molecular assemblies targeting the metal-ligand fingerprint region (200–500 cm^{-1}, 24–60 meV). This requires high average spectral flux to distinguish weak signals, with <10 meV resolution and high time resolution (i.e., close to the Fourier transform limit of time and frequency bandwidth), at photon energies resonant with transition-metal K-edges and L-edges (up to 12 keV) [7].

Inelastic X-ray scattering also has tremendous potential for characterizing ground-state and low-energy collective modes that are hallmarks of complex matter.

For example, in strongly correlated materials, collective modes of the electronic continuum determine the material properties and are described by the momentum-dependent dynamic structure factor $Se(\boldsymbol{q},\omega) \sim \chi(\boldsymbol{q},\omega)$. However, comprehensive direct measurements of this fundamental description of quantum matter at meV energy resolution that spans the entire Brillouin zone have remained largely inaccessible. In soft matter and biological systems, low-energy collective dynamics mediate self-assembly and the development of non-equilibrium forms of matter. Our ability to understand and ultimately control such processes relies on developing much deeper insight into the low-energy potential landscape.

The full scientific potential of inelastic X-ray scattering (both resonant and non-resonant) has yet to be realized, owing to the low average spectral flux of present hard X-ray sources. The high spectral flux from the new high repetition rate sources, at close to the transform limit and in a uniform pulse structure to facilitate sample replacement/recovery, will transform this field. Importantly, the hard X-ray regime provides bulk sensitivity and access to sub-nm length scales (large momentum transfer), and is compatible with diamond anvil cells for pressure-dependent studies. The new source capabilities will support the highest-resolution spectrometer designs [5] with unprecedented throughput.

16.2.3 Coherent Hard X-ray Scattering: Materials Heterogeneity and Spontaneous Fluctuations

Spontaneous fluctuations and heterogeneity are pervasive in complex matter, and are often central to their functional properties. X-ray photon correlation spectroscopy (XPCS) is a powerful time-domain approach for characterizing ground-state fluctuations and heterogeneity, as illustrated in Fig. 16.4. The coherent X-ray analogue of dynamic light scattering, XPCS produces a speckle pattern that is a Fourier projection of the material structure. Changes in the speckle pattern map the evolving material structure in momentum space ($\boldsymbol{q}$) through the dynamic structure factor $S(\boldsymbol{q},t)$. At a fundamental level, the ability to discern structural changes at short time intervals scales inversely with the square of the X-ray source brightness ($\Delta t \sim B^{-2}$). At a practical level, accessing these timescales depends on multiple-pulse measurement schemes (with moderate peak brightness and high average brightness) and/or detector speed and sensitivity. The high average coherent power of LCLS-II-HE, with programmable pulses at high repetition rate, will enable studies of spontaneous fluctuations at the atomic scale from microseconds (or longer) down to fundamental femtosecond timescales, thus opening up entirely new measurement possibilities.

XPCS approaches will likely include the following:

- "Sequential" XPCS measures a real-time sequence of individual speckle patterns, limited either by the repetition rate of the XFEL or by the 2D detector readout rate [13].

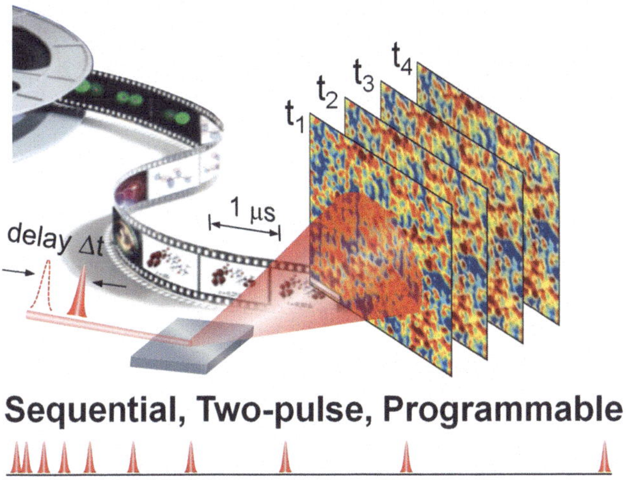

Fig. 16.4 Multi-pulse X-ray photon correlation spectroscopy of materials heterogeneity and fluctuations

- "Two-pulse" XPCS [14, 15] relies on superimposed pairs of speckle patterns created by time-delayed X-ray pulses. The decay in speckle visibility as a function of pulse delay is directly related to $S(\boldsymbol{q},t)$. This approach will probe ultrafast atomic-scale statistical dynamics in materials for the first time, with a time resolution limited only by the X-ray pulse delay.
- A "programmable" XPCS mode is uniquely enabled by the CW-SCRF Linac of LCLS-II-HE. In the simplest configuration, X-ray pulse pairs can be generated directly from the X-ray FEL with time delays of <100 fs up to milliseconds or longer. In addition, potentially much more powerful XPCS schemes are enabled, in which the entire time structure is encoded in the X-ray pulse sequence (e.g., in a logarithmic distribution of time intervals).

The new source capabilities will enable qualitative advances in emerging approaches such as time-domain inelastic X-ray scattering [16] and X-ray Fourier-transform spectroscopy using Bragg crystal interferometers [17], which are potentially powerful complements to IXS and XPCS.

For example, FT-IXS is an important new approach for probing low-energy collective modes in condensed matter that exploits the information content of diffuse scattering and the ultrashort pulses of X-ray FELs to map the dynamic structure factor in the non-equilibrium regime. Initial LCLS experiments using FT-IXS yielded new insight into time- and momentum-dependent phonon–phonon correlations in model bulk materials [16] (Fig. 16.5). The high repetition rate of future sources will enable this powerful approach to be applied to functional materials, nanostructures, and assemblies.

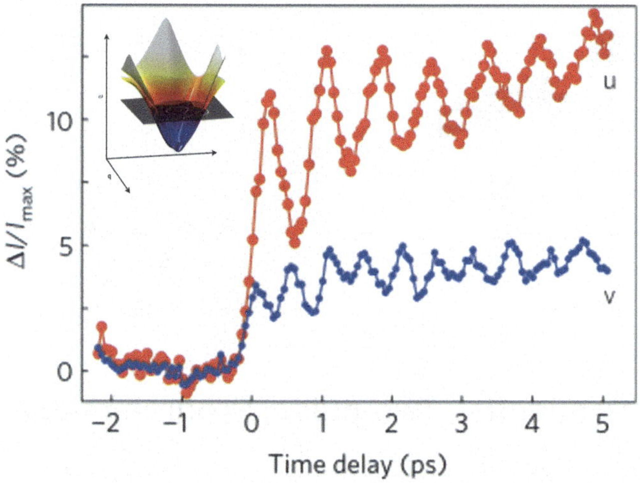

Fig. 16.5 Coherent phonon–phonon correlations and acoustic phonon dispersion function (inset) measured in bulk Ge via Fourier-transform inelastic X-ray scattering (FT-IXS) [16]

16.2.4 Coherent X-ray Scattering: Complete Time Sequences of Structural Dynamics at the Atomic Scale

A major early success of X-ray FELs is the determination of high-resolution structures of biological systems and nanoscale matter before the onset of damage. The combination of serial femtosecond crystallography (SFX) and "diffraction before destruction" [18] has provided atomic-resolution structures of clusters, nanomaterials, and many biologically important classes of molecules [1], often from micro- to nano-sized crystals that are unsuitable for study with the best conventional synchrotron X-ray sources, and with higher resolution or with structural details more representative of the room-temperature native state. However, studies to date have been limited largely to a few snapshots of static structures.

An important scientific challenge for structural biology and the broader area of functional nanomaterials and self-assembly is the ability to study structural *dynamics*—at the atomic scale, and under operating conditions or in environments that are more physiologically relevant (e.g., aqueous, room temperature). Dynamics are an important key to understanding the function of macromolecular complexes and nanomachines whose structures change over a range of time and length scales, often in response to changes in local environment (temperature, pH) or interaction with small molecules. Determining the conformational states of molecular machines and the low-energy landscape that determines their function are important new frontiers for both biology and nanomaterial self-assembly.

Recent time-resolved SFX studies of photo-detachment dynamics in CO-myoglobin [19], and isomerization dynamics in photoactive proteins [20, 21] illustrate the potential of dynamic pump–probe structure studies. However, these results are limited to just a few time snapshots of model systems at high concentration and at very high excitation fluences. Hard X-rays at much higher repetition rate are needed in order to extract complete time sequences from small differential scattering signals that originate from more relevant complexes, where active sites are often in dilute concentrations and where low photolysis levels are essential in order to provide interpretable results.

While crystallography is an extremely powerful tool for elucidating atomic structures, many complex biological machines defy crystallization due to weak interactions among constituent components and their intrinsic flexibility. Time-resolved solution X-ray scattering [22], fluctuation X-ray scattering (fSAXS), and single particle imaging (SPI) [18, 23, 24] can provide alternative paths toward understanding dynamics of non-crystalline samples at low to medium resolution. SPI continues to evolve toward higher resolution, complemented by fluctuation X-ray scattering techniques that are enabled by the combination of ultrafast X-ray pulses and high repetition rate. These techniques are briefly discussed further below.

16.2.5 Single Particle Imaging: Toward Atomic Resolution

As discussed in a prior chapter of this book, the scientific community using LCLS developed a comprehensive road map to advance the development of single particle imaging (coherent diffractive imaging), with a goal of reaching 3 Å resolution of biological objects [25].

Key X-ray source requirements identified in this road map include:

- Target photon energy range: 3–8 keV.
- Pulse duration <20 fs with maximum flux per pulse.
- Clean X-ray focus of 100–200 nm (to minimize beamline scattering while maximizing signal from a small object).

Detector development for single particle imaging will focus on:

- Optimum quantum efficiency in the tender X-ray range.
- Single photon counting at high-q and sub-Poisson noise elsewhere.
- $\sim 10^4$ dynamic range, <100 μm pixel size
- High read out rates (>1 kHz).

In addition to the development of high-intensity *single-shot* X-ray imaging in the tender X-ray regime, the European XFEL and LCLS-II will enable the development of new approaches to biological imaging from *large data sets* of relatively low-contrast scattering patterns collected at high repetition rate (limited by the detector read out).

16.2.6 *Fluctuation X-ray Scattering: Interacting Complexes and Assemblies in Natural Environments*

Fluctuation X-ray scattering (fSAXS) [26–29] has emerged as a method bridging SPI and crystallography, and is a promising route for 3-D imaging of anisotropic ensembles of interacting complexes in natural (solution) environments. fSAXS is an extension of well-developed SAXS/WAXS techniques, but in a regime in which the X-ray exposure is much shorter than rotational diffusion times. This gives rise to anisotropic scattering patterns (with annular correlations or fluctuations) which contain ~100 times greater information content than typical isotropic SAXS/WAXS patterns, thus enabling the reconstruction of 3-D objects.

The requirements (and development road map) are quite similar to those for single particle imaging as described above, with the following differences and additional requirements:

- Adjustable X-ray focus from ~1 μm to ~50 μm (depending on sample concentration etc.)
- Photon energy range: Both tender X-rays (2–8 keV) and soft X-rays (0.3–1.2 keV) are needed for different applications, and to exploit larger scattering cross sections, resonant scattering, and transmission in the water-window.
- Moderate initial goal of better than 1 nm resolution.
- High repetition rate (limited by detector read out) to exploit the highest average X-ray flux.

In the fSAXS approach, annular correlations from many scattering patterns (at modest signal-to-noise) can be summed. Thus, fSAXS will exploit the high repetition rate of European XFEL and LCLS-II/LCLS-II-HE (limited by detector read out) to achieve the highest average X-ray flux.

16.2.7 *High Repetition Rate, Heterogeneous Sample Ensembles, and Rare Events*

The high repetition rate of the new sources provides a transformational capability to collect 10^8–10^{10} scattering patterns (or spectra) per day with sample replacement between pulses. By exploiting revolutionary advances in data science (e.g., Bayesian analysis, pattern recognition, manifold maps, and machine learning algorithms as outlined below) it should be possible to characterize heterogeneous ensembles of particles or extract new information about rare transient events from comprehensive data sets of X-ray scattering patterns and/or spectra.

For fluctuation X-ray scattering and single particle imaging, high-intensity pulses in the 5–10 keV range hold tremendous promise for characterizing conformational heterogeneity of biological machines at different stages in their work cycle. For example, recent cryo-EM studies of the ribosome have demonstrated that it is possible to extract conformational movies from a sufficient number of snapshots of nonidentical, unoriented biomolecules cryo-trapped at random points in their work cycle [30]. This approach of reconstructing a movie from random individual snapshots is directly applicable to X-ray FEL studies, with the significance and impact determined by how finely the conformations are sampled with respect to the underlying conformational heterogeneity (i.e., a large number of snapshots are required). Among the hybrid approaches that are increasingly important for biology, the key advantages provided by XFELs include access to dynamics in near-native environments (room temperature, liquid); the ability to probe thick samples or dense media; and an unprecedented number of independent snapshots provided by high repetition rate.

16.2.7.1 Advanced Computation, Data Science, and Synergy with X-ray FEL Sources

Dramatic advances in computational capability and advanced algorithms, coupled with massive data sets, are creating profound opportunities for *data science*, that is, for scientific extraction of new knowledge and insight from data that goes far beyond what was previously possible. This is not simply doing the same analysis faster or on a larger scale, but spawning entirely new methods and establishing new paradigms for analysis. Advanced computation and data science will be integral to high-repetition-rate X-ray science, and massive data sets will further fuel advances in new algorithms and computational approaches. This powerful combination will allow scientists to envision entirely new experiments and analysis methods that will accelerate many fields. Here we highlight just three of many classes of computational techniques that the new sources will enable and exploit.

16.2.7.2 Mapping Reaction Landscapes and Kinetic Relationships between Conformations

A sufficiently fine-grained statistical sampling of configurational space can be used to map the kinetic topology or reaction pathways of a dynamic system. By analogy, measurement of the GPS locations of an ensemble of cars every minute as they randomly traverse the interstate highway system will provide a good representation of the metropolitan layout of a country—where cities are and how they are connected. GPS measurements at much less frequent intervals (e.g., once per week) would not provide the same insight. Similarly, the high repetition rate of new X-ray FEL sources will enable the mapping of the configurational space of complex systems such as proteins, viruses, or chemical reactions. For instance, extensive sampling will enable researchers to determine the energy landscape of a protein (i.e., the local potential energy minima and the connecting transition states) by analyzing the mutual information content between proximate samples. Because protein systems are high dimensional, consisting of many thousands of atoms (each with multiple degrees of freedom), it will be necessary to develop techniques for inferring lower dimensional representations of protein motion on which to map these dynamics [31]. Manifold embedding, diffusion maps, kernel methods and other dimensionality reduction techniques are under intense development for application in a broad range of data science tasks. These methods will be of particular importance in the study of the structure and dynamics of molecular complexes, exploiting large datasets of scattering patterns or spectra.

16.2.7.3 Automatic Pattern Recognition for Distinguishing Groups of Data or Events

Large datasets will enable users to rapidly and robustly identify what features of their data distinguish two or more sample configurations or experimental conditions—for example, single-particle versus many-particle hits. Such approaches are already being employed at LCLS, for example to exploit the two-color self-amplified spontaneous emission (SASE) operating mode where the presence of either X-ray color from LCLS is a stochastic process (output may consist of either pulse-color individually or both together). Neural networks have been employed to learn to distinguish between two single-color events and a two-color event (Fig. 16.6). The result is a model capable of both classifying events by the X-ray FEL colors present, and understanding the patterns within the data that reveal the temporal and spectral content of the X-ray FEL beam. The data rate and volume of experiments in the future will make manual analysis increasingly burdensome and

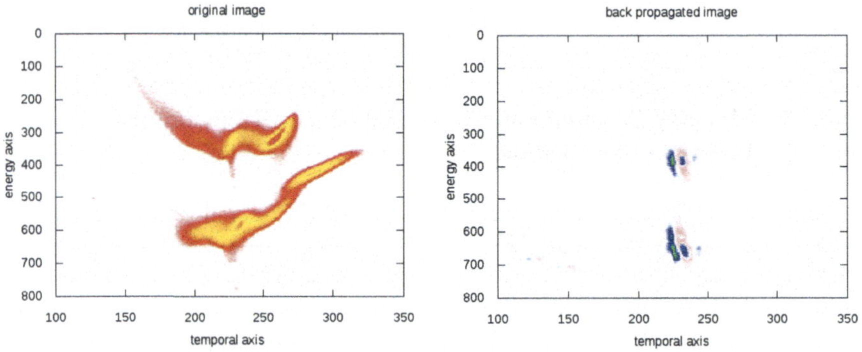

Fig. 16.6 Neural networks are employed to discover new structure within LCLS data. Left: Raw output image of an electron-beam diagnostic, the X- ray transverse cavity (XTCAV), which fluctuates substantially from shot to shot. Images such as these are analyzed to determine if one or both of two electron beams contribute to X-ray lasing. Right: By starting from known results, the neural network algorithm was able learn and discover this subtle pattern within the raw XTCAV image. This pattern highlights the specific regions that reveal the lasing conditions for that X-ray shot. The result is sufficiently nontrivial that discovery of this subtle pattern was only possible through machine learning

increase the need for high- performance machine learning algorithms that perform robustly in the presence of large datasets. Similar machine learning concepts will be required for the analysis of large X-ray data sets.

16.2.7.4 Adaptive Scanning of Higher-Dimensional Parameter Spaces

The experiments enabled by X-ray FELs have grown increasingly complex over the course of the first few years of operation, and this trend will accelerate with the new sources. For instance, pump–probe studies of photosystem-II [32, 33] have exploited a times sequence of three different pump pulses prior to the X-ray FEL probe. The European XFEL and LCLS-II-HE will enable experiments where scanning combinations of many controlled parameters—for example, pump laser or FEL delays, polarizations, intensities, perhaps multiplied by many perturbations and samples—are necessary for success. Each additional experimental variable increases the required data size exponentially (for fixed detection statistics).

Advanced computational techniques will not only be necessary for inference in the high-dimensional parameter spaces that result from such experiments, but will further allow closed feedback-loops to drive these experiments to combinations of input parameters that provide the greatest scientific insight.

16.3 Development of New XFEL Capabilities

One of the hallmarks of the early operation of X-ray FELs has been the rapid development of new modes of operation. For example, over the first few years of operation, the pulse duration at LCLS has been shortened from over 100 fs to <0.2 fs; the energy reach has been extended from an initial 0.8–8 keV to 0.25–13 keV; full polarization control has been introduced for soft X-rays; longitudinal coherence has been provided via self-seeding in both the soft-X-ray and hard-X-ray domains; and a wide array of dual pulse options have been developed with independent control of pulse timing, energy and polarization.

In the future, it is to be expected that a similar rate of progression with high repetition-rate sources will be seen. This section provides a brief overview of some of the current and upcoming developments in source characteristics of high repetition rate X-ray FELs. Near-term areas include increasing the power of sub-femtosecond X-ray pulse for fundamental charge migration and structural biology studies; advancing from double-pulse to multi-pulse capability to follow the evolution of materials undergoing rapid changes; exploring pathways to the highest possible peak powers; and delivering fully quantitative measurements of the beam properties to further enable precision science.

16.3.1 Sub-Femtosecond Pulses

An exciting new horizon is opened by the development of methods to deliver sub-femtosecond pulses, ideally close to the Fourier-Transform limit (for optimum flexibility in spectral bandwidth). The provision of such ultrashort pulses with high bandwidth (5 eV or above) will coherently excite multiple electronic states in a molecule, thereby creating a localized nonstationary electronic wavepacket and opening up new areas of atomic and molecular physics. For the biological sciences, the delivery of sub-femtosecond sources will ensure that signals are generated before any meaningful structural motion (damage) can occur. The key here is to obtain sufficient flux in each pulse to deliver the required scattering intensity.

Sub-femtosecond pulse generation was recently demonstrated in the hard X-ray regime. To achieve this, only a small part of the electron bunch was selected to lase, either using nonlinear compression of the electron bunch [34], or by passing the bunch through a slotted metal foil [35]. This resulted, for example, in pulse lengths inferred from the spectral bandwidth to be ~180 attoseconds (with a 14.4 eV coherent bandwidth at 9 keV).

A further technique under development is optical manipulation of the electron bunch [36, 37], which holds the potential to generate higher peak power pulses, and also operate in the soft-X-ray domain, as required for many applications. This takes advantage of the high field intensities of lasers to generate a short, high-current spike in an electron bunch via electron/photon interaction in a magnetic undulator. The short current spike is subsequently used in the FEL undulator to generate a short X-ray pulse. An alternative is to use infrared light generated by the electron beam itself, which then provides the ability to have fully locked time delays between two separate sub-femtosecond pulses. Recent studies at LCLS indicate that pulses as short as a few hundred attoseconds at photon energy of about 1 keV can be generated. An experimental program (known as X-ray Laser-Enhanced Attosecond Pulses (XLEAP)) is underway to develop this approach, with nominal performance as follows:

- 30–50 uJ per pulse
- 0.5 fs FWHM pulse duration
- 4–8 eV FWHM bandwidth.

16.3.2 Bandwidth Control Via Seeding

A significant number of envisioned science opportunities require exquisite control of the FEL coherent bandwidth, and utilization of a well-defined spectrum. The potential for creating large coherent spectral bandwidth, and pulse durations ~1 fs or less will enable nonlinear X-ray spectroscopies in a similar manner to that achieved for optical nonlinear spectroscopy.

At present, bandwidth selection is achieved by inserting a monochromator into the undulator chain to collect the SASE radiation produced by an initial set of undulators. This selects a narrower bandwidth and uses the resulting monochromatic beam to "self-seed" the electrons in the following undulator section.

In the soft X-ray domain, the resolving power of the system used at LCLS is roughly 5000. Since the resolving power is dominated mainly by the diffraction limit contribution of the grating (the number of illuminated lines), one can increase the resolving power by moving the source point further upstream, for example by opening the gap of the last few undulators just before the monochromator. In principle this approach can increase the resolving power almost arbitrarily. However, one must also increase the power delivered to the grating, to preserve the power required to seed the electron beam. Within the projected damage limit of the grating, a resolving power of the order of 30,000 or slightly higher is expected.

The single crystal self-seeding approach used to date at LCLS for 5–9 keV X-rays has difficulty in the tender X-ray regime, or much beyond 10 keV. An alternative approach is to use a two-bunch concept, letting the first bunch provide the seed for the second bunch. This requires development of a chicane with sufficient delay and optimization of the location due to the reduced seed power compared with the single crystal approach. Recent work using a "dechirper" device prior to the undulator chain has provided exquisite control of the electron bunch trajectory, enabling lasing of a "fresh slice" of the bunch in the latter half of the undulator. Combining this with a self-seeding chicane allows substantial (>2×) increase in peak brightness using short (10 fs) pulses.

While these self-seeding approaches have provided much-needed capability, the development of an *external* seeding system would provide far greater flexibility, if it can be realized—as demonstrated in the EUV domain at the Fermi@Elettra facility in Trieste. There are presently a variety of methods under investigation in the community for the 100–1000 eV regime including High Gain Harmonic Generation (HGHG) [38], Echo-Enabled Harmonic Generation (EEHG) [39, 40] and direct High Harmonic Generation (HHG) [41] injection schemes. Alongside this, there are other options to produce narrow bandwidth pulses, including harmonic lasing [42], purified SASE (pSASE) [43], and improved SASE (iSASE) [44], as well as novel emittance exchange schemes that may be able to operate in the tender or hard X-ray domain. There is no clearly preferred solution, and so the immediate task is to undertake a quantitative trade-off analysis; learn from recent experience at Fermi@Elettra, FLASH, SACLA, and SLAC.

16.3.3 Two-Pulse (Variable Delay)

X-ray pump with X-ray probe experiments open a very broad array of new scientific possibilities. At LCLS, extensive machine development effort has recently been devoted to developing 2-pulse, 2-color techniques, with notable success. Table 16.2 provides a snapshot of current capabilities at LCLS.

Table 16.2 Summary of two-pulse, two-color modes of operation at LCLS

Multi-color pulse mode table—SHORT FORM—Status 4/10/2017						
Technique	Pulse separation	Min pulse duration	Energy separation	Max energy/pulse	Mode	Comments
SOFT X-RAYS						
Fresh slice						Modes with the dechirper + orbit control
Two SASE pulses	$\sim -15 - +850$ fs	~5–8 fs	±2.5%	200–500 uJ (20 fs duration)	SASE	Probe intensity is higher if the max delay req'd is 35 fs. Pump pulse intensity is higher if the min delay req'd is +15 fs or more (no zero delay).
Linear SASE + polarization controlled SASE	$\sim -15 - +850$ fs	~5–8 fs	±2.5%	300 uJ	SASE	Only pump polarization can be controlled. See also comments re: Fresh-slice, two SASE pulses
One pulse self-seeded, one SASE	0–50 fs	~15–20 fs	±2.5%	100 uJ seeded, 200 uJ SASE	SASE SEEDED	Only probe polarization can be controlled. See also comments re: Fresh-slice, two SASE pulses. Requires longer setup
Three SASE pulses	0–900 fs (first to 2nd), 0–50 fs (second to 3rd)	~5–8 fs	2.5% range for all	100 uJ	SASE	Second pulse has lowest intensity, weak if $E > 700$ eV
Split undulator SASE	0–50 fs	40 fs	±2.0%	30 uJ	SASE	Minimally invasive, easy to maintain
Double slotted foil	15–70 fs	~10 fs	±1.5%	100–300 uJ	SASE	Minimally invasive, easy to maintain. Delay and energy separation are not independent, minor tuning needed between changes
Two bucket (ns spacing)	350 ps increments, ±38 ns	40 fs	±2%	0.5–1 mJ (100 fs duration SASE)	SASE SEEDED	Under development
Twin bunches (fs spacing) w/o slotted foil	125 fs max	70–100 fs	±2.5%	1.3 mJ	SASE	Intensity performance comparable to fresh-slice. Max time
Twin bunches (fs spacing) w slotted foil	~70 fs (bunch duration)	3–10 fs	±2.5%	50 uJ	SASE	Slice going forward

HARD X-RAYS						
Twin bunches						Requires long setup (laser stacker/injector tune)
Two SASE pulses	0–125 fs	~10 fs	0.2–3%	2 mJ (30 fs duration)	SASE	1st/probe pulse always higher photon energy
Twin bunches + V slotted foil	±50 fs	−5 − 10 fs	~3%	50 uJ	SASE	
Twin bunches + HXR self-seeding	0–100 fs	~10 fs	~1%	150 uJ per pulse	SEEDED	Both colors or a single color can be seeded. Requires longer setup time (hours)
Double slotted foil	7–20 fs	−10 fs	±1.5%	100–300 uJ	SASE	Minimally invasive, faster setup than twin bunches. Delay/energy separation not independent, minor tuning needed between changes
Two bucket (ns spacing)	350 ps increments, ±38 ns	20 fs	~2%	1–2 mJ (40 fs duration SASE)	SASE SEEDED	Under development

See: https://lcls.slac.stanford.edu/parameters

An important advance beyond two pulses with the same energy and variable delay has been two pulses with different colors and variable delay. Accelerator-based techniques have addressed this within a finite parameter space: two colors with separations on order of 1% have been demonstrated with delays ranging from 50 fs in soft X-rays to ~150 fs in hard X-rays. Also, two pulses separated by one or more RF buckets (providing up to >100 ns separation in steps of 0.35 ns) can be provided with different colors. The challenge is the delay range between 100 fs and a few hundred picoseconds. Solutions for hard X-rays using crystal split and delay methods are being implemented, but the soft X-ray regime is challenging.

X-ray FEL sources with variable gap undulators (such as SACLA) can produce two well-separated pulses, more than a keV apart [45], although results to date have not yet reached saturation for either color. Here too, the delay from the accelerator is from a magnetic chicane and is of order a few tens of fs. There is the possibility in hard X-rays again to use a crystal based split and delay to extend this range.

Finally, LCLS-II will offer a unique potential: to bring the soft X-ray (SXR) and hard X-ray (HXR) undulator beams to one experiment with variable delay and high timing precision. A crystal split and delay unit can also be used to vary the HXR arrival time for X-ray pump–probe experiments. There are many variants of this scheme, for example using the SCRF Linac for both the SXR and the HXR undulators and "combine" the third harmonic of the SXR undulator with a similar energy HXR undulator pulse at repetition rates in the 100 kHz scale.

16.3.4 High Repetition Rate Experimental Technology

There is a broad technology development program required to enable high repetition rate operation, with some key examples listed here:

16.3.4.1 X-ray Detectors

The requirements for pixelated large area detectors can be summarized in two main categories: large dynamic range (e.g., for crystallography) and low noise, small pixel (e.g., for spectroscopy)—in both cases with frame readout rates as high as possible. In the first case the detector should be able to detect a single photon and provide linear response up to at least 10,000 photons. This type of detector is needed for coherent scattering, imaging, diffraction and crystallography. The second type of detector requires modest dynamic range but very low noise and small pixels, as required for XPCS experiments, for example.

Multi kHz frame-rate cameras can be achieved by moving the signal digitization into the readout chip, with auto-ranging to provide high dynamic range and tiled systems to provide both small detectors (comprised of a single chip) as well as tiled multi-megapixel systems for large area forward scattering studies. There are a number of such detector systems currently under development, including JUNGFRAU and MOENCH at the Paul Scherrer Institute [46, 47] (with capability up to 2 kHz and 8 kHz); the ePix family of detectors at the SLAC National Accelerator Laboratory [48] (with the potential to achieve tens of kHz operation); and the High Dynamic Range–Pixel Array Detector (HDR-PAD), currently being developed at Cornell University [49], which will be capable of a few kHz continuous frame readout combined with a very large dynamic range capability.

Detectors working in burst mode, capable of storing a limited number of frames, for example JUNGFRAU [46], AGIPD [50], LPD [51], FASPAX [52], could provide viable options for implementing detection schemes that effectively utilize the available machine rate, that is, using only "good data" or on-chip data reduction.

Finally, it is worth noting that for this range of photon energies in the case of photon-starved experiments such as XPCS, where each frame has a very sparse occupancy, detectors based on conventional counting chips could be used. New generations of counters provide small pixels, multiple thresholds, and sophisticated on-system algorithms for handling charge sharing. For example, the VIPIC-L, developed in collaboration between Brookhaven, Fermilab, and Argonne national labs, will provide sparsified readout for pulses as close as 400 ns and full frame at 55 kHz [53].

16.3.4.2 X-ray Optics

Cooled mirrors are required to handle the average heat load of high repetition rate systems such as LCLS-II-HE. There is a tradeoff between the achievable clear aperture and the highest reflected photon energy, with an example compromise being meter-scale mirrors coated with a heavy metal such as Rh, to provide a cutoff energy of 16 keV.

Similarly, water-cooled refractive focusing lenses are needed (in a so-called transfocator system), particularly for refocusing into secondary interaction chambers, as is currently the design at LCLS (for the CXI instrument) and European-XFEL (for SFX/SPB).

Recent work has allowed detailed wavefront measurements to be made of the focal profile to provide quantitative information on focused intensity (as required for the interpretation of structural damage studies). The combination of Talbot interferometry with a high-resolution scintillator microscope provided single shot, high sensitivity, non-iterative wavefront recovery in 2D. Furthermore, this can be implemented with a long working distance from interaction point ($\sim$1 m) to allow in situ measurements concurrently with diffraction studies. For studies that require the tightest focal spot ($\sim$100 nm), these systems can be used to provide live feedback to remove high-order aberrations via mirror pitch, and astigmatism removal via mirror translation.

16.3.4.3 Sample Environment

In order for a sample to remain undamaged by X-rays preceding arrival to the interaction region, the sample must move further than the zone of damage surrounding the interaction region in less than the time between X-ray pulses. This damage zone may be as large as 100 μm at the highest photon fluence [54]. Focused aerosols and micron sized liquid jets can travel at several tens of m/s making them compatible with repetition rates of ~1 MHz. Room temperature supersonic atomic and molecular beams move even faster at several hundred m/s. The situation becomes more difficult for larger, slower moving liquid jets used in recirculating systems and for fixed targets, such that the sample environment system may ultimately limit the highest usable repetition rate.

Continuous jet sources can also be driven to break up into well-defined droplets at regular intervals with the current state of the art at about 2 MHz. Controlling the breakup and timing the drop arrival to that of the X-ray pulse could allow data collection further downstream of the injector. This would increase injector lifetime and allow a greater range of time points to be accessible for time-resolved studies.

In all cases, advances in the management of debris will be required. Debris generated from FEL pulses ablating or vaporizing samples can coat the injector or in-vacuum apertures, necessitating periodic cleaning, and can sometimes interrupt beam time operations even at low (~100 Hz) repetition rates.

16.3.4.4 Data Systems

The management of the extreme data rates produced from megapixel detectors operating at high repetition rate is perhaps one of the most taxing problems to be addressed, leading to an increase in the data throughput from the ~10 GB/s level seen today to >TB/s. This will lead to storage requirements in excess of 100 PB and processing rates approaching the exaflop level (10^{18} floating point operations per second).

The data systems will need to:

- Apply on-the-fly data reduction to reject unnecessary data (e.g., a veto that assesses when the beam missed the sample, or extraction of only the relevant features from a sparse dataset) on a timescale of <1 s to prevent unacceptable data buildup;
- Allow "fast feedback" analysis codes to be run in bursts of short jobs, with very fast startup time (contrary to the mode of operation of most supercomputers). This is needed to extract the critical information content from the data on a timescale of minutes—To enable real-time changes to the experimental configuration, and to identify when a rare or important event occurs;
- Access high-end offline computing, typically during the "off-shift" of an experiment, to deliver reasonably complete data analysis prior to the next shift.

- Ensure high availability of all data management and analysis systems throughout the duration of the experiments (noting that the facility will typically be operating 24 h per day);
- Enable a highly flexible development cycle for the analysis codes, due to the wide variety of experiments, the rapid turnaround required, and the need to modify data analysis during experiments.

This substantial increase in the scale of the data being produced will necessitate sociological change within our community, both with regard to retention of raw data, and in terms of our approach to the application of numerical modeling (to simulate expected signals) and real-time data analysis. Coordinated development of such tools in preparation for the operation of high repetition rate sources is essential, and already underway with projects such as EUCALL (https://www.eucall.eu) and ExaFEL (https://www.exascaleproject.org/video-exascale-free-electron-lasers-project/).

16.4 Conclusions

The results achieved to date on hard X-ray FEL sources such as SACLA and LCLS have been transformational, with particular impact in the field of structural biology, as outlined in the bulk of this book. The transition to high repetition rate sources, which is now upon us with the start of operations of the European XFEL, provides a highly complementary suite of new capabilities for this field. Fulfillment of this potential will require parallel development of new X-ray detectors, pump/probe laser sources, synchronization systems, high power optics, data acquisition systems and novel algorithms for real-time data reduction and analysis—as well as the development of the sources and X-ray techniques outlined above.

At high repetition rates, serial femtosecond crystallography (SFX) will advance from successful demonstration experiments to addressing some of the most pressing challenges in structural biology for which only very limited sample volumes are available (e.g., human proteins); or only very small crystal sizes can be achieved (<1 μm); or where current structural information is significantly compromised by damage from conventional X-ray methods (e.g., redox effects in metalloproteins). In all of these cases, high throughput and near-physiological conditions of room temperature crystallography will be qualitative advances. X-ray energies spanning the Se K-edge (12.66 keV) will further enable de novo phasing via isomorphous replacement and anomalous scattering. Time-resolved SFX and solution SAXS will advance from the present few time-snapshots of model systems at high photolysis levels, to full time-sequences of molecular dynamics that are most relevant for biology. Hard X-rays and high repetition rates will further enable advanced crystallography methods that exploit diffuse scattering from imperfect crystals, as well as advanced solution scattering and single particle imaging methods to map sample heterogeneity and conformational dynamics in native environments.

As we say in the X-ray FEL community, the future is bright.

Acknowledgements This chapter describes the work of a very large number of people at the SLAC National Accelerator Laboratory, the users of LCLS, and the wider community. Use of the LCLS is supported by the US Department of Energy, Office of Science, Office of Basic Energy Sciences under Contract no. DE-AC02-76SF00515.

References

1. Bostedt, C., Boutet, S., Fritz, D. M., Huang, Z., Lee, H. J., Lemke, H. T., et al. (2016). Linac Coherent Light Source: The first five years. *Reviews of Modern Physics, 88*(1), 015007.
2. Kjær, K. S., & Gaffney, K. J. (2017). Finding intersections between electronic excited states with ultrafast X-ray scattering and spectroscopy. In *Frontiers in optics 2017*. Washington, D.C.: Optical Society of America.
3. Wernet, P., Kunnus, K., Josefsson, I., Rajkovic, I., Quevedo, W., Beye, M., et al. (2015). Orbital-specific mapping of the ligand exchange dynamics of Fe(CO)(5) in solution. *Nature, 520*(7545), 78–81.
4. Zhang, W., Alonso-Mori, R., Bergmann, U., Bressler, C., Chollet, M., Galler, A., et al. (2014). Tracking excited-state charge and spin dynamics in iron coordination complexes. *Nature, 509*(7500), 345–348.
5. Shvyd'ko, Y. (2015). Theory of angular-dispersive, imaging hard-x-ray spectrographs. *Physical Review A, 91*(5), 053817.
6. Shvyd'ko, Y., et al. (2014). High-contrast sub-millivolt inelastic X-ray scattering for nano- and mesoscale science. *Nature Communications, 5*, 4219.
7. Sutter, J. P., Baron, A. Q. R., Ishikawa, T., & Yamazaki, H. (2005). Examination of Bragg backscattering from crystalline quartz. *Journal of Physics and Chemistry of Solids, 66*(12), 2306–2309.
8. Kukura, P., McCamant, D. W., & Mathies, R. A. (2007). Femtosecond stimulated Raman spectroscopy. *Annual Review of Physical Chemistry, 58*, 461–488.
9. Harada, Y., Tokushima, T., Horikawa, Y., Takahashi, O., Niwa, H., Kobayashi, M., et al. (2013). Selective probing of the OH or OD stretch vibration in liquid water using resonant inelastic soft-X-ray scattering. *Physical Review Letters, 111*(19), 193001.
10. Hennies, F., Pietzsch, A., Berglund, M., Föhlisch, A., Schmitt, T., Strocov, V., et al. (2010). Resonant inelastic scattering spectra of free molecules with vibrational resolution. *Physical Review Letters, 104*(19), 193002.
11. Sinn, H. (2001). Spectroscopy with meV energy resolution. *Journal of Physics-Condensed Matter, 13*(34), 7525–7537.
12. Yavas, H., et al. (2007). Sapphire analyzers for high-resolution X-ray spectroscopy. *Nuclear Instruments & Methods in Physics Research Section A-Accelerators Spectrometers Detectors and Associated Equipment, 582*(1), 149–151.
13. Rumaiz, A. K., et al. (2016). First experimental feasibility study of VIPIC: A custom-made detector for X-ray speckle measurements. *Journal of Synchrotron Radiation, 23*, 404–409.
14. Grubel, G., et al. (2007). XPCS at the European X-ray free electron laser facility. *Nuclear Instruments & Methods in Physics Research Section B-Beam Interactions with Materials and Atoms, 262*(2), 357–367.
15. Gutt, C., et al. (2009). Measuring temporal speckle correlations at ultrafast x-ray sources. *Optics Express, 17*(1), 55–61.
16. Trigo, M., et al. (2013). Fourier-transform inelastic X-ray scattering from time- and momentum-dependent phonon-phonon correlations. *Nature Physics, 9*(12), 790–794.
17. Tamasaku, K., Ishikawa, T., & Yabashi, M. (2003). High-resolution Fourier transform x-ray spectroscopy. *Applied Physics Letters, 83*(15), 2994–2996.
18. Neutze, R., et al. (2000). Potential for biomolecular imaging with femtosecond X-ray pulses. *Nature, 406*(6797), 752–757.

19. Barends, T. R. M., et al. (2015). Direct observation of ultrafast collective motions in CO myoglobin upon ligand dissociation. *Science, 350*(6259), 445–450.
20. Coquelle, N., et al. (2017). Chromophore twisting in the excited state of a photoswitchable fluorescent protein captured by time-resolved serial femtosecond crystallography. *Nature Chemistry, 10*, 31.
21. Tenboer, J., et al. (2014). Time-resolved serial crystallography captures high-resolution intermediates of photoactive yellow protein. *Science, 346*(6214), 1242–1246.
22. Arnlund, D., et al. (2014). Visualizing a protein quake with time-resolved X-ray scattering at a free-electron laser. *Nature Methods, 11*(9), 923–926.
23. Bergh, M., et al. (2008). Feasibility of imaging living cells at subnanometer resolutions by ultrafast X-ray diffraction. *Quarterly Reviews of Biophysics, 41*(3-4), 181–204.
24. Huldt, G., Szoke, A., & Hajdu, J. (2003). Diffraction imaging of single particles and biomolecules. *Journal of Structural Biology, 144*(1-2), 219–227.
25. Aquila, A., Barty, A., Bostedt, C., Boutet, S., Carini, G., dePonte, D., et al. (2015). The Linac Coherent Light Source single particle imaging road map. *Structural Dynamics, 2*(4), 041701.
26. Kam, Z. (1977). Determination of macromolecular structure in solution by spatial correlation of scattering fluctuations. *Macromolecules, 10*(5), 927–934.
27. Kam, Z., Koch, M. H. J., & Bordas, J. (1981). Fluctuation X-ray-scattering from biological particles in frozen solution by using synchrotron radiation. *Proceedings of the National Academy of Sciences of the United States of America-Biological Sciences, 78*(6), 3559–3562.
28. Malmerberg, E., Kerfeld, C. A., & Zwart, P. H. (2015). Operational properties of fluctuation X-ray scattering data. *IUCrJ, 2*, 309–316.
29. Saldin, D. K., Shneerson, V. L., Howells, M. R., Marchesini, S., Chapman, H. N., Bogan, M., et al. (2010). Structure of a single particle from scattering by many particles randomly oriented about an axis: Toward structure solution without crystallization? *New Journal of Physics, 12*, 035014.
30. Hosseinizadeh, A., et al. (2014). High-resolution structure of viruses from random diffraction snapshots. *Philosophical Transactions of the Royal Society B-Biological Sciences, 369*(1647), 20130326.
31. Dashti, A., et al. (2014). Trajectories of the ribosome as a Brownian nanomachine. *Proceedings of the National Academy of Sciences of the United States of America, 111*(49), 17492–17497.
32. Kern, J., Tran, R., Alonso-Mori, R., Koroidov, S., Echols, N., Hattne, J., et al. (2014). Taking snapshots of photosynthetic water oxidation using femtosecond X-ray diffraction and spectroscopy. *Nature Communications, 5*, 4371.
33. Kupitz, C., et al. (2014). Serial time-resolved crystallography of photosystem II using a femtosecond X-ray laser. *Nature, 513*(7517), 261–265.
34. Huang, S., et al. (2017). Generating single-spike hard X-ray pulses with nonlinear bunch compression in free-electron lasers. *Physical Review Letters, 119*(15), 154801.
35. Marinelli, A., et al. (2017). Experimental demonstration of a single-spike hard-X-ray free-electron laser starting from noise. *Applied Physics Letters, 111*(15), 151101.
36. Marcus, G., Penn, G., & Zholents, A. A. (2014). Free-electron laser design for four-wave mixing experiments with soft-X-ray pulses. *Physical Review Letters, 113*(2), 024801.
37. Zholents, A. A., & Fawley, W. M. (2004). *Proposal for intense attosecond radiation from an x-ray free-electron laser. Physical Review Letters, 92*(22), 224801.
38. Allaria, E., et al. (2013). Two-stage seeded soft-X-ray free-electron laser. *Nature Photonics, 7*(11), 913–918.
39. Hemsing, E., et al. (2016). Echo-enabled harmonics up to the 75th order from precisely tailored electron beams. *Nature Photonics, 10*(8), 512–515.
40. Stupakov, G. (2009). Using the beam-Echo effect for generation of short-wavelength radiation. *Physical Review Letters, 102*(7), 074801.
41. Ackermann, S., et al. (2013). Generation of coherent 19-and 38-nm radiation at a free-Electron laser directly seeded at 38 nm. *Physical Review Letters, 111*(11), 114801.
42. Schneidmiller, E. A., et al. (2017). First operation of a harmonic lasing self-seeded free electron laser. *Physical Review Accelerators and Beams, 20*(2), 020705.

43. Xiang, D., et al. (2013). Purified self-amplified spontaneous emission free-electron lasers with slippage-boosted filtering. *Physical Review Special Topics-Accelerators and Beams, 16*(1), 010703.
44. McNeil, B. W. J., Thompson, N. R., & Dunning, D. J. (2013). Transform-limited X-ray pulse generation from a high-brightness self-amplified spontaneous-emission free-electron laser. *Physical Review Letters, 110*(13), 134802.
45. Hara, T., et al. (2013). Two-colour hard X-ray free-electron laser with wide tunability. *Nature Communications, 4*, 2919.
46. Mozzanica, A., et al. (2016). Characterization results of the JUNGFRAU full scale readout ASIC. *Journal of Instrumentation, 11*, C02047.
47. Ramilli, M., et al. (2017). Measurements with MONCH, a 25 mu m pixel pitch hybrid pixel detector. *Journal of Instrumentation, 12*, C01071.
48. Blaj, G., Caragiulo, P., Carini, G., Dragone, A., Haller, G., Hart, P., et al. (2016). Future of ePix detectors for high repetition rate FELs. *AIP Conference Proceedings, 1741*, 040012.
49. Shanks, K. S., Philipp, H. T., Weiss, J. T., Becker, J., Tate, M. W., & Gruner, S. M. (2016). The high dynamic range pixel array detector (HDR-PAD): Concept and design. *AIP Conference Proceedings, 1741*, 040009.
50. Allahgholi, A., et al. (2015). AGIPD, a high dynamic range fast detector for the European XFEL. *Journal of Instrumentation, 10*, C12013.
51. Veale, M. C., et al. (2017). Characterisation of the high dynamic range large pixel detector (LPD) and its use at X-ray free electron laser sources. *Journal of Instrumentation, 12*, P12003.
52. Shin, K. W., Bradford, R., Lipton, R., Deptuch, G., Fahim, F., Madden, T., et al. (2015). Optimizing floating guard ring designs for FASPAX N-in-P silicon sensors. In *2015 IEEE nuclear science symposium and medical imaging conference (Nss/Mic)* (pp. 1–8). New York: IEEE.
53. Deptuch, G. W., et al. (2014). Design and tests of the vertically integrated photon imaging chip. *IEEE Transactions on Nuclear Science, 61*(1), 663–674.
54. Stan, C. A., et al. (2016). Liquid explosions induced by X-ray laser pulses. *Nature Physics, 12*(10), 966–971.

Index

S. Boutet et al. (eds.), *X-ray Free Electron Lasers*,
https://doi.org/10.1007/978-3-030-00551-1

M

N

The manufacturer's authorised representative in the EU is Springer Nature Customer Service Centre GmbH, Europaplatz 3, 69115 Heidelberg, Germany. If you have any concerns regarding our products, please contact ProductSafety@springernature.com

Printed and bound by CPI Group (UK) Ltd, Croydon, CR0 4YY
30/09/2026
02225651-0016